HANDBOOK OF

IS MANAGEMENT

FIFTH EDITION

HANDBOOK OF
IS MANAGEMENT

FIFTH EDITION

ROBERT E. UMBAUGH
EDITOR

Auerbach
RIA GROUP

Other books edited by Robert E. Umbaugh:

Productivity Improvement in IS
Quality and Control in IS

Contributors

EILEEN BIRGE, *Director of Information Technology, BSG Alliance/IT, Inc., Houston TX*

BIJOY BORDOLOI, *Associate Professor, Department of IS and Management Sciences, The University of Texas at Arlington*

DAVE BRUEGGEN, *Systems Analyst, Union Central Life Insurance Co., Cincinnati OH*

YAOWALUK CHADBUNCHACHAI, *Senior Manager, TelcomAsia Corporation Public Co., Ltd., Bangkok, Thailand*

LESLIE S. CHALMERS, *Vice-President, Bank of California, San Francisco CA*

AMY Y. CHOU, *Data Administrator, Corporate Systems, Ltd., Amarillo TX*

DAVID C. CHOU, *Associate Professor of CIS, West Texas A&M University, Canyon TX*

JIM COCHRANE, *Director, AT&T Business Broadband Networks Breakthrough Offers, Naperville IL*

PAUL CULLEN, *EDP Audit Specialist, Norwest Audit Services, Inc., Minneapolis MN*

WILLIAM J. EGENTON, *Manufacturing Engineer, Raychem Limited, Swindon, UK*

LARRY P. ENGLISH, *President and Principal, Information Impact International, Inc., Brentwood TN*

VATCHARAPORN ESICHAIKUL, *Assistant Professor, School of Advanced Technology, Asian Institute of Technology, Bangkok, Thailand*

G.A. FLANAGAN, *Managing Consultant, IBM Consulting Group, Raleigh NC*

EDWARD A. FORBES, *Computer Systems Specialist, The Vanguard Group, Malvern PA*

LOUIS FRIED, *CIO, TELLUS Corp., Edmonton, Alberta, Canada*

LEORA FROCHT, *Vice-President, Fixed-Income Capital Market Systems, Smith-Barney, New York NY*

MALCOLM A. FRY, *President, Fry Consultants, Essex, UK*

MICHAEL L. GIBSON, *Professor of Management, Auburn University, Auburn AL*

HAL H. GREEN, *Vice-President, Business Development, Aspen Technology, Inc., Houston TX*

STEVE GUYNES, *Professor, College of Business Administration, University of North Texas, Denton TX*

RON HALE, *Senior Manager, Deloitte & Touche LLP, Chicago IL*

IAN S. HAYES, *Principal, Clarity Consulting, Salem MA*

GILBERT HELD, *Director, 4–Degree Consulting, Macon GA*

IRA HERTZOFF, *President, Datagram, Columbus OH*

RUDY HIRSCHHEIM, *Professor, Information Systems Research Center, Houston TX*

SHEILA M. JACOBS, *Assistant Professor of Management Information Systems, Oakland University, Rochester MI*

R.C. JOHNSON, JR., *Consultant, IBM Consulting Group, Piscadero CA*

KEITH A. JONES, *Certified Quality Analyst and Senior Data Services Consultant, Dun & Bradstreet, Palm Harbor FL*

MARTIN KANSKY, *Information Engineering Consultant and Facilitator, Martin Marietta, Oak Ridge TN*

KEITH G. KNIGHTSON, *President, KGK Enterprises, Kanata, Ontario, Canada*

LYNN C. KUBECK, *Manager, Computing and Telecommunications Services, Purdue University Calumet, Hammond IN*

JAMES J. KUBIE, *Vice-President, IBM Consulting Group, Charleston SC*

MARY LACITY, *Assistant Professor of MIS at the University of Missouri, St. Louis MO*

SOOUN LEE, *Associate Professor in the Department of Decision Sciences and Management Information Systems, Miami University, Oxford OH*

GREGORY R. MADEY, *Associate Professor, Department of Administrative Sciences, Kent State University, Kent OH*

L.A. MELKUS, *Consultant, IBM Consulting Group, Dallas TX*

N. DEAN MEYER, *President, NDMA, Inc., Ridgefield CT*

NANCY BLUMENSTALK MINGUS, *President, Mingus Associates, Inc., Williamsville NY*

NATHAN J. MULLER, *Consultant, The Oxford Group, Huntsville AL*

JOHN P. MURRAY, *Technical Resource Manager, Compuware, Madison WI*

KATE NASSER, *President, CAS, Inc., Piscataway NJ*

STEFAN M. NEIKES, *Data Analyst, The Tandy Corporation, Fort Worth TX*

STUART NELSON, *Partner, Government and Financial Services, CSC Consulting, Minneapolis MN*

PAUL NISENBAUM, *Software Engineer, Candle Corp., Santa Monica CA*

DAVID O'SULLIVAN, *Research Manager, Quality Assurance Research Unit, University College, Galway, Ireland*

MILES H. OVERHOLT, *Principal, Riverton Management Consulting Group, Palmyra NJ*

JOHN PETERSON, *Manager, Technology Strategy, AT&T Global Public Networks' Advanced Engineering Concept Center, Naperville IL*

YANNIS A. POLLALIS, *Assistant Professor of Information Resources Management, Syracuse University, Syracuse NY*

KENNETH P. PRAGER, *Principal, Riverton Management Consulting Group, Palmyra NJ*

E. TED PRINCE, *Founder, Perth Ventures, Inc., New York NY*

RICHARD ROSS, *Principal, CSC Index, New York NY*

HOWARD A. RUBIN, *President, Rubin Systems, Inc., Pound Ridge NY*

RANDALL H. RUSSELL, *Senior Manager, Ernst & Young Center for Business Innovation, Boston MA*

DONALD SAELENS, *CIO, Minnesota Department of Revenue, St. Paul MN*

JOHN SATZINGER, *Assistant Professor of Management, University of Georgia, Athens GA*

A. PERRY SCHWARTZ, *President, Computer Associates, Inc., Flowery Branch GA*

S. YVONNE SCOTT, *Manager of Business Systems and Planning, GATX Corporation, Chicago IL*

JAMES A. SENN, *Director, Information Technology Management Group, Georgia State University, Atlanta GA*

SUMIT SIRCAR, *Director, Center for Information Technologies Management, The University of Texas at Arlington*

IL-YEOL SONG, *Associate Professor, College of Information Science and Technology, Drexel University, Philadelphia PA*

RALPH R. STAHL, JR., *Chief Global Information Security Officer, AT&T, Dayton OH*

STANLEY H. STAHL, *President, Solution Dynamics, Los Angeles CA*

DJOEN S. TAN, *Director, Tanconsult, Information Management, Hilversum, The Netherlands*

JOHN TEMPLE, *Partner, Deloitte-Touche, Paris, France*

ROBERT E. TYPANSKI, *Manager of Data Access Services, Bayer Corp., Pittsburgh PA*

ROBERT E. UMBAUGH, *Principal and Head, Carlisle Consulting, Carlisle PA*

MARY VAN SELL, *Associate Professor of Management Information Systems, Oakland University, Rochester MI*

RAY WALKER, *Senior Consultant, DuPont Engineering, Wilmington DE*

TRENTON WATERHOUSE, *Marketing Manager, LAN Switching Systems, Cabletron Systems, Rochester NY*

HUGH J. WATSON, *C. Herman and Mary Virginia Terry Chair of Business Administration, University of Georgia, Athens GA*

MICHAEL E. WHITMAN, *Assistant Professor of MIS, University of Nevada, Las Vegas NV*

LESLIE WILLCOCKS, *Fellow in Information Management and University Lecturer in Management Studies, Templeton College, University of Oxford, Oxford UK*

STEVEN M. WILLIFORD, *President, Franklin Services Group, Inc., Columbus OH*

JOHN WINDSOR, *Interim Chairman, College of Business Administration, University of North Texas, Denton TX*

LEO WROBEL, *President, Premier Network Services, Inc., Dallas TX*

Contents

Contents

Introduction

THE GREAT CHALLENGE IN MANAGING information technology continues to be balancing scarce resources and increasing demands for service. There are so many opportunities to apply IT to crucial business problems that the list of wants always exceeds the can-do list. IS managers at all levels need every bit of help they can get to survive and prosper in this environment. Often the key to success is the skillful combination of business know-how and an intimate familiarity with exceedingly complex technology issues.

Discussions with today's IS managers raise recurring themes: outsourcing, insourcing, client/server computing, process reengineering, data warehousing, downsizing, rightsizing, customer satisfaction—the list goes on and on. It is very much in vogue to talk and write about the strategic use of information systems in business. The idea is for organizations to gain competitive advantage by using information in some way that competitors cannot match quickly enough.

The business press has glorified select information systems that gave (in some cases retrospectively) a competitive advantage to the organizations that developed and used them. As a result of these impressive examples of profiting from the use of information-based systems, the entire IS industry is encouraged to search for competitive systems. IS managers trying to heed this advice, however, have found that it is easier said than done.

In many cases, organizations neglect the basics. Most of the organizations that have succeeded in creating information systems of substantial value can attest that the basics remain as important as ever. Successful companies use thorough, practical business planning, treat information as a resource, and integrate the information advantage with other sound business systems.

PARTNERING WITH USERS

We work in an age where the computer is a common fixture in business offices, and increasingly, in the home. Easy-to-use software is the reason computers are available to a wide base of employees and educators. Advances in the application of computer technology will continue to be paced primarily by the development of software that is simple for novices to use.

Chief information officers (CIOs) and other senior IS executives should realize that there will be an overwhelming need for training and consulting even if hardware and software is extremely easy to use. Users will always be testing the limitations of commercially available productions.

With so much software in the hands of users, chances are that strategically important systems will evolve even faster. Many people in the IS industry acknowledge that few strategic systems find their origins in IS; these systems almost always originate in a department outside IS, and they rarely evolve from a strategic planning process. They are likely to arise from a business need or opportunity identified spontaneously or, equally likely, to evolve from a sophisticated operational system within which someone recognizes even further opportunity.

NEW CHALLENGES

For many years, the IS function has been considered a support organization—often one with a minor role in the enterprise. Now, more and more organizations view the IS function in a new light. In some cases (though, admittedly, still relatively few), IS management plays a truly strategic role as part of the organization's policy group. In many other cases, IS management is not involved at the policy level but plays a greater role in strategic enterprise issues than it has in the past.

A second important way that the IS role is changing involves the nature of the assignments given to IS staff. Closer and more frequent ties to user departments are common. The IS department is involved in training and consulting tasks, and users are more likely to involve members of the IS staff in their planning sessions.

Providing user support demands an ever-increasing proportion of IS resources. The issue of what to support and to what extent occasions much debate. Some IS managers believe that user support can easily consume more than half of the IS budget. The contradiction is that at the same time, the pressure to reduce IS expenditures has been extreme, which makes budgeting a difficult task.

Another stressor is data management. Many users are reluctant to acknowledge that the information resident on their microcomputers is owned by the organization and not by them. The IS department should know what information is being stored and how well it is being stored. Decisions about the most efficient method of cataloging, protecting, and retrieving stored data remain a preeminent responsibility of the IS group.

THE NEED FOR STRATEGY

We live in an era rich in new technology. PCs with more capabilities seem to be introduced each week. Networking capabilities are expanding. Data base technology is being enhanced rapidly—almost every aspect of information processing technology is being improved. In fact, the availability of new technology exceeds the ability of most organizations to assess and assimilate that technology. Even research universities have trouble keeping up. One serious risk we all run is being so eager to try new technology that we institutionalize varied technologies before we can properly integrate them. This leads to even more so-called archipelagos—islands of technology with no bridges in sight.

ACQUIRING TECHNOLOGY AND BUSINESS SKILLS

Managers who climb the corporate ladder are almost always challenged by complex problems with which they lack familiarity. IS departments face similar challenges in their organizational development.

As the IS function becomes involved at the higher levels of the organization (e.g., through strategic systems, executive information systems, and decision support systems), IS personnel may find that they must keep abreast of rapidly changing technology while learning to understand business issues with which they have less than intimate knowledge. This book aims to help IS professionals meet these challenges.

As managers of information systems, we are well into the Age of Integration, having left the Age of Implementation a few years ago. Industry leaders have long recognized the need to integrate information systems—integration has been the driving force of data base management efforts as far back as the late 1960s. It has only been recently, however, that we've had all the technology necessary to achieve that integration. This fifth edition of the *Handbook of IS Management* focuses on those aspects of technology and business management necessary for integrating the application of information technology.

There are many books that cover management and perhaps just as many on technology. However, there are very few books on the management of technology. We have been applying computing technology to business problems for nearly 40 years, yet in my consulting practice and in my long-standing role as consulting editor of the journal *Information Systems Management*, I continue to see examples of failure in IS management. There are, of course, many who do an excellent job as IS managers, and from them we can learn better ways to do our jobs.

HOW TO USE THIS HANDBOOK

This book is an excellent reference for the IS management team. It is a source of ideas to improve the effectiveness and productivity of the entire IS organization.

We call this book a handbook because it is written for the practicing IS manager. Practitioners will recognize that the book's subjects reflect every area of IS where major change is occurring. Among the areas where we've given special emphasis are support for internal customers, outsourcing, client/server computing, network strategies, and improved methods of delivering IS products and services. The emphasis in all the chapters is on practices and methods IS managers should be adopting (or adapting) if they hope to remain effective and responsive in today's business environment. There is an extensive index to help readers easily find specific subjects and other subjects of related interest. Some managers use this handbook as a training tool for subordinate managers and project managers, often by leading a discussion of various chapters. However you use this book, it is my wish that it will help guide you in your pursuit of a rewarding career.

Throughout this handbook, you will find the work of many experts in the field. These authors bring a wealth of experience and education to their writing in an attempt to share with readers successful techniques for managing IS. I trust that you will benefit from the time you spend reading this book.

RObert E. Umbaugh
Carlisle PA
May 1997

Section I

Management Planning, Policy, and Oversight

THE INCREASING INTEGRATION OF BUSINESS direction and information technology is changing the role and stature of the IS organization. The adoption of technology, particularly the seamless integration of IT in business processes, is leading to a true information age much more advanced and more sophisticated than what we experience now.

For these reasons, IT managers must refocus their strategy, rethink their role, and assume the leadership role that has so often been described and so infrequently achieved. This section is intended, in part, to offer some guidance for the changing role of IS and its managers.

Organizational trends that feature process modification, reduced staffs, closer ties to the customer, and shorter product lead times are transforming the strategic use of information. Chapter I-1, "Transforming Information Management," introduces what IS must do to further automate processes to support rapid change and better support knowledge workers. It includes practical steps for accomplishing the transformation of the information management function.

The risks of large-scale business redesign are well known, but they can be reduced through enterprise modeling, a tool that allows strategic planners and IS management to assess an organization's current position before trying to set objectives for improvement. Chapter I-2, "Enterprise Modeling for Strategic Support," shows how to analyze the potential of enterprise modeling as a diagnostic tool in support of business reengineering.

Chapter I-3, "New Directions in Strategic Planning," introduces the concept of the core mission to IS planning. This chapter categorizes key core missions and shows how an understanding of the enterprise's core mission leads IS planning in the right direction.

Whenever organizations try to implement new technology there is risk, and risk must be managed. One of the most significant risks is that people and their organizations will not accept change and will therefore resist and block it. In its discussion of change management and people-centered organizations, Chapter I-4, "Organizational Impact of Technology and Change," offers IS managers a perspective on more effectively aligning the IS organization to better manage change.

One strategy that many organizations are considering is outsourcing. Drawing from the firsthand experiences of US and British companies, Chapter I-5 summarizes the expectations and results of outsourcing experiences. "Outsourcing Realities" explains what went right and what went wrong—and why—as seen through the eyes of both IS and senior executives. Successful outsourcing experiences are then used to outline a prescription for ensuring that expected benefits are fully realized.

Outsourcing brings with it the promise of great reward and regret, depending on how and when the strategy is used. Unfortunately, the risks and costs of outsourcing are sometimes lost amid the rhetoric about outsourcing's benefits. Chapter I-6 presents "A Strategy for Outsourcing" that weighs supposed benefits against realistic outcomes and provides guidelines for cases when outsourcing makes sense. It also clarifies the advantage of retaining the IS function in-house rather than outsourcing this important group to vendors whose agenda is likely to differ from that of the enterprise.

Chapter I-7, "Organizational Architecture for IS," addresses key issues that accompany business reengineering and introduces the concept of IS management plateaus—development stages in the application of information technology. Five management plateaus characterize the development of IT in organizations. On each successive plateau, the costs and risks of the IT investment are higher, but IT's potential to add value to the business and to successfully support business redirection also increases.

Gaining budgetary and management approval to bring in a new technology or start a project to enhance the capability of the organization to use information—something that can truly help the organization compete—more often than not involves a tremendous amount of convincing and cajoling. Chapter I-8, "Economics of Advanced Information Technology," details an approach to assessing the adoption of new technology.

It is perplexing that although most organizations have been automated and computerized for more than a quarter of a century, the idea of customer service has only come to light in the past decade. Chapter I-9 presents a history of the business culture's shift from simply being unavailable to customers to today's customer service-driven operations. "Moving from Availability to Service" includes an analysis of the three primary driving factors behind this transition and discusses the obstacles that must be overcome to achieve service excellence.

Long-term planning has never been easy. In today's fast-paced technology, regulatory, and economic environment, it is even more difficult. Chapter I-10 offers an approach to better planning—helping to discern the shapes and shadows of things to come and guiding organizations accordingly. In the process, it offers some forecasts of how one industry—the communications industry—can prepare to meet its future and how the industry's customers may be affected by new communication products and services. The approach of "Forecasting Technology Needs: An Example" can be easily used as a model for other industries.

I-1

Transforming Information Management

Larry P. English

THE ROLE INFORMATION SYSTEMS and data plays in the information age of the twenty-first century will be different than it is today. Enterprisewide trends emphasizing business processes, teamwork, reduced time to market, and customer service are transforming IS. Information systems must do more than automate processes; they must empower and "informate" knowledge workers. The failure of IS to make this transformation may jeopardize the enterprise itself. This chapter reviews the role of IS and data in the transformed IM function and concludes with practical steps to making the transformation a reality.

REDEFINING IM

Information management (IM) applies sound management principles (e.g., planning, organizing and staffing, directing, and controlling) to information as a key enterprise resource. It includes three components: data resource management, process management (i.e., business activities that collect or present information to knowledge workers), and information technology management.

The premise underlying these components is that the value of data is optimized when data is managed to be shared by many applications and knowledge workers, processes are managed to maximize value-adding activities and eliminate non-value-adding activities, and technology is exploited to enable just-in-time delivery of information. This synchrony requires a cross-functional approach.

IM's mission is simple: to enable the fulfillment of the business mission of the enterprise through managed information, managed processes, and managed information technology.

EVOLUTION OF DATA RESOURCE MANAGEMENT

Data resource management (DRM) applies management principles to the data and information leg of the IM triangle. Data resource management has three components: data management, data base management, and data technology management.

From Data Administration to Data Management

Data administration evolved as a function to address the problem of uninte-grated application data bases. As such, data administrators developed common data models and data definitions using data standards, eliminated unnecessary redundancy and managed required redundancy, improved data integrity, and increased data access and sharability. Studies conducted by the author revealed that a small minority of enterprises have achieved phenomenal success with their data administration implementation, but the vast majority of organizations have had only limited success. The common thread in the most effective organizations is that they manage data and develop applications from a horizontal and inte-grated, business-centric resource approach.

Data administration must evolve into a true data management and leadership role. Rather than simply taking data definitions from an applications develop-ment scope, data management must facilitate common data architectures with data definitions and domains that support the entire enterprise. Data manage-ment coordinates information stewardship throughout the enterprise.

Information stewardship in turn supports business process reengineering and total quality management because it establishes single definition across the busi-ness value chain and information accountability. Data management enables the creation of a common business chart of facts that becomes the enterprise's ev-eryday language. It introduces quality programs for the information product along with metrics to measure data quality and ensure the reliability needed by knowledge workers.

From Data Base Administration to Data Base Management

Data base administration arose in the 1970s to manage data stored in so-called sharable data bases. However, applications defined by functional areas, using a sys-tems approach, generally resulted in application-specific data bases not usable by other functional areas. As a result, the number of nonshared sharable data bases proliferated, and data base systems management became an expensive access method. The most significant benefits of data base technology—reusability, sharabil-ity, minimal redundancy—were mostly unrealized.

Data base administration must evolve into a true data base management role that maximizes the sharability of the data stored in the enterprise data bases. Data bases that contain data stored redundantly, whether owing to distribution, client/server replication or download, or an information warehouse, must be designed with a common format and definition that supports the information requirements of all knowledge workers and eliminates the need for interfaces.

From Data Technology Administration to Data Technology Management

Data technology management establishes a minimalist suite of data manage-ment technologies such as data modeling, repository, data base management, object data base management, and data quality management tools. Repository or data dictionary management lets knowledge workers and data producers access the definition of data and the data needed to perform their jobs.

Some organizations, as they elevate the scope and business impact of data resource management, have changed the function's name. Weyerhaeuser, the forest and paper company of Federal Way WA, calls its DRM program "Information Excellence."

EVOLUTION OF PROCESS MANAGEMENT

The information age challenges the systems approach paradigm. Although the systems approach has many benefits, it also has a spectacular weakness. By breaking problems into smaller, isolated components, it fragments business processes into nonintegratable parts. An organization has only to examine its own portfolio of applications to discover hundreds, perhaps thousands, of interface programs that copy data from one data base and transform it into a format usable by another system. Billions of dollars have been spent on so-called systems integration, signaling that something is badly wrong with the systems development process. Outsourcing as a solution is a further sign that the processes used to build systems and manage data have not worked.

One major corporation discovered that 40% of its applications development staff was dedicated to developing and maintaining redundant data create/update programs (capturing data that existed elsewhere) and interface programs. An insurance company discovered that one of its core business facts was captured and entered in 43 different application programs by 43 different businesspeople who stored it in 43 different data bases. Any total quality management program or business process reengineering methodology will confirm that interface programs and redundant data create programs that do not add business value. Rather, they add cost, virtually guarantee the creation of a significant potential source of error, and create information float, which delays timely access to information.

The systems approach fails to understand the real product of a system. It focuses on the functions the application is automating and sees the applications system as the product. The information age demands a resource management approach that sees the products of applications development as managed information and informed and empowered knowledge workers.

From a Systems Approach to a Resource Approach

In the information age, applications development must balance the processes automated with the information product created or managed and the people being empowered. As such, the information systems process must emphasize a resource approach to the life cycle of the information product.

The resource approach applies the management principles used in human and financial resources to information. Exhibit I-1-1 and Exhibit I-1-2 compare the similarities of the resource life cycle of the human resource and the information resource. A managed resource life cycle comprises five basic processes required to minimize the resource's cost and maximize its value. They are:

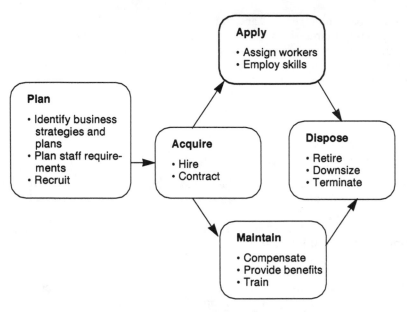

Exhibit I-1-1. Resource Life Cycle of the Human Resource

- Planning.
- Acquisition.
- Maintenance.
- Disposition
- Application.

The first four resource processes constitute the cost basis of the resource. The economic feasibility of an enterprise directly correlates to its ability to apply (the fifth resource process) a resource in a way that generates value greater than the resource's costs. The enterprise must apply its human resources by assigning workers to roles that use their skills in ways that provide value greater than the cost of planning, hiring, compensating, and retiring or terminating them.

Information has a similar resource life cycle. To accomplish the objectives of the enterprise, planning must include what the enterprise needs to know to carry out its strategic initiatives.

This strategic planning yields enterprise information, process and application, and technology models that are integrated with business objectives. Tactical planning results in reengineered processes, operational applications, and populated data bases.

The information model reflects the enterprise's knowledge base organized around the enterprise's fundamental resources, called subject areas, and the

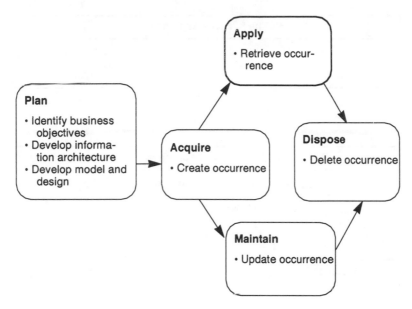

Exhibit I-1-2. Resource Life Cycle of the Information Resource

fundamental entity types about which the enterprise must know information. The information model must be as understandable to the business executive as the organization chart, which models the organization of human resources, and the chart of accounts, which models financial resources in the form of sources of revenue and expense.

The process and application models reflect the fundamental business value chains and are integrated through shared information commonly defined. Developing applications from a common model reduces the costs of developing similar nonintegratable applications and interfaces.

Acquisition of the information resource occurs through applications that create occurrences of each entity type. Maintenance updates occurrences of each entity type, either in the form of changing values over time, such as customer addresses, or changing states of existence, such as when a placed order is verified for credit worthiness and becomes a validated order. Disposition occurs when knowledge workers no longer need to know the data; for example, when an order is paid, it is deleted.

These processes represent the cost basis of the information resource. As with other resources, the application of information—the retrieve processes—provides the value basis of the information resource. Data derives its value when it is retrieved by knowledge workers and used intelligently to accomplish business objectives, perform business processes, support business decisions, and create competitive advantage.

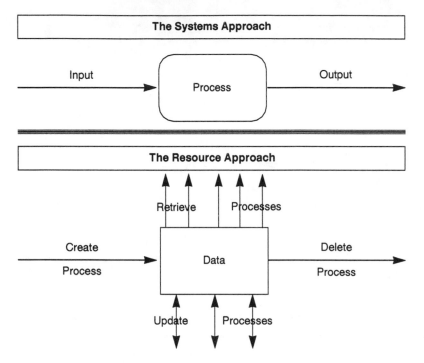

Exhibit I-1-3. A Comparison of the Systems Approach and Resource Approach to Information

Managing information as a business resource represents a 180-degree shift from the systems approach. Rather than building data around the processes, the resource approach builds the processes around the data, as illustrated in Exhibit I-1-3. This is the basis for the object paradigm that encapsulates processes, called methods, around data types.

Development methodologies must provide for cross-functional modeling and definition of the information resource. Those that do not will add unnecessary cost to applications development and maintenance and will not maximize the information resource's value. Today's explosion of prototyping tools and rapid development products and methodologies have the potential to be misused; they may further the proliferation of redundant data create/update programs and the fragmentation of applications and information.

In his book *Enterprise Engineering* (Lancashire UK: Savant Institute, 1994), James Martin illustrates that both data bases and applications must not be modeled and built within the scope of functional business areas. Rather, both data and process must be designed to support cross-functional value streams (e.g., business processes or value chains), "end-to-end collections of activities that create a result for a customer." The cross-functional value stream or process becomes the basis for modeling the data; that is, data is modeled, defined, and built in a way that supports all processes requiring it.

The resource approach identifies the point (e.g., business event, business activity, and data producer) at which a fact of information becomes known and develops the create application. Business events that change occurrences of those entity types have update applications; new data is not created elsewhere. Information management places the create and update applications as close to the source (data producers) and time (process of origination) as is feasible, eliminating information float and enabling just-in-time information.

The resource approach defines data so that every entity type and every attribute have a common and consistent meaning across the enterprise. Common data definition is required so that all knowledge workers interacting with a given fact of information have a common understanding of its meaning.

From Process Automation to Process Management

The role of applications development must evolve into a true process engineering and management function. This function applies management principles to the process leg of information management. It includes process evaluation as well as process and applications design that includes the creation, updating, and presentation of information. Process management has three components:

- Process engineering.
- Process development and human factor design.
- Process technology management.

Process management establishes process and applications architectures and develops common definitions for reusable processes. It assesses and evaluates processes based on value-adding versus cost-adding criteria. It develops an applications architecture against the business value chains that places create applications at the process of original point-of-data capture.

Applications design must be based on an assemble-to-order philosophy that minimizes duplication of code around common processes. Applications are designed and implemented within a business value chain and integrated across functions. They operate on commonly defined shared data and therefore require minimal interface programs. Interface programs are used only for temporary coexistence with existing legacy systems until such time as they can be completely replaced. Permanent interface programs are used only to interface required software data base packages that are commodities and when no competitive advantage would be gained by internal development. (Of course, organizations may have a significant portfolio of legacy systems that cannot be replaced in the short run, if at all. Although some legacy systems contain data of significant value to a business, others may be so ill-defined as to cost far more than their value to the organization. Techniques do exist that allow for an architectural approach to the new development of enterprise models with a phased, and temporary, interface to obsolete legacy applications and data bases.)

Object technology (OT) and, more important, object methods are radically changing how applications are developed. OT is the Copernican technology that literally forces code (process) to be written around data types (the resource). However, the technology is not the solution. The solution is making the paradigm shift to a resource approach to development processes that will realize the potential benefits of using object technology.

Applications software packages are evaluated against the enterprise data architecture as well as on functional requirements. Costs of packages are reduced when a minimum of translation is needed in the required interfaces.

User interface design (i.e., screen layouts and reports) must evolve into human factor design, which includes designing the work processes, dialog, and presentation of information in human-machine interactions.

EVOLUTION OF INFORMATION TECHNOLOGY MANAGEMENT

Information technology support must evolve into an information technology management role. This role applies management principles to information technology to develop a common IT architecture with a minimalist set of tightly interoperable technology components. It must establish guidelines for the evaluation of IT across the enterprise to prevent unnecessary duplication of disparate technologies that provide the same basic functions and drive up the support and technology interface costs without adding value. Information technology management increases the value of information technology through maximum use with minimum support costs.

ORGANIZATIONAL STRUCTURE OF THE NEW IM

Just as the larger business enterprise is reshaping itself around its business processes with flattened, more horizontal organizations, so too must the new information management (IM) organization reconfigure itself. As depicted in Exhibit I-1-4, in this new organization, multifunctional teams manage the IM processes.

At center stage, conducting the IM organization, is the chief executive officer (CEO), with the chief information officer (CIO) as section leader. Architecture development, the result of cross-functional architecture planning teams, provides the framework.

The interaction between IM and the business occurs through multifunctional teams that cooperate with multifunctional business teams responsible for business value chains. As data requirements are discovered in a business value chain, data resource management identifies other interested knowledge workers and ensures that all significant business events affecting the entity types have been identified. Data definition supports all knowledge workers' views. Process management ensures that point-of-origin events and processes are discovered for the create applications along with all processes and events that change the state for update applications. Human factor design ensures efficient and ergonomic work design and information presentation.

In a distributed or decentralized enterprise, depicted in Exhibit I-1-5, additional multifunctional teams are required to manage the information and pro-

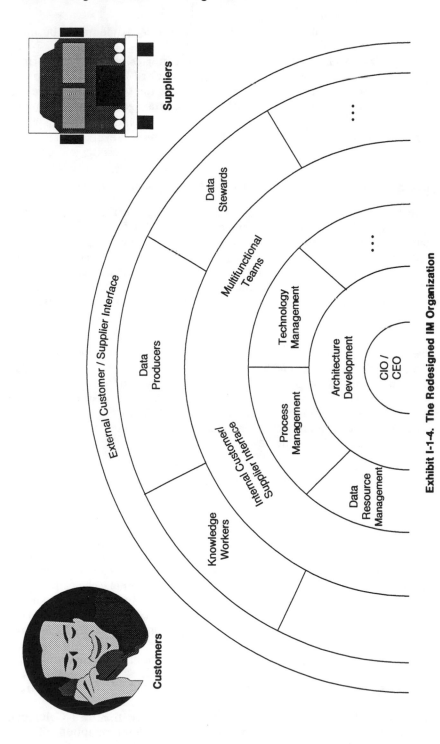

Exhibit I-1-4. The Redesigned IM Organization

cesses of the business value chains that cross multiple, autonomous business units.

Some organizations have organized to address strategically common information, process, and technology management. ITT Hartford Insurance Group has established an information strategy unit accountable for strategic initiatives to provide for common information architecture and management, process management and human factor design, technology architecture and management, and object management across business units.

INFORMATION STEWARDSHIP: THE BUSINESS ROLE IN IM

Strong business involvement in information management characterizes the information age. Unlike in the past, when all responsibility for information management was either intentionally or inadvertently delegated to IS departments, information can no longer be considered a technical resource of the organization. It is a business resource used by business personnel (i.e., data consumers or knowledge workers), created by business personnel (i.e., data producers), and defined and guided by business personnel (data definers or tactical and strategic information stewards).

Knowledge-Worker Steward: The Data Consumer. The industrial-age organization holds workers accountable for the work they perform. This is true also for knowledge workers, who use information as a raw material in their work. Operational knowledge workers are on the front lines processing insurance claims or filling customer orders. Strategic and tactical knowledge workers make decisions such as determining customer satisfaction of existing products, analyze potential products, or determine new business directions.

Operational Information Steward: The Data Producer. Knowledge workers who use information depend completely on the accuracy and quality of the information created by data producers. If the data is inaccurate (e.g., a diagnosis code or shipping address is incorrect), the knowledge worker's result may likewise be incorrect.

Therefore, the information-age organization will empower and hold the data producer, or operational information steward, accountable for the accuracy and quality of the data produced. Furthermore, data producers may be called on to capture facts that may not be needed in their jobs or business units but are required by knowledge workers in downstream activities. Data intermediaries are data producers who simply transcribe data from one form to another, such as key data from a paper form into a data base, without adding value. The real data producer is the individual who first knows the facts. Capturing data at this point of origin is a much more cost-efficient and reliable process than are downstream processes that attempt to rediscover or re-create the facts or transcribe them from some manual form.

Companies such as Cominco, Ltd., a mining company in British Columbia, Canada, have such data accountability written into the job descriptions of their

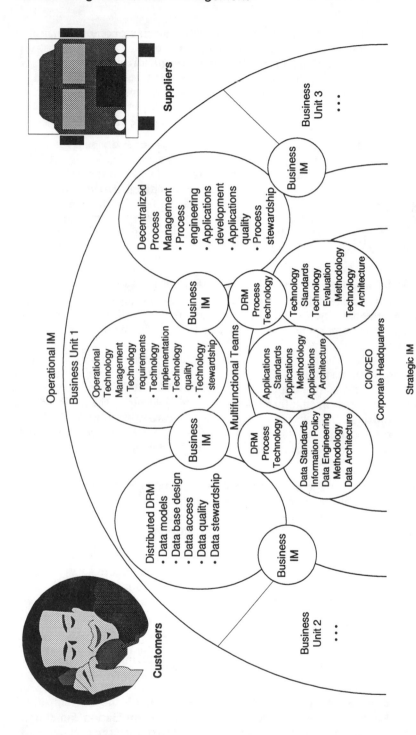

Exhibit I-1-5. IM in a Distributed Enterprise

miners. Because the safety of all miners depends on accurate information about drilling activities, the miners are accountable not just for how much ore they dig, but also for the accuracy of the information on their time sheets describing where they have drilled. If this information is incomplete or inaccurate, they do not get paid—regardless of whether they can read or write. This is an industrial company of the information age.

Tactical Information Steward: The Business Manager or Process Owner. Every process owner or manager in an information age organization is a tactical information steward accountable for the data produced as a result of the business processes for which the manager is accountable. Managers' accountability also includes implementing and enforcing the information policies developed by the CIO and issued through senior management, and providing information quality training to data producers. In the information-age organization, this accountability is included in every manager's job description.

Strategic Information Steward: The Business Information Domain Expert. Because data producers are held accountable for the accuracy of the information they produce, the definition of information must be clear and precise. While the DRM function is responsible for facilitating the development of information models and definition of information, the enterprise must identify the high-level business personnel with expertise in some business subject who will serve as strategic information stewards. Generally, this stewardship role is part-time. However, some organizations, such as Denver-based US West Communications, have full-time stewards in an information stewardship group in the business area.

Strategic information stewards verify or authenticate the definition of data in their scope of expertise, establish data quality levels, approve access to secured information, and resolve business issues and conflicts over data.

Senior Business Management. Senior business management issues the information policy, establishes information accountability in the business, resolves issues concerning information sharing across business units, and sponsors and creates cross-functional organizational change with its new information-sharing culture.

TRANSFORMING IM INTO A BUSINESS ENABLER

Preparation for the twenty-first century cannot happen tomorrow. It must occur now. The following steps will help IS managers exploit the power of both today's information technology and tomorrow's emerging technology and transform IS into a strategic information management function.

Developing a Quick but Explicit Self-Assessment of the Organization's Information Health. Change will occur only when an organization can quantify the cost of the status quo and realize it is more expensive than the cost of change.

IS managers should take a small set of critical core business facts, such as customer address or product ID, through the following process:

1. Counting the number of redundant data bases these facts are stored in; the number of redundant applications maintaining the facts; the number of interface programs copying the facts from one place to another; and the number of redundant data producers required to sustain duplicate data entry of these same facts.

2. Estimating (if actual metrics are unavailable) the cost of developing all the redundant create, update, and interface applications and redundant data bases. The level of redundancy in all other areas should also be estimated and its total cost extrapolated.

3. Conducting a data audit to identify some sample occurrences, such as a specific customer, then extracting the data values from the redundant data bases and comparing consistency. Customers should be called to see how accurate the data actually is, and the business costs of rework and lost business opportunity should be estimated.

Developing a Vision. IS managers should develop strategic thinking about the business and about how technology can solve business problems, beginning with the business mission. Everything IM does must be geared toward enabling the realization of business objectives. Managers should mentally step out of the day-to-day role in IM and think of radically new ways that information and information technology can be used.

Today's information professionals must embrace the larger purpose of their work. Instead of thinking in terms of just defining data or building a data base, they should focus on enabling knowledge workers to be empowered and informated for competitive advantage.

Developing Rapport with Senior Management. Senior managers are IM's most important customers. IS managers must earn their commitment by listening to their concerns, a process that includes identifying their strategic objectives and developing proposals that quantify cost savings and revenue generation in concrete terms. This puts senior managers' decision to commit to change in the form of an evaluation of return-on-investment rather than in the form of a request for additional expenditures.

Enabling a Paradigm Shift. The transition to the information age requires two paradigm shifts. First, the business must transform from an industrial and hierarchical management style to an information-based and ensemble management style. Second, the IS function must transform itself from a procedural, function-driven systems approach to an event-driven information resource approach.

Paradigm shifts are usually created by outsiders, who see things in ways not clearly recognizable to those who have a vested interest in the status quo. Once a vision is established, it should be articulated to the managers who have the authority to create the paradigm shift. A vision has to be expressed in concrete terms that people can understand. This entails:

1. Quantifying the cost of the status quo. The cost of applications development should be determined, as should the cost of maintaining create/update programs, interface programs, and redundant data bases. The level of redundant create/update programs, interface programs, and data bases should also be determined. Finally, IS managers should quantify the cost of the development and maintenance of these redundant (i.e., non-value-adding) applications and their operational costs, including the cost of business personnel (i.e., data intermediaries) entering data that exists elsewhere.

2. Identifying the benefits of the information resource approach. One benefit is that it eliminates redundant applications and data bases, inconsistency of redundant data bases, and non-value-adding information intermediary time. Another centers around improving information quality through point-of-origin information capture with stewardship accountability. The information resource approach is also useful in identifying new opportunities created by timely information (as a result of reduced or eliminated information float and reduction of lost business owing to inaccurate, out-of-date, or missing information) and new business opportunities enabled through integrated, shared information that can be combined and analyzed in new ways.

3. Overcoming paradigm paralysis by providing continuous education that encourages new thinking and new behaviors among managers and IS professionals.

4. Actively involving those whose habits are changing in the decision-making process and in the design of the new development methods. Managers should be sensitive to the needs of IS professionals whose applications development habits are changing and allow for risk taking and mistakes.

Analyzing Key Strengths and Weaknesses. IS managers should identify the internal strengths and weaknesses of the IM function and its external opportunities and threats. This includes planning a course of action to neutralize weaknesses and threats, maximize strengths, and exploit opportunities.

Reviewing Current IM Processes to Identify and Define Core Processes. IS managers should review current processes in terms of the value they add to the business, emphasizing those that contribute value. Processes discovered during the transition to the new business paradigm, with its new rules and regulations, should be integrated within the paradigm's framework.

Eliminating Non-Value-Adding Processes. IM processes that do not contribute or add value to the strategic business objectives should be reviewed and deleted.

Redefining the Applications Development Process. IS managers should evaluate the development methodology from a resource perspective that includes:

- Developing information, process, and technology architectures to model the business.

- Identifying cross-functional business processes or business value chains.
- Identifying business events that trigger the process activities within the value chain.
- Applying business process reengineering principles to a process before automating the process.
- Defining data to support all business processes that use that data.
- Building data models and data bases with common data definitions that can be fully sharable or replicable to other server data bases.
- Developing a single create/update program for each discrete entity type and locating it at the process activity at the original source of create.

Using Multidisciplinary Teams for Development. When development occurs, IS managers should ensure that representatives from data resource management, process management, and technology management are appropriately involved at all levels of the process.

Investing In and Exploiting the Right Technologies. Guidelines for evaluating and selecting information technology within the enterprise should be developed. IS managers should evaluate technology from a best-fit approach and choose a single type of technology (e.g., a DBMS , transaction application tool, decision-support application tool, or object-oriented programming language) for each class of business problem. Minimal-value tools and their support costs should be eliminated. All parties to be affected by a technology decision should be included in the evaluation of IT products.

Developing a Plan to Move to a Shared Information Resource. IS managers must develop a high-level enterprise information model that identifies the business subject areas (i.e., business resources) and fundamental business entity types. They should use that model in planning and developing applications and data bases. Legacy data bases in which data currently resides should be identified. An orderly migration should be planned to eliminate unnecessary data bases and applications as new development supports the processes and data required by the legacy applications and data bases.

Managing Change Effectively. IS managers must prepare for change and develop plans to institute change in a positive way. This includes identifying and soliciting support from the change sponsor, who can authorize change. Change agents should be chosen who can communicate with and involve the audience affected by the change. New procedures and standards should be prototyped in a way that encourages improvement. After standards and procedures are prototyped, they should be modified based on feedback from the pilot team, and thorough training to subsequent teams should be provided. Managers should continue to improve and solicit improvements to standards and procedures from the people who use them.

Changing Reward Mechanisms. Changing reward mechanisms can solidify new habits. IS managers should reevaluate the reward mechanisms of the old paradigm, such as rewarding individuals, meeting target dates when scope is reduced and shortcuts are taken, and solving problems that are really symptoms of a larger problem that should have been prevented. They should then identify the reward mechanisms of the new paradigm: teamwork and meeting target dates without compromising the architecture and with common cross-functional definition of data using customer satisfaction surveys.

CONCLUSION

There is no question that the successful enterprises of the early twenty-first century will look and behave differently than their counterparts of the late-twentieth century. Indeed, the transformations have already begun. The questions are: Who will make it and who will not? And who will be the catalysts for the successful enterprises?

This is the most exciting time to be a part of the information management field—the catalyst for society's transformation to the new economic paradigm of the information age. Although the information age was triggered in 1946 when the ENIAC computer came online, only now have the principles for successful use of information technology become clear. This is the decade of the maturing of the information age. As Peter Drucker advised, "Put your resources on tomorrow, where the results are—and not on yesterday, where the memories are."

I-2

Enterprise Modeling for Strategic Support

Michael E. Whitman
Michael L. Gibson

SUCCESSFUL STRATEGIC BUSINESS ENGINEERING, whether a reactive effort to regain competitive advantage or a proactive effort to maintain and improve performance, depends on an organization's ability to accurately and methodically analyze its internal and external environments, people, processes, organizational structure, information uses, and technology. Enterprise modeling (EM) greatly enhances strategic business engineering by providing a structured, diagrammatic framework for depicting the myriad interconnected and changing components addressed in large-scale efforts to implement change. Its representative models of the organization serve as baseline features against which all subsequent change is measured and provide a basis for strategic planning. Using EM as a forecasting tool fosters a more effective and efficient planning process that dramatically increases the probabilities of success.

IS professionals are uniquely situated to apply enterprise modeling technology to overall business change. Enterprise modeling originated in data processing departments as a software development tool and plays a critical role in computer-aided systems engineering, allowing systems designers to map current and proposed information systems as a predecessor to development. IS professionals have direct access to the organizational information and the discipline and training in modeling and designing processes necessary to support information flow. It is, therefore, a logical extension for the IS function to aid business engineering strategists achieve the revisions inherent in strategic business engineering.

ENTERPRISE MODELING

Enterprise modeling has been described by E. Aranow as "a combination of diagrammatic, tabular, or other visual structures, which represent the key components of the business that need to be understood." More simply put, enterprise

19

modeling consists of representing complex objects in easy-to-understand diagrams. Complex modeling tools facilitate EM, but the basis for understanding a large, complex organization relates to the ability to represent it in a series of elementary graphs that allow modelers to view components individually without losing the contextual overview of the entire organization.

Enterprise modeling is referred to as one of many qualitative models which can be used to represent quantitative measures of particular facets of an organization in a higher level of abstraction that represents the organization in a more holistic manner. Qualitative modeling allows the modelers to view the organization's synergistic existence as a whole entity versus the sum of its parts, in supporting the organizational mission, objectives, and functions. This systems view provides critical analysis of the organization as a preparation for strategic business engineering, as organizational modelers must encapsulate the interaction of the components of the organization depicted as part of the conceptual view of the organization.

The enterprise model itself can be decomposed into two functional models, a business model and a systems, or information, model (see Exhibit I-2-1). The business model depicts business functions, events, and activities, and organizational structures, geographic locations, and interrelationships. The information systems model comprises the information needed, produced, and used by business functions.

ENTERPRISE MODELING CONSTRUCTS

Business models are concerned with what processes are needed to run a selected business area, how processes interact, and what data are needed. Systems models present a high-level overview of the enterprise—its functions, data, and information needs. Whereas the business model represents a view of the entire business, the information systems model represents that portion of the business targeted for computer support.

Modeling an organization begins at the strategic or corporate level and progresses through decomposition of corporate-level objects, activities, and associations down to the tactical and operational levels. Objects (business) are those things internal and external to the business for which it is important that the business retain information. Activities (enterprise) represent what the enterprise does in terms of major functions, processes, and procedures. Associations are relationships, dependencies, shared characteristics, or other connections between organizational objects and activities.

As shown in Exhibit I-2-2, a top-down, comprehensive approach to modeling the business divides the conceptual (i.e., strategic) objects, activities, and associations into greater and greater detail. In the intermediate levels, these objects, activities, and associations are logical (i.e., functional), at the lowest level they are physical (i.e., operational). This process spans the business life cycle, continuing down through the layers of the organization until they are at a primitive level and need no further decomposition to be clearly understood. At this lowest level, the objects, activities, and associations inform and are used by the employees (who are also some of the business objects) in conducting the daily activities of the business.

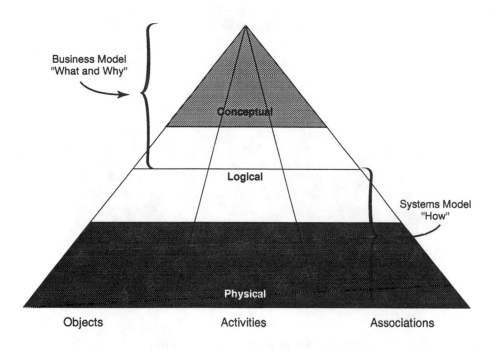

Source: M.L. Gibson and C.A. Snyder, "Computer Aided Software Engineering: Facilitating the Path for True Software Knowledge Engineering," International Journal of Software Engineering and Knowledge Engineering (January 1991).

Exhibit I-2-1. Functional Models of the Enterprise Model

The concept of organizational analysis resembles the business evaluation techniques presented in most current management literature. Just as a formalized structure is required for collecting and representing environmental scanning data; in organizational analysis, modeling methodologists responsible for modeling the enterprise must use a structured framework to facilitate accurate and precise analysis. The resulting models should accurately represent what the business is, does, and uses, and with whom it interacts.

A CASE tool that has high degrees of graphic and text (dictionary and repository) support often helps the modeler show how things are related (associated). A good CASE tool should have some type of intelligence. Modeling with case tools imposes formality on all phases of the methodology and provides repositories and extensive capabilities for cross-checking and storing of data.

The Enterprise Modeling Methodologist

As the CIO of a large oil company has said, "The business line managers within the organization must be the individuals responsible for any change effort, and for reengineering specifically." For any intervention strategy to succeed, the

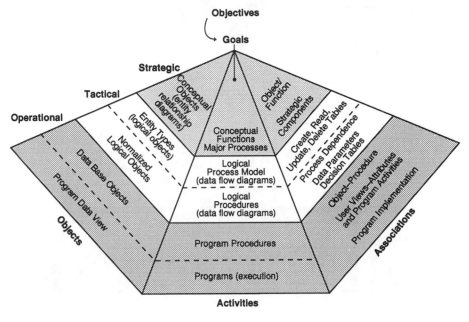

Exhibit I-2-2. Enterprise Modeling Paradigm

managers and employees within an organization targeted for strategic business engineering must become an integral part of the effort and continually be informed and educated regarding the process. In fact, for a successful strategic business engineering (SBE) effort, managers and employees must be the individuals actually performing the change, empowered by their supervisors to use their personal innovation and ingenuity to achieve desired improvements.

The complexity of EM, however, necessitates that a methodologist trained in EM techniques instruct managers and employees in the EM process, if not to actually conduct the modeling. Using a methodologist does not detract power or responsibility from the business personnel for the end project; it provides them with an experienced knowledge base from which to draw ideas and suggestions on overall design improvements.

ENTERPRISE MODELING IN SBE

As depicted in a contextual SBE model (Exhibit I-2-3), enterprise modeling plays a significant role in supporting the various processes of a typical engineering effort. The model begins with the strategic planners' mental model of the organization: a conceptual image of what the organization is, how it operates, and how it interacts with its environment. The mental models of

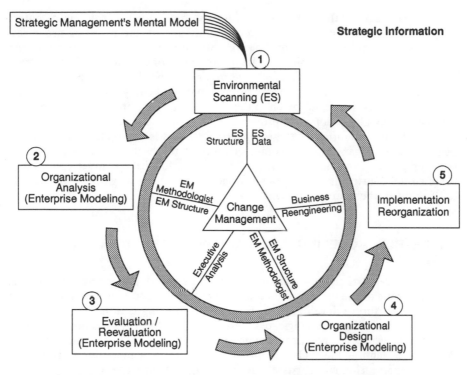

Exhibit I-2-3. Contextual Paradigm of Strategic Business Engineering

executives are ever-striving attempts to match the physical with the logical that drive the organizational restructuring and change management evident in today's organizations.

The subsequent stages in the SBE process are as follows:

- *Environmental scanning.* Selectively retrieving information from the organization's internal and external environments to support the development of an organization's strategy.

- *Enterprise modeling.* Accurately reflecting the current processes and information usage within an organization and mapping out the desired end result of the SBE effort.

- *Evaluation and reevaluation of the enterprise model.* Developing a blueprint of the organization as it currently exists to allow strategic planners to compare the model of what the organization is to their conceptual models.

- *Blueprinting the new and improved organization.* Creating a revised organizational design that models what should be instead of what currently exists.

The final stages of the SBE paradigm involve implementing the revised enterprise model and continuously monitoring and revising the previous processes. The central controlling factor of the entire SBE process embodies the structured change management principles that lie in the center of the model. Current practices in organizational development and organizational behavior theory are extremely useful in SBE. In fact, some projects undertaken by IS professionals fail because due attention is not given to the change intervention strategies proposed by organizational development professionals.

Once the first iteration of the engineering process is completed, the next one begins. As an organization becomes committed to the kinds of revolutionary changes and improvements resulting from successful SBE, it must continue to practice the constructs and lessons learned, or else again begin to stagnate and fall behind in the competitive race.

ENTERPRISE MODELING, REENGINEERING, AND STRATEGIC PLANNING

The true value of an enterprise model lies in its ability to support a conceptual understanding of the present situation of an enterprise and to aid in mapping out a strategy for future developments. Enterprise modeling is thus a dynamic strategic tool that allows strategic planners to assess the organization's position before establishing the means to accomplish organizational goals and objectives. A strategic planning life cycle that incorporates enterprise modeling and strategic business engineering encompasses the following several steps.

Goal Development

Directly related to the business profile analysis is the identification and evaluation of business opportunities and threats present in the company's external environments. These opportunities and threats help develop strategic goals and objectives. Business engineering, as a logical extension of the change process, should be a cornerstone for strategy development. The enterprise models created should be used to supplement the goals developed within this process.

Strategy Formulation

The second step in a typical strategic planning life cycle is the development of specific strategies for the organization, as well as a single context-level strategy that addresses the focus and mission of the organization. As strategy emerges, a key component is the continuous process of evaluation and engineering. Again, enterprise modeling provides a useful tool in evaluating the organization to determine the feasibility of the various alternative strategies.

Strategy Implementation

The next phase of the strategy life cycle involves implementing the developed strategy, focusing on achieving results, and relying heavily on change management, organizational behavior analysis, and performance measures. SBE during implementation concentrates on organizational structure, relationships, and processes, and on the behavior of the firm's top leadership. Modeling the structures and associations within the organization coupled with an aggressive implementation strategy provides a fundamental blueprint for a successful implementation.

Strategy Assessment

During the implementation stage, the simultaneous evaluation of strategy development and strategy implementation directly affect final strategies. The engineering phase of implementation and reorganization is ideally suited to support the integration of the strategy into the business environment. As the strategy is implemented, the engineering constructs are ingrained in the process, creating an atmosphere conducive to the ongoing change that characterizes reengineering and strategy implementation.

Strategy Control and Maintenance

Success in strategic planning is a relative concept. Reengineering does not ensure the success or failure of a strategy; rather it serves to report the state of the organization as it responds to the planning process. Strategy maintenance is a continuous looping process whereby the organization continues existing strategies and develops new ones throughout the strategy life cycle. The continuing analysis, design, and implementation steps in enterprise modeling and engineering facilitate strategy maintenance. The relationship between SBE and strategic planning is symbiotic—the constructs behind each support the other.

The following example illustrates the application of enterprise modeling.

ENTERPRISE MODELING AT STATE UNIVERSITY

State University (a pseudonym) is a major public university in the southeast with an enrollment of approximately 25,000 students. As a land grant university, State pursues its charter missions of research, instruction, and extension. The university currently manages its financial operations through a single functional division, known as the Business and Finance Division. The department within the business office primarily responsible for the information systems and financial reporting procedures supporting the university is known as Financial Information Systems (FIS). The director of FIS contacted an enterprise modeling group to develop a series of models of FIS operations in preparation for business process redesign efforts.

History of Computing in the Business Office

The administrative computing function at State University was originally part of the Business Office (now known as the Business and Finance Division). In the mid-1970s, administrative computing was supported by an IBM 370 mainframe located in the basement of the administration building. Around the same time, the Division of University Computing (DUC) was formed and the function of administrative computing support was moved to it. The office of Financial Information Systems was formed primarily as the Business Office's central data-entry office.

After the departure of the administrative computing function from the university's administration building, the Division of University Computing set up a remote batch station to handle the transmission of the data that was key punched by FIS. The remote batch station also printed output generated by the administrative mainframe. At that time, FIS consisted of four data-entry clerks, a production supervisor, and the director. The responsibility for the remote batch station was taken over by FIS in 1981. The batch station operator was then transferred to FIS. In 1982, FIS purchased an IBM System/38 to meet the growing demand for business office computing support. Within two years the network grew to over 150 users, requiring that FIS upgrade the System/38 to the most current and largest model available.

From 1981 until 1986, the director handled all mainframe ad hoc programming requests. The director also handled the coordination of projects involving the installation and upgrade of mainframe-based systems used by the Business Office. In 1988 the System/38 was replaced by an IBM AS/400 B60, which in turn was replaced by an AS/400 model E60 providing 120 megabytes of main memory and approximately 20 gigabytes of direct-access storage devices. At the time of the study, FIS offered administrative support computing services on the IBM AS/400. The system supports two high-speed printers and more than 260 local and remote terminals, personal computers, and printers located in administrative offices around campus. The primary operations of FIS are presented in Exhibit I-2-4; an organizational chart of staff is presented in Exhibit I-2-5.

Existing Strategic Plans

The existing plan of FIS centered around directives from the Office of the Vice-President of Business and Finance that attempted to integrate the university's long-range goals with this office. The strategic planning and implementation process was, at best, ad-hoc and informal. At the time of the study, the university and the business office were undergoing a change in administration and organizational structure, which resulted in less emphasis on long-range planning in the area of information technology.

The ad-hoc nature of the planning process and the lack of integration with the rest of the university created a static environment for FIS. The implementation and maintenance of the strategy planning process was the same as the plans themselves, ad hoc and temporary, and resulted in a short-term focus that forced FIS staff to adopt a fire-fighting approach.

Objective of the Enterprise Model

Creating an enterprise model for FIS and the university's central computer center would allow the university to formulate a long-term plan for decentral-

1. Provide computing and technical support to the various departments within the Buisiness Office.
2. Act as computing liaison between the Business Office and the Division of University Computing for mainframe computer application problems.
3. Coordinate the purchase of computing equipment and software.
4. Coordinate the maintenance and repair of computer equipment.
5. Maintain computer data communications network between offices within the Business Office and between the AS/400 and the campus network.
6. Coordinate the installation of new mainframe-based systems or upgrades of existing systems.
7. Coordinate the scheduling of all production program runs.
8. Print and distribute reports generated by production systems.
9. Provide traiing to end users of new systems or applications.
10. Coordinate the requesting of new reports or applications with DUC; this includes establishing programming priorities for all outstanding requests at DUC that belong to the Business Office.
11. Develop new department AS/400–based applications when requested by departments.

Exhibit I-2-4. FIS Operations

---

izing the activities that FIS currently performs and making constituent groups self-sufficient. This includes examining mainframe activities and replacing them with AS/400 applications and client-server processing, integrated with PC networks. Modeling FIS supported a structured analysis of the functions and processes that can be redesigned with an overall IT focus. The underlying objective of this goal set was to lessen the load on FIS to create a more efficient operation that still meets the financial reporting needs of the constituent groups.

The FIS director and his staff would use the enterprise model for the long-term strategic planning process and as a blueprint for process redesign. By shedding additional light on the operations and expertise of FIS, the models would also allow the Office of Business and Finance and the Office of the Vice-President of Academic Affairs to support their long-range planning.

The various constituency groups served by FIS are also affected by the project. These include the bursar's office, the bookstore, the police department, risk management and insurance functions, and property control functions of the university.

Methodology

The first step in developing the model was to delineate the scope of the study. FIS has three primary functions:

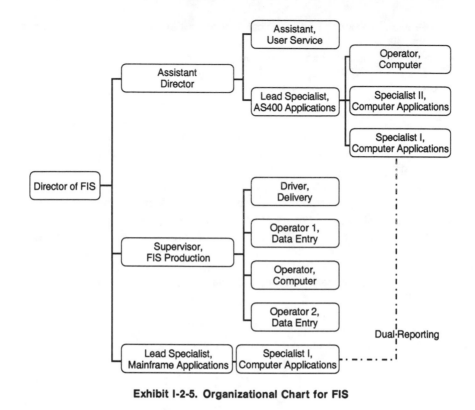

Exhibit I-2-5. Organizational Chart for FIS

- To provide computer support to various other departments of the university.
- To provide mail services within the university.
- To administer the department.

Clearly, the main focus of FIS is in the provision of computing support. The development team was fortunate to have the director of FIS as a member of the team. As director for more than ten years, he was intimately acquainted with operations and organization. His input was invaluable in accurately modeling this portion of the enterprise.

The enterprise modeling itself was performed on a 486-33 microcomputer running Knowledgeware's Information Engineering Workbench (IEW), a planning workstation software. It was formulated on James Martin's information engineering modeling methodologies with some modifications.

Procedure

Four aspects of FIS were entered into the project encyclopedia and decomposed in diagrams. These were organizational units, functions, goals, and critical

1. Goal: To increase Computing Self-Sufficiency

Subgoals:
- Maximize end-user services.
- Decentralize computing on the State University campus.
- Improve operational quality.
   2. Goal: To shift all users on the campus away from mainframe computing toward computing on mini- and microcomputers

Subgoals:
- Install additional microcomputers where appropriate.
- Increase promotional materials for user of micromachines.
- Enforce a policy to promote the use of microcomputers.
- Provide application training for all related computer areas.
- Outsource training for all related computer areas.
- Decrease dependence on the mainframe by:
  - Shifting toward miniapplications from mainframe applications where feasible.
  - Streamlining ad hoc reporting.
  - Streamlining output generation.
- Decrease dependence on minicomputers by:
  - Increasing dependence on network usage across campus.
  - Reviewing minicomputer use.

Exhibit I-2-6. FIS Goals and Subgoals

---

success factors. Decomposition proceeded in a top-down fashion. Units were decomposed into subunits and, in some cases, into individual employees. Functions were decomposed into subfunctions down to the highest level of processes. Goals were decomposed into subgoals. Through an iterative sequence of interviews, the top two goals and their corresponding subgoals were identified (see Exhibit I-2-6).

The functional decomposition was examined for responsibility and correspondence with the organizational goals. Then the corresponding subfunctions were identified and their subfunctions, and so on down the line until major processes were identified.

Once the functions and goals were identified, the critical success factors (CSFs) were identified and associated with functions and goals. Examples of CSFs for FIS are presented in Exhibit I-2-7.

Next, association matrices were created. These matrices relate two aspects to one another: functions to organizational units, goals to organizational units, and functions to goals. The purpose of doing this is to see which organizational units

1. Attract, train, and retain high-quality IS personnel.
2. Communicate service quality and reliability to top mangement.
3. Continually align IS priorities with university goals.
4. Deliver reliable, high-quality IS service.
5. Effectively use current technology.
6. Maintain professional IS operating performance.
7. Maintain close contact with all users, especially those in top management.
8. Maintain effective IS planning and leadership.

Exhibit I-2-7. Critical Success Factors for FIS

and subunits and which functions and subfunctions support which goals. At this point, factors critical to the success (CSFs) of FIS were entered into a series of matrices to determine which activities of FIS supported the achievement of success for the department.

Data model (entity-relationship) diagrams were created for each subfunction. These depict the various entities tracked by the organization, and the relationships between them. Thus, for example, the first diagram clearly shows what entities are involved in establishing priorities to guide the installation of applications on the university's mainframe computer. Also shown on this diagram are the ways in which the entities relate to one another in the course of establishing these priorities. The encyclopedia tracks these relationships to facilitate analysis of which entities support which subfunctions and processes of FIS.

The resulting matrix of entities and business functions was augmented by showing the involvement of the entities that have responsibilities for the functions. The classification included: direct management responsibility, executive or policy-making authority, functional involvement, technical expertise, and actual execution of the work. The association matrices resulted in the following comparisons (the order of mention is vertical axis, then horizontal axis):

- Subject area supports goal.
- Goal is cited by organizational unit.
- Entity type is responsibility of organizational unit.
- Function is responsibility of organizational unit.
- Function supports goal.
- Critical success factor is cited by organizational unit.
- Function supports critical success factor.
- Goal is affected by critical success factor.

Following the creation of these association matrices, a collection of property matrices was developed based on entity type (either fundamental, associative or attributive); organizational units; critical success factors; and goals.

Throughout the modeling process, several interactive sessions were conducted to integrate the director of FIS and selected staff into the development of the models. The director acknowledged that he really hadn't comprehended the department's complexity until it was laid out in the models. Integrating various members of the modeling team and end users resulted in a much more comprehensive analysis of the business operations. These interactive sessions are better known as joint strategic planning (JSP) sessions.

Joint Strategic Planning

JSP sessions involve end users in collaborative strategy development and planning sessions. IE methodology is used to provide an overview enterprise model of the organization, including its strategies and plans. CASE systems, particularly an upper-case system, may be used to support interactive and more productive sessions. During these sessions, executive and top-level systems personnel interact to develop or modify strategies and plans and subsequently develop the enterprise model for the functional activities of the organization. These sessions can also be used to determine how to use information technology strategically.

The earlier stages (statement of objectives, development of user requirements, and logical design) of the traditional systems development life cycle are the most important for user participation. From the IE perspective, these stages correspond roughly to the strategy, planning, analysis, and design stages.

There are three compelling reasons for stressing user participation during these phases/stages. First, when CASE is employed, the use of JSP sessions followed by joint application development sessions permits strategic specifications to be integrated with analysis and design specifications. This allows corporate strategy to be integrated with systems development. Second, during the early stages, the users' experience and contributions are strongest and the IS professionals' are weakest. Third, the potential costs associated with detecting and correcting errors in the later stages of the systems development cycle can be minimized by concentrating effort in the development of complete and accurate user requirements.

JSP may require many participants. To be most effective, each participant (both user and IS professional) must have the authority to make decisions concerning user requirements that will not later be overruled by higher-level management. For this reason, it is important that participants be experts in the areas they represent.

During sessions with the modeling team, the director of FIS helped evaluate and define the functional relationships incorporated into the CASE-driven enterprise model. The JSP sessions employed by this project enabled the business modelers to fully and quickly engage the director in the accurate description and subsequent modeling of the business. As the director of FIS was able to see the model being built interactively, he was able to suggest changes from previous sessions, thus deriving a better model in the earlier stages. The Information Engineering Workbench (IEW) software enabled interactive changes to be made in the model that automatically made subsequent changes in all related parts of the model. Use of JSP improves requirements analysis and subsequently results in more effective systems that enable and enact the strategic plans of the organization.

CONCLUSION

The ideals of enterprise modeling as an integrated component of strategic business engineering provide a competitive alternative to the continual decline of businesses and organizations. Organizations continually focus their attention on reducing existing cost structures and resources in a vain attempt to counteract the increasing gap between mature organizations and newer, more innovative organizations. Only by fundamentally changing not only how business is performed, but also how business is defined, can businesses expect to survive and adapt in the new information age.

I-3

New Directions in
Strategic Planning

E. Ted Prince

BY IDENTIFYING THE CORE MISSION, planners can immediately understand the kinds of issues on which they must focus. In using the concept of the core mission, classical planning approaches are not discarded. Instead, they are modified by adding the concept of core mission. In particular, such new planning concepts as critical success factors (CSFs), total quality management (TQM), gap analysis, and key competencies analysis can be integrated tightly into core mission approaches.

CHANGING GOALS

The fundamental issue in IS planning is dynamism. Systems are becoming larger, the cost of failure is higher, cycle time of products and markets is decreasing, and technological change is increasing. In the midst of all of these changes, IS must still keep current systems operating, continuously enhance them, and decide how to cope with the many paradigm shifts in the business and technological spheres.

The types of change IS must cope with are both business- and technology-based. Both business and technology goals are experiencing new emphases. These changing goals can be categorized as:

- Traditional or financial.
- Classical or production.
- Contemporary or sales and service.
- Avant-garde or innovation.

In each of these categories, objectives are constantly changing. All of these changes affect IS—ranging from application goals, design, skills, relationships between applications, and relationships with users and corporate executives. For example, IS now must add quality measurement at the design stage of an appli-

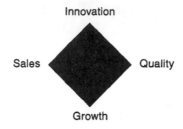

Exhibit I-3-1. The Mission Diamond

cation because it is required from the business end. IS cannot expect to ignore this issue, or attempt to retrofit it into the application.

Similarly, technological goals are changing, This affects not only IS, but in turn the business areas that may need to change their business models and processes as a result of new opportunities presented by these technological advances and changed emphases.

The fundamental problem for IS is that all of these changing business and technological goals must be incorporated into its plan if it is to respond to real world needs. With limited resources, IS must assign priority to where it expends its resources. What process can it adopt to ensure that it chooses correctly? The answer lies in the concept of the core mission.

THE CORE MISSION

The theory behind the concept of the core mission is that all organizations focus on a core group of goals. These goals are determined by several factors, including history, a concept of mission, ideology, leadership type, and culture. All organizations have a core mission. The core mission is usually implicit, however, and often does not line up with the explicit mission. An analogy in the political sphere is that a politician cannot be judged by what he or she says but by his or her actions only. In the economic sphere, this is known as the doctrine of revealed preference (i.e., what consumers want can only be judged by what they actually choose, not by what they say they would choose).

Four major types of core mission exist: innovation, quality, sales, and growth. These should not be regarded as mutually exclusive, but as four poles where an organization's core mission may be at any point bounded by the mission diamond (see Exhibit I-3-1). Nor does the core mission exclude the possibility of multiple goals—rather it focuses on the weighting of various goals, and the cultural preference between them. A large organization can also have multiple core missions, as in the case of several subsidiaries or divisions of a large company or the acquired company of a larger company. The core mission is a construct that assists in identifying the main cultural direction preferred by the organization.

Innovation. The core mission of innovation is to create new products and services. The emphasis is on research and development. The culture is research driven. The focus is revolutionary rather than evolutionary. Differentiation comes from product innovation. The CEO is often a former scientist, and researchers set the major thrust of the company. Examples include Wang (under An Wang), Apple Computers (under Steve Jobs), Xerox Corp., Polaroid, and 3M.

Quality. The quality core mission is to produce high-quality products through engineering and constant improvements. The culture is engineering-driven and evolutionary. Differentiation is achieved through product quality and performance rather than newness. The CEO is often a former engineer. These organizations often end up creating excellent products that become technically obsolete. Examples include Digital Equipment Corp., Mercedes Benz, and AT&T.

Sales. The sales core mission is to maximize market share. Emphasis is on sales and marketing and contact with clients. Salespeople tend to set the thrust of the culture and the CEO is often a salesperson. Differentiation is achieved through attention to the customer and service. Examples include IBM, ADP, General Motors, and Wal-Mart.

Growth. The growth core mission is to build a large company, particularly through acquisition and diversification, often in several completely different markets. Such core missions are typically found in conglomerates, and the CEO is often an accountant. Because the company operates in different markets, no one can be an expert. The culture tends to be controls and planning driven. Differentiation comes from low costs, tight controls, and efficiency. Examples include ITT (under Geneen), General Electric, the old Gulf and Western, and United Technologies.

In each of these different types of companies, the prevailing culture selects the core mission from many different alternatives. The core mission then guides all business choices. It determines strategies in all parts of the organization. It provides the implicit criteria for hiring and dismissing. It produces a cadre of similar-thinking people that in turn reinforces the core mission. All other types of business goals are dealt with by how they affect the core mission. Therefore, priorities for resources in an innovative organization tend to favor research and new products, as opposed to expanding sales. In an engineering organization, there will be a preferential shift towards product improvement as opposed to new products.

To be relevant to an organization, plans and strategies must take the core mission into account. In so doing, the plans take account of the particular cultural and organizational factors and preferences in the organization. IS planning must be based on this approach.

CORE MISSION AND IS STRATEGY

IS strategies and plans, like those of the other divisions in any organization, must be consistent with, and promote, the core mission of the corporation. IS

IS Innovation

IS Marketing   IS Quality

IS Planning and Growth

Exhibit I-3-2. IS Mission Diamond

	Business Focus	**Technology Focus**
IS Innovation	Avant Garde	Avant Garde
IS Quality	Classical/Production	Classical/Production
IS Planning and Control	Traditional/Finanacial	Traditional/Finanacial
IS Marketing	Contemporary/Sales and Service	Contemporary/Sales and Services

Exhibit I-3-3. IS Missions

priorities must match the cultural priorities of the organization in its strategic planning. The core mission of the organization will also dictate the types of people, skills, processes, and applications in IS. Simply put, the IS mission is to treat the organization of which it is a part in the same way that the organization treats the outside world.

IS must therefore strive to situate itself roughly at the point on the mission diamond occupied by its host organization. This means it must strive to emulate the core mission within IS. Therefore, IS can be seen to have four polar missions (analogous to that of the organization): IS innovation, IS quality, IS planning and control, and IS marketing (see Exhibit I-3-2).

Each of the four IS missions corresponds to a particular type of business or technology focus (shown in Exhibit I-3-3). IS marketing corresponds to contemporary sales and service goals. IS planning and control to traditional or financial goals, IS quality to classical or production goals, and IS innovation to the avant garde goals. In attempting to answer the question as to where IS should focus, this exhibit can be used as a starting point. It can help decide what specific types of strategies IS will follow given the different core missions of the host organizations.

IS Innovation. Innovative organizations focus on the front end of the supply chain (e.g., invention, design, and creation). They tend to be weak at the back end in such areas as enhancement, sales, service, and operational efficiency. IS departments in such organizations would benefit from focusing on such areas as formal, creative needs-finding processes, and new product creation and planning. Their application focus may be in such areas as building research and development repositories, enhancing shareware, and enhancing communication and creativity.

Quality-Oriented Organizations. Quality-oriented organizations focus on enhancement, continuous improvement, or the middle, production part of the supply chain. In such organizations, the process focus should be on formal software quality assurance, formal measurement of quality, formal testing methodologies, and defect measurement programs, both in IS and in the production and service areas. The application focus is likely to be on production systems, particularly in companies using computer-aided manufacturing. The technology focus is likely to be the classical or production focus (i.e., unlikely to be leading-edge or too old to provide the quality desired by the engineering and actuarial culture).

Planning and Control Organizations. The planning and control core mission is interested in the cost of the entire chain. IS organizations in their process focus must concentrate on formal measurement systems for IS costs and productivity, and formal standards for resource allocation. Such organizations will be impressed by development of a planning and control systems architecture, and methods for cheaper and quicker development (e.g., RAD). In their application focus, these organizations will focus on financial systems, planning, accounting, budgeting and analysis, executive information systems, and communications, the lifeblood of such control-oriented organizations. The technology focus here will be traditional.

IS Marketing. The IS marketing core mission will include formal standards for user service, formal measurement of user and customer satisfaction, formal presentation standards for ease of use, formal disaster recovery standards, and measurement and improvement of systems responsiveness. The application focus will be on sales and marketing systems, customer-response systems, and post-sale service systems. The technology focus will be contemporary, or sales- and service-oriented to the customer rather than the developer or producer.

The assessment of such strategies makes only generic suggestions. The precise mix of strategy must be tailored to the particular circumstances of the organization, taking into account its core mission, its position in the mission diamond, and the level of maturity of the organization. The suggestions provided in this chapter, however, are examples of the truism that if the precise strategies are not tailored to the core mission, IS will be seen as out of step with the organization.

Formulating the Core Mission-Correct Strategy

The approach of using the core mission allows classical planning methods to be modified to allow planners to control the two key issues of dynamism in the environment, and culture within the organization. New strategic planning ap-

Core Mission	Business Processes
Innovation	New Product
Quality	Production
Growth	Planning and Control
Sales	Customer

Exhibit I-3-4. TQM Process

---

proaches have emerged, however, and they should be viewed together with the core mission approach.

There have been four main new entrants to strategic planning methods over the past few years. All derive from non-IS areas and are therefore being increasingly used in an effort to more closely align business and IS goals. These methods are: critical success factors (CSF) analysis, total quality management (TQM), key competencies analysis, and gap analysis.

CSF Analysis. This is an important planning tool. By examining those factors on which business success critically depends, management can focus scarce resources for maximum impact. Nonetheless, CSF analysis has flaws. The principal problem is that many CSFs may be identified, all of which appear to be important, but only some of which are actually important. It is here that the core mission approach can be crucial in identifying which of the many CSFs are important to that particular organization and the way in which they should be pursued. By more closely aligning CSFs with the cultural preferences of the organization, their number can be reduced, the most important ones identified, and the potential for success correspondingly increased.

TQM. Although TQM has been popular in recent years, its efficacy has been widely questioned. Much of this stems from the fact that TQM is typically applied without regard to the particular core mission of the organization so that inappropriate implementations are prescribed, leading to frequent failure and loss of credibility.

Exhibit I-3-4 points out the relationship between business processes on which TQM typically concentrates and the four types of core mission. TQM takes as its aim the application of process methods to all areas of the supply chain, ranging from the creation of the new product, through production planning, and customer processes. It is clear that the core missions relate to four points along the supply chain and that what is regarded as a core mission simply represents the propensity of an organization to have particular competencies in one part of the supply chain.

If no regard is paid to the core mission, it is easy for TQM to be applied to all parts of the chain, even parts that an organization is less interested in. The core

Exhibit I-3-5. TQM and Strategy

mission approach tells executives that, in introducing TQM, they should look at the core mission, and apply TQM first to those parts of the supply chain in which the organization has most interest. This maximizes the perceived payoff and increases the potential internal support for the TQM process.

One significant implication of this in the IS area is that, armed with this core mission approach, IS can unbundle the development lifecycle and focus TQM efforts on those parts of the lifecycle having most cultural significance. Too often, lifecycle approaches fail because too much is required of the organization at one time in introducing it. The core mission approach allows executives to cope with constraints on organizational learning. Because IS is in a constant battle to implement some sort of developmental approach and standards, this is an important issue.

A further implication is depicted in Exhibit I-3-5. Typically, in most organizations, quality assurance, TQM, strategic quality planning, and strategic planning are all carried out as different processes by different units within the same IS organization. TQM should be regarded as part of strategic planning, as should strategic quality planning. In addition, quality assurance, usually regarded as an organizational orphan, should also be tightly linked to these activities.

Key Competencies Analysis. This has become more popular in recent years. It is easy to identify key competencies in isolation, however, without regard to what the organization is really interested in achieving. Core mission approaches allow competencies to be identified that overlap with the core mission of the organization.

Gap Analysis. This has been practiced for many years. In its usual form, gap analysis identifies gaps in technical skills in IS that must be filled for the department to achieve its technical goals. This is too narrow a focus, and the real effort should be in identifying gaps in systems, technology, skills, and processes relative to the core mission.

In summary, the core mission approach does not have to be applied in isolation from classical and newer strategic planning approaches. It can and should be used as a template for these approaches, to filter out those issues of real importance to the organization. In so doing it can act as a short cut strategic approach, to be used as a quick credibility check on the results of more conventional approaches.

Dealing with Changing Core Missions

Even where a core mission is constant, many employees may not understand what it is, or may not look behind the organization's rhetoric to decide what is really important to it. Organizations constantly have staff turnover at all levels, so even if some people understand the core mission, new employees may not. Very often, an organization may not be introspective enough to know what its core mission really is. In all of these cases it is important to identify it as a key component of strategic planning, not only in MIS but in all parts of the organization.

More interesting is the case where the core mission changes. This may occur in a variety of ways (e.g., merger, acquisition, divestiture, change of CEO or board, or an unusual or abrupt shift in the organization precipitated by massive change in its environment). These events occur continuously, but it is rare for most members of an organization to realize that an abrupt change in core mission invalidates many of the assumptions on which planning was previously based. Although change is accelerating, there is often failure to see this type of transformation either by employees at the working level or, just as easily, by senior management.

Changes in the core mission, however, can be life and death issues in the marketplace, and matters of corporate, individual and professional survival. Neither corporations, their managements, senior executives, or workers themselves can afford to ignore such changes. In IS, where technological change at huge rates compounds all the usual business factors, this issue is even more acute. All participants must constantly look to whether the core mission has changed as the key input to their strategic planning.

CONCLUSION

In many organizations, the leadership of IS is a revolving door. The initial problem may well be that IS has no concept—let alone understanding of—the core mission. A lack of understanding on the part of IS, however, probably reflects a lack of understanding also on the part of corporate management. To the extent that there is a lack of understanding on both sides, the issue is ultimately one of process. Unless a process is put into place to educate both sides, the problem will continue, to the overall organization's detriment.

Much of the problem in IS reflects a wider problem in the organization. The concept of the core mission can be a major benefit to IS in countering its lack of understanding, but by itself it may do little to solve the wider problem. The concept of the core mission will be most useful only in the context of an organizationwide process to share knowledge concerning the core mission, changes in it, and how to ensure that all members of the organization share this knowledge to maximize the efficiency of its efforts.

I-4

Organizational Impact of Technology and Change

Kenneth P. Prager
Miles H. Overholt

AT SOME POINT , almost everyone has been part of a failed project. Sometimes the projects fail technically because the system is not correctly designed, the hardware malfunctions, or the software does not do what everyone hoped it would do. However, projects often fail not because of technical flaws, but because the people in the organization reject them.

Some of these failures occur because the systems do not accurately address a real business need. Other systems fail because they contradict the expected reporting relationships and communication paths within the organization, or because they do not harmonize with the way managers look at "who we are as a company." Many systems fail because they require that people in separate functional areas work together. A few systems even fail because of rumors about the new system's impact on people's jobs. Typical symptoms of a failed project are:

- The intended users are not using it.
- Managers and employees are using it, but feel it does not meet their needs—made obvious by constant complaining about the system, about the IS group, or about the difficulty of using the system.
- There may be open animosity toward IS.

Whatever the reasons, a failed effort to implement new technology is always a failure to understand and adequately manage the change process.

Whenever organizations try to change, there is risk. One of the most significant risks is that people within the organization do not accept the change and therefore block or defeat it. Because new technology represents radical changes to people and their jobs the risks are great. Changing peoples' lives is a serious issue that must be managed with care.

For most of the twentieth century, the primary strategy of the successful corporation was to create stability, institutionalize routine, and manage procedures. Management did not have to know how to guide people through significant, life-style altering change. Today, because of restricted capital, growing competition, and demands for quality, every manager needs to understand how to orchestrate change.

PEOPLE-CENTERED ORGANIZATIONS

One way that managers begin to gain this expertise is to view organizations differently. Executives often refer to people as human capital or "our greatest asset," as if people were entries on the balance sheet. Unintentionally, this implies that organizations exist without people and that, in fact, organizations simply use people like any other piece of machinery or equipment.

A new focus is needed that views people as the core of the organization and recognizes that all other aspects of the organization are built on people. The people-centered organization is this new paradigm.

Seven Components

People-centered organizations consist of seven interactive and interdependent components. In this context, organization means not only the company but also each division, function, and team, because each of these organizations is itself people-centered and has its own seven components.

- The genetic core is the center of the organization and the locus of power for all decisions. The more people the genetic core includes, the more powerful the organization.
- The philosophy comprises the publicly articulated beliefs of the organization that emanate from the genetic core.
- The organizational behaviors are the set of interactions between individuals and groups within the organization.
- The formal organization is the structure, reporting, and reward system of the organization.
- The information systems and technology are the structures, constraints, and demands of the information, manufacturing, and service delivery systems within the organization. This component deals with acquiring the information and using it to make decisions.
- The informal organization is the network that binds the employees together. The grapevine is its primary communication channel and the informal leaders are some of the most powerful in the company. Frequently, the informal organization counterbalances a weakness in the formal organization.
- The culture is the set of beliefs about how the organization runs day to day. It includes the organization's history and myths, artifacts and symbols, as well as expectations of such things as how meetings are to be run and how individuals are to dress.

In flexible, responsive organizations, the seven components are aligned and congruent, supporting and reinforcing employees. If one or more of the components is out of alignment, the resulting imbalance creates a blockage that hinders employees in their work and in implementing change.

REQUIRED CHANGES

For example, if top management decides to implement total quality management (TQM), the organization undergoes a fundamental philosophical change from which management can map the required changes in each of the other six segments. The genetic core, for example, must focus on quality and customer-oriented strategies. The formal organization must be redesigned so compensation systems reward teams, communication channels broaden, and work processes and tasks change. New systems and technologies must be put in place to measure, support, and enhance the changes. Users need greater access to more customized data. The informal leaders need to adopt and support the new philosophy, because if they reject the change, then TQM becomes just another program of the month. The culture must foster customer-oriented thinking and open communication among all the levels of the organization.

Successful change requires that the seven segments are congruent to support people in their new way of operating. The greater the balance and the fit among the seven components, the fewer the barriers to effective change by people.

TECHNOLOGY, NEW SYSTEMS, AND CHANGE

In many situations, IS professionals and IS management are the driving force behind change. They are designing, creating, and implementing massive system changes that force the company to change its culture. IS enables teams to become self-directed and empowers individuals to redesign work-flows.

This push from the information and technology component of the people-centered organization also causes reactions in the other six components. The IS manager must design implementation processes that ensure that the components are realigned to support the people as they use the new technology. Consequently, IS professionals are now in the change business and must become experts in managing organizational and human change.

Designing the Process

To maximize employee support and ownership of new information technology and ensure the best fit, the change process should:

- Guide work groups at different hierarchical levels through the process of aligning the seven components of the people-centered organization.
- Assist people in removing the organizational and personal barriers that block the change.
- Make the alignment process obvious so that the people can replicate it.

IS managers should view the redesign as a participative process that ripples down through the organization. Each team designs a clear framework at his own hierarchical level that allows teams at the next lower level to redesign their work processes to match the requirements of the new system.

For example, first the executive team uses the people-centered organization framework, as described in this chapter, to redesign the entire organization on a very broad level. Then the functional management teams participate in a similar process to plan how their areas fit this new design. Next, natural work groups within the functions analyze the required changes. In this way, each successive level is empowered to make the appropriate detailed changes at its level. The process begins at the top, cascades down the levels, and loops back and forth among the levels as strategies and issues need to be reexamined.

Using the people-centered organization as a framework, IS managers can lead groups through this redesign process, identifying barriers to implementation and mapping the strategies changing to the new technology.

The process consists of analyzing each component of the people-centered organization. For each component, and at each hierarchical level, the facilitator leads the group as it develops a vision of what the future state of that component will be after the new technology is implemented. Then, the group profiles the current state of the component. Finally, the group compares the two to analyze what needs to be changed and what the group perceives as potential barriers to the change. The facilitator asks the participants to use phrases that describe their perceptions of each of the components.

Action Plans

The obvious product of this process is an action plan at each level and within each work group that incorporates the groups' decisions. A not-so-obvious but more important product of this process is buy-in to the new technology because the people have participated in the decisions that affect their own jobs—they have devised their own plans for changing to the new way. Finally, because people know what the facilitator is doing, people can apply the same process themselves to later changes.

IS managers need to decide if everyone in the organization should be included, which would be ideal, or if a set of representatives is sufficient. Leading the process does not require extensive theoretical knowledge, but it does require significant group leadership skill.

ALIGNING THE INFORMATION SYSTEMS FUNCTION

When the information systems and technology component is the focal point of the change, then IS management must first use the people-centered organization model to align itself with the new technology. IS management and staff must answer such questions as:

- *Genetic Core.* Is the IS management team fully on board?
- *Philosophy.* Does everyone in the IS department agree that supporting the

new technology fits with "who we will be" as an IS department after it is in place?

- *Behavior.* How will the new technology affect interaction within the IS department? How should IS personnel behave when they interact with people from other departments?
- *Formal organization.* Is the IS function appropriately organized to support the new technology? Are new positions needed (e.g., for help desks)? Should the department create new communication mechanisms within IS and between IS and the rest of the company (e.g., E-mail or an IS newsletter)?
- *Information and technology systems.* Are the IS department's internal systems congruent with the proposed change? Does the IS staff have access to the information needed to make decisions?
- *Informal organization.* Are the informal leaders within IS supporting this new technology? Are there rumors circulating within IS that may create resistance? For example, will people believe they will lose their jobs if they are not properly trained?
- *Culture.* Do the normative beliefs of the IS employees fit the new technology? Does the actual daily running of the department support the change?

Only after IS management and staff have ensured alignment within the IS department can the IS manager attempt to gain the necessary buy-in from the people in the rest of the organization.

Issues Affecting the Process

The organizational redesign process must tackle hierarchical, level-specific strategic issues surfaced by the technological change. IS management and the functional or work area participants must address the issues if there is to be successful implementation. IS management may expect to encounter issues similar to the following examples.

Genetic Core Issue. Gaining support from management for the new technology is key. Without management support, employees perceive the change as unimportant and choose not to support it. Frequently, IS management decides to sell new technology to executive or functional management on benefits that impress IS professionals. Non-IS management will not be influenced to change to a client/server environment because the technology encompasses different platforms. Rather, management must perceive a real benefit, a solution to a problem that it is experiencing.

Example. An IS vice-president for a growing $150 million manufacturer learned this lesson the hard way. During the strategic planning process, IS management and staff recognized the need to shift to an online, real-time system to solve certain scheduling and production problems. Accustomed to unthinking agreement by executive management, the IS vice-president recommended implementation of the new technology. The IS strategic plan was accepted and the IS department committed hundreds of working hours developing specifications and researching hardware and software. When the IS staff presented the findings to executive management, it disagreed with the premise that there was an organizational need and wondered why IS had spent all this time on a "star wars" project. Two months later the IS vice-president was looking for a new position.

Preferred Action. The IS vice-president failed to identify executive management's needs and link them to the needs of other functions. If IS had facilitated an organizational redesign session with executive management, it would have discovered management's perception of future problems and been able to address its particular needs. Then IS would have been able to project several future scenarios illustrating the need for an online, real-time system.

Philosophy Issue. The primary issue in the philosophy component is how well the new technology supports what the belief systems need to be after a new technology has been implemented. IS management must match the philosophy of new technology with the corporate philosophy to ensure acceptance. Technologies with philosophies that conflict with those of the corporation are barriers to implementation.

Example. A large hospital system installed electronic data interchange (EDI), creating major changes in employees' roles. IS management and staff were overwhelmed with the conversion from the old operating system to EDI. To complete the conversion on time, the department focused on providing comprehensive technical training. Almost all the trained employees understood how to use EDI and demonstrated great proficiency in using the system. However, everyone in the purchasing department was unhappy with EDI and avoided linking suppliers into the system. Proficiency and familiarity with the system was not a problem, and IS management could not understand what was wrong. Finally, a supplier casually informed management that the purchasers were afraid of losing their jobs and did not understand that EDI was intended to free their time to do more sophisticated activities.

Preferred Action. IS management was in a familiar double bind. The deadline pressures caused IS staff to forget that functional management and employees needed to participate in a process so they would understand how the system would change their jobs. In addition, employees needed an opportunity to create new roles and work processes to enhance the use of EDI.

Formal Organization Issue. The introduction of new technology alters the hierarchical structures, workflows, communication process, communication processes, and compensation systems. To gain full commitment and ensure that the new technology is used, IS management must involve management and employees in the redesign process.

Example. In a large financial services corporation, IS and executive management were pleased with the company's new system. It had increased productivity and was accepted by users throughout the corporation. Serendipitously, a task force working on improving communications discovered that cross-functional communication required by the new system was blocked by the traditional hierarchy. Users who needed to communicate cross-functionally had to send information upward, then wait for it to be communicated laterally and downward before it arrived to the intended receiver. The delay and the inevitable distortion of the information was blocking further increases in productivity.

Action. IS management conducted an organizational redesign process throughout the corporation. Even though the process was time-consuming and lengthy, lasting 19 months, the results were that work groups realigned communication channels to match the system, cycle time decreased, and employees felt empowered to work more productively.

Behavior Issue. New technologies demand new behaviors between users and the system, between functions, and between individuals. Everyone must perform different tasks with the computer, new and more information is available to functions, and employees no longer rely on the same individuals for information. IS management must assist everyone in identifying what new behaviors are required, in acquiring the behaviors, and in practicing the behaviors.

Example. The manufacturing division of a Fortune 500 company had been operating on a mainframe with no server system for 10 years; in fact, the technology was dubbed "no-line, past time." When the division started to change from the old system to a WAN, the IS vice-president created a cross-functional pilot team. Its assignment was to determine the behavioral changes necessitated by the WAN, document those changes, then create a training lab that simulated the experience for all employees. The lab was deliberately nicknamed "WAN-a-be" and became a major factor in ensuring successful implementation in a resistant culture.

Action. Successful organizational redesign requires anticipating changes rather than reacting to them. If employees have the opportunity to anticipate how their work areas will change, they will shift their behavior to match the new realities.

Information Systems and Technology Issues. The key issue when technology is the driving force behind large-scale organizational change is creating a reciprocal feedback and adjustment process that enables the IS staff to improve the system. Most new technologies must be adapted to fit specific user needs. The organizational redesign process is an excellent opportunity to jointly consider the best way to make the adaptations.

Example. A small professional services company engaged a consulting firm to create a new reporting system to track costs and billable time more accurately. Once the system had been implemented, the staff complained about wasted time and useless information. The managing partners were upset because the staff had been interviewed and involved in the design of the system. However, after the system had been implemented the staff had no opportunity to refine it. The consulting firm was no longer available and the IS department was overwhelmed by efforts to train new users of the system and manage the increase in data. Before a reciprocal feedback channel could be established, staff discontent with the system was so high that most returned to the manual reporting system.

Preferred Action. The managing partners and the consulting firm were so preoccupied with ensuring that the system was perfect that they were unable to plan for user feedback after implementation. If managing partners had simply announced that the new system would be helpful but would also require much feedback and joint participation in improving the details, the staff attitude would have been cooperative. Instead, the staff refused to use the system.

Informal Organization Issue. Informal leaders can make or break the acceptance of new systems. IS management wants to avoid a negative review of a change being broadcast through the company's grapevine. IS management must take advantage of the power of the informal organization to gain support for new technologies. To manage the informal organization you must know who the informal leaders of the organization are.

Example. Senior management in a large corporate services department decided to install a departmental LAN. Working closely with the IS staff, departmental management carefully developed a training and installation schedule, involved key supervisors in the design phase, and was poised to install the new system. Just as the first staff members started training sessions, a rumor began to circulate that all part-time staff members would lose their jobs when the implementation was completed. The training instructors reported to management that the staff appeared disinterested in learning the new system and that some individuals were asking pointed, angry questions that had nothing to do with the training.

Action. To address the issue, IS management was asked to facilitate employee discussion sessions on the impact of the new system. The staff remained quiet through the first two sessions. Finally, a part-time employee asked when someone was going to tell them that their jobs were being eliminated. The IS facilitator had the opportunity to explain that no jobs were to be lost. To prove the point, the facilitator conducted a brief redesign session on the role of part-time workers. Shortly thereafter, the rumors stopped. Not only had management successfully stopped the misinformation from circulating, but a more productive role for the part-time employees had been created.

Culture Issue. The organization must ultimately accept and institutionalize the new values that are created by new technologies. Employees must learn to value the technology and to integrate it into the mainstream of the organization. Status symbols must change to reflect this acceptance, oral history should include the new system, and the new technology must become just another part of the way people work. This is necessarily a slow process, but one that IS management can influence.

Example. When a Fortune 500 corporation began to downsize, secretaries were among the first employees to be laid off. Senior management decided that directors, managers, and supervisors could share secretaries, enabling the corporation to lay off 50% of the secretaries. Twelve months later, senior management authorized the purchase of PCs so that directors, managers, and supervisors could do their own word processing. Senior management learned that few directors or managers ordered the PCs. When managers seeking promotion would not use a PC, senior management realized that it had failed to see the connection between the status of having a secretary and the lack of status of using a PC for word processing.

Action. At a break in an executive meeting, one senior manager shared this problem with a peer, who in a staff meeting had mentioned the problem to corporate directors. The senior manager immediately had a desktop PC installed and started attending word processing training.

CONCLUSION

Developing expertise in and becoming comfortable with managing change can help the IS manager avoid projects that fail for nontechnical reasons—failures caused by people and organizational problems. To develop this expertise, there are some specific actions the IS manager can take.

First, the IS manager should embrace the people-centered organization model. This recognizes that organizations are built on people and that people need to have an active role in redefining and redesigning their organizations to align with the new technology.

Second, the IS manager must develop new skills and gain a new understanding of people issues. Some of these skills are:

- Team building and teamwork skills to help work groups function effectively.
- Facilitation skills to lead work groups through the redesign process.
- Communication skills, because communication within a work group, within a department, among departments, and throughout the organization is absolutely critical for successful change.
- Conflict resolution skills to deal with the inevitable struggles as people change.

Finally, using these skills, the IS manager must ensure that all components are aligned—not only with each other, but also with the new technology and the new organization. The IS manager can act as a facilitator to help work groups examine each of the seven components of the people-centered organization. For each component, the work groups must define what that component will look like with the new technology, what the component looks like today, and what the strategy is for moving the component from today to where it needs to be to support the new technology. This redesign occurs at a macro level where the executive team broadly redefines the organization; at a mezzo level where functional managers redefine their own departments; and at a micro level where individuals in the organization define their own strategies for changing to the new environment. By applying these concepts, IS and users together can anticipate and manage the many changes that need to be made to ensure successful implementation of new technologies and systems.

I-5

Outsourcing Realities

Mary Lacity
Rudy Hirschheim
Leslie Willcocks

WHEN KODAK OUTSOURCED its information technology operations to IBM Corp., Businessland, and Digital Equipment Corp. in 1988, it triggered a renewed interest in outsourcing. Although selective outsourcing of certain IT functions—programming, training, documentation, and disaster recovery—has existed since the beginning of data processing, Kodak legitimized the use of total outsourcing in which companies dismantle internal IT departments by transferring IT employees, facilities, hardware leases, and software leases to third-party vendors.

Kodak's success sent a message to senior executives that IT had matured into a commodity service best managed by an external supplier. A bandwagon effect resulted as other senior executives sought to duplicate the strategic alliances enjoyed by Kodak and its IT outsourcing vendors. Other large companies, among them Enron, Freeport-McMoran, Continental Airlines, General Dynamics, and Continental Bank, signed long-term outsourcing contacts and publicly announced the anticipated benefits: 10% to 50% cost reductions, increased service levels, and access to new technologies and technical expertise.

In the years that have passed since the resurgence of outsourcing interest, companies that leapt on the outsourcing bandwagon have had time to evaluate whether their expectations were realized. Did IT costs drop by as much as 50%? Did service levels increase? Did vendors introduce new technologies? This chapter attempts to answer these questions by reporting on the experiences of senior executives, IT managers, and vendor account managers from 14 US and 15 British companies involved in outsourcing decisions.

THE OUTSOURCING STUDY IN BRIEF

In all, more than 100 interviews were undertaken. Each interview (typically lasting between one and two hours) was tape-recorded, transcribed, and analyzed. In each company, at least two people in different roles (e.g., CEO, CFO, business unit manager, IT director, account manager, or IT staff member) who might have different perspectives on the outsourcing decision process was interviewed.

The companies were from a variety of industries. The study included a broad spectrum of outsourcing arrangements: selective (partial) versus total outsourcing; use of multiple vendors versus a single vendor; short-term versus long-term contracts; detailed versus non-detailed contracts. Also included were organizations that had been involved with outsourcing for a considerable length of time (3 to 10 years) as well as those that had more recently signed outsourcing deals that were less than three years old. In total, 40 outsourcing decisions were examined (several companies evaluated outsourcing on multiple occasions). Of these, 14 decisions resulted in total outsourcing where at least 80% of the total IT budget was outsourced; 26 decisions resulted in selective outsourcing where 5% to 30% of the IT budget was outsourced.

Although it is dangerous to make sweeping generalizations based on only 40 outsourcing decisions, a number of patterns emerged, common to both sides of the Atlantic. This chapter discusses these patterns by:

- Summarizing the expectations people had before outsourcing.
- Explaining what went wrong—and why—when the expectations were not met.
- Suggesting mechanisms, based on successful outsourcing experiences, to help ensure that all parties have sensible outsourcing expectations that can be realized.

Before going into detail on these three points, one broad theme arises from the data that needs to be expressed at the outset. Based on the participants' outsourcing expectations and experiences, the participants who were most disappointed with outsourcing followed a total outsourcing strategy, whereas participants most pleased with outsourcing generally pursued a less publicized—yet more controllable—selective sourcing strategy. Although some companies have been satisfied with their total outsourcing arrangements (mainly those that had signed airtight contracts), the same cannot be said for a significant portion in our study (particularly those that did not sign detailed contracts).

Companies that engaged in total outsourcing often suffered service degradation and, in some cases, increased IT costs. Indeed, some companies have threatened to sue their outsourcing vendors for nonperformance. Three companies proclaimed their total outsourcing decisions to be outright failures—they terminated their contracts early despite significant penalties—and rebuilt internal IT departments from scratch.

While it might be too strong to say these outsourcing disappointments offer evidence that total outsourcing is a fad surrounded by hype, false hopes and empty promises, it at least appears that there is some semblance of truth to this criticism. In contrast, companies that engaged in selective outsourcing often realized their outsourcing expectations. With selective outsourcing, however, initial expectations were often much more modest than the 50% saving hoped for by the total outsourcing participants. However, modest, selective outsourcing participants have generally been able to avoid the potentially negative conse-quences of outsourcing by signing tight, short-term contracts for a definable subset of IT services.

WHERE DO EXPECTATIONS COME FROM?

The main goal of this chapter is to contrast participants' expectations before outsourcing with their actual outsourcing experiences. Where did their expecta-tions come about? The trade press, discussions with peers, and consultants' forecasts all portray outsourcing in a highly rational way—management looks for the best way to deliver a cost-efficient IT service to the organization. In evalu-ating outsourcing, management uses objective criteria assuming a common (agreed) set of beliefs and values about the need for having the best IT service at the lowest possible cost. But is this the way outsourcing evaluation is actually done?

In addition to the typical financial, business, and technical expectations about outsourcing, several political motivations for outsourcing were uncovered (see Exhibit I-5-1). More than a few participants viewed outsourcing as a way to demonstrate their corporate citizenship, to enhance their careers, or to eliminate a troublesome IT function. We recorded numerous cases where an outsourcing decision was perceived as a political battle to justify the existence of IT. When IT was perceived by senior management as a cost pit, it often initiated outsourcing evaluations. Similarly, IT managers often used outsourcing evaluations to justify their existence—either to demonstrate that in-house performance was superior or to show a business orientation to senior managers—by outsourcing "commodity" IT functions in order to reduce costs or focus on higher value-added IT work.

Outsourcing becomes a vehicle for subtly managing (some might say manipu-lating) the perception about the value of IT. The inclusion of political expecta-tions and behaviors to the more rational expectations provides a more realistic assessment of sourcing decisions.

Financial Expectations

Many participants, especially senior managers, cited financial reasons for out-sourcing, usually to cut costs, improve cost control, and restructure the IT budget.

Cost Reductions. Many participants expected that outsourcing would save them money. They perceived vendors to enjoy economies of scale that enable them to provide IT services at a lower cost than internal IT departments. In particular, participants believed that a vendor's unit costs are less expensive because of mass production efficiencies and labor specialization. This perception was based on the fact that vendors submitted lower bids than current IT costs.

After participants engaged in outsourcing, however, many failed to fully re-alize their anticipated savings, and indeed, IT costs rose in some cases. The participants who failed to realize cost savings were surprised by hidden costs in the contracts. The disappointments were associated with loose contracts that

Initial Expectations/ Reasons for Outsourcing	Determinants of Failed Expectations	Determinants of Realized Expectations
Financial		
Reduce Costs	Hidden costs	Create specific contracts
Improve Cost Controls	Users bypassed vendor bureaucracy	Involve business managers in rating/ranking user requests
Restructure IT Budgets	Overtime, IT unit costs out of sync with price/performance improvements	Sign short contracts
Business		
Return to Core Competencies	Loss of business expertise resulting from treating the entire IT department as a utility	View IT as a portfolio of core and noncore activities
Facilitate Mergers and Acquisitions	Vendor unable to absorb new assets and people because of short notification	Select a vendor with expertise in mantling and dismantling systems
Start-Up Companies	IT costs and service expectations not met because of the new company's inexperience with IT	Sign short contracts that are renegotiable as soon as IT needs are better defined
Technical		
Improve Technical Service	Service degrades because of lack of service-level agreements, contention with the vendor's other clients, overworked vendor staff	Specify service-level agreements, vendor account manager, and penalties for nonperformance in contract
Access to Technical Talent	No change in technical talent when staff transfers to the vendor; new vendor talent is possible but expensive	Make explicit what technical talent is desired in the contract and its cost
Access to New Technologies	Possible but expensive; vendors motivated to run old technologies as long as possible	Sign contracts for duration of the expected life of current technology and renegotiate as new technologies emerge; for longer contracts, include incentives to share the benefits of new technologies, negotiate planned obsolescence and replacement of technology

Initial Expectations/ Reasons for Outsourcing	Determinants of Failed Expectations	Determinants of Realized Expectations
Political		
Prove Efficiency	Senior management views evidence of IT efficiency with skepticism	Ensure credibility of IT efficiency through senior management buy-in; senior management sponsors outsourcing evaluation
Justify New Resources	Senior management views justification of new resources with skepticism	Ensure the justification for new resources through senior management buy-in; senior management sponsors outsourcing evaluation
Duplicate Success	Senior management initiates outsourcing based on exaggerated claims appearing in the literature	Involve outside expertise to help temper expectations; analyze the "real" experiences of outsourcing successes
Expose Exaggerated Claims	IT manager sponsors outsourcing evaluation to expose exaggerated claims, but it is perceived by senior management as a ploy to simply buy additional time; IT manager fails to establish credibility	Involve outside experts in developing objective evaluation criteria and site visits to best-of-breed companies
Eliminate Troublesome Function	Cannot outsource the management of IT	Outsource IT only from a position of strength (i.e., a company cannot outsource a problem)
Break the Glass Ceiling	Initiating of outsourcing evaluation perceived as ploy to enhance reputation of IT manager	IT manager must possess business savvy rather than technical capabilities

Exhibit I-5-1. Determinants of Failed and Realized Outsourcing Expectations

merely stipulated that vendors perform the same services previously performed by the internal IT department. Consequently, the vendor billed participants for services the participants assumed were in the contract. In one case, a participant was charged $500,000 for so-called extra services. Another participant warned that exceedingly low bids may indicate hidden costs.

Another source of hidden costs were the vendors' standard change of character clauses. For example, if a participant changed from one electronic spreadsheet package to another, the vendor charged an excess fee to support the new package. From the participant's perspective, an electronic spreadsheet is an electronic spreadsheet; why should there be a charge for changing packages? From the vendor's perspective, a new spreadsheet package triggers a need to retrain staff and also creates more customer queries as new users adapt to the package. Participants who realized cost savings generally had signed airtight contracts that fully documented the services and service levels performed by the previous IT department. In some cases, they prudently demanded that the vendors demonstrate how they planned to reduce costs by identifying where exactly the economies of scale exist. Although some vendors could clearly demonstrate economies of scale, others provided intangible reasons, such as "We are technical experts."

Some vendors refused to divulge their cost structures because it would potentially sabotage future bids with other customers. In general, those participants whose cost expectations were met were able to fully define their IT needs and were able to assess the vendor's true cost advantage over the internal IT department.

Improved Cost Control. Another financial rationale for outsourcing was gaining control over IT costs. As any IT manager can attest to, IT costs are directly related to IT user demands. In most organizations, however, IT costs are controlled through general allocation systems that motivate users to excessively demand and consume resources. General allocation systems are analogous to splitting a restaurant tab—each dinner guest is motivated to order an expensive dinner because half the cost will be shared by the other party.

Participants saw outsourcing as a way to contain costs because vendors implement cost controls that more directly tie usage to costs. In addition, users can no longer call their favorite analysts to request frivolous changes but instead must submit requests through a formal cost control process.

Some participants failed to realize cost control expectations through outsourcing. Instead, users began to use their discretionary budgets to bypass vendor bureaucracy. Users reason that they shouldn't have to waste time justifying something that is critical to their job. Rather than control costs, some participants claimed outsourcing created islands of hidden IT costs as users dipped into departmental budgets to satisfy their IT demands. Some participants succeeded in realizing their cost control expectations. In these cases, senior business managers worked with the vendor to prioritize user request and to prevent users from circumventing cost control mechanisms. Although companies could have theoretically implemented cost control mechanisms without outsourcing, the formal outsourcing relationship bypassed the internal politics that had previously prevented internal IT departments from implementing cost controls.

Restructuring IT Budgets. Some participants wanted to use outsourcing to restructure their IT budgets from lumbering capital budgets to more flexible operating budgets. For example, rather than retain a $15 million mainframe on the books, participants could sell the asset to the vendor and merely buy the number of MIPS they need each year from the vendor. The sale of the asset also generates cash up front, which increases the participants' cash flow. In addition, some vendors will purchase stock and postpone the bulk of IT payments to near the end of the contract, making the overall net present value extremely attractive to participants. In return for these financial incentives, vendors require long-term contracts, typically 10 years in duration.

Participants often failed to understand the consequences of a long-term contract. How can vendors offer such sweet deals? Vendors know that the unit costs of IT drop exponentially over time. Although IT costs decrease from year-one perspective, over time participants did not share in the benefits of price/performance improvements because they are obliged to pay the same fee for the duration of the contract.

Participants who most fully realized their expectations about restructuring the IT budget signed shorter-term contracts. Although many vendors may fail to submit bids for shorter contracts, participants usually found at least one vendor willing to sign a two- to five-year contract, or a contract incorporating frequent staging points for reassessment. This allowed participants to reevaluate changes in the underlying cost structure of IT and to renegotiate contracts based on current price/performance ratios.

BUSINESS EXPECTATIONS

Participants expressed three business rationales for outsourcing: return to core competencies, facilitate mergers and acquisitions, and start-up new companies.

A Return to Core Competencies

During the 1990s, many large companies abandoned their diversification strategies—once pursued to mediate risk—to focus on core competencies. In other words, executives have come to believe that the most important sustainable competitive advantage is strategic focus (i.e., concentrating on what an organization does better than anyone else while subcontracting everything else to vendors).

As a result of the focus strategy, IT came under scrutiny: Is IT a competitive weapon or merely a utility? Even within companies, perceptions over IT's contribution to core activities varied. In general, senior executives frequently viewed the entire IT function as a noncore activity, whereas IT managers and some business unit managers contended that certain IT activities are core to the business.

Those participants most disappointed with outsourcing tended to view the entire IT department as a utility and thus pursued a total outsourcing strategy. Problems arose, however, when participants realized that certain IT functions—such as strategic planning, development of business-specific applications, support of critical systems—should have remained in-house because they require detailed business knowledge. Although vendors are fully capable of providing technical expertise, they often lack such knowledge.

Those participants most pleased with outsourcing view IT services as a portfolio containing both core and noncore activities. Before outsourcing, participants evaluated the contribution of each IT activity. Noncore activities, such as PC maintenance, data center operations, or run-of-the-mill accounting software, were outsourced while core activities, such as development of new strategic applications, remained in-house. In this way, the company's focus strategy was successful—internal IT resources are focused on business-critical applications whereas the more routine IT activities are outsourced.

Facilitating Mergers and Acquisitions

Because the participants were from large companies, as indicated by their presence on the US Fortune 500 or Europe's Times 1000 lists, many of the companies pursue a growth strategy through mergers and acquisitions. Mergers and acquisitions create many nightmares for IT managers, who are required to absorb acquired companies into existing systems. Participants expected outsourcing to solve the technical incompatibilities, absorb the excess IT assets (e.g., additional data centers), and absorb the additional IT employees generated by mergers and acquisitions.

Some participants found that outsourcing failed to solve their merger and acquisition problems. IT problems associated with mergers and acquisitions can be readily solved by involving IT in the decision process. But because IT departments rarely participate in non-IT strategic decisions, IT managers are usually informed about mergers or acquisitions only after deals are consummated, leaving them with little time to migrate systems. With outsourcing, the IT problems caused by mergers and acquisitions can worsen because senior executives are even less likely to share these plans with vendors than they are with their own IT managers.

One participant, however, successfully used outsourcing to reduce IT problems caused by mergers and acquisitions, having selected a vendor that was an expert at mantling and dismantling data centers. In the outsourcing contract, the participant developed specific service-level measures to accommodate mergers and acquisitions. Because the company in question had acquired 18 companies in a span of 2 years, it had enough knowledge of the effects of mergers on IT to detail a merger/acquisition clause in the contract.

Providing IT for Start-Up Companies

Some participants explained that they outsourced IT when the company was first incorporated. At the time, participants expected that outsourcing was a quicker and less expensive way to provide IT services. Start-up companies simply could not afford the capital investment required to erect internal IT departments because they had neither the technical expertise present nor the business desire to hire such talent internally.

For start-up companies, the primary danger in outsourcing IT is that customers cannot specify their IT needs very well to vendors. Participants were uncertain about the IT services, service levels, and volumes needed to support the new business. Over time, participants realized that they grossly underestimated their IT needs and were tied to a long-term contract that strongly favored the vendor.

To ensure that outsourcing expectations are realized, start-up companies should sign short-term contracts. By signing short-term contracts, start-up companies minimize their outsourcing risks because they are allowed to renegotiate the contract much sooner. Participants stated it took as little as six months to understand their IT needs well enough to sign a fair contract.

TECHNICAL EXPECTATIONS

Companies may determine that outsourcing can either improve technical services, allow them to gain access to technical talent not currently available in the organization, or provide access to new technologies.

Improving Technical Service

Some participants were dissatisfied that their in-house IT departments delivered systems late and over budget and did not respond quickly enough to user requests. They viewed outsourcing as a way to improve technical service, reasoning that outsourcing vendors possess a technical expertise lacking in internal IT departments. Yet some participants were disappointed that outsourcing failed to improve their technical service; in some instances, service levels actually degraded after outsourcing. Participants cited three reasons for failed expectations:

- Service levels were not fully documented in the contract.
- Contention with the vendors' other customers compromised IT service.
- The vendors' staff was overworked and thus often made mistakes.

When service levels were not fully documented in the contract, participants often found severe service degradation. For example, one vendor took 17 working days to implement security requests for new logon IDs and access to data sets. From the vendor's viewpoint, it needed time to verify requests, obtain approvals, and obtain the appropriate signatures. From the customer's perspective, a service that previously took 5 days to deliver now took 17 days.

Some participants complained that the level of service degraded because the vendor overworked its staff. Vendor analysts and programmers working overtime were prone to make errors and subsequently provided a level of service that led to lower overall user satisfaction with IT.

The participants' disappointments in technical service stem from a lack of understanding of the IT cost/service trade-off. To contain costs, vendors implement cost control measures, such as centralized functions to submit and prioritize user requests, standardized packages to obtain economies of scale, and reduced staff to lower personnel costs. If not for these measures, the vendors may not have been able to submit low bids.

Participants who were pleased with the vendors' level of technical service understood the IT cost/service level trade-off. By creating specific service levels in the request for proposals—which are later incorporated into the contract—participants may be willing to accept a more modest savings of 10% to 20% to more drastic savings of 40% to 50%. With high cost savings expectations, participants are more likely to experience service degradation as the vendor attempts to slash internal costs to meet bid specifications. Some senior managers we interviewed specifically said that they understood that IT services would likely degrade through outsourcing but that this is the price they were willing to pay to reduce the costs of IT. Summing up this sentiment, one CEO was overheard to say, "They [complained] about IT before outsourcing, they [complain] now—but at least it's costing me a lot less."

Other ways to ensure quality technical service are to stipulate significant vendor cash penalties for nonconformance of service-levels agreements and to specify the staff size in the contract. For example, some companies stipulate cash penalties for prime-shift downtimes, late reports, or late delivery of newly developed systems. Companies might also specify the number of personnel assigned to specific tasks to ensure service levels. One company specified that five people work the help desk. Another company in the study went so far as to name the vendor account manager it wanted to oversee the account in the contract, choosing an individual who had the respect of senior management, knew the business, and was generally perceived to be critical in making the contract work. The arrangement was highly successful in this particular case but may not be implementable as a general rule.

Access to Technical Talent

Some participants expected outsourcing to provide access to technical talent. Many of them found it difficult to find or retain staff with the desired state-of-the art technical skills. This is the primary reason one large UK retailing and distribution company outsourced telecommunications in 1990. An international manufacturing company outsourced its systems development work on several projects to gain access to IT expertise not available in-house. Numerous companies consider outsourcing partly for the access to greater IT expertise it would bring.

Our study showed, however, that such talent is not all that easy to come by. Many participants complained that the technical expertise remained the same because their internal staff simply transferred to the vendor. One industrial equipment manufacturer outsourced because managers felt the IT staff did not have the technical skills needed to implement a new computer architecture. The conversion (not to mention the entire outsourcing arrangement) failed because the same IT people (now vendor employees) performed the installation.

In another case, the access to new technical talent was made available to the company but at significant expense. The only way to ensure that access to technical talent will occur is to have it specifically noted in the contract. Informal understandings and appeals to "strategic partnerships" are ineffectual. Instead, customers should detail the price of additional technical skills they need. We have found that the contract is the only mechanisms to ensure wishes are fulfilled.

Gaining Access to New Technologies

Some participants viewed outsourcing as a way to hedge bets on emerging technologies providing them with access to the products of the vendors' large research and development departments. Participants were most interested in client/server technology, expert systems, new development methodologies, and CASE tools.

Some participants were disappointed that outsourcing did not automatically provide access to new technologies. Rather, outsourcing contracts often motivated vendors to run older technology as long as possible. For example, one company signed a 10-year contract for all mainframe applications. Even though the company wants to migrate to client/server technology, this shift—considered a change in character—is subject to additional fees. The participant complained that the vendor brought in the desired new technology and the technical skills to use the technology, but at enormous expense. In addition, the contract does not provide a discount for discontinuation of the mainframe applications, thus the customer is contractually bound to old technology. In hindsight, the participant should have signed a shorter contract for the duration of the expected life of the mainframe.

Participants most pleased with their access to new technology created specific contracts for new technologies. One company specifically hired a vendor to help implement its first client/server application. The company maintained managerial control over the development but used the vendor to provide technical expertise. In this way, the company learned about the technical aspects of client/server technology, which enabled it to implement future client/server applications on its own if it wished.

Other participants used outsourcing for just the opposite reason: they outsourced their old technology to a vendor to refocus internal resources on new technology. Sun Microsystems, for example, signed a $27 million deal for CSC to handle all of Sun's mainframe operations for up to three years. Meanwhile, Sun will rewrite its mainframe-based manufacturing and financial applications to run on a new client/server architecture. The new systems will not be outsourced but run by the Sun Microsystems staff.

A third option to ensure that access to new technologies becomes a reality is to include incentives in the contract to share benefits of new technologies. Customers and vendors can collaborate to develop valuable business applications to harness the power of new technologies. For example, one company hired a vendor to create an image-based customer-tracking system that was subsequently sold on the market. Both the customer and vendor share in the profits.

In short, outsourcing provides access to new technologies only if the contract specifically addresses the issue.

POLITICAL EXPECTATIONS

The political dimension of outsourcing involves the behavior of the various parties involved in the decision-making process and how they shape senior management's perception about IT and its value. Political rationale for why organizations outsource includes: proving efficiency, justifying new resources, duplicating the outsourcing success of others, exposing exaggerated claims, eliminating a troublesome function, and breaking the so-called glass ceiling.

Proving Efficiency

Because many companies account for IT as an overhead function, senior managers frequently evaluate the function solely on cost efficiency. Because no concrete measures of actual efficiency exist, senior managers formulate only a perception of efficiency. Some participants, especially IT managers, expected that an outsourcing evaluation would demonstrate to senior management that the internal IT department was cost-efficient. By comparing internal IT costs with vendor bids, IT managers can hold up their reports and say, "See, no one can provide services cheaper than us." The hard numbers appear objective and, therefore, add credibility to their efficiency claims.

Some IT managers were disappointed that their senior managers viewed the outsourcing evaluation results with skepticism. Senior managers may suspect that IT managers biased the evaluation by selectively picking functions that they knew vendors could not underbid—that is, IT managers select one of their best-managed functions as a candidate for outsourcing while eliminating poorly managed functions from the scope of the evaluation. One IT manager conducted an outsourcing evaluation for data center operations that had exceedingly low costs and excellent service levels. The IT manager's real problem was with systems development—users were complaining that new systems were late, over-budget, and defective. By limiting the evaluation to what IT performed well, the IT manager hoped that the positive outsourcing evaluation would have a knock-on effect for the whole IT department.

Another source of skepticism came from vendors. Some vendors view outsourcing requests initiated by IT managers with suspicion, claiming that many IT managers merely inquire about outsourcing for a free assessment. Vendors have responded by charging as much as $20,000 to submit a bid. Other vendors may bypass the IT manager by contacting a more senior manager in the company.

The most effective way to ensure credibility of IT efficiency is to involve senior management in the evaluation process. Senior management may serve to verify the scope of the evaluation, help develop bid criteria, and review the bid analysis. IT, however, must be integrally involved in the evaluation and is needed to specify the technical (and often financial) details in the request for proposal and to interpret the technical aspects of the external bids. Senior management sponsorship does not replace IT involvement, but merely serves to minimize politics.

Justifying New Resources

Some IT managers initiate outsourcing evaluations in order to acquire new resources, such as machine upgrades and additional personnel. Because senior management many times views the IT department as a cost burden, it is reluctant to provide additional funds for new IT investments without substantial justification. By showing that growth cannot be satisfied more efficiently through outsourcing, the IT managers expected that their resource requests would be granted.

One IT director wanted to upgrade to a more sophisticated operating system on the AS/400 and move to relational data bases. Knowing that senior management would question whether the additional capacity could be acquired without purchasing a new machine, the IT director bundled resource requests with the results of an outsourcing evaluation that demonstrated it was cheaper to purchase the machine than to outsource.

Such a request may be perceived to be biased. Senior management may be reluctant to buy in to the need for additional resources argument, choosing instead to view the justification as not credible. Management may simply continue to fund IT at the minimum level possible. In one case, the IT manager initiated what might be termed a cursory outsourcing evaluation. Without any formal bidding process, the IT manager claimed that outsourcing was not a viable alternative and that the IT department needed the money for hardware and software upgrades. In this case, senior management was not persuaded—less than a year later, it outsourced all of the IT department.

It must be reemphasized that the most effective way to handle such a crisis is to have senior management involved in the outsourcing evaluation exercise. In addition, the involvement of outside experts may help the evaluation of IT alternatives.

Duplicating Success

Favorable outsourcing reports trigger managers, particularly senior managers, to initiate outsourcing evaluations to duplicate the success stories at other companies. Because these senior managers often do not truly value the IT functions anyway, they hope to at least reduce costs to the levels their competitors allegedly achieved through outsourcing. These managers jump on the bandwagon after hearing good reports from other companies, even though these early successes are usually reported during the so-called honeymoon phase of the contract. Actual savings often fall short of anticipated savings.

Nonetheless, several participants initiated outsourcing evaluations because they wanted to duplicate the outsourcing success stories they read in the trade press or heard about at seminars or from colleagues. A senior vice-president of operations at a bank who participated in our study considered outsourcing after talking to several colleagues in other banks that had outsourced and reported significant anticipated savings. With little formal analysis and little understanding of outsourcing, the bank selected a vendor and hastily signed an agreement in hopes that outsourcing would bail it out of its financial troubles. As in many cases, the anticipated savings were not realized.

The only way to guard against such optimism is to educate senior management of the often exaggerated claims by analyzing the real experiences of the so-called success stories. The successes are neither as prevalent nor as dramatic as reported. Bringing in outside (and hopefully neutral) experts often helps temper expectations.

Exposing Exaggerated Claims

IT managers, on the other hand, usually fear that the favorable reports will seduce their managers into outsourcing. IT managers will often conduct their own outsourcing evaluations in an attempt to paint a more realistic picture of outsourcing than the one created by public information sources.

IT managers face a major risk if they conduct a cursory outsourcing evaluation simply to expose exaggerated claims: Senior management may perceive that the IT department is unwilling to make a rationale decision. As a result, senior management may conduct its own outsourcing evaluation without including the IT department in the process. For example, one IT manager in our study commissioned the IT staff to generate a white paper on outsourcing after hearing senior management quote a business magazine article on the subject.

In summary, if senior management perceives that the IT manager is only trying to buy time by "researching" outsourcing claims (i.e., trying to disprove them), it may simply move ahead while hiding the evaluation from the IT group. Once again, the only way to guard against such negative interaction is to conduct a formal evaluation process involving benchmarking, often against best-of-breed companies.

Eliminating a Troublesome Function

It is not uncommon for IT to be perceived by senior management as a troublesome function—in a word, a headache. IT administrators receive few accolades for managing the function. When the function runs smoothly, senior executives do not notice. When the function experiences problems, senior management screams. Some people wonder, who needs the aggravation? Why not outsource the function and let the vendor worry about it?

Some senior managers may consider outsourcing as a way to eliminate a burdensome function. On the other hand, is it really possible to outsource the management of IT? It is likely that by eliminating one headache, all that management may be doing is exchanging one headache for another.

Clearly, if outsourcing is considered, it must be evaluated from a position of strength. A problem cannot be successfully outsourced. Successful outsourcing cases involve outsourcing well-understood IT environments. Similarly, most unsuccessful outsourcing cases involve management turning over to an outside vendor something that they did not understand or want to understand. The latter case is a recipe for disaster.

Breaking the Glass Ceiling

Sometimes IT directors see the purpose of outsourcing evaluations as a means to enhance their personal or departmental credibility. Because senior managers appear not to fully value the services of the IT department, they may not value the contribution of the people who run the function. IT personnel rarely break into the upper echelons of management. Hence, IT managers may initiate outsourcing decisions for the purpose of enhancing their credibility. By showing that they are willing to outsource their kingdom for the good of the company, they prove to senior management that they are corporate players.

IT managers often undertake outsourcing evaluations to alleviate the misconception that IT managers are myopic. They demonstrate to senior management that they are business professionals committed to corporate goals, not techno-crats attempting to build technology empires. According to IT managers, in-creasing their personal credibility benefits the entire IT organization—upper management support means support for the entire IT team, not just the depart-ment head.

The fact that IT directors are willing to sell off their empire for the good of the company appears to have very little bearing on credibility. On the other hand, it seems very clear that IT managers must become more business-minded or be-come vulnerable to outsourcing. Internal IT managers' knowledge of the business may be the only factor that differentiates them from outsourcing vendors. In order for IT directors to become part of the true corporate team, they must be, first and foremost, business savvy.

GUIDELINES FOR SUCCESS

Although it is imprudent to make superficial generalizations, several guide-lines can be inferred on the relative success of certain outsourcing practices.

Both Senior Management and IT Management Involvement Is Required to Conduct a Rational Outsourcing Evaluation. Otherwise, politics from either side may drive an outsourcing evaluation, with senior management possibly attempting to negotiate with a vendor without IT involvement and IT managers attempting to conduct a cursory evaluation to temper exaggerated claims, justify new resources, or prove efficiency. When both senior management and IT man-agement are involved, each assumes a role that helps reduce political behavior. Senior management assumes the role of defining the objectives—whether they are financial, business, or technical—defining the scope of the outsourcing evalu-ation, developing bid analysis criteria, and verifying the bid analysis. IT assumes the critical role of creating the detailed request for proposal, evaluating the legitimacy of vendor economies of scale, estimating the effects of price/performance improvements, and providing insights on emerging technolo-gies that may effect the business.

Selective Outsourcing Is Often Preferred. Participants' successful experi-ences with total outsourcing were usually limited to cases in which the company's internal politics prevented the IT department from initiating cost savings; the company needed to focus energies on more strategic issues than IT; or the company needed the cash infusion from the sale of information assets. In most cases, however, selective sourcing was the preferred option. Although vendors may be technical experts, critical business skills are required to align IT with overall business strategy. Participants agree it is unwise to outsource IT activities that require extensive knowledge of business needs, such as the development and support of strategic systems, IT planning and strategy, and IT architecture. In addition, it is unwise to outsource any IT activity that is perceived as a problem. Outsourcing IT activities that are not understood leaves the customer in a poor position to negotiate a sound contract. Companies cannot outsource a problem—they merely exchange one set of problems for another.

Companies Need a Solid Contract. Depending on the precise sourcing objectives, customers should negotiate the appropriate contract type, contract length, and contract detail. In general, customers can sign one of three types of contracts:

- Resource buy-in, in which the customer retains managerial control.
- Resource buy-out, in which the vendor assumes management control.
- Strategic partnership, in which the customer and vendor jointly manage IT and share in the risks and rewards.

In general, customers prefer short-term contracts because it is very difficult to predict future IT demand and supply—few participants felt confident predicting their own IT needs or the impact of future technologies past five years. Vendors prefer long-term contracts for precisely this reason. The longer the contract, the more likely the vendor can exploit price/performance improvements to increase their profits and sell additional products and services. In general, specific contracts with detailed service level requirements and penalties for nonperformance are preferred to nonspecific contracts. If customers follow the prescription to outsource only what they already understand, then creating a detailed contract should not be a problem.

CONCLUSION

All organizations will undertake outsourcing evaluations—it is not a passing fad. It is simply common sense for organizations to consider whether outside vendors can better meet sourcing objectives. If they can, then outsourcing may make sense. This, of course, assumes that IT is a commodity and that there is no intrinsic difference between the provision of an IT service from outside or inside. The arguments used to support the alternative positions are equally convincing and strongly held. Yet irrespective of the argument, it is interesting to speculate why in-house IT departments could not reduce their costs to be competitive with an outside vendor.

First, an internal IT department does not have to make a profit. Vendors not only have to make a profit, but they have significant overhead expenses that increase their costs. Second, an in-house IT department has knowledge of the business (and industry) that the vendor will likely not have. If an outside vendor can provide service less expensively than an internal IT department, then why shouldn't its strategies be adopted by the in-house IT department? So-called insourcing arrangements can work very effectively if the in-house IT unit can cut costs—for example through data center consolidation, the implementation of improved chargeback systems, worker empowerment, the purchase of used equipment, and offering a reduced software portfolio. Although insourcing may not be the preferred solution, it should be considered alongside the options of total and selective outsourcing.

I-6

A Strategy for Outsourcing

N. Dean Meyer

OUTSOURCING SOME OR ALL of IS activities is an option being considered in many US organizations today. Outsourcing—that is, paying other firms to perform all or part of the IS function—is not new, but there is renewed interest in it. It can be a viable business option when used in the right circumstances and for the right reasons. Outsourcing has been used very effectively in a few cases; it has had disappointing results in many other cases.

In many cases, an IS organization's interest in outsourcing originates with pressure from top executives who use outsourcing as a threat to force change. Outsourcing vendors promise dramatic savings and enhanced flexibility so that line executives have more time to focus on their core businesses. On the surface, these claims seem plausible, but they often do not hold up well under closer scrutiny.

RHETORIC VERSUS REALITY

Each of the claimed benefits of outsourcing has underlying assumptions, each of which must be considered in a decision to outsource.

Cost Savings

Economies of scale seem to reduce costs. The vendor, however, must earn a profit at its customer's expense. Furthermore, external contracting brings added sales and transactions costs.

Most of the savings from outsourcing generally come from data center consolidation. Once a firm has consolidated its data centers on its own, outsourcing is generally more expensive. The only lasting cost savings occur where there are true economies of scale across corporate boundaries—for example, in long-distance communications. In some cases, interorganizational sharing is possible, but such cases must be examined carefully. Hardware no longer presents economies of scale, and many software licenses are corporation specific.

In a few cases, outsourcing has been viewed as a source of near-term cash, because IS assets may be sold to the outsourcing vendor. Selling a strategic resource is an extreme way to save a sinking firm. Furthermore, selling off IS cripples all remaining business units and increases long-term costs.

Better Access to Technology

Equipment vendors suggest they can provide customers with better access to new technologies. In practice, vendor sales representatives are quick to bring new products to the attention of internal IS staff in any case. The firm must judge for itself when to adopt a new vendor offering rather than leave that decision to a vendor that has a vested interest in selling new products. Vendor-owned outsourcing services are also less likely to tap opportunities presented by competitive vendors (e.g., more cost-effective, plug-compatible products).

Increased Flexibility

Some people say that outsourcing converts fixed costs (or such relatively fixed costs as people) into variable costs, giving the firm greater financial flexibility. In fact, most outsourcing vendors require long-term contracts that provide them with stable revenues over time. Renegotiating these contracts may be more expensive than changing internal commitments.

If flexibility is the goal, the contract must be carefully negotiated to allow variability in demand and cost. Flexibility, however, comes at a relatively high price.

Greater Competence

It can be argued that vendors are more experienced than internal staff at running an IS function. This situation can be remedied by hiring competent IS managers as readily as by hiring an outsourcing vendor.

In some cases, outsourcing is simply a matter of paying someone else to experience the pain of managing a dysfunctional IS function, rather than trying to figure out how to make the function healthy again. This costly form of escapism sacrifices a valuable component of business strategy for a short-term convenience.

Downsizing

In organizations that must downsize, it may seem that the outsourcing vendor can move surplus people to other jobs serving other companies. If those other jobs exist, surplus staff can compete for them on the open market with or without the outsourcing deal. When staff members have the qualifications for those other positions, they will get them—regardless of whether the firm pursues outsourcing. If they are not qualified enough to win other jobs on their own merits, it is unlikely that a highly competitive outsourcing vendor will retain them in these positions for long. Ultimately, then, outsourcing does little to change the employment picture for surplus IS professionals.

More Time for Business Issues

Some organizations are attracted to outsourcing because it relieves senior management of having to worry about managing the IS function. The argument is that outsourcing reduces the demand on senior management because a contract is substituted for direct authority. This rarely proves to be the case. In fact, managing an outsourcing vendor is no easier (and is often more difficult) than managing an internal IS executive.

If senior managers become less involved in managing IS, outsourcing may actually be counterproductive. Those who understand the strategic value of IS argue that managers should spend more, not less, time thinking about IS. Without management involvement, the IS function may do no more than it has done in the past. That is, it may continue to invest in administrative systems but fail to find breakthroughs in strategic applications.

Another argument is that when a firm outsources IS, business managers have more time to focus on the corporation's main lines of business. This is only true if IS managers are transferred into other business functions. If the IS managers are fired or transferred to the outsourcing vendor, there are not likely to be more business managers to focus on the organization's business issues than there were before outsourcing. In other words, the business only gets more attention if line management is expanded (and costs are increased). Of course, line managers can be added regardless of whether IS is outsourced.

INSIDERS' ADVANTAGE

There are two key reasons why insiders have an advantage over outsourcing vendors for some key functions within the department: continuity and vested interests. For both of these reasons, insiders are more likely to be invited to clients' key meetings and will be in a better position to play a role in the firm's strategic imperatives.

Continuity

Internal staff members have a history and an expectation of continuity with the organization that may pay off in a better understanding of the business and improved partnerships with clients. By contrast, outsourcing vendors may rotate their staff more easily, because individuals develop loyalties to the outsourcing vendor rather than one customer organization. The insiders' improved partnership advantage pays off in client satisfaction and more meaningful strategic alignment. Long-term employees better understand the clients' culture, strategies, and politics; they also know they will be around to deal with the consequences of their actions.

Vested Interests

Outsourcing vendors may be sincere about partnership, but ultimately they work for different shareholders and ethically must (and will) place their shareholders' interests first. For example, what would happen if, in a needs assessment interview with a client, the IS consultant sees an opportunity for either a $200,000 administrative application that should save clerical time, or a $200 end-user computing tool that could significantly affect the client's personal effectiveness? Although the latter choice may provide a higher payoff and more strategic value, the outsourcing vendor has a strong incentive to recommend the more lucrative administrative system project because it generates more revenues, which are added costs rather than cost savings.

In one extreme case, an automobile rental company wanted to acquire software that would help it make better use of its fleet. An expert in the industry offered to license a state-of-the-art yield management package, but because the company had outsourced its IS function, the outsourcing vendor saw no profit in this arrangement. Instead, the outsourced IS department spent hundreds of thousands of dollars—nearly the price of the package—on a study of alternatives, and then even more money replicating the package to ensure that it had a role in both development and support.

Insourcing

The insiders' advantage may also be used to bring in new revenue. *Insourcing* is a term that refers to sales made by a staff function to clients outside the corporation. A staff function may sell directly to external clients if two conditions are met:

- The staff function has a distinctive competence in a particular area that ensures success in a well-defined niche, in spite of competition.
- Insourcing will improve internal client satisfaction, at a minimum by building a critical mass of specialists in an area that otherwise might not warrant permanent headcount. (Size permits a higher degree of specialization, which in turn reduces costs, improves quality, speeds time to market, and accelerates the pace of innovation.)

Before insourcing is considered, the IS function must be sure that internal clients are completely satisfied with its work. Unless the corporation is pursuing a strategy that takes it into the IS business, it is more important that the IS function help internal clients to succeed than that it makes money on its own.

In general, it is preferable to sell products and services to internal clients, who in turn may add value and sell them outside the corporation. This ensures consistency and coordination with line management's customer-oriented strategies. It also keeps the focus of the staff function on its primary mission—serving internal clients.

WHEN OUTSOURCING MAKES SENSE

Upon scrutiny, outsourcing is usually found to carry with it certain risks. After investigation, many firms have shied away from outsourcing all or major portions of an IS function. Instead, these organizations are pursuing a selective outsourcing strategy.

Outsourcing selected functions can be quite valuable, especially in the case of functions where an insiders' advantage—continuity and appropriate vested interests—are not so important. Indeed, the use of contractors and consultants is hardly new and generally represents a healthy form of outsourcing.

The goals of selective outsourcing include:

- Minimizing fluctuations in staffing that could result from rising and falling demand.
- Maximizing the development of employees by outsourcing less interesting work.
- Minimizing costs by using relatively less expensive employees whenever possible or sharing costs with other corporations.

For this discussion, the term *external consultants* refers to people who are hired to transfer their skills and methods and improve employees' effectiveness. These consultants are distinct from contractors who do work in place of employees.

Generally, consultants may be used by anyone whenever justifiable. Contractors are considered extensions of the staff of employees. Employers should decide when and whom to hire, supervise all contractors, and be accountable for their work.

STRATEGY FOR OUTSOURCING, FUNCTION BY FUNCTION

At a more detailed level, the pros and cons of outsourcing vary by function within the IS department. The five basic functions found in any IS department are:

- Machine-based service bureaus that own and operate systems for use by others (e.g., computer and network operations).
- People-based service bureaus providing routine services that are produced by people (e.g., client support, training, and administration).
- Technologists who design, build, maintain, and support systems (e.g., applications developers, platform experts, information engineers, and end-user computing specialists).
- The architects who coordinate agreements on standards and guidelines that constrain design choices for the sake of integration.
- The consultancy that works with clients to define requirements, set priorities, measure benefits, and coordinate relations with the rest of the IS department.

The issues related to outsourcing can be examined within each of these functional elements.

Machine-Based Service Bureaus

Outsourcing machine-based service bureaus is equivalent to buying computer and network time from vendors. This is a common practice and should be evaluated for each new increment of capacity. It is a form of financing when capital is short and a way to satisfy temporary or otherwise limited needs for specialized platforms. Outsourcing a major portion of the machine-based service bureau may also be useful for a limited period of time during a migration to a new platform.

Permanent outsourcing may be cost-effective when there are economies of scale and multiple corporations can share infrastructure; a common example is in the field of communications, where few companies run their own private long-distance networks. However, this usually does not apply to software licenses, which are generally corporation specific.

Outsourcing should only be used when the same economies of scale cannot be attained internally. If outsourcing the entire function appears to save money, internal consolidation of data centers should also be considered as an alternative.

People-Based Service Bureaus

Some people-based service bureaus can also be outsourced. For example, installation and repair services are commonly outsourced, and training courses can be purchased for many common end-user computing packages.

Vendors offer commercial hot lines that support most of the common end-user computing products, but these are best used in combination with internal support functions. Inside staff should be the first line of support, calling on external resources for a specific class of questions. This approach ensures the required quality of service and handles internally developed packages and configurations. In general, the following guidelines apply:

- Hire enough staff to satisfy the steady demand; outsource to satisfy peak loads.
- Outsource only commodity and end-of-life services; keep new growth opportunities inside.

Technologists

No matter how big the IS organization, it can never afford to hire a specialist in every possible discipline. Contractors are therefore a valuable source of specialized expertise in less frequently used areas.

A preference to buying over making is another variant on outsourcing. Turnkey packages are attractive because they free scarce internal talent from the ongoing burden of maintaining systems, allowing applications technologists to focus on new requirements. Packages should be used whenever the requirements for customization are limited. In most cases, clients appreciate receiving a proposal that offers a choice between custom code and a package—often this amounts to a choice between 100% of what they want for 100% of the cost, versus 80% of what they want for 20% of the cost.

The guidelines here are similar to those for people-based service bureaus: Hire enough staff to satisfy basic demand; outsource to satisfy peak loads. Outsource only commodity and end-of-life skills; keep new growth opportunities inside.

In every case, employees should be used to manage outsourced technologists to ensure quality, systems integration, architectural compliance, and responsiveness to the business.

Architects and Consultancies

The functions of architects and consultancies are highly strategic and require an insider's deep understanding of the corporation. They not only require an intimate knowledge of the business, but success in these functions requires close relations with clients and the rest of the IS function. Although many management consulting firms are willing to sell these two types of planning, such high-leverage functions should never be outsourced.

Architect. Even in the smallest of operations, the architect function is extremely important. Architects facilitate a consensus on standards so the IS function can be responsive to clients' strategic needs—tailoring solutions to their unique missions rather than blindly following a static top-down plan. They are also needed for a firm to successfully evolve toward integrated systems.

Where staff is lacking, the architect function should not be outsourced. It should be part of the responsibility of the chief technologist or perhaps the head of the department.

Consultancy. The consultancy function must not be ignored, however scarce personnel may be. It is essential to ensure a strategic return on IS investments and healthy client relations. (Return on investment is, of course, every bit as important to a small organization as it is to a large one.) Consultants are the primary liaison to clients, diagnosing the client's strategic needs without any bias for particular solutions and setting up contracts with the rest of the IS function. Consultancy requires an insider's understanding of the business and carefully cultivated relationships with key clients—the result of continuity over time. Outside consultants can train and assist internal consultants, but contractors should not be used to do their jobs for them.

Even in the smallest organizations, responsibility for strategic consultancy should be placed somewhere within the IS department. The responsibility may be assigned to one person or, in groups of only a few people, it may be a part-time responsibility of the top IS executive. Because this function depends on relationships with clients and an understanding of the corporation's strategy and politics, the consultancy function should never be left to outsiders.

HOW TO MANAGE OUTSOURCING VENDORS

Any degree of outsourcing necessitates clear designation of responsibility within the organization for acquiring these outside resources. For example, in the case of external technical specialists, the appropriate technologist is expected to know where to find the right people and how to manage them. The internal manager of outsourcing contracts (in any area) is responsible for:

- Shopping for the best deal, negotiating the contract, and managing contractor performance.
- Resolving problems in the relationship and maintaining healthy collaboration between the two parties.
- Generating entrepreneurial ideas within the established charter and domain and deciding whether to make or to buy.
- Establishing clear contracts with internal customers and suppliers and retaining responsibility for fulfilling those contracts (whether outsourcing vendors are involved or not).

For all practical purposes, outside contractors should be considered part of the group that hired them. The fact that their paycheck is written by a different corporation does not change the nature of their work. All outsourcing vendors will automatically live within the bounds of the existing organizational structure, and clients need not worry about who is chosen to staff their projects.

CONCLUSION

In general, outsourcing the entire IS function should only be done in severe cases. Doing so risks higher costs, less flexibility, and a loss of strategic alignment. Instead, each group within an IS department should cultivate contacts with outside contractors and package vendors in its area of expertise and manage them as part of its staff. Each internal entrepreneur should proactively decide whether to make or to buy in the course of every project.

I-7

Organizational Architecture for IS

Djoen S. Tan

THE ROLE AND IMPACT of information technology in organizations have changed significantly during the past decade. The application of IT has evolved from an administrative support function to a more strategic role. The costs of IT have also evolved from an overhead expense to a business investment. Yet, there is serious concern that the expected value of the investment in IT will not be achieved.

According to the Management in the 1990s research program of the MIT Sloan School of Management (the MIT90s framework), N. Venkatraman noted that the inability to realize value from IT investments is mainly due to the lack of strategic alignment. His strategic alignment model showed that alignment should involve business strategy, business organization, IT strategy, and IT organization. Accordingly, this alignment can be performed on several levels of IT-enabled business reconfiguration.

The above-mentioned research program conceptualizes an organization as consisting of five sets of forces that are in dynamic equilibrium as the organization is subjected to influences from the outside socioeconomic environment and the external technological environment (see Exhibit I-7-1). General management's task is to ensure that all five forces move through time to accomplish the organization's objectives. This MIT90s framework is a so-called consistency model.

The MIT90s framework can be extended to a more complete consistency model of IT management. This chapter defines five IT management plateaus by combining this consistency model with a stage model of the application of IT, based on the levels of IT-enabled business reconfiguration of Venkatraman. The model can also be applied to coalitions of organizations.

CONSISTENCY MODEL OF IT MANAGEMENT

A consistency model of an organization views the organization as an open system with a limited number of basic subsystems (aspect systems) or "forces."

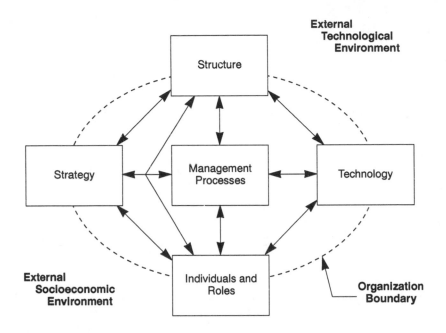

Exhibit I-7-1. The MIT90s Framework

To achieve internal and external stability of the organization, these subsystems strive for harmony and consistency with each other by mutual adaptation and adjustment.

A consistency model of IT management in organizations is shown in Exhibit I-7-2. The model consists of three domains:

- The environment.
- The organization domain.
- The IT domain.

In accordance with the MIT90s framework, five subsystems are distinguished in the organization domain:

- *Strategy.* The organization's objectives and the ways the organization tries to realize these objectives; the selection of markets and products and the marketing mix (the business plan).
- *Structure.* The organization structure, roles and responsibilities, decision-making procedures, and planning and control systems.
- *Technology.* The technology used for the business processes.
- *People.* The knowledge, skills, ambitions, attitudes, and social relations of the people in the organization (human resources).

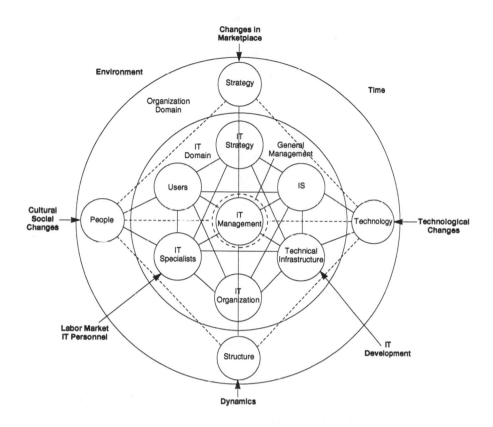

Exhibit I-7-2. Consistency Model of IT Management

- *Management.* General management (senior and line management) responsible for the selection and realization of the business strategy and for adjusting the subsystems in the organization to each other.

All the subsystems influence each other by mutual interaction. The first four subsystems are influenced, respectively, by the changes in the marketplace, the dynamics (the rate of change), the technological developments, and the cultural and social changes in the environment.

The IT domain is a subdomain of the organization domain. In the IT domain, the subsystems of the organization should be focused on the problems concerning the application of IT. For example:

- Strategy focuses on IT strategy.
- Structure focuses on the IT organization structure.

- Technology focuses on information systems and the technical infrastructure.
- People focuses on IT specialists and the users.
- Management focuses on IT management.

In the IT domain, the technology subsystem is divided into information systems (i.e., applications) and technical infrastructure (i.e., hardware, operating systems, and network and data base management systems) because the technical infrastructure has more long-term and common characteristics than the applications. The people subsystem is divided into IT specialists and users because these groups usually have a different view of information systems and their managers often have different priorities.

The five subsystems of the organization domain therefore interact with the following seven subsystems of the IT domain, which also influence each other:

1. *IT strategy.* The objectives, rules, selected standards, and planning of the application of IT.
2. *Information systems.* The applications, including manual procedures.
3. *Users.* The knowledge, skills, ambitions, and attitudes of the users of the information systems.
4. *Technical infrastructure.* The hardware and systems software used by the applications.
5. *IT organization structure.* The structure, roles, responsibilities, procedures, and planning and control systems of the IT organization.
6. *IT specialists.* The knowledge, skills, ambitions, and attitudes of the IT specialists.
7. *IT management.* The management of the application of IT by senior, line, and IT managers, who are responsible for an optimum application of IT by adjusting the aforementioned subsystems to each other.

Subsystems 1 to 3 express the need of the organization for information systems and the knowledge and skills of the users to make use of these systems (the demand side). Subsystems 4 to 6 concern the available IT organization and the capacities of the IT staff and the technical infrastructure (the supply side). IT management should balance the supply and the demand sides at an acceptable cost level.

Environmental forces also directly influence the subsystems of the IT domain within an organization—for example, the impact of IT developments on the technical infrastructure and the labor market for IT personnel on the IT specialists. Finally, in Exhibit I-7-3, time is shown as a third dimension and is dealt with in the following discussion.

IT MANAGEMENT PLATEAUS

Within a continuously changing environment, an organization is continuously searching for new situations in which the subsystems are in equilibrium with each other. Time is a critical factor. The consistency model of the IT organization

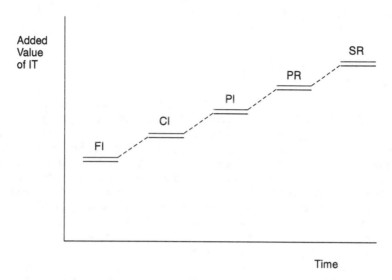

Notes:

FI Functional integration
CI Cross-functional integration
PI Process integration
PR Process redesign
SR Scope redefinition

Exhibit I-7-3. The Five IT Management Plateaus

should therefore be combined with a stage model to take into account time and the changes in the environment and the organization as time passes.

A stage model of an organization assumes that the development of an organization occurs stage by stage, much like human development from childhood through adolescence to maturity. A stage or phase cannot be passed over but can be passed through faster.

A well-known stage model of the development of automation in organizations is Nolan's Stages Theory. Nolan described six stages: initiation, contagion, control, integration, data administration, and maturity. Later on the last two stages were changed to architecture and demassing.

In this chapter, an IT management plateau is defined as a period of time during which the subsystems of an organization and its IT domain are in balance with each other. An IT management plateau is regarded as a development stage of the application of IT. The term "plateau" has previously been used in a framework to plan and control the realization of an IT architecture.

Based on the levels of IT-enabled business reconfiguration of Venkatraman, the following five IT management plateaus are identified:

1. Functional integration (FI).
2. Cross-functional integration (CI).
3. Process integration (PI).
4. Business process redesign (PR).
5. Business scope redefinition (SR).

On each successive IT management plateau the organization is further tailored to the possibilities of IT. Each successive plateau therefore has a higher potential of IT to add value to the business but also requires more complex information systems, more knowledge to build these systems, bigger organizational changes to implement these systems, and higher costs and risks (see Exhibit I-7-3).

On each plateau, a dynamic equilibrium between all the subsystems of the consistency model should exist before the next plateau can be reached by a harmonic development of the subsystems. In this way, periods of relative stability alternate with periods of change. A harmonic development preserves unity by simultaneous and equivalent growth of the subsystems, which is only possible through mutual interaction and adjustment.

The mutual adjustment of the subsystems occurs by means of communication and coordination within and between the organization domain and the IT domain. Each successive plateau requires a higher degree of communication and coordination.

A higher IT management plateau cannot be reached by, for example, merely changing the organization structure and building new information systems. If the IT strategy is not changed, the technical infrastructure not adapted, and management, users, and IT specialists not properly trained, the operation will fail.

Subsystems of the IT Domain on Each IT Management Plateau

Functional Integration. On this first IT management plateau, the information systems support the existing workflow within the business functions (functional departments), such as manufacturing, sales, and finance. Apart from some adjustment of procedures, no organizational changes are needed. Each business function often exploits different technical platforms.

The objective of the application of IT is operational efficiency. The IT plan is an inventory of the users' demands. The IT organization mirrors the business organization and can vary from a completely centralized IT organization for a functional organization to a completely decentralized IT organization for a divisional organization. The information systems are developed and built by the IT specialists according to the functional specifications of the users. Within the IT budget, the IT decisions are made by the IT management.

Until the beginning of the 1980s, most organizations were on this IT management plateau.

Cross-Functional Integration. The information systems integrate the flow of work across several functional departments, such as purchasing, inventory control, and accounting. These systems are more complex to develop and to implement compared to those on the first plateau. Minor organizational adjustments may be required, but the functional structure and the working methods remain largely unchanged. The technical infrastructures of the functional departments concerned must work together. There should at least be some common standards.

The objective of the application of IT is to improve the effectiveness of the business. The IT plan is derived from the business plan. The IT organization cannot be completely decentralized because some form of central coordination is required. The information systems are developed and built by the IT specialists according to the functional specifications of the users. IT decisions are made by a steering committee of line and IT managers.

Most organizations today are situated on this plateau.

Process Integration. The information systems enable the work of a complete business process to be carried out as a single entity. A business process is a set of logically related work activities that achieves a tangible and important business result, such as customer order fulfillment, new product development, or after-sales service. Usually these activities are done by several (functional) departments. The implementation of these information systems requires extensive organizational changes and an integrated technical infrastructure.

The objective of the application of IT is to achieve competitive advantage in existing product-market combinations. The IT plan and the business plan should be adjusted to each other. The IT organization on this plateau cannot be completely decentralized. On the other hand, a centralized organization is not in accordance with the responsibility of the line managers for their information systems. Therefore a selective decentralization (a combination of centralization and decentralization) should be chosen—for example, decentralized systems development and centralized management of the technical infrastructure.

Together with the users, the IT specialists identify and analyze the most important business processes, using for example, the value-chain method of Michael Porter. A project manager (process owner) for each main business process should be appointed. The information systems are built by the IT specialists under the supervision of line management. IT decisions are made by a steering committee of line and IT managers with a member of senior management as chairperson.

Many organizations for which IT is of strategic importance, such as financial institutions and transportation companies, are now situated on this plateau.

Business Process Redesign. Just as on the former plateau, information systems enable the work of the most important business processes to be carried out as single entities. The existing business processes are not taken for granted, however, but are redesigned or reengineered to make use of IT in new ways. So business processes are performed in ways that were previously impractical or impossible. A customer order fulfillment process, for example, can be eliminated by arranging for customers to place their orders electronically. Dramatic organizational changes are necessary to implement these information systems.

This is, of course, only possible with an integrated technical infrastructure. Furthermore, this technical infrastructure should include CASE tools and standard application modules to speed up systems development. A successful business process redesign often depends on the speed of implementation.

The objective of the use of IT is to realize competitive advantage in existing and new product-market combinations. The IT plan and the business plan should be one integrated strategic plan. Because of the consequences of IT for the entire corporation, the IT organization should be selectively decentralized with sufficient involvement of senior management. Users and management consultants of user departments select and redesign the business processes together with the IT specialists. The information systems are developed by the IT specialists together with the users under the supervision of line management. IT decisions are made by senior management after consultation of a steering committee with the same composition as on the former plateau.

Business process redesign or reengineering (BPR) is a hot item. As yet, only a few pioneering companies have experience with BPR. Most companies today face the challenge of reaching this IT management plateau by the end of the 1990s.

Business Scope Redefinition. Here again, information systems enable the most important business processes to be carried out using an integrated technical infrastructure. Optimum use of IT, however, requires more than the redesign of business processes; even the business scope is enlarged or shifted, if necessary. The redefinition of the business scope gives rise to new business processes and redesign of existing processes. This implies radical organizational changes.

The objectives of the application of IT on this plateau is to realize innovative product-market combination. The business plan and the IT plan form one integrated strategic plan. A highly selective decentralization of the IT organization is required. Users and IT specialists develop the new business processes and the corresponding information systems in small teams (fewer than 10 persons). Line management is completely responsible for the new systems. The new systems must be realized quickly by using appropriate tools and standard application modules as part of the integrated technical infrastructure. IT decisions are made by senior management, because of the consequences for the business scope. The strong involvement of senior management and the close cooperation between users and IT specialists require the organization to be a networked organization.

A networked organization is also called an information-based organization. The essence of a networked organization is the integration of thinking and doing. The main characteristics are:

- A diamond-shaped instead of a pyramid-shaped organization.
- Few hierarchical levels and a minimum of lower-level functions (these functions are automated or outsourced).
- Knowledge workers (managers and specialists) as the dominant force in the organization, working together in varying multidisciplinary teams (a highly selective decentralization).
- Intensive use of IT on all levels of the organization (a networked organization cannot exist without IT).
- Self-organizing and self-learning capabilities, because of the integration of thinking and doing.

Corporations will not reach this IT management plateau before the year 2000; for many, it may well take longer.

This highest plateau does not signify that there will be no further developments after this level. The application of IT has, however, become as standard as the use of paper and pencil, and IT processes have become ordinary business processes.

Exhibit I-7-4 summarizes the seven subsystems for each plateau. The first plateau covers Nolan's first three stages. The second plateau corresponds to Nolan's fourth stage (integration). Stage 5 (architecture) and 6 (demassing) of Nolan have characteristics of the third IT plateau.

The IT management plateaus show the natural development of the application of IT in organizations. Owing to the learning processes of managers, users, and IT specialists, different sections of an organization can be situated on different plateaus.

A faster transition to a higher plateau can be achieved by adequate planning and control of the required activities. The higher plateau is then regarded as the goal of the IT strategy. Intermediate plateaus can be defined, if necessary, to realize the transformation in smaller steps and so enlarge the manageability of the migration.

It is not always necessary to reach the highest plateau as soon as possible. This is only true if IT is of strategic importance for the survival of the organization. The conditions for a successful transition to a higher IT management plateau are:

- All subsystems of the organization domain and the IT domain on the present plateau are in balance (i.e., consistent) with each other.
- Senior management has a vision that IT can be used as a lever for competitive advantage.
- A clearly defined IT strategy exists, supported by senior, line, and IT management.
- A climate of change is created within the entire organization.
- The IT strategy is worked out into concrete action plans or projects for each subsystem.
- Adequate control exists for the realization of the projects.
- Senior management provides sufficient commitment and support.

IT MANAGEMENT PLATEAUS IN COALITIONS OF ORGANIZATIONS

Organizations form together with their suppliers, distribution channels, and customers to create a value system. By working together, the value chain of each of the participating organizations can be improved.

Interorganizational information systems enable organizations to extend their reach and capabilities outward to customers and suppliers. They can support or trigger the redesign of business processes across the boundaries of multiple organizations and have the potential to transform a business, a marketplace, or even an entire branch of industry.

The consistency model of IT management shown in Exhibit I-7-2 can also be applied to a group of cooperating organizations or a coalition of organizations. A

IT MANAGEMENT PLATEAUS / SUBSYSTEMS	FI	CI	PI	PR	SR
IT Strategy	Increase efficiency. IT plan based on user demands.	Increase effectiveness. IT plan derived from business plan.	Competitive advantage in existing PMCs. IT and business plan adjusted to each other.	New PMCs. Integrated strategic plan.	Innovative PMCs. Integrated strategic plan.
Information Systems	Support existing workflow in business functions.	Integrate workflow across business functions.	Enable complete business process.	Enable complete business processes after redesign.	Enable complete business processes after scope redefinition.
Technical Infrastructure	Platform for each function	Common standards	Integrated infrastructure	Integrated infrastructure	Integrated infrastructure
IT Organization	From completely centralized to completely decentralized	From completely centralized to federal	Selectively decentralized	Selectively decentralized	Highly selectively decentralized
Users	Functional specifications	Functional specifications	Responsible for IS realization	Redesign business processes and IS	Design new business process and IS
IT Specialists	Realize IS	Realize IS	Design and build IS	Build IS	Build IS
IT Management	IT managers decide within budget	Steering committee of line and IT managers	Steering committee chaired by senior management	Senior management decides after consulting with steering committee	Senior management decides

Notes:

FI Functional integration   PR Process redesign
CI Cross-functional integration   SR Scope redefinition
PI Process integration   PMC Product-market combination

Exhibit I-7-4. Summary of the Subsystems on Each IT Management Plateau in Organizations

coalition or a strategic alliance is defined as a long-term agreement that is more than a usual business transaction but less than a merger or an acquisition.

The organization domain comprises all the participating organizations. The IT domain refers to their joint application of IT. For a coalition of organizations, the same five IT management plateaus in Exhibit I-7-3 can be distinguished as development stages of their joint application of IT.

The subsystems "IT specialists" and "users" comprise all the participating organizations. The interorganizational systems are developed and built by joint project teams of IT specialists and users of the participating organizations. The involvement of users is stronger on each successive plateau. However, close cooperation between IT specialists and users of the different organizations is more difficult to achieve in this case because of differences in culture, among other things. A short description of the other subsystems of the IT domain for each IT management plateau follows.

Functional Integration. Within a coalition of organizations, functional integration implies transaction automation. High-volume, repetitive paper transactions (e.g., orders, receipt notes, invoices) that flow between the participating organizations are replaced by electronic messages (electronic mail or basic EDI transactions).

The objective of the common application of IT is to increase the speed and reliability of the transactions between the organizations. Usually the messages do not link directly into the main applications of the participants. Apart from minor alterations in the working procedures at the boundaries of the organizations, nothing has to be changed.

The IT strategy, the technical infrastructure, the information systems, and the IT organization of the business participants can remain the same. There is no overall IT management. The situation is similar to a completely decentralized IT organization at a single company.

Cross-Functional Integration. On this second IT management plateau, the interorganizational systems enable all transactions of a procedure to take place in the form of electronic messages. A procedure consists of several related transactions often performed by multiple functional departments, such as an invoice payment procedure. Procedure automation requires a direct link into the application systems of the participants, because transactions have to trigger responses.

The objective of the common application of IT is cost reduction. Working methods within the participating organizations are affected by procedural adjustments, but no change of internal organization structure is necessary.

The IT strategies and technical infrastructures of the participating organizations should enable the realization of these systems. This implies at least an agreement on common IT standards. A steering committee of the involved line and IT managers is needed to define these standards and to supervise the realization of these interorganizational systems. The situation is similar to one decentralized IT organization with cooperation on some common systems. Most present-day EDI applications are situated on the first or the second plateau.

Process Integration. The interorganizational systems enable the execution of a main business process between the participants in a value chain—keeping shelves stocked, for example. This is, of course, only possible when process integration has been implemented within the individual organizations and a common technical infrastructure is available. Process integration requires adjustments of the internal structure and the working methods of the participating organizations.

The objective of the common application of IT is competitive advantage in existing product-market combinations. A joint IT plan should exist that is consistent with the individual business plans and IT plans. The definition and the implementation of the joint IT plan is controlled by a steering committee with members of senior management of the participating businesses. Businesses such as department stores and supermarkets are situated on this plateau.

Business Process Redesign. As on the former plateau, the interorganizational systems enable all the transactions of a main process and are fully integrated with the corresponding process-supporting applications of the participating organizations. In addition, however, the potential of IT is exploited to fundamentally alter the way the business process is carried out. This gives rise to radical changes of the internal structure and working methods of the participants. Sometimes even the boundaries between the organizations have to be adjusted.

The objective of the common application of IT is competitive advantage in existing and new product-market combinations. A joint strategic plan, which integrates a business and an IT plan, is necessary to realize the interorganizational systems and the integrated technical infrastructure. The definition and implementation of this plan is controlled by a steering committee with members of senior management of the participating organizations (a federal IT organization). Only a few pioneering companies have reached this fourth IT management plateau.

Business Scope Redefinition. Business scope redefinition can also occur within a coalition of organizations as a next step after the redesign of business processes across the boundaries of the individual organizations. This implies boundary corrections and dramatic organizational changes of the participating organizations.

The objective of the common application of IT is to realize innovative product-market combinations. Entirely new business and market opportunities may be created. This also requires a joint integrated strategic plan. The IT activities are functionally coordinated by senior management. Because of the strong interdependence of the participants, the steering committee on this highest IT management plateau takes the form of a board of directors. The coalition has in fact become a single (networked) organization. It is expected that coalitions of organizations will reach this highest plateau in the twenty-first century.

Exhibit I-7-5 summarizes the seven subsystems of the IT domain on each IT management plateau for a coalition of organizations. Also within a coalition of

IT MANAGEMENT PLATEAUS / SUBSYSTEMS	FI	CI	PI	PR	SR
IT Strategy	Increase speed and reliability of transactions. IT plan based on users demands.	Increase efficiency. IT plan derived from individual business and IT plans.	Competitive advantage in existing PMCs. IT plan adjusted to individual business and IT plans.	New PMCs. Joint integrated strategic plan.	Innovative PMCs. Joint integrated strategic plan.
Information Systems	Transaction automation	Procedure automation	Enable common business process	Enable common business processes after redesign.	Enable common business processes after scope redefinition.
Technical Infrastructure	Different infrastructures	Common standards	Integrated infrastructure	Integrated infrastructure	Integrated infrastructure
IT Organization	Project organization	Project organization	Federal	Federal	Functional coordination
Users	Functional specifications	Functional specifications	Responsible for IS realization	Redesign business processes and IS	Design new business processes and IS
IT Specialists	Realize IS	Realize IS	Design and build IS	Build IS	Build IS
IT Management	IT managers decide within budget	Steering committee of line and IT managers	Steering committee chaired by senior management	Steering committee of senior managers	Steering committee of senior managers

Notes:

FI	Functional integration	PR	Process redesign
CI	Cross-functional integration	SR	Scope redefinition
PI	Process integration	PMC	Product-market combination

Exhibit I-7-5. Summary of the Subsystems on Each IT Management Plateau in Coalitions of Organizations

organizations, all subsystems on each plateau should be in balance with one another. On each successive plateau, the impact of IT on the individual organizations and the relationships between the organizations is more severe, the costs and risks are higher, but the potential competitive advantage generated by IT also increases.

Adequate planning and control of the required activities is necessary for a purposeful migration to a higher plateau. An essential condition for a successful migration is mutual trust within the coalition. Furthermore, the conditions mentioned in process integration also apply to coalitions of organizations.

CONCLUSION

All subsystems of the organization domain and the IT domain should be in balance with each other for the successful application of IT to occur. Good communication between users and IT specialists must exist, as well as adequate coordination by senior, line, and IT management. Each successive IT management plateau requires a higher degree of communication and coordination.

In organizations, as well as in coalitions of organizations, the subsystems of the IT domain on the lower plateaus are adapted to the subsystems of the organization domain. These reactive applications of IT can offer productivity improvements of 10% to 20%. On the higher plateaus, the subsystems of the organization domain are adapted to the subsystems of the IT domain. With these proactive applications of IT, productivity gains of more than 80% are possible.

Depending on the business organization structure, the IT organization on the lower plateaus can vary from a centralized to a decentralized organization. On the higher plateaus, only a selectively decentralized IT organization is appropriate. On the highest plateau, an organization (as well as a coalition of organizations) transforms into a networked organization.

Because a single organization is formed on the highest plateau, the IT management plateaus can also be used as stepping stones to integrate a group of businesses after mergers or acquisitions. This applies especially to organizations with data processing as the primary business process, such as financial institutions. IT is then used as a catalyst for the integration of businesses.

I-8

Economics of Advanced Information Technology

A. Perry Schwartz

IN COMPANY AFTER COMPANY advanced information technology is a hot topic; in many of these same companies, however, there is a tremendous gap between talk and action. That gap exists as a direct result of the failure of too many information systems projects to have immediate, tangible payoffs. In fact, many projects to reduce costs that were approved by management have actually increased costs. By now, every company has experienced its share of these projects. On the other hand, projects that have produced tangible increases in revenue or profitability are few and far between. Indeed, expectations of increased revenue or profitability are often so low that many firms fail to keep even the most basic data for determining whether such increases have occurred.

Consequently, the typical executive has been preoccupied with the cost side of the equation, ignoring the benefit side. This means that the executive is essentially ignoring the fact that the fundamental economic purpose of introducing new technology is not really to reduce costs but to increase a firm's growth and profitability potential. Cost savings are only one part of this equation and arguably not even the most important part.

ENHANCING GROWTH AND PROFITS

Some firms have already discovered the connection between information technology and profitability. For example, one manufacturing firm saw the possibility of using computers to store and disseminate information it was already gathering in the normal course of business. After a large initial investment in new technology, this company now derives a significant share of its revenue from selling this information to business and individual subscribers. Costs rose, but revenues increased much more.

In a second manufacturing firm, five years' worth of financial experience were tracked after a large investment had been made in a new information system. In this case, the additional fixed cost for the new system was substantial: nearly $1

million for improving the support of 30 engineers. As a result of the new technology, however, these engineers were able to speed the pace of key development
projects. After five years, the additional net revenue attributable to this department's work exceeded $30 million.

During the past several years, the implementation of substantial upgrades and
new features—costing upward of $2 million—in an old information system in a
division of a large financial services corporation was tracked. This division had
a historical growth rate of 2.5% per month. In the last three years of the study,
following the installation of the new system modules, this growth rate rose to
more than 4% per month. At the same time, the number of staff members has
remained the same. The result has been a substantial increase in staff efficiency
and division profitability. These companies' philosophies are typical of firms that
have experienced great success in exploiting advanced information technology;
they give cost savings low priority while seeking to maximize growth and profit.

PACED AND PACING ACTIVITIES

In working with these successful firms, it was found that the key to making
their philosophy work is to place the initial emphasis on understanding the
activities of the business rather than on understanding the technology. As most
companies are beginning to realize, this is because technology itself does not have
a payoff; payoff can be realized only to the extent that the firm's activities are
amenable to enhancement by application of the technology.

On the basis of these experiences, a model was developed that explains the
choices these firms made, and guides the successful targeting of activities for
improvement by advanced information technology. For this model, it is useful to
distinguish between two types of activities by introducing two new terms: *paced*
and *pacing*:

- *Paced activities.* Provide support for the firm's operation, but do not drive
  revenue.
- *Pacing activities.* Improves performance and increases revenue.

Some examples help to distinguish between paced and pacing activities. In a
bank, prospecting for new corporate account is a pacing activity; credit file
maintenance is a paced activity. More prospecting generally increases revenues.
More credit file maintenance, above the level required for the current amount of
business, does not contribute to an increase in revenues. For firms that depend on
product innovation to attract customers, research and development is pacing and
production is paced. In most firms, the sales activity is pacing while accounting,
personnel, maintenance, legal, and staff functions are generally paced.

By definition, an asymmetry exists between paced and pacing activities as
illustrated in Exhibit I-8-1. For a pacing activity, an increase in activity enhances
revenues, whereas a decrease in activity reduces revenues. Restrictions or limits
on pacing activities constrain growth. A decrease in a paced activity may also
reduce revenue, but this is a result of the paced activity acting as a constraint on
a pacing activity. For example, if a bank's computer cannot handle additional
accounts, there is no point in making any extra effort to attract new accounts.

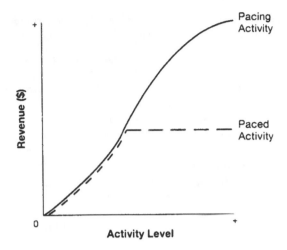

Exhibit I-8-1. Relationship Between Revenue and Activity Level for a Paced and a Pacing Activity

On the other hand, an increase in the level of a paced activity does not increase revenue or constrain growth. If the sample bank's computer system capacity is increased but no new accounts are opened, the change in computer capacity has only increased costs and reduced profits.

Although the difference between pacing and paced activities seems obvious, a distinct tendency is observed among all but the most senior managers to overlook pacing activities and to focus on paced activities when they are asked to consider the prospects of advanced information technology projects. This is understandable, because most of these managers are more concerned with cost control, as measured by their departmental budgets, than growth, and they have a correspondingly narrow view of the firm. On the other hand, higher-level executives have a more comprehensive view of the firm and can distinguish between paced and pacing activities.

Selecting Target Pacing Activities

Investments in advanced information systems are usually made for the purpose of enhancing economic outcomes. If pacing activities are targeted, increased revenues are likely. Alternately, if paced activities are targeted, decreased costs are usually the only reasonable goal. Ordinarily, the incremental increase in revenues from leveraging pacing activities far exceeds the potential cost savings from improvements in paced activities. Therefore, pacing activities represent the greater opportunity for advanced information systems, and as the first step in developing a technology strategy, these activities should be identified.

Once the pacing activities have been identified, four criteria can help select the ones that are likely to be successful advanced information technology projects:

- *Will increased emphasis on the activity actually result in more business?* This confirms whether the activity is really one that is pacing. If the sales staff believe that the market is saturated, for example, new product development rather than sales may really be the critical pacing activity.

- *Will the additional capacity for quality or quantity of work be directed toward the pacing activity?* For example, if freed time will be soaked up by administrative work rather than by increased sales efforts, the investment in advanced information technology is not likely to pay off.

- *Will the additional business resulting from enhancing the pacing activity actually yield more profit?* This may not be true in cases in which a costly production and manufacturing infrastructure must be created to support a large increase in sales or in which prices must be lowered to induce additional purchases.

- *Does the pacing activity contribute to a line of business that is a high priority?* This firm should have a fundamental interest in enlarging the aspect of its business that will be targeted by the advanced technology.

- *What is the technical feasibility of any proposed solution?* This assessment must be made in light of the firm's current level of information technology. Solutions requiring the firm to make technological leaps have high failure rates and should be considered research and development efforts.

- *What is the level of support for the introduction of new technology by the managers of the activity?* If the managers are not enthusiastic, successful implementation will be jeopardized.

- *What will the level of acceptance of the proposed solution be by those who will be asked to use it?* Strong resistance by the target users will increase the risk of failure.

In general, the more positive the answers to these questions, the greater the likelihood for significant payoff.

After the key pacing activities have been identified, it is time to determine whether opportunities, problems, or bottlenecks exist that can be resolved by advanced information technology. Those activities that cannot be improved through the use of advanced information technology may still be identified as targets for improvement but are not relevant to the technology strategy of the firm. Those pacing activities amenable to enhancement by technological means should be subjected to thorough systems analysis.

ADDING PEOPLE OR COMPUTERS

Insightful executives were asked whether they can get the same increase in output just by adding more staff, rather than more technology. The question reflects a course of action every prudent executive should consider. After all, if increased activity is the goal, it can certainly be generated by additional staff. A three-part answer is offered to the question of people versus computers:

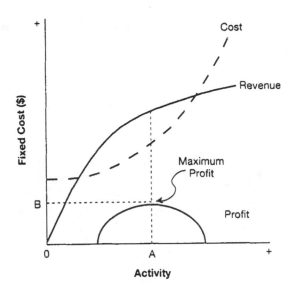

Exhibit I-8-2. Profit Maximization Based on Activity Level

1. One of the major benefits of advanced information technology is the capability to perform tasks more quickly and thoroughly. An increase in staff size often degrades responsiveness because of the increased need for communication and management control within the organization. The hidden costs of organizational control can seriously drain profits.
2. Nevertheless, if an increase in activity through the addition of staff will increase revenues at a much greater rate than it increases costs, yielding increasing incremental profits, a staff increase is likely to be desirable.
3. On the other hand, if the resources committed to the pacing activity have been set to maximize profit, expanding the resource commitment along conventional lines by adding staff will only decrease profits. Because advanced information technology can change the underlying cost structure of the business, use of technology to increase effective activity without a corresponding increase in staff may provide the basis for increasing profits.

Exhibit I-8-2 helps explain these last two points. In this figure, the x-axis is the level of activity and the y-axis is dollars. The revenue curve is typical for a pacing activity. The cost curve, showing fixed and variable costs, reflects the general concave upward pattern that the standard economic theory of the firm projects for costs. The profit curve is simply the difference between the revenue curve and the cost curve.

The maximum profit occurs at point A on the x-axis in Exhibit I-8-2. If the level of activity is less than A, it may be appropriate to add staff, thereby increasing profits. This is the circumstance described in item 2 in the preceding

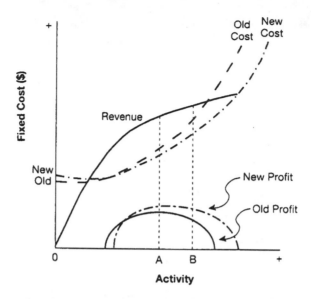

Exhibit I-8-3. Effect of Change in Cost Structure on Profit

list. If the level of activity is greater than A, however, adding staff is a poor strategy. This is the circumstance described in item 3.

USING IT TO INCREASE PROFITS

Effective executives ask tough questions and demand sound answers. For advocates of new information technology projects, one of the tougher questions is: How is the new technology going to increase profits? The answer is rooted in the basic economies of the firm. In fact, the economics are much the same as those used as a basis for the introduction of most industrial automation in factories.

The basic economics are illustrated in Exhibit I-8-3. This figure superimposes two new curves of those in Exhibit I-8-2 (i.e., new cost and new profit) to illustrate one possible effect of the introduction of advanced information technology. In this case, the new technology increases fixed costs while it decreases the variable costs of performing an activity. Thus, the new cost curve starts above the old cost curve but has a lower slope. The break-even point occurs where the two cost curves intersect. From this point on, the advanced information technology will produce profits that are uniformly higher than those that would have been achieved without it, and the new, higher maximum profit occurs at activity level B rather than level A.

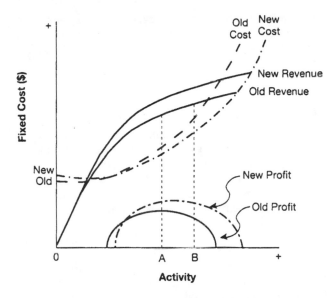

Exhibit I-8-4. Acceleration of the Firm

In addition to this long-run cost saving, advanced information technology holds out the promise of increased revenues. For example, technology might be used to create new features or services. The firm may anticipate additional revenue per unit of activity to the extent that the basis of competition is altered.

Exhibit I-8-4 illustrates the combined impact of both decreased costs and increased revenues. In this case, the effect of introducing advanced information technology is reflected by two structural curves. The first is the new cost curve. This reflects the increase in fixed costs and the decrease in variable costs associated with the new technology and is the same curve as shown in Exhibit I-8-3. The second curve reflects new revenue that would result from new or improved products and services. The new profit is then the difference between the new revenue and the new cost; there is an associated new profit-maximizing activity level B, reflecting profit levels substantially higher than those achieved at the previous maximizing level A. The overall result is a change in the cost structure of performing the activity.

The alteration of the cost/revenue structure illustrated in Exhibit I-8-4 is referred to as acceleration. The acceleration effect is the most desirable outcome, occurring only when the firm takes advantage of both the cost-reducing and the profit-enhancing potential of a new technology.

The typical executive is not likely to have reviewed the benefits of advanced information technology proposals or project outcomes in terms of acceleration. Experience suggests that the reason for this is not the absence of acceleration but the inadequacy of systems for measuring the economic contribution of information technology projects. Yet a complete program tracking the impact of dollars spent on information technology in terms of cost savings and revenue increases would certainly yield invaluable data. It would also be the basis for more sound technology investment decisions.

CONCLUSION

In the face of an explosion in advance information technologies, many firms are approaching new information systems opportunities without a guiding set of principles for leveraging the results. Traditional strategies focus on costs and savings related primarily to paced activities. However, the potential for payoff is far greater if the firm can establish a strategy for applying advanced information technology to the improvement of pacing activities.

Nevertheless, the choice of pacing activities as targets for advanced information technologies is only the start. Over the long term, a successful strategy for the introduction of technology must be driven by fundamental economic considerations. Obviously, no firm can continue for long sacrificing profits to buy advanced information technology. On the other hand, the inclination to apply advanced information technology will be strengthened greatly as information technology initiatives clearly and demonstrably result in economic acceleration of the firm. The challenge is to ensure that projects are chosen to maximize this possibility and that project tracking systems are adequate to measure the economic outcomes.

I-9

Moving From Availability
to Service

Malcolm A. Fry

IT SEEMS STRANGE TO NOTE that although many large organizations have been computerized for more than 25 years, customer service has only been fashionable since the mid- to late 1980s. Why is this? Did the pioneers of industrial technology not care about their customers? It is only by analyzing the driving factors behind the dramatic emergence of customer service that service can be successfully implemented within an organization. Therefore, the question that must first be answered is, why has service suddenly become the vogue?

Many factors are behind service becoming the vogue, but the three prime factors are the technology profit triangle, the new age of technology, and the performance to service transition. Individually each of these factors has a tremendous effect since the late eighties but it is the accumulative effect of the three that has provided the driving force.

THE TECHNOLOGY PROFIT TRIANGLE

The technology profit triangle is a theory developed by the author in 1985 that concentrates on business applications, rather than hardware and software, to plot the effect of technology on business processes. Although the development of new technologies has allowed the penetration of automation deeper and deeper into the fabric of industry, it is the application of that technology that has played the primary role in underpinning the blossoming of customer service. Exhibit I-9-1 shows the development of business applications through three major eras of progression: support, mechanics, and profit.

The bottom left-hand corner of the exhibit shows that all of the early business application development was in the area of business support. Business support applications are the basic business activities that occur in every organization (e.g., payrolls, credit management, statements, invoicing, and other accounting practices). In fact, it is estimated that more than 90% of all mainframes were originally installed for these applications, with payroll being the leading appli-

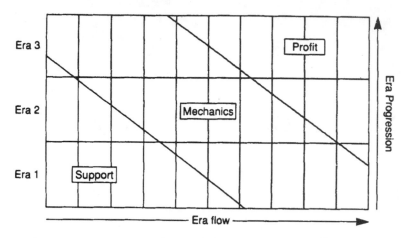

Exhibit I-9-1. The Technology Profit Triangle

cation. Support processes are often called back office functions because they do not directly affect the business products but are essential to the successful running of any organization.

During these halcyon early days most computer systems were batch, or possibly simple networks, with a very small number of internal customers and, critically, did not directly affect products or corporate customers. As a result of the small customer base, the low problem impact, and unsophisticated technology, customer service was not perceived as a high priority. The driving factors for investment were reliability and performance.

As era 1 progressed, technology became both more sophisticated and reliable, which encouraged progression and investment into more critical and vulnerable business areas. In Exhibit I-9-1 these are called mechanics, and are all of the business functions in an organization excluding the support activities. Therefore, if all support and all mechanics functions are computerized, the result is a fully automated organization.

It was during the automation of business mechanics that the requirement for customer service began to become an issue. This was a result of three factors: pain, awareness, and sophistication. The pain originated because for the first time technology failures could adversely affect the normal business activities and, in exceptional cases, lose clients. In addition the pain was fueled by the increasing number of staff members dependent on technology to perform their daily duties. Awareness became an issue because technology began to require much higher levels of investment, which in turn generated the demand for a return on that investment. Finally, technology became much more sophisticated, requiring more skill to solve problems coupled with an increase in the range of problems. So customer service was born, still a low priority but the flame had been ignited!

As sophistication increased and mechanics became more automated, the demand for service began to grow—the next level of business automation, however, caused a major shift in the emphasis of technology. Toward the end of era 2, technology began to be harnessed to create profit by allowing companies to expand into new markets, provide new services for their customers, compete in new markets, and trade internationally from a single location. This led to the technology profit triangle. It is probably easier to use some simple examples than to provide a detailed explanation:

- *Case 1: Automated teller machines.* ATMs allow 24-hour, 7-days-a-week access to withdraw cash, deposit cash or checks, check account balances, and perform numerous other banking functions. How could a bank provide this service without technology? How long would a bank last without ATM machines? What would people do if their bank took actions in person in the future?

- *Case 2: Airline boarding passes.* In the US, all major airlines now issue boarding passes, complete with preallocated seat numbers, as opposed to tickets. The result of this practice? No check-in lines and less delay at airports. How could an airline provide this service without technology? How long would an airline last if they removed this function and began to issue old-style tickets and insisted that customers had to stand in line for check-in?

- *Case 3: French hotel chain.* A hotel chain in France operates automated hotels overnight. These hotels are situated on French highways and are not staffed after 7pm (time varies). Guests perform their own check-in from a computer screen using a credit card for payment and gain access to a hotel and their room by way of a secret code issued by the computer. The hotels cost 130 French francs ($21) per night for three people compared with prices of about 400 francs ($65) for a traditional hotel. Which hotel would most customers stay in? Fast check-in and low prices versus slow check-in and higher prices? No contest! But how long would this hotel chain survive if their technology failed?

These three cases illustrate the new direction that technology is following and it is easy to see how important customer service is in each of these cases. As soon as many companies in the same industry are using similar elements from the technology profit triangle, others are almost forced to follow to remain competitive (e.g., banks and ATMs). Entry into the technology profit triangle is a gamble, because nobody is certain that the customer will buy the product.

Success brings all of the rewards of being one step ahead of the competition, however, and it would be extremely embarrassing for such an advantage to be lost because of poor support and service. In addition, entry into the technology profit triangle increases the technology awareness of corporate executives who will watch the project constantly to verify that they have made a worthwhile investment. Therefore, once an organization enters the triangle, customer service becomes a vital element.

New Age of Technology

It is hard to believe that personal computing did not start to have a significant effect in most organizations until after 1985! The revolution in technology since that date has been dramatic and has been a significant factor in the emergence and subsequent growth in customer service. Personal computers, however, are not the only component that is included in the new age. The following paragraphs explain other technological advances making an impact.

Personal Computers. Personal computers have provided the basis upon which much of the new age has been built. Their assets are obvious: fast, offline operation, graphics, full color, common look and feel, powerful, cost effective, and connectable to other platforms. They have become the standard technology delivery agent used by most organizations. But a complicated beast is hidden behind the pleasing facade. They are highly complex units that are often in the hands of complete technology novices, which is a sure-fire recipe for trouble. Customers using personal computers require a high level of support for both problem solving and question answering. Problem diagnosis can be very difficult from a distance and many of the questions are difficult to answer because the programs resident on the personal computers were not written in-house.

Office Automation. The use of electronic mail, word processors, spreadsheets, electronic calendars, and presentation graphics has created a whole new category of customer support and service. In the past, most of the need for customer service was driven by problems, but it is now being increasingly driven by how-to questions. Most office automation tools are highly sophisticated and extremely function-rich, which is great for the expert but can be a nightmare for the novice. Unfortunately, most office automation users are undereducated, which obviously increases the number of how-to questions. In fact, many help desk staffs consider that they are becoming the company online training service. This problem is exaggerated by the fact that the office automation products are all purchased, rather than developed in-house, so the help desk is becoming the last line of defense.

Single Point of Delivery. Organizations use their networks, local servers, and personal computers to provide a single point of delivery. This means that all of the systems they are using, irrespective of platform, are delivered to one single point so customers can access all of their systems from one source. Often this is achieved by using a common front end (e.g., Microsoft Windows), to provide common look, feel, and functional ability. A seamless single point of delivery is wonderful—exactly what the customer wants—but the implications are tremendous. If the delivery personal computer fails, the customer has not only lost contact with all platform connected systems but also with PC resident office automation systems! In earlier times, the loss of a mainframe terminal would not have stopped the customer from writing reports, sending E-mail, or issuing memos. The loss of the PC now often means that the customer cannot do any productive work.

Intercompany Connectivity. As if single point of delivery was not enough, more organizations are connecting their technology to provide inter-company communications (e.g., electronic data interchange, or EDI)! One example is travel agents accessing airline networks to book flights. Failure to connect to these EDI networks results in the loss of revenue and reputation—if the company is not connected, it cannot sell its products (e.g., flights).

Corporate Dependency. All of the previous topics illustrate just how dependent most organizations are becoming on technology, and as this dependency is being recognized, so is the need for investment in customer service being appreciated.

Portable Technology. Since 1990, a new revolution has begun in the use of portable technology, especially with the reduction in size but increase in capacity and performance of notepad personal computers. There are now cordless networks providing mobile connectivity to servers. One company in the US with 1,000 notepads has noticed a major change in the working habits of its owners and of the level of callers to its help desk. Portable technology is changing the timetable of customer support while introducing customer support at a much higher level in the organization.

New age technologies have forced customer service and support to become more professional. Many service departments and help desks are finding the transition difficult, however, which is why they must concentrate on the fundamentals of service rather than haphazardly attempting to do their best.

The Availability-to-Service Transition

In the past, much effort has been expended by IS departments to achieve high levels of performance but little emphasis has been placed on service. Now that performance is beginning to reach an acceptable level, more concentration is being focused on service. Performance without service is pointless. For example, what is the point of buying a performance car if every time its owner drives it he or she gets a backache? Car designers learned a long time ago that performance and comfort must be given equal treatment. Likewise, IS departments are beginning to learn that performance and customer service are mutually complementary. This shift in emphasis has occurred gradually, but will continue to grow towards the end of the century, by which time the investment in service will exceed the investment in performance.

The technology profit triangle, the new age of technology, and the performance-to-service transition have been the driving forces, but these must be considered in context with many of the other factors that have also contributed to the shift toward service. These other factors include: growing technology awareness, the end of technophobia, continual development of new technologies, improvements in the performance of telecommunications, cost effectiveness of personal computers and mid-range systems, a young generation of technology aware users, increasing corporate dependency on technology and the fact that it is becoming impossible to compete in the market place without the support of technology.

WHAT IS EXCELLENT SERVICE?

If only this question could be answered with a simple constant that could be adopted as a standard. Unfortunately, service is driven by a set of ingredients that are dependent on many variables, including the following:

- Corporate structure.
- Technology supported by IS.
- Technology installed at customer location.
- Corporate culture.
- Business peaks and valleys.
- Financial investment.
- Business agreements (e.g., service level agreements).
- Software and business applications.
- Technology platforms installed.
- Personalities.
- Differing concepts of good service.

Of course many other ingredients will apply, but the preceding list shows how the perceived level of service can vary not just from company to company but also from individual to individual. The approach to excellent service can be summarized as follows: Excellent service requires concentrated effort to meet precise targets agreed upon with the customer community, coupled with a policy that guarantees to continue to improve those targets on a regular basis. At the same time, it must be ensured that the required infrastructure and skills exist to meet and measure those targets.

Excellent service is, and always will be, subjected to personal interpretations. For example, Why does my computer always fail when I need it most? Only the customers can determine whether the service they receive is excellent or poor, and excellent service can be achieved only by consulting and involving the customer. To summarize, excellent service depends on many variables but will be a success only if the customer is involved in the process of determining excellence and thereafter in the measuring and improvement process.

BARRIERS TO SERVICE EXCELLENCE

Most of the barriers that exist today are a result of attitudes rather than tools or technologies. Service is a volatile mixture of technology and attitude, but unfortunately it is easier to change technology than it is to change attitude. Only by attacking those attitude barriers can organizations improve service.

Service Perception. For many years, technicians have been proud of achieving 99+% availability without realizing that this level of service proves to many customers that they are receiving poor service. For example, who would purchase a television if the salesperson told them that only 99% of the televisions worked? Who would fly on an airline if only 99% of their flights arrived safely? Nobody would spend their own money on 99% goods. Customers understand this and relate these figures to technology. For example, if there are 100 people in a department and 99% availability, does this mean that the headcount can be reduced by one? After all, only 99 will be working at any one time if they are all dependent on technology to perform their job functions. Technicians must learn to match their expectations with those of the customer.

Isolation. For many years, technicians have led an isolated life with minimum contact with the customer community and as a result may have lost the ability to communicate with the rest of the organization except under specific circumstances (e.g., when developing a new system). IS departments must learn to build new relationships with their customers.

No Business Knowledge. Isolation is compounded by the fact that many technicians are graduates who have never worked in the industry that they are supporting, but came straight from school into computing. Given this background, how can they be expected to communicate with customers unless they are given corporate exposure and specific training?

Disbelief in Perfection. Technologists have always argued that it is impossible, too expensive, or unnecessary to achieve 100% availability. This sense of imperfection can also be applied to all of the other areas of service. This sense soon transmits itself to customers who in turn begin to accept that technology will regularly fail. This attitude is usually transmitted when technicians are fixing faults and emit the aura of it-is-just-another problem. IS departments must train their staff to believe that perfection must be the overall goal and that achieving 99.5% is not a success but a .5% failure.

Poor Quality Routines. Too many technicians have a poor attitude toward quality when attacking problem elimination. Customers do not want help desks, what they want is zero problems! A problem must be first located, then solved, and finally eliminated. Too often IS departments only work on a locate-it, solve-it basis, allowing repetitive problems to exist simply because they can be easily solved. In fact, the ability to solve repetitive problems is sometimes one of the reasons that some technicians are promoted. How many are promoted for eliminating problems? IS departments must install processes that incorporate the eliminate-it element, possibly by the use of improved problem and change management.

Lack of Customer Consultation. Too often, the only contact technicians have with customers is when problems occur. A positive working relationship cannot be built when most communication is negative. It is amazing how many help desks and service level agreements have been created without involving customers in the building process—no wonder many are not as good as they could be! Is this oversight a result of arrogance, bad planning, or simply a mistake? Whatever the reason, IS must learn to involve customers in the design of any customer service functions.

Each of these barriers can be a major obstacle, but all can be easily overcome if technicians can learn to change their attitude and involve customers in the development and support process of customer service rather than attempt to continue in isolation.

EFFECTS OF POOR SERVICE

Although it is not possible to provide general quantification of poor service, the potential costs are high. The effects of poor service include the following:

- *High revenue wastage.* Large amounts of revenue are wasted because of poor service. For example, 99% availability equals 1% of the total IS revenue that is wasted as a result of nonproductivity. 99% also means that on average 1% of the corporate workforce cannot use their technology. In the case of a company with 10,000 technology-dependent employees, this means that on average, 100 are not working. Given an average cost of $50,000 ($25,000 average salary plus another $25,000 for overhead costs), this represents a loss of $5 million in direct costs alone. In addition, each help desk staff member costs $100,000 per year (same formula, but two people—one on each end of the telephone). This hard revenue can be partially recovered by improving service support.

- *Lost business opportunities.* When customers do not trust their technology, they will not attempt to maximize that resource but will simply perform only the necessary actions. Technology should be embraced to improve business opportunities, not simply used as a large corporate calculator. Poor service will reduce customer confidence even if performance operates at an acceptable level. Lost business opportunities cannot be quantified, but can contribute to reduced competitiveness in the open market.

- *IS staff frustration.* Working in permanent confrontation with customers adversely affects those front-line staff who by the very nature of their job must come into constant contact with the customer community. These staff members see their job as firefighting rather than rewarding, as frustrating rather than satisfying, and as negative rather than positive. Excellent service support would alleviate those frustrations and create more satisfied and productive staff.

- *Poor new developments.* Lack of service will contribute directly toward poor new developments because no good data will exist concerning the perception and quality of current systems.

BACK TO FUNDAMENTALS

The fundamentals required for customer service are not remarkable nor are they original, but they are often ignored. Most of the fundamentals exist—to some degree—in many IS departments, but are often seen as low priority and low importance rather than as vital to meet today's and more importantly, tomorrow's technology challenges. The key lies in recognizing the fundamentals. Once recognized, they can be addressed and a new challenge will emerge for IS—one that is more demanding than any previous challenge.

A Joint Venture

It is essential to attack the basic fundamentals in conjunction with the customers from the beginning, it is not sufficient to involve them only toward the completion of any service initiatives. If the customers do not like the final product they will, quite rightly, complain and find their own solution. Therefore, customers must be involved from the very beginning of service reconstruction and both fully involved and consulted at all subsequent steps.

Customer Technology Officers

One of the prerequisites for achieving an IS and customer partnership is the establishment of customer technology officers (CTOs) within the customer community. CTOs are responsible for local technology issues within the customer community (e.g., help desk liaison, problem solving, local training, change requests, media ordering, local installations, and IS contact point). The number of CTOs depends on the shape and nature of the organization and in most cases would not be a full-time job. Once established, the CTOs provide a vital conduit between the customers and IS to monitor the progress of service excellence restoration. Incentives for appointing a CTO within each department include fewer problems, more control over IS, better-educated staff, faster problem solving, control over resources, better purchasing, better new systems, and independence.

The Vision

Service crosses all of the IS boundaries and requires a common direction that all staff—from operators to programmers—can strive to reach. There is no point in stating that excellent service is the objective without quantifying those objectives. Everybody must be harnessed to the same objectives and not their personal objective of excellent service (i.e., to successfully row a boat all of the rowers must be pulling in the same direction). A vision statement should be created and endorsed by senior IS management to provide the common objective for all IS staff members. This statement should clearly specify what the deliverables of excellent service will be.

Management Buy-In

Many IS managers claim to believe in customer service, but in reality are not interested in putting any time, effort, or financing into achieving excellent service. It is often more a case of doing only enough to satisfy the short-term needs. The further away a manager is from front-line services (e.g., development), the more likely it is that the manager will regard service as a lower priority. This is quite logical, because those managers are not in daily contact with the customer community. Therefore, excellent service cannot be achieved until all managers are fully committed to actively support customer service.

The Service Culture

Even if a vision exists and management has committed to excellent service, the next stage (i.e., to convert all IS staff into believers), is probably the most difficult. It is often difficult to convince technicians who have no contact with customers that they need to change their attitude to become more service oriented, but even a programmer can contribute by thinking service when writing a new program, by empathizing with the end user. All IS departments must embark on an educational and promotional program to convert staff members into believing and following a service culture. In other words, staff members may follow different religions but they must all pray to the same god.

Service Elements

Excellent service has many elements that require investment and attention before any level of success can be guaranteed, including the following:

- Business-level agreements.
- Service monitoring.
- Help desks.
- Problem management.
- Change management.
- Asset management.
- Customer documentation.
- Customer liaison.
- Customer education.
- Quality assurance.

Apart from business-level agreements, none of these elements are new or unique but most IS departments do not concentrate enough investment, effort, or time in these areas. In fact, most staff members believe that they can ignore them if the circumstances suggest, in their opinion, that they can be bypassed. So staff members take calls that should go to the help desk, enter late changes, do not bother to close solved problems, only train customers when a system is installed, provide paper manuals for online systems, and deliver systems with known quality weaknesses (e.g., poor operational documentation). These practices must stop;

management must make it clear that these practices are unacceptable and take appropriate action against members of staff who ignore them. Each of the elements should be reviewed as part of the drive towards excellent service with an emphasis on state-of-the-art products to provide the basis for future progression.

Service- and Business-Level Agreements

Service-level agreements (SLAs) have existed for a considerable time and, in most cases, were never acceptable to the customer community. They were seen as an IS contrivance that was forced on them. A device over which they had no influence, no input, could not measure, and did not meet their business requirements. Therefore, SLAs often meant service level arguments rather than service level agreements.

Customers were being asked to sign service-level agreements stating figures that they did not agree to (e.g., 99% availability). When they wanted higher levels of performance, 100% availability, and could not prove, or confirm, the figures when they were issued because only IS could perform the measuring, they had to accept any figures issued. Not so much an agreement, more of a decree. SLAs may have been good for traditional mainframe applications, but are not so for modern multiplatform environments.

Business-level agreements (BLAs) use a completely different philosophy by working from the customer toward IS, rather than the opposite as employed by SLAs. Customers are given a set of headings and asked to specify their requirements, which are then discussed with IS and an appropriate agreement established. Once established, the agreed information should be dynamically available to the customers so they can check the performance against the agreement at any time from their own technology resources. BLAs require a completely different approach from IS and it is the approach rather than the content that is important.

CONCLUSION

Historically, performance has been the benchmark against which IS has been measured. By the year 2000, service will become the benchmark. IS departments with vision will begin now to lay the foundations for a successful business partnership with their customer community so that by the year 2000 they will be providing a world-class service that will leave their business competitors trailing behind.

I-10

Forecasting Technology Needs: An Example

Jim Cochrane
John Temple
John Peterson

THE FUSION OF INFORMATION and communications technologies is changing the world in which communications services and equipment suppliers compete. New technology and services requirements are creating whole industries, restructuring others, radically changing the nature of competition, and redefining how customers behave. Service suppliers are working to provide widely encompassing information systems and service strategies, including market-specific solutions. Technological advances are making such solutions both technically possible and affordable.

However, in the short run, CIOs and IT strategists find themselves trying to forecast their organizations' technological needs and business opportunities in the midst of technical uncertainty and conflicting market signals. To find their way through this confusion, some strategists are trying to foresee the future and create current business strategies and plans according to that vision.

By working back from a business vision to trends, directions, and other weak but real signals of business and regulatory developments, planners can monitor and sort through the clutter. They can determine whether accumulating signals are consistent with the scenarios and business plans their organizations have developed, and they can confirm or adjust their planning assumptions as appropriate. An example of how this is done is provided in the following sections. (See Exhibit I-10-1.)

THE SHAPE OF THE COMMUNICATIONS FUTURE

The following example traces expected developments in the communications industry and in the process suggests a strategy organizations can implement to take advantage of those developments. The developments are described in terms

111

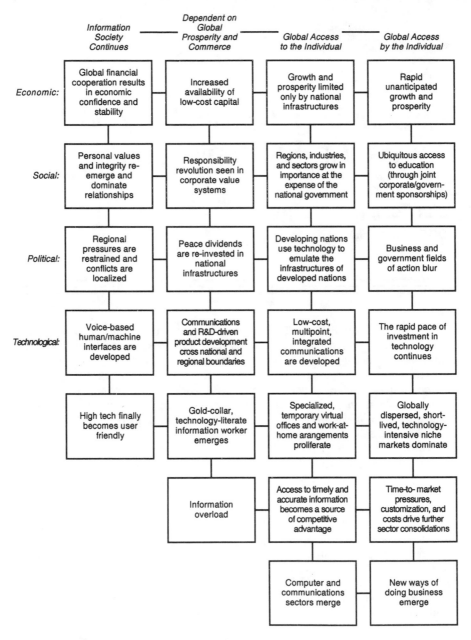

Exhibit I-10-1. A Business-Driven Scenario (A Twenty-Five Year, Linear, Economic-Driven View)

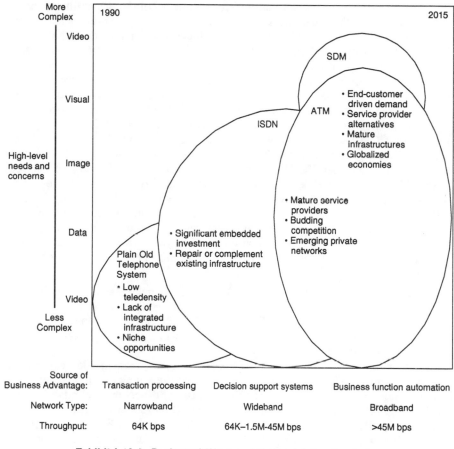

Exhibit I-10-2. Projected Waves of Network Infrastructures

of waves of changes in global networking. Each wave develops as technology providers find new ways to enable businesses (and nations and regions) to compete in an era of new patterns of business transactions. Although described in a sequence, the waves are not time bound. Applications and technology offerings overlap, and some solutions exist in all the waves, though their implementations may differ within them. (See Exhibit I-10-2.)

Looking Back from the Future

By the year 2015, an era of ubiquitous information access will be well underway. Throughout the world, businesses will compete, based not on long-term product and service differentiation but on their information management capabilities. True competitive advantage will come from a company's ability to match distributors (and other customers with cash) with significant volumes of customized high-quality product in near real time. Successful businesses will automate their globally distributed business functions, real-time transaction-based decisions, and coverage of global niche markets.

Efficient telecommunications infrastructures will become a source of short-term competitive advantage. Major digital upgrades to the world's public switched voice/data networks will have been virtually completed. Analog-to-digital upgrades in the 1980s and 1990s will have allowed frame relay and ubiquitous CCITT (International Telegraph and Telephone Consultative Committee) signaling system seven (CCS7) services such as the CLASS offerings (calling and called identification information and features). Integrated Services Digital Network (ISDN) services will be at last globally transparent and widely deployed. Such traditional services, or their local equivalents, including Centrex and custom calling, ISDN, voice mail, electronic mail, and fax store and forward, will be basic services around the world. Therefore, telecommunications service providers seeking a competitive advantage will have to provide custom applications in addition to high quality service and responsiveness. (See Exhibit I-10-3.)

Wave 1. Business advantage, initially found in inexpensive the end-to-end digital connectivity provided by the Synchronous Digital Hierarchy (SDH) [Synchronous Optical Network (SONET) in the United States], will foster the introduction of optical transport, which will allow network elements to communicate on very reliable and low-cost bandwidth.

Communicating intelligence in the network switches, transmission equipment, and intelligent networking products will allow the networks to become self-aware. They will be able to monitor their status and become self-adapting and self-healing. Networks will be able to recognize the type and quality of services being delivered and to predict failures and interruptions based on historical trends. They will monitor the status of key network elements and products and current traffic and congestion patterns. They will be programmed to provide for automatic recovery.

Initially, business users will receive transaction summaries for use in decision making; these summaries will report on the accuracy of the information carried and its timeliness. Combining the traditional network intelligence of the switched network, "very smart" terminals and adjunct elements will further enhance the value of the data through advanced decision support services and applications. The complexity and number of services and applications options available through the network having increased, the needs for even greater transparency and flexibility for network access will accelerate. New user interfaces, including multilingual voice response menus, will be developed to provide flexibility and simplicity for managers increasingly relying on network-delivered information for critical decision making.

In North America, the Internet will be replaced with more effective commercial alternatives. These replacements will evolve when local access transport area (LATA) restrictions are lifted to accelerate competition in the local access and distribution markets. The upgraded, self-healing, public switched networks will support transparent interoperability and globally distributed business operations,

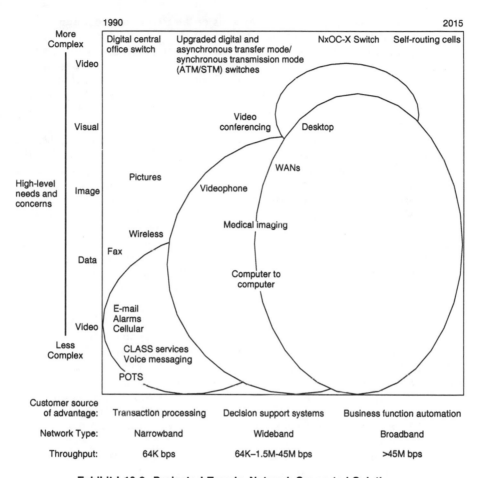

Exhibit I-10-3. Projected Trends: Network-Supported Solutions

remote testing, diagnostics, and administration as well as facilitate access to the individual using networked user-terminal locator sensors.

Wave 2. To facilitate transaction data review and real-time decision discussions, hybrid, or private overlay networks, have been used since the 1970s. These virtual private information networks will become increasingly important to businesses that use them because they (and the real-time information they carry) will initially provide a source of global competitive advantage. The very specific needs of their paying customers will discourage the creation of public network-oriented solutions. Private customers will not be concerned with the time delays and deployment requirements of ubiquity, versatility, standardized services and allocation of costs that are the major concerns of the traditional communications services providers (hereafter, services providers).

Customers will use public network transport (at SDH/SONET transport speeds from point to point) but will support encryption and private proprietary interfaces using specialized end systems. Developed to satisfy LAN-to-LAN connection requirements and the need to support secure globally dispersed private videoconferencing, these public networks will provide a mix of Ethernet, FDDI, frame relay, and switched multimegabit data services.

Customers' willingness to pay for high margin services will encourage service providers to develop a range of broadband services. By the late nineties, the traditional services providers will begin to respond to private hybrid-overlay networks by providing many narrowband and some wideband data-handling services, including wireless access to assuage many data users' initial needs. The traditional service providers will offer these basic public switched data networks and services offerings to begin competing with the new services providers' private hybrid-overlay networks. Such services will help foster partnerships between large data customers and the traditional services providers. Over time, public service providers will add to their voice services private line, medium-speed packet data, and visual (first image and then full-motion video) applications.

Wave 3. As increasing numbers of businesses implement wideband or broadband access architectures, major geographical clusters of public narrowband and wideband data networking will become increasingly common. This will help service providers to expand greatly the number and nature of their services. Initially, these platforms will provide more robust virtual channel switching capabilities. Sophisticated (and still paying) customers will encourage the development of more specialized virtual private networking by leveraging switched wideband public transport. Such switched networks will be developed to meet the business user's demand for broadband services.

A second impetus to the evolution of this public switched wideband network will be the increase in bandwidth (i.e., spectrum, fiber, and coaxial cable) to the desk top, conference table, and the home. Customers will use increased bandwidth to more fully integrate the synchronous bit rate streams of visual and voice services with the synchronous bit rate streams of specific data requirements. The gradual obsolescence of the centralized office environment will help to increase demand as the cost of linking multiple team members or tasks is dwarfed by the costs of transportation and maintaining separate, globally dispersed, central work locations.

The large scale adoption of asynchronous transfer mode (ATM) packet technologies in the late 1990s will provide high-speed broadband transmission and switching technologies to the services providers. Economical broadband access in the office will finally become a reality, and broadband residential access will begin accelerating. Users will have access to full-motion video and other multimedia applications over fiber. The increased speeds first encouraged by ATM applications will lead to widespread demand for business and residence access to multiparty full-motion videoconferencing and other multimedia applications.

As broadband applications spur the demand for more bandwidth and higher switching speeds, optical switching components will be introduced into the services providers' networks, first in cross-connects and then in electronically controlled packet switching fabrics. Business's absolute cost of communications services will increase, despite ever decreasing unit costs, causing paying customers to seek more and more economical bandwidth. Volume and margin skimming by entrepreneurs and private network owners will increase steadily. Loose networks of virtually linked parallel processing computers will further destroy the economies of scale and other advantages of the large, centralized, bureaucratic, services providers. Within a decade, broadband services will become ubiquitous for global companies and no longer a source of commercial advantage. It will be simply a prerequisite for competitive parity.

Wave 4. In response, the upstart services providers will develop new network topologies and classes of value-added services using both wireless and fiber-based access and a new generation of digital optical switching vehicles. As bandwidth dramatically increases with the development of terabit switching capabilities, the virtual corporation will emerge. In such corporations, nonhierarchical management teams will implement business strategies by sending management decision data, packaged and provided in real time to globally dispersed locations. Intracompany communications will have to cross national boundaries. Consequently, political boundaries will have to become increasingly transparent, and large bureaucracies, in all industry sectors across the world, will crumble under the cost and time advantages of the flat virtual corporations.

By the second decade of the millennium, the aging of the populations in the western countries and the multiplier effect on costs of regional, cultural, and language differences will stimulate investment in research to radically simplify human access to ever more complex devices and services. Identifying individuals, locating them, and gaining access to them anytime will be a major function of global networks. The use of network-based virtual personal assistants will gradually become widespread. These systems will take on increased workloads in service industries and in those centers of excellence that will provide staff support to the flat, globally dispersed management teams. Investment in passive sensor technology to track and locate registered network users will also become widespread. Experiments with the ultimate user friendly network interface, alphawave controlled devices, will become a necessity as the complexity and sophistication of the services and options continues to grow exponentially. Finally, globally dispersed businesses will have to maintain competitive parity based on the information advantage provided by new terabit network topologies, which will also aid in the development of a truly global economy.

SOME CURRENT SIGNS OF THE GROWING WAVES

In the very early stages of a revolution, change is rarely immediate or obvious. Most observers simply do not recognize the signals until they become overwhelming. However, in the business world, organizations must look for changes before they occur. This willingness to anticipate changes can decrease an organization's response time, giving it a competitive advantage. The following sections describe some signals that may foreshadow the waves just described in this chapter.

Mobility

In Italy, the Autostrada has provided drivers with remote, high-speed automatic debiting for toll payment capabilities for years. The system uses existing sensor technology, video, communications, and radio equipment, to query and approve access to high-speed entry and exit lanes of the highway. It assists in managing traffic volumes and speeds. The system queries a data base and debits the driver's bank account in real time when the vehicle passes electronic toll gates. The sensors and data base can be used to identify users on the highway system and to provide a general location for simple message delivery by means of car radio.

By the end of the decade, cars in Europe are expected to have an intelligent road transport informatics (RTI) visor that will project both symbols and text several meters in front of the driver. The unit will include European route guidance, parking guidance, automatic debiting, and a transmitter and receiver for communications with roadside beacons. Public transportation and commercial vehicles are also expected to use RTI. This system will improve dispatch planning, vehicle tracking and location, and load and fleet management, encouraging the creation of trains of commercial vehicles in dense formations directed and timed to facilitate control of the road networks.

Transportation system improvements that further the optimization of commercial performance will occur throughout Europe. The improvements will all be made possible by intelligent mobile terminals interworking with wire line networked intelligence.

Wireless Access

In Quitaque TX an entire village of just under a thousand people has foregone wired local loop technology for radio-based voice and low speed data communications. This may well be a model for accelerating teledensity in rural and remote areas throughout the world.

Wireless voice, cellular, and PCS-type systems provide mobility at both the pedestrian and vehicular level. The cellular voice low-speed data (by means of cellular modems) connects to the wire line network at the local mobile telephone switching station. Radio technologies for wireless networks can accommodate heavy traffic with acceptable quality input and a minimum of interference from external sources. Using such technologies, however, involves trade-offs between encryption costs and privacy—and system costs, power usage, spectrum usage, the complexity of the terminal device, and communications range. Consequently, different wireless systems are used for premises, local, and global communications.

Medical Advances

Two technical advances, image-guided therapy and robotics, are accelerating changes in healthcare. Image-guided therapy allows surgeons to see into and through their patients' bodies. Robots are used in hip replacement surgery. In a hip replacement, a specially shaped cavity is made in the thigh bone to accept the replacement prosthesis. A milling robot linked to a computer positions itself using information from a computer tomography scan of the patient. A separate computer attached to the robot monitors its movements against the CT scan and ensures that the drill is in the right place. This approach has proven to be 20 times more accurate than that of a human doctor.

The United States Department of Defense (DOD) is exploring electronic means of bringing a remote doctor to combat casualties. (Significant exploration of remote surgery is also taking place at the Technical University of Karlsruhe in Germany.) Colonel Richard Satava of the Advanced Research Projects Agency of the DOD says "with medical informatics and networking, and mechanically driven doctoring, the physician can see, feel, and interact at a remote site from the patient. Surgeons will be able to operate in dangerous and inaccessible areas, from war zones to third-world countries, without the expense and time of traveling there." Remote surgical operations have already been conducted on animals.

As early as June of 1993, the United States Army Medical Corps conducted a field training exercise that included remote surgery. Medics on the simulated battlefield first held a teleconference with experts to determine whether immediate surgery were required for a wounded soldier. It was. To bring in a qualified surgeon, an ambulance, in which the tele-operation system had been installed, drove to the casualty. The surgeon, watching on a three-dimensional monitor, consulted with the medics in the ambulance that had prepared the patient for the surgery. The remote surgeon then performed the operation using a central master controller that manipulated the forceps, scalpels, and needles.

Information Superhighway

The fourth signal is an interesting article by Robert Benjamin and Rolf Wigand in the *Sloan Management Review*. The authors maintain that the national information infrastructure will one day provide consumers full access to the goods and services provided by businesses. This will force distributors to change their distribution channels, which in turn will spur restructuring and a redistribution of profits throughout the business value chain.

They predict that one or more organizations within the traditional business value chain will be by-passed as the information infrastructure provides the links for new patterns of transactions. The result will be a reconfiguration of the economic system. Successful players in the reconfigured economies will have to learn how to leverage the emerging information infrastructures. The players must identify which technologies are or will complement their own competencies and which are crucial for their business objectives, and they must carefully monitor and respond to the emergence of those technologies.

TECHNOLOGY VALUE CHAIN

If the waves this chapter describes are reasonable, suppliers of both products and services will face more competitors and a greater diversification of products than they ever have before. Traditional industry players face a decade of market discontinuities and opportunities. Opportunities will be found in a mix of customers for traditional and new service providers, new network topologies, new products or architectures, new technologies, and shorter market and opportunity windows. All this will happen as the services providers continue to interject intelligence into both network elements and really intelligent terminal devices. Relationships among customers, network service providers, and value-added resellers, not technology, will determine where intelligence resides in a network (i.e., at the premise, local exchange, or interexchange).

Business's current base of communications equipment is a direct descendant of 120 years of telephony. As the communications industry develops networks with increasingly distributed architectures, industry players will have to compete with an ever expanding range of communications offerings. For example, it is already possible to overlay high-speed, interconnected, synchronous transfer-mode WANs on the current generation of 64K bps switched networks. The incremental changes required to provide traditional voice services over interconnected server-based ATM LAN networks are easy to envision. If accompanied by the evolution of end-user terminal equipment to cell/packet/sender/receiver/router products, the product could provide a cross-elastic fully distributed server-based alternative to the current hierarchical switch-based network offerings.

Technology Value Chain Concept Model

Glimpsing shadows of potential futures is an interesting exercise, but these shadows lack commercial value unless they can be linked to technological realities and business strategies. A successful strategy development process that includes both scenario development and threat assessment is essential for planning for the long term. Part of evaluating the impacts of alternative factors is creating a business-specific technology value chain. It is a planning tool that helps business planners to link and balance the push of technology and the pull of the business strategy. (See Exhibit I-10-4.)

At a very high level, the technology value chain model includes:

- The business's strategic direction, defined in terms of specific business objectives (i.e., growth, revenues, profits).
- Key elements of the strategy (i.e., ten or so key policies that relate to business performance against the strategic business objectives).
- Identification of strategic investments, offerings, and platforms.
- Systems-level competencies, technologies, and skills required to support the strategic investments, offerings, and platforms.
- Device-level technologies (both hardware and software) that are available or necessary to support systems-level competencies, technologies and skills; and potential sources of technology (both internal and external).

Device-Level Technologies

Exhibit I-10-5 illustrates how planners can link technology evolution requirements to planned solutions. The steady movement in hardware toward the physi-

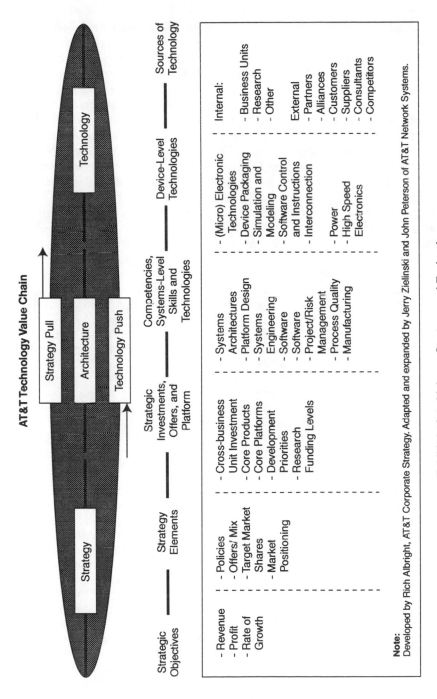

AT&T Technology Value Chain

Strategic Objectives	Strategy Elements	Strategic Investments, Offers, and Platform	Competencies, Systems-Level Skills and Technologies	Device-Level Technologies	Sources of Technology
- Revenue - Profit - Rate of Growth	- Policies - Offers/ Mix - Target Market Shares - Market Positioning	- Cross-business Unit Investment - Core Products - Core Platforms - Development Priorities - Research Funding Levels	- Systems Architectures - Platform Design - Systems Engineering - Software - Software - Project/Risk Management - Process Quality - Manufacturing	- (Micro) Electronic Technologies - Device Packaging - Simulation and Modeling - Software Control and Instructions - Interconnection - Power - High Speed Electronics	Internal: - Business Units - Research - Other External - Partners - Alliances - Customers - Suppliers - Consultants - Competitors

Note:
Developed by Rich Albright, AT&T Corporate Strategy. Adapted and expanded by Jerry Zielinski and John Peterson of AT&T Network Systems.

Exhibit I-10-4. Linkage: Strategy and Technology

cal limits of single-chip microprocessors, combined with the rising costs of fabrication necessary for ever-faster parts, are pressuring designers to seek alternative ways to get increasing levels of systems performance. Improvements in very large scale integration (VLSI) have improved the cost performance ratios by using a collection of inexpensive mass-produced processors rather than single-purpose custom-designed devices. Throughout this decade and early into the next, silicon will be the primary integrated circuit material, and the silicon process of choice for the design and manufacture of memories, processors and application-specific integrated circuits will continue to be CMOS (complementary metal-oxide semiconductor). The need for minimum chip count and improving speeds will drive increasing amounts of intellectual content onto the microprocessor.

Hardware is important because of the intellectual property in system firmware and software. Software engineering has evolved in response to the needs to design, configure, document, and maintain tens of millions of lines of software code over the installed life of network equipment. Managing basic (or device-level) software technology requires standardization, structure, support (tools and testing), process, and most importantly discipline. The systematic approaches, tools, and standards that allow the development of large complex systems are a key competency that assures the attainment of acceptable performance.

One result of the increased interdependence of device-level technologies, hardware, and software, will be a growth in the number of strategic partnerships exploiting the breadth of intellectual content that can be produced on a single chip. Intellectual property ownership and trades will play an ever increasing role in forming successful teaming. This will occur as existing functions are performed by one or a few chips, and new system-level functions, including network architectures, are delivered in ASIC technologies.

Systems-Level Technologies

Device-level evolution does not easily translate into complex networking offerings and applications. New device capabilities support new architectures and improved hardware and software platforms; in turn, new architectures and platforms support new or improved applications. The systems-level skills and technologies become the engines for development. It is here that differential value can be added to components and assemblies. These are the specific competencies that differ from company to company and are among the key differentiators that customers will buy, sometimes at a premium.

Systems-level skills and technologies are not addressed in this chapter because of the strategic importance of such unique competencies to individual companies. The systems-level skills and technology assessment process, however, can be used to benchmark strategic competitors (including the user). The results can be used to generate a matrix of capabilities and tendencies that can be analyzed to determine a relative industry positioning by specific companies or selected technology. Besides the generic descriptions provided in the concept model, these skills and technologies might include such competencies as applications and systems software environments, software tools and visualization techniques,

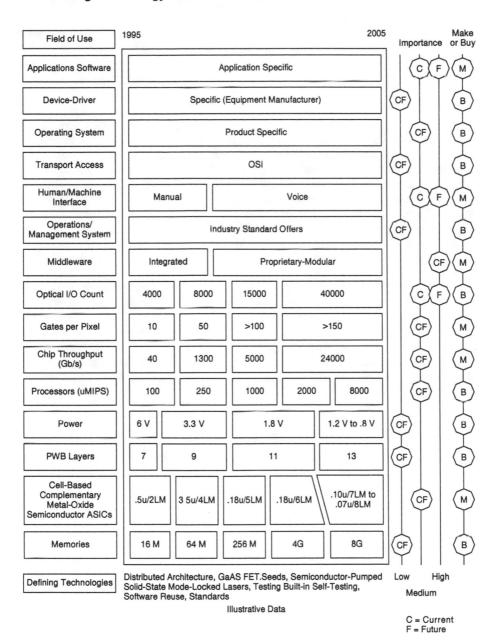

Exhibit I-10-5. High-Level NxOc-X System Roadmap

compression techniques, architecture skills, object-oriented technologies, and formal quality processes and disciplines. (See Exhibit I-10-6.)

INNOVATION

Because balancing technological advances and business strategy is critical to an enterprise in an industry in transition, it important to try to understand how technology will develop in the future. Technology can significantly affect the enterprise and how it competes. Some enterprises become so successful in the way they use technology (or the way they generally operate) that they refuse to change. Their focus is incremental, self-limiting, and bounded by near-term needs. They avoid considering the advantages of radical, longer term changes. This is the effect of "killer competency."

Exhibit I-10-7 illustrates one approach that organizations suffering from killer competency—or those simply struggling to compete in a changing world—can use to evaluate their technology portfolios. Technology strategists can try to determine the effects of technology changes on their enterprises by mapping their organizations' current technology portfolios against technological innovations now underway and their enterprises' critical competencies. Such maps show whether the enterprises' critical mass of the technology investment supports their business objectives and whether their plans for future technology investment should be changed. The following sections describe some technological developments that strategists should consider when developing their advanced technology development portfolios.

Human-Machine Interface: Niche Innovation

Small, fault-tolerant systems first designed to automate communications between hearing-impaired and normal-hearing customers can provide systems with text-to-speech and automatic speech recognition to help automate communications services and minimize operational costs for services providers. After the turn of the century, systems will accommodate large vocabularies (i.e., around 20,000 words) and most normal language usage. Sub-word recognition techniques and language models of natural English and other languages will support fluent interactions in which task-specific semantic controls will still apply. Systems will be able to make rudimentary language translations, allowing them to be used eventually as computer-based personal assistants, interactive text processing machines, and voice-operated intelligent terminals.

Niche Innovation—Cable Modems

Cable modems promise to break the bottleneck of the last mile of twisted pair, especially in densely populated metropolitan areas. They allow the end user to access online services through the coaxial cable that delivers cable television signals to the home (or business). In a client/server architecture, the cable modem takes a datagram generated in the network server and broadcasts it over the

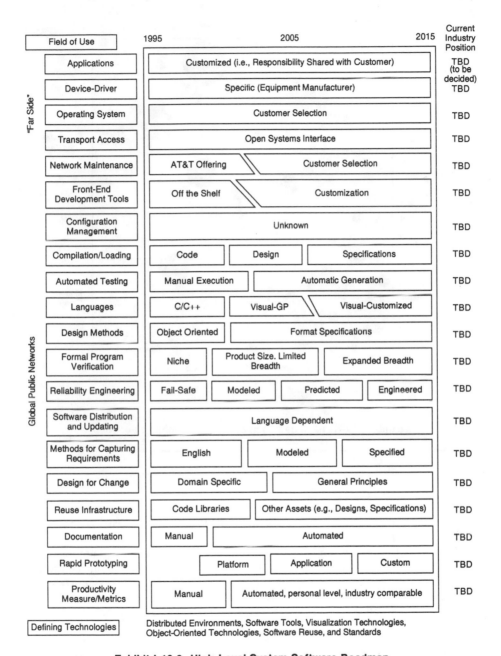

Exhibit I-10-6. High-Level System Software Roadmap

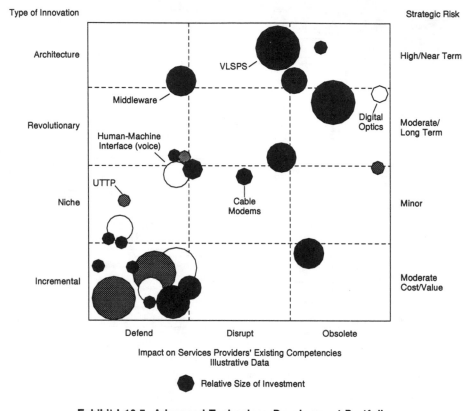

Exhibit I-10-7. Advanced Technology Development Portfolio

network to a microcomputer tuned to the right frequency for distribution on the cable television network.

Typically, the device sends and receives data on two different channels. The receiving channel usually supports LAN transmission data rates but communicates to the network only at narrowband data rates. Their primary shortcoming is the opportunity cost that the cable service provider must bear to support the online services delivery. They must forego a 6 MHz wide channel for the data path from the provider to the homes.

The inherent limitations of the coaxial capacity restricts the number of computers to about 500 on any downstream frequency at any one time. Although a return path is dedicated to a single customer, it remains a narrowband channel. The upstream frequency is usually relegated to the 5 MHz to 40 MHz spectrum, where, in the US, it must contend with the clutter of marine and air traffic radio and broadcast shortwave. Although not a strategic threat to the traditional service providers on a national basis, cable modems do present a potentially significant information technology alternative for local access, especially in areas of high population density with existing cable television service.

Niche Innovation—Utility Telemetry Trunk Protocol (UTTP)

The telephone network was designed from its beginnings to alert a called party when a call has been placed. Simultaneous dial tone is provided to alert the calling party that the network is available to process his or her call. The network has also been engineered and programmed to connect the calling and called parties based on the routing number (or telephone number) dialed. However, the complexity of the existing network topology and software has limited the ability to add intelligence to modify the routing and call termination processes.

These two facts make it difficult to use the local access network for telemetry. A telemetry call must go through, despite individual line feature assignments (e.g., call forwarding), to its intended end point. The second problem that confronts the telemetry user is that the network is designed to alert the end point that an incoming call should be answered. For a telemetry application, this alerting, which is fundamental to voice networks, is actually an unwanted intrusion.

The utility telemetry trunk protocol (UTTP) makes an incremental change in call handling technology that has broadened the basic network platform to allow it to provide telemetry services. UTTP has overcome the roadblocks to telemetry that were inherent because the network was designed to work according to the fixed principles of routing and alerting. It has opened new communication services without requiring separate networks. Now smart homes (with smart appliances, security devices, and electric consumption management) can be monitored with a telemetry call.

Revolutionary Innovation—Digital Optics

Many other factors influence the evolution toward ubiquitous broadband access. Primary among them is the deployment of optical fiber to the home, the desk top, and the conference table. If end users can supply or receive hundreds of megabits per second, networks must provide switching capabilities in the tens of terabits per second. Today, only parts of the infrastructure for an all optical wire line network are in place. The transmission part of the network routinely uses optical transport, and optical LANs are commercially available. For optical switching systems to become economically justifiable, evolution to broadband services must occur. In communications networks, the most significant advantages of digital optics can be realized in the routing and switching of very large bandwidths of traffic.

Free-space photonics appear to provide differential advantages over electronics when the number of connections, reduced connection energy, and greatly reduced skew in the arrival times of bus signals are an issue. These three technical advantages promise novel and improved architectures as the communications networks are beginning to move from offering only voice and limited data communications to real-time broadband multimedia services.

When taken together, the technical advantages of digital optics offer the system designer the flexibility of entirely new architectural options. For a switching system, an entire high-speed network, capable of total throughput of hundreds of gigabits per second, can be built on a single circuit pack. In a computer architecture, separate busses may be employed from the processor to the memory, I/O, cache, or peripheral controllers. In addition, the busses can be very wide to accommodate the natural width of the data being moved.

Architectural Innovation—Middleware

Modular middleware interfaces are a software environment that reside above the operating machine kernel. They provide an object-oriented rather than machine-oriented view of system capabilities. The function of middleware is to improve the portability of applications over a wide range of hardware platforms and to improve the productivity of applications development. High-level environment constructs for applications developers provide some productivity improvement; the rest is achieved by employing a high-level object paradigm, its supporting software development tools, and the reuse of implementations.

The modular middleware software interfaces are part of a distributed object system that supports both real-time and time-shared (and mixed) scheduling. They operate over multi-computer architectures and support incremental updates to the running system. This promotes highly productive software development and integration processes. It also provides machine independent cross-machine object messaging and automatic memory management based on self-defining memory structures. Middleware promotes extensive reuse of interfaces and implementations by providing libraries of modular building blocks. Applications programming interfaces (APIs) define the boundary between platform software and application software. All platforms and APIs can evolve in place to reduce flash-cut software upgrades. Included in the widely used platform APIs are standard protocol stacks, interfaces to peripheral device control, and OAM&P support functions.

Architectural Innovation—Very Low Cost Programmable Switches (VLCPS)

Appearing now in private data networks, these products provide a simple but potentially robust method of connection that is physically separated from its source logic. They use off-the-shelf hardware and software components that are integrated into a programmable platform suitable for wireless, cable, ATM and SDH interfaces. This approach, referred to as 10G bps in a box, may prove to be for the telecommunications industry what distributed processing has been to the computer industry.

Such architectural innovations represent the greatest near-term threats to the traditional equipment manufacturers. Their ability to rapidly assemble inexpensive, off-the-shelf components and assemblies and some systems software in an architecture that solves a services providers' needs is impressive. Given access to reasonable business value chain elements and some good fortune, these new competitors could force accelerated change in the traditional communications industry.

Innovation Wild Cards

A review of the technology portfolio shows what exists, not those gaps that should be addressed. In *Technotrends: How to Use Technology to Go Beyond Your Competition* (New York: HarperBusiness, 1993), Daniel Burris offers some interesting insights along these lines. He identifies four major engines (he calls them rivers) of change: technological innovation, globalization, decentralization of authority, and global demographic shifts. (These rivers are similar to the underlying assumptions of this chapter's business-driven scenario.) He also offers 30 new rules for businesses. They include: think 10 years out and plan back to the present; provide customers with the ability to do what they cannot but would have wanted to if they thought they could; and leverage time with technology. It is difficult to determine what future technologies will provide organizations such capabilities. Nevertheless, it is necessary to try when making very long-term technology plans.

CONCLUSION

If services providers and equipment suppliers are to survive, the industry must begin to shift emphasis from cost-reduced traditional products to more encompassing communications systems applications and services, especially those that provide competitive advantage to the service providers' customers. The equipment world is changing because of competitive pressures at two levels. First is the developing worldwide competition based on technologies developed in alliances between traditional and new services providers. Intelligent peripherals, distributed network intelligence, and massively parallel processing are all trends adding to the pressure. Second is the fierce global competition among and between the service and the equipment providers. The historic price-performance relationships of networking services are becoming obsolete. Traditional wire line services such as voice and low-speed data can be implemented inexpensively using broadcast technologies. In a similar manner, traditional broadcast technologies, such as television, are being provided with improved sound and picture quality over coaxial networks.

Communications players and the companies that use their services will find that one prerequisite to survival is the forming of teams to take advantage of the intellectual property and unique competencies of individual players. Successful companies will have to learn to create teams with other companies to take advantage of specific opportunities, while at the same time remaining fierce competitors for other opportunities. The prerequisite core competency of the next century could well be the ability to continually form, dissolve, and then reform effective teaming arrangements that allow all the partners to benefit.

Section II
User-Directed Computing

ALL COMPUTING SHOULD BE USER-DIRECTED, of course. The title of this section aims to draw attention to information processing that is done under the direct and immediate control of users instead of centrally. With each passing week, more computing resides in the hands of the user. This is a good thing if the user is fully competent to handle the matter. Many users, however, need support in varying degrees. What was called end-user computing in the past has grown to be user-directed computing in every sense of the word.

Taking advantage of opportunities offered by the Internet is a high priority with many users. Electronic markets are rapidly emerging alongside interorganizational systems as a vehicle for electronic commerce. Chapter II-1, "Capitalizing on Electronic Commerce," helps IS managers evaluate the business potential of electronic markets and the characteristics that make the Internet a highly effective forum for such markets. Principles followed by firms already capitalizing on the Internet's electronic market potential are outlined as well.

The increased mobility of computing has paved the way for telecommuting. Users work from virtual offices enabling them to access data wherever and whenever necessary. "Telecommuting: An Overview for the IS Manager," Chapter II-2, provides suggestions for establishing a successful telecommuting program and for managing telecommuters.

Data warehousing offers users access to most, if not all, corporate data for the first time in computing history and is bringing the promises of the data base management systems of the 1960s and 1970s finally within reach. Chapter II-3, "Data Warehousing Concepts and Strategies," describes data warehousing in depth, including the managerial and organizational impacts of the technology.

Information engineering is a very popular methodology in the IS management field. It is designed to improve the integration of technology and to develop a top-down plan for an organization's information systems. Chapter II-4 describes a methodology consisting of four stages. This methodology differs from others because it considers an organization's information to be a significant asset. "A

Practical Approach to Information Engineering" asserts that when an organization understands and manages its data more effectively, it can increase its competitive advantage.

Chapter II-5 addresses the management of advanced information technology, which encompasses the range of issues encountered when introducing and deploying information technology in support of business operations. "Managing Advanced Information Technology" discusses how IS technology planners can determine what technologies will be appropriate for the organization and how they can achieve alignment between technology and business strategies.

As client/server and other approaches to distributed computing take hold, more and different challenges confront IS management. Distributed environments require that IS managers be able to shift from being operators of a central utility to facilitators who can recognize the synergy among loosely related groups of users. Success depends on the development of new ways of visualizing and measuring the service delivery process. Chapter II-6, "Shifting to Distributed Computing," discusses the concerns of senior IS management as computing technology moves further out into the user community.

Chapter II-7, "Leveraging Developed Software," describes an approach to making an investment in software even more valuable to the enterprise. Leveraging is the reusability or portability of application software across multiple business units. The chapter uses an example from the manufacturing sector to show how leveraging software works and the conditions necessary for its success. Leveragability is the extent that the application can remain unchanged as it is installed and made operational at each location. Because leveraging can reduce the cost of acquiring and maintaining software, IS managers should make it an important part of their IS strategy.

Organizational change is the primary force behind client/server computing. IS managers have felt the brunt of organizational change, especially that caused by technology. Chapter II-8, "Client/Server as an Enabler of Change," is a case study in organizational change at the Minnesota Department of Revenue. The chapter shows IS managers how strategic change was fueled by the need to serve customers better and improve revenue—the very objectives that drive desired change in the business world—and how it was implemented through the use of client/server technology and business process redesign.

Chapter II-9 addresses the political issues related to moving to client/server computing. "Client/Server Computing: Politics and Solutions" defines some of the benefits and drawbacks of the technology and offers a short checklist for when it comes time to implement a company's first client/server system.

II-1

Capitalizing on Electronic Commerce

James A. Senn

ELECTRONIC COMMERCE is the handling of business transactions over communications networks. Although already an established form of business for a large number of firms, electronic commerce is evolving rapidly, moving well beyond its origins in electronic data interchange (EDI) and other forms of interorganizational systems toward the creation of electronic markets. Two examples illustrate how the broad participation of many of the world's leading companies has given impetus to these markets:

- In 1995 the US Chamber of Commerce, Dun & Bradstreet, Chase Manhattan Bank, and publishing giant Simon and Schuster, among others, launched the International Business Exchange as an international electronic marketplace that can be used for trading virtually any product or service. Anyone with a personal computer will be able to join the Exchange and post messages describing the products or services they are offering or seeking. An electronic virtual agent will match buyers and sellers, giving them a chance to negotiate their own terms (anonymously, if desired). Credit histories can be verified and other pretransaction services used to ensure that the sale and purchase meet each party's requirements. The base of 50,000 business participants in the first year is expected to grow to more than 1.5 million by the end of the decade.

- CommerceNet is a consortium of companies and organizations formed to facilitate the use of an Internet-based infrastructure for electronic commerce. More than 100 commercial, education, not-for-profit, and government entities composed the early membership group. CommerceNet supports business services that normally depend on paper-based transactions. From their desktop computers, buyers browse multimedia catalogs, solicit bids, and place orders. Sellers respond to bids, schedule production, and coordinate deliveries. All necessary financial and transaction support services are available through the network. Yet CommerceNet is much more than a place to strike a deal. Its working groups provide a forum for industry leaders to discuss issues, deploy pilot applications, and, from these experiences, define standards and best business practices for using the Internet to conduct electronic commerce.

The sheer magnitude and speed of change in business is one factor causing forward-looking executives and managers to consider new business forms, including electronic markets, as well as the role of information technology in delivering products and services. Skyrocketing use and continuing development of the Internet is creating new opportunities for businesses, large and small, and for entrepreneurs seeking to utilize their ingenuity to deliver new products and services in innovative ways.

This chapter explores the two forms of electronic commerce:

- Interorganizational systems, which already have become an integral part of business processes in so many firms.
- Emerging electronic markets.

The rationale for participating in electronic markets is new opportunity to create a product, deliver a service, or get in touch with potential customers. Internet commerce is one highly visible forum for electronic markets. As this chapter illustrates, the characteristics of the Internet merit its careful evaluation by executives and managers alike.

INTERORGANIZATIONAL SYSTEMS

Through interorganizational systems, buyers and sellers arrange for routine exchange of business transactions without the necessity of direct negotiation. Because information is exchanged over communications networks using prearranged formats (see Exhibit II-1-1), there is no need for telephone calls, paper documents, or business correspondence to create and carry out the transactions. Although interorganizational systems may involve proprietary communication links, firms are, at an increasing rate, evaluating the desirability of using public networks for the systems.

Emergence of Interorganizational Systems

Interorganizational systems were driven by business needs and facilitated through information technology's continuing advances. The systems are a direct result of the growing desirability of interconnecting business partners to streamline business processes by:

- Reducing the costs of routine business transactions.
- Collapsing cycle time in the fulfillment of business transactions, regardless of geographic distance.
- Eliminating paper and the inefficiencies associated with paper processing.

Pursuit of these objectives was facilitated by networks that interconnected the diverse desktop and data systems used by business partners. Both proprietary

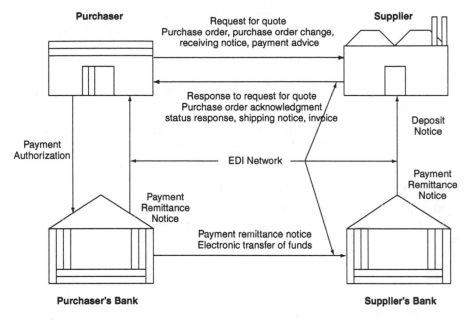

Exhibit II-1-1. Interorganizational System for Electronic Data Interchange (EDI)

---

network solutions and the services of value-added network carriers ensured that any firm wishing to link up could do so.

Types of Interorganizational Systems

The term *interorganizational system* describes a variety of business activities rather than a single entity. Following are five of the most prominent types of interorganizational systems:

- *Electronic data interchange (EDI).* Computer-to-computer (or application-to-application) exchange of standard, formatted business documents transmitted over computer networks where translation systems overcome differences in information technology used by trading partners.
- *Electronic funds transfer (EFT).* Automated exchange of money between parties in a commercial transaction or between banks representing businesses responsible for conducting the settlement portion of a business transaction.
- *Electronic forms.* Online completion and transmission of business forms (e.g., claims forms and contracts, complete with electronic signature) that the recipient can route to the appropriate in-house destination for proper handling.

- *Integrated messaging.* Delivery of electronic mail and facsimile documents through a single electronic transmission system; it may include the combining of EDI, electronic mail, and electronic forms for transmission.
- *Shared data bases.* Information stored in repositories shared between trading partners and accessible to both; such data bases are often used to reduce elapsed time in communicating information between parties as well as to arrange cooperative activities.

Other types of interorganizational systems will undoubtedly evolve as businesses refine and capitalize on their IT capabilities.

Scope of Interorganizational Systems

All interorganizational systems share common characteristics (see Exhibit II-1-2). The principal activities of the systems are business-to-business or business-to-government in nature. In many instances, intermediaries operate the networks that carry the information or provide transaction processing services or data base access.

The communications infrastructure of an interorganizational system is predetermined. All parties know the links over which transactions will be transmitted and where and how they will be received, including the use of electronic mailboxes. Whether public or private networks are used varies from situation to situation.

Parties participating in electronic commerce interact on the basis of a relationship that is defined and preestablished. Terms and conditions of that relationship are often set forth either as contracts or in briefs that specify the expectations and responsibilities of each party.

Interorganizational systems are firmly established in business. The transfer of funds electronically is becoming the norm for such systems, both nationally and internationally. In the US alone, some 50,000 firms conduct business by way of electronic data interchange. Although businesses use the terms *electronic commerce* and *EDI* synonymously, electronic commerce encompasses capabilities much broader than EDI. All forms of interorganizational systems promise to continue growing at an accelerating rate.

THE BUSINESS CASE FOR ELECTRONIC MARKETS

Electronic markets are rapidly emerging alongside interorganizational systems as a vehicle for conducting business. A market is a network of interactions and relationships where information, products, services, and payments are exchanged. When the marketplace is electronic, the business center is not a physical building but rather a network-based location where business interactions occur. The interactions themselves are managed by a broad array of IT applications (see Exhibit II-1-3).

In electronic markets, the principal participants—transaction handlers, buyers, and sellers—are not only at different locations but they seldom even know one another. Nor are relationships between buyers and sellers likely to be predetermined

Interorganizational Systems

Customer/supplier relationship is determined in advance with the anticipation it will be an ongoing relationship based on multiple transactions.

Interorganizational systems may be built around private or publicly accessible networks.

When outside communications companies are involved, they are typically value-added carriers (VANs).

Advance arrangements result in agreements on the nature and format of business documents that will be exchanged.

Advance arrangements are made so both parties know which communication networks will be integral to the system.

Joint guidelines and expectations of each party are formulated so each knows how the system is to be used and when transactions will be submitted and received by each business partner.

Electronic Markets

Two types of relationships may exist:

- Customer/seller linkage is established at time of transactions and may be for one transaction only (i.e., purchase transaction).

- Customer/seller purchase agreement is established whereby the seller agrees to deliver services or products to customer for a defined period of time (i.e., a subscription transaction).

Electronic markets are typically built around publicly accessible networks.

When outside communications companies are involved they are typically online service providers (which function as market makers).

Seller determine, in conjunction with the market maker, which business transactions they will provide.

Customers and sellers independently determine which communication networks they will use in participating in the electronic market. The network used may vary from transaction to transaction.

No joint guidelines are drawn in advance.

Exhibit II-1-2. Distinguishing Features of Interorganizational Systems and Electronic Markets

by agreements. The means of interconnection varies between parties and may change from event to event, even between the same parties. Exhibit II-1-2 summarizes how electronic markets differ from interorganizational systems.

Executives and managers should evaluate the potential of electronic markets on the basis of five business benefits:

- Extending the firm's reach.
- Bypassing traditional channels.
- Augmenting traditional markets.
- Boosting service.
- Advertising.

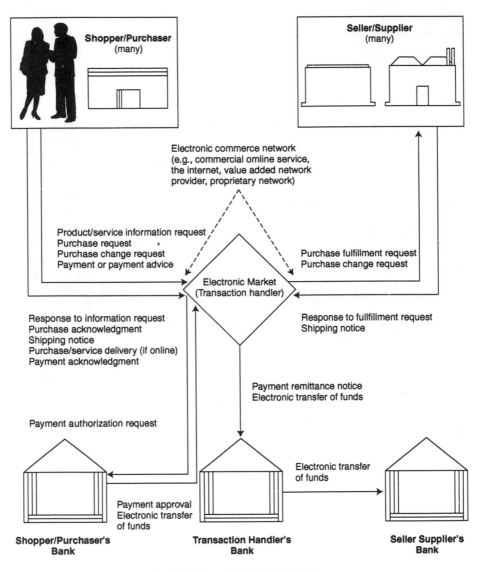

Exhibit II-1-3. Electronic Markets

Extending the Firm's Reach

The ability of a firm to interact with customers or with business partners is defined by its reach. The ultimate objective is to be able to reach any potential customer, regardless of location, without the need for prior arrangement. Even though they are valuable business tools, interorganizational systems cannot achieve this objective because they depend on predefined relationships and communications paths.

Firms are often limited in their ability to reach customers by their sales and marketing processes. The size and location of their sales force, the breadth and depth of their distributor network, the extent of their dealer chain, the number of business locations, or the size and effectiveness of the mailing list all determine a firm's reach. These factors also determine the nature and extent of information exchange. On the other hand, the innovative use of communications networks for electronic markets can create the most dynamic form of reach: anyone, anytime, anywhere.

Bypassing Traditional Channels

Heightened competition and shareholder push for return on investments make it increasingly important for a firm to assess the value added, as well as the costs incurred, in working with its business partners. This is particularly true for distribution channels. If the services of a broker, representative, or distributor do not add value, firms will seek to bypass them to eliminate costs, delays, and other inefficiencies.

Largely for this reason, a growing number of firms are attempting to deal directly with manufacturers, passing along savings to customers in the form of lower prices. Electronic markets facilitate bypass if they enable firms to deal directly with actual and potential customers. Moreover, firms can enter the market even when they do not have, do not wish to create, or cannot establish access to traditional channels.

Augmenting Traditional Markets

Catalog companies have competed successfully against traditional retailers for many years by bypassing both traditional channels and markets where items are bought and sold (i.e., retail stores and other types of sales centers). Electronic markets are a natural evolution of catalog selling and direct dealing, except that both the catalog and order entry process, and in some cases actual fulfillment, are online. In fact, the best known catalog companies, including L.L. Bean, Lands End, and Spiegel, are expanding well beyond their traditional markets to compete in electronic markets.

Among the most effective electronic market alternatives are:

- *Electronic shopping malls (cybermalls).* Emulations of traditional malls that encompass a variety of stores, services, and information guides.
- *Direct retail sales outlets.* Electronic storefronts where customers deal directly with the retailer to create and carry out a sales transaction.

- *Online catalogs.* A special case of the sales outlet where companies create data bases that can be browsed by customers and used by the firm to distribute information.
- *Direct service centers.* Electronic locations from which customer service, advertising, marketing, and technical support are provided.

Boosting Service

Service knows no boundary when markets are electronic. Time windows are eliminated because online services can be delivered 24 hours a day. Important customer and supplier information is available around the clock yet inquirers need not wait for an assistant to provide the details. Careful and creative use of information technology also means that the information can consist of much more than narrative explanations, for drawings, photographs, animated descriptions, and full multimedia presentations are all within the scope of service support in the electronic market.

Other important service options include online sections that provide answers to the most frequently asked pre- and post-sale questions. Support can go well beyond troubleshooting concerns. Organizations have found that their descriptions of product updates or new service features can be much more detailed and offer better explanations when provided in this manner. Of course E-mail and fax-back responses are easily provided as well.

Even if a company chooses to never make a single sale by way of electronic commerce, it can still build its business. Boosting service by way of electronic markets has the potential to be much more than just another business tactic.

Advertising

Awareness, visibility, and opportunity, all important benefits of advertising, take on special importance when markets are electronic. Firms are not constrained by the boundaries of a printed document or by the length of a time slot, both common constraints of advertising through conventional broadcast media.

Carefully chosen listings in online catalogs and data bases enable a firm's consumers to learn about the company and its products even when they lack prior knowledge of their existence. Electronic links make it possible for shoppers to jump from the advertising spot to the firm's location in the market. There a seemingly unlimited range of alternatives can be used to inform, educate, and perhaps convince the customer of the company's capabilities. Product samples and colorful demonstrations, delivered electronically, are highly effective vehicles for gaining attention and garnering good will while building the business.

CREATING ELECTRONIC MARKETS USING THE INTERNET

Firms seeking to pursue development and cultivation of electronic markets can choose from among a variety of communications networks alternatives, including proprietary networks, online services (e.g., America Online, CompuServe, and

Prodigy), and the Internet. All are elements in the frequent discussions calling for a national information structure in the United States. Similar discussions are ongoing in other nations.

There is little doubt that both the expanding reach of the Internet and the accelerating international interest in national information infrastructures will stimulate creation of electronic marketplaces. As more and more firms take steps to move the electronic marketplace from concept to reality, a broad array of innovations will emerge, making it possible for firms and individuals to capitalize on communications networks and overcome the traditional business barriers of time and distance.

Because the Internet has captured the attention of many IT users and observers, it is useful to examine the Internet's value in terms of electronic markets. These following sections explore the reasons why firms may want to include the Internet in their electronic market plans.

Internet Features

The characteristics of the Internet are widely documented (see Exhibit II-1-4). Eight key features are of greatest importance to businesses interested in participating in electronic markets.

Public Resource. The very public nature of the Internet is among its most important distinguishing features. Thus the vast majority of business practitioners are aware of the Internet and its widespread accessibility, even though most do not yet consider its business value. However, the skyrocketing attention to the Internet by the print and broadcast media is certain to fuel the growth in public awareness. Potential customers and business partners will expect firms to be accessible on the Internet.

Because virtually anyone can participate in the Internet as a business by making only a modest start-up investment, the number and diversity of firms participating will continue to increase rapidly. Moreover, the opportunities to announce new products and services, to source materials and services, and to reach potential customers or partners (television home shopping services pale by comparison) are abundant.

Global Reach. Approximately 20% of all interaction on the Internet originates from outside of the United States (see Exhibit II-1-5). In addition, a substantial number of host sites reside in non-US cities, making it a truly international network. Both sectors are growing rapidly.

The broad international reach of the Internet means much more than business access to individuals and firms in developed countries, even though that alone is sufficient for many firms to integrate Internet commerce into their businesses. For the first time, individuals and shops in many underdeveloped countries can interact online as telephone links to the Internet make it possible to span vast geographic distances. No one knows how large this vastly undeveloped market will be.

Span

Connected Networks Worldwide	30,000
Connected Countries	75
Connected Computers	2 million
Rate of Monthly Growth	7% to 10%

Source: US Department of Commerce, Washington DC; Internet Society, Reston VA.

Usage: US and Canada

Individuals with Internet Access	37 million
Internet Users	24 million
World Wide Web Users	18 million

Typical Internet Use (based on the 24 hours preceding survey)

Purpose

World Wide Web Access	5 million
E-Mail	4.5 million
Discussion (noninteractive)	2.5 million
Downloading software	2.1 million
Remote Computer Use	2.1 million
Discussion (interactive)	1.5 million
Audio or Video (real-time)	1.3 million

Location (includes use in multiple locations)

Work	66%
Home	44%
School	7%

Source: CommerceNet Consortium/A.C. Nielsen.

Exhibit II-1-4. Internet Characteristics

Capability to Link. The Internet's capability to link firms has not been fully discovered. Most business users of the Internet still view it primarily as an electronic mail system—that is, a communications tool. Hence, only a fraction of companies connected to the Internet have sought to capitalize on its vast capabilities.

When viewed as a connection tool, rather than as a communications network, many other intriguing possibilities emerge. A variety of different business-to-business transactions can be passed through the network, and there is the growing likelihood that EDI documents will also be transmitted through the Internet.

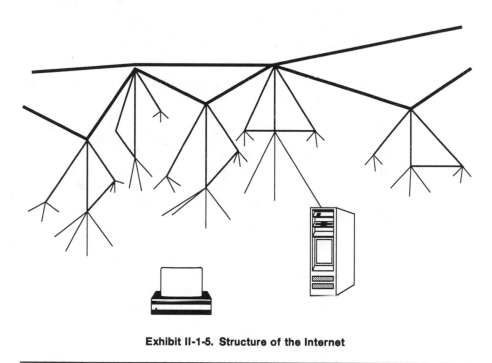

Exhibit II-1-5. Structure of the Internet

Several traditional EDI vendors have developed capabilities to support Internet EDI, although not all have actually announced their capabilities.

Shared Ownership. No company, society, association, or individual owns the Internet. Rather, some thousands of independently owned and operated networks are interconnected to form the Internet. As a result, the Internet is distinguished by collaboration, not proprietary designs. The broad base of public participation means that new initiatives can be successful only if the majority of participants are interested in using them. Even more, it means that virtually every individual and firm, large or small, has the opportunity to participate.

Shared ownership does mean, however, that the Internet has some awkward features, especially in the areas of security and reliability.

Platform Flexibility/Diversity. There are few limits on the nature of the computing and communications that can be interconnected with the Internet. Companies are thus free to use the systems of their choosing (e.g., UNIX, DOS, Windows, OS/2, Macintosh, and Sun). Yet the choice of system platform places no restriction on others using the system or wishing to interconnect with them.

In many instances, the computing systems attached to the Internet are less sophisticated than those used in proprietary systems. Networking and applications software compensate for differences in systems capabilities even as they accommodate the diverse computing and communications platforms.

Cost Advantages. The cost of conducting business on the Internet is quite modest. The principal requirement is to create a business site, typically on the World Wide Web (WWW) portion of the Internet. Getting on the WWW may cost as little as $100. Low-cost kits are readily available to construct the necessary features (such as home pages, online catalogs, and communication links). For a modest fee, the development of such features can be contracted.

Because of the many companies that have emerged to provide access to the World Wide Web or other portions of the Internet, it is not necessary to even operate a computer network to be able to participate in Internet commerce. These companies, which in effect function as "on ramps" to the network, will provide all services, at a cost that depends on the frequency of use for the service. Representative Internet companies are listed in Exhibit II-1-6.

Some companies are investing heavily in their Internet resources, dedicating several staff members and a significant hardware and software investment into supporting their presence on the network. They are choosing to do so because, compared to other alternatives, including developing and maintaining a proprietary computer network or supporting a direct dial-up bulletin board, they view the Internet as a cost-effective resource.

CAPITALIZING ON THE INTERNET FOR ELECTRONIC MARKETS

Ongoing monitoring of firms using the Internet provides growing evidence that those who are capitalizing on the network's electronic market potential appear to follow several principles:

- They treat the Internet as a new medium.
- They use the Internet to leverage existing business and support capabilities.
- They formulate clear business objectives for Internet use.

The Internet as a New Medium

Many businesses tend to consider the Internet's features as supplementary to what they already do. Although this approach may offer attractive possibilities, greater opportunities may be found by taking a fresh approach to the Internet as a medium for reaching out, linking up, and delivering something entirely different. Hence, management should raise stimulating discussion by asking questions that will unleash creative possibilities, such as:

- What are the current limitations in linking up with business partners or supporting customers? What is the impact of those limitations?
- What new products, services, or supports can be offered?
- What opportunities exist to aid the firm's customers in being more successful with their customers?

Firm	World Wide Web Address
Best Internet Communications, Inc. Mountain View CA	http://www.best.com
CyberGate, Inc. Deerfield Beach FL	http://www.gate.net
CTS Network Services San Diego CA	http://www.cts.com
Engineering International, Inc. Albuquerque NM	http://www.olworld.com
Global Shopping Network, Inc. Pasadena CA	http://www.gsn.com
The Internet Access Company, Inc. Bedford MA	http://www.tiac.net
MindSpring Enterprises, Inc. Atlanta GA	http://www.mindspring.com
Open Market, Inc. Cambridge MA	http://www.openmarket.com
Primenet Corporate Center Phoenix AZ	http://www.primenet.com
Teleport International Services Portland OR	http://www.teleport.com
Web Communications Santa Cruz CA	http://www.webcom.com
XMission Internet Access Salt Lake City UT	http://www.xmission.com
Whole Earth 'Lectronic Link (The Well) San Francisco CA	http://www.well.com
Zilker Internet Park Austin TX	http://www.bizpro.com

Exhibit II-1-6. Representative Commercial Electronic Market Providers on the Internet

- How can the firm's current competitors turn the Internet into a competitive weapon that is detrimental to the firm?
- What new businesses can be developed as a means of offering Internet capabilities to others?

Leveraging Existing Business and Support Capabilities

Firms creating value through the Internet are doing so because they are able to leverage resources and expertise already present in the firm. Hence, it is vital that firms directly address these important questions:

- What is it that the firm does best—the products or services it delivers—and how can they be leveraged into new business arenas or as new products and services to a different market?

- What important resources is the firm underutilizing and how can they be put to new or extended use by making them available through electronic markets?
- How can the knowledge-base of the firm be enhanced through access to new customers or business partners who are willing to share their insights and needs in an interactive environment?
- How can the knowledge-base be leveraged into a product or service that will be accessible to virtually any individual or firm through the power of electronic markets?

Formulating a Business Case

Unless a company's journey onto the Internet is designed to be nothing more than an exploratory adventure or distraction, any rationale for moving onto the network should be formulated as a business case. This means establishing and then measuring against clear objectives, preferably with a timetable describing expected milestones. The business case should identify points of success, whether they be potential customer contacts, information inquiries, revenue generation, or profit margins. It should clearly answer two key questions:

- What will the company gain?
- How will success be measured?

It is all too easy to seek to justify new initiatives through such nebulous terms as visibility, public relations, advertising, and public awareness. Yet if these are important reasons for joining the network, as they often are, they should be cast in measurable business terms that will enable even the strongest (or weakest) supporter to gauge success.

CONCLUSION

Electronic markets and the Internet are in their infancy. Although it is not clear how either will evolve, both represent fundamental shifts in electronic commerce with significant implications for business in general. An ever-greater portion of business will be conducted online, with extensive reliance on communications networks.

Waiting to see how the promise and possibilities of electronic markets will evolve may appear the safest strategy in the short-term, particularly for managers averse to high risk. Yet, organizations must have ample opportunity to gain insight into the potential of electronic markets and to create the necessary experience and knowledge to capitalize on the opportunities that may emerge. Organizations that do so may gain long-term advantages that latecomers will never overcome.

II-2

Telecommuting: An Overview for the IS Manager

Sheila M. Jacobs
Mary Van Sell

TELECOMMUTING—a term coined in 1973—enables office employees to work effectively in nontraditional settings. Telecommuters work in remote locations, such as their homes or neighborhood satellite offices, one or more days a week. Using personal computers, these employees can link to their companies' computer systems via telecommunications lines.

The IS manager may be involved with telecommuting in the organization in three major ways:

- Helping functional area managers establish and manage telecommuting in their departments.
- Deciding which products or technology the organization should use for telecommuting arrangements.
- Establishing and managing telecommuting in the IS department.

SOCIETAL BENEFITS OF TELECOMMUTING

For millions of people in America and other high-technology nations, the work day is bracketed by stressful rush-hour commutes over clogged and often decaying highways or railways to congested urban centers where the costs of parking and office space rise continually. Mass transportation systems, where they are available, are overcrowded and often unpleasant.

Recently, the efforts of many nations to ease traffic congestion and air pollution, or to curb spending on public roadways, have included the promotion of telecommuting programs. In 1989, for example, four southern California counties began requiring companies with more than 100 employees at one location to develop plans for cutting commuter traffic. Similar laws exist in other states, including Arizona, Hawaii, Texas, and Washington.

If telecommuting could replace 10 to 20% of US road trips, it could save as much as $23 billion per year in energy, transportation, and environmental costs. A telecommuting program can also help a company lower real estate costs in urban areas by allowing employees to share a smaller office space by spending different days in the main office.

Telecommuting also facilitates employment for workers who have physical difficulty getting to an office, including employees who are disabled, recovering from illness or injury, or are on maternity leave.

Why Employees Want to Telecommute

Most telecommuters are successful professionals who want more from life than traditional corporate mobility. Employees who request telecommuting often do so because telecommuting enables them to combine work with another valued goal. This goal may be related to personal enrichment, such as travel or enrollment in a graduate degree program, or balancing work and family commitments more effectively. Telecommuters value the flexibility to attend their children's special needs (e.g., medical appointments, school visits) during normal work hours. Telecommuting gives employees the freedom to spend more time with their families.

The Benefits of Autonomy. Most telecommuters feel that working at home, rather than in the office, enhances their ability to concentrate on their work because there are fewer distractions. Other motivations for telecommuters are autonomy and control over their time. For example, telecommuters in IBM's nine-month telecommuting pilot program said that telecommuting had given them time to experiment, as well as the ability to increase job turnaround time, by scheduling their work at faster, off-shift times.

Many telecommuters report that their primary motivation for working at home is that they can get more work done. The reason for this is not apparent. It may be that successful telecommuters are individuals with "night owl" circadian rhythms, who work after normal office hours when they are most energetic.

ORGANIZATIONAL BENEFITS OF TELECOMMUTING

Organizations, whether or not they allow telecommuting, want to minimize their labor and operating costs. At the same time, they want to increase levels of productivity. Having a telecommuting program makes it easier for an organization to attract top-quality employees who may not wish to move or commute. Telecommuting also helps reduce labor costs by lowering the rates of absenteeism and employee turnover.

Increased Productivity

Telecommuters are not only more productive at home than they are in the office; they are also more productive than their in-office counterparts—by approximately 30%. Telecommuters are generally more effective because they:

- Work at times of day when they are most productive.
- Are more likely to finish projects ahead of schedule.
- Work for longer periods of time without interruptions.
- Experience improved communication with the work group.
- Are more available for consultation with clients and supervisors at home by phone than when in the office.
- Are more creative because they concentrate better at home.

Reduced Costs

Although 30% of cost savings attributed to telecommuting are due to decreased labor costs, 70% of cost savings result from reduced overhead expenses. Organizations save between $1,500 and $6,000 per telecommuting employee on reduced office space and related overhead expenses. Telecommuting also lowers the costs of using and maintaining equipment, such as mainframe computers that can be used by telecommuters at off-peak hours.

Telecommuter Job Satisfaction

Several studies have found that the majority of telecommuters are satisfied with their arrangement and prefer telecommuting to working full time in the office. Satisfied telecommuters also report:

- Higher self-rated productivity.
- Satisfaction with the performance appraisal system for telecommuters.
- Technical and emotional support from managers.
- A lack of family disruptions.
- Greater loyalty to the organization as a result of being trusted by managers.

Predictions that telecommuting employees would become isolated from the organization and overlooked for promotions have not been supported. Telecommuters are usually more visible in their companies because they are almost always more productive than office workers. Some telecommuting employees report a feeling of missing daily social and professional interaction. The establishment of a neighborhood shared workspace, referred to as a "telecenter" or a satellite work center, can alleviate this problem.

PREPARING THE ORGANIZATION FOR A TELECOMMUTING PROGRAM

Not all companies are good candidates for telecommuting programs. The structure of the organization must be considered. Companies with little work autonomy that use time-based methods of work supervision, and whose decision-making processes are centralized and hierarchical, would find it virtually impossible to supervise the work of telecommuters without changing all these aspects of organizational structure.

If the company is flexible and results-oriented, then the organizational structure of the company is suitable for telecommuting. However, although most companies have more volunteers for telecommuting than they can use in a pilot program, there may still be resistance to telecommuting from managers who fear losing control of their employees or who do not trust employees. Telecommuting presupposes supervisors managing employees by results, communicating expected outcomes clearly, and controlling output and quality rather than time spent and processes used.

Jobs Suitable for Telecommuting

Job tasks that are suitable for telecommuting have been classified as output tasks. Output tasks typically produce discrete pieces of work, such as project reports produced by one person working alone. Jobs that consist of information processing, that result in measurable output, and that do not involve physical contact are candidates for telecommuting programs. Telecommuting enhances productivity in jobs requiring creativity and analysis, where the need to interact with others is not critical.

Many categories of jobs are suitable for telecommuting programs. Several of these jobs are in the information systems field, such as:

- Computer programmer.
- Software engineer.
- Computer systems analyst.
- Technical writer.
- Data-entry clerk.
- Consultant.
- Technical supporter.
- Word processor.

Outside the IS department, the range of jobs that lend themselves to telecommuting includes:

- Translator.
- Sales representative.
- News reporter.
- Public relations professional.
- Stockbroker.
- Lawyer.
- Accountant.
- Engineer.
- Architect.
- Real estate agent.
- Travel agent.
- Writer.

- Insurance agent.
- Purchasing agent.
- Claims processor.
- Marketing manager.
- Customer service representative.

Successful telecommuters tend to be technically skilled and often have substantial professional experience. Telecommuters often have to perform tasks that would be done for them by others in the office (e.g., quality testing, job completion time estimation), so they need a variety and depth of skills. Because telecommuters will have limited face-to-face contact with their supervisors and co-workers, good communication and organization skills are important. Telecommuters should also be adept at scheduling, preparing, and documenting their work.

Technology Requirements

An organization does not need a large capital investment in equipment to initiate and support a telecommuting program. An employee can telecommute with a personal computer, a modem, and the appropriate software. Many prospective telecommuters already own these products.

Computer Equipment. Personal computers are the foundation of telecommuting. Laptop computers, portable computers that can send faxes, and client/server computing equipment are critical. Modems and printers are important, and fax/modems are desirable for additional flexibility.

LANs. Remote access to the company's LAN is important. Some telecommuters need remote access to a LAN only for short time periods (e.g., for E-mail); others need their home computers to behave as nodes on the network. Remote software packages facilitate this arrangement. Telecommunications technology and services make it possible to network home computers to office computers.

Sophisticated Phone Systems. A customer or a co-worker should be able to reach a telecommuter at home or at the office by dialing the same phone number. Today's phone systems can make the physical location of the telecommuter irrelevant. The phone technology in the telecommuter's home (or satellite office) should include the same features found at the office, such as speed dialing, redialing, return dialing, caller ID, priority call, and select forward. Interactive voice services, which provide telephone recordings such as, "If calling for billing information, press 1," may also be important.

Voice Mail. Voice mail, like E-mail, enables a telecommuter to remain informed about office activities and to communicate with the office effectively.

Voice mail messages can be recorded and retrieved at any time of day, so tele-commuters on flexible schedules are not disadvantaged.

Videoconferencing. Videoconferencing allows people at geographically sepa-rate locations to have face-to-face meetings. As the cost of videoconferencing drops, remote conferences will increase. A telecommuter may be able to partici-pate in a videoconference with the company office by going to a telecenter or satellite office that has video equipment or by going to a videoconferencing facility at another location, such as a phone company office that provides this service. Desktop videophones, with small screens, are also available and may be placed in the homes of telecommuters.

Other Products. Other products, such as fax machines and cellular phones, can facilitate telecommuting programs. An electronic imaging system is another useful product. With this equipment, an employee at the office can scan a docu-ment onto the main computer. The telecommuter can then access this document using a home computer and a modem.

Future Needs. Some of the future needs for products and services to support telecommuting include:

- Equipment that is light, compact, and easily portable, such as portable fax machines and printers weighing less than five pounds.
- Better paging and remote access technologies.
- Client/server operating environments that use applications programming interfaces to better support remote access technologies.
- Remote access technologies with built-in security controls.

IMPLEMENTING A TELECOMMUTING PROGRAM

A company's first telecommuting program should be a pilot program, lasting six to eighteen months. The pilot program will help the company determine needed training and equipment, program costs and benefits, and program man-agement. At the beginning and end of the pilot program, levels of productivity and other outcomes, such as morale and overhead expenses, should be formally measured.

When the actual program is launched, mandatory core times should be estab-lished. These are times when the telecommuter is to be available by phone or E-mail to customers or supervisors. Usually, telecommuting employees have scheduled days in the office to keep them in touch with their co-workers, man-agers, and the day-to-day affairs of the company, and to prevent feelings of isolation.

Telecenters

There is a growing trend to establish neighborhood telecenters as alternatives to employees working at home. A telecenter is somewhat like a branch office, except that its location is convenient for employees rather than for customers. Employees from a variety of departments, or even from a variety of companies, may work together in one telecenter. The telecenter may be set up by the company or by an independent organization that rents space to several companies.

Telecenters can give a telecommuting employee a structured work environment while still saving on transportation, real estate, and overhead costs. A telecommuting employee can work at a convenient location without feeling isolated and without missing social interaction. Employees who do not have room at home to set up office equipment can still telecommute. A telecenter is advantageous economically because telecommuting employees share equipment. The telecenter may also serve as a pilot program before the company fully invests in telecommuting.

Managing Telecommuters

Although careful preparation for a telecommuting program is essential, the ultimate success of the program depends on the way it is carried out by the telecommuting employees and their managers. Managers and telecommuting employees should agree on what work is expected, how it should look, and when it should be completed. A written agreement, signed by both parties, may be helpful. The most common reason for failure of a telecommuting program is inadequate communication between managers and employees.

Managers of telecommuting employees need training in techniques of results-based management, job analysis, work specification, and performance appraisal. They may also need training in communications skills.

IS managers must know how to prepare organizations for the changes induced by telecommuting. They should help functional managers establish telecommuting in their departments, and determine the technology requirements for telecommuting. Finally, IS managers should know how to implement telecommuting programs, meeting the information and communications needs of the telecommuters while monitoring and controlling telecommunications costs for the company.

Suggestions for Successful Implementation

Tips for implementing a successful telecommuting program include:

- Start with a pilot program.
- Decide in advance what the desired outcomes are (e.g., increased productivity, reduced turnover) and how to formally measure them.
- Select employees who are motivated and qualified to work independently.
- Consider both the personality of the employee and the needs of the company when deciding how many days per week the employee can telecommute. Decide each case on an individual basis.
- Make sure the telecommuters can be reached by those working in the main office.
- Schedule regular office visits for the telecommuters.

- Be sure the supervisors of telecommuters have (or learn) management skills that focus on results and communication.
- Give telecommuters regular feedback on quality control.
- Try to avoid impromptu office meetings; telecommuters may feel excluded.
- Do not assign telecommuters tasks that could cause delays or backlogs for in-office workers.

II-3

Data Warehousing Concepts and Strategies

Stefan M. Neikes
Sumit Sircar
Bijoy Bordoloi

MANY IT ORGANIZATIONS are increasingly adopting data warehousing as a way of improving their relationships with corporate users. Proponents of data warehousing technology claim the technology will contribute immensely to a company's strategic advantage. According to the Gartner Group, by 1998, the $2 billion data warehouse market will quadruple to an incredible $8 billion.

Companies contemplating the implementation of a data warehouse need to address many issues concerning strategies, type of data warehouse, front-end tools, and even corporate culture. Other issues that also need to be examined include who will maintain the data warehouse and how often and, most of all, which corporate users will have access to it.

After defining the concept of data warehousing, this chapter provides an in-depth look at design and construction issues, types of data warehouses and their respective applications, data mining concepts, techniques, and tools, and managerial and organizational impacts of data warehousing.

HISTORY OF DATA WAREHOUSING

The concept of data warehousing is best presented as part of an evolution that began about 35 years ago. In the early 1960s, the arena of computing was limited by punch cards, files on magnetic tape, slow access times, and an immense amount of overhead. About the mid-1960s, the near explosive growth in the usage of magnetic tapes increased the amount of data redundancy. Suddenly, new problems, ranging from synchronizing data after updating to handling the complexity of maintaining old programs and developing new ones, had to be resolved.

The 1970s saw the rise of direct access storage devices and the concomitant technology of data base management systems (DBMSs). DBMSs made it possible to reduce the redundancy of data by storing it in a single place for all processing. Only a few years later, data bases were used in conjunction with online transaction processing (OLTP). This advancement enabled the implementation of such applications as automated teller machines and reservations systems used by the travel and airline industries to store up-to-date information. By the early 1980s, the introduction of the PC and fourth-generation technology let end users innovatively and more effectively utilize data in the data base to guide decision making.

All these advances, however, engendered additional problems, such as producing consistent reports for corporate data. It was difficult and time-consuming to accomplish the step from pure data to information that gives meaning to the data. One reason for this was the variety of applications inherent in any large organization and a lack of integration across applications. Poor or nonexistent historical data only added to the problems of transforming raw data into intelligent information.

This dilemma led to the realization that organizations need two fundamentally different sets of data. On the one hand, there is so-called primitive data, which is detailed, can be updated, and is used to run the day-to-day operations of a business. On the other hand, there is summarized or derived data, which is less frequently updated and is needed by management to make higher-level decisions. The origins of the data warehouse, as a subject-oriented collection of data that supports managerial decision making, are therefore not surprising.

Many companies have finally realized that they cannot ignore the role of strategic information systems if they are to attain a strategic advantage in the marketplace. CEOs and CIOs throughout the US and the world are steadily seeking new ways to increase the benefits that IT provides. Data is increasingly viewed as an asset with as much importance in many cases as financial assets. New methods and technologies are being developed to improve the use of corporate data and provide for faster analyses of business information.

Operational systems are not able to meet decision support needs for several reasons. First, most organizations lack online historical data. Second, the data required for analysis often resides on different platforms and operational systems, which complicates the issue even further. Third, the query performance of many operational systems is extremely poor, which in turn affects their performance. Fourth, operational data base designs are inappropriate for decision support.

For these reasons, the concept of data warehousing, which has been around for as long as data bases have existed, has suddenly come to the forefront. A data warehouse eliminates the decision support shortfalls of operational data bases by storing the current and historical data from different operational systems in a single, consolidated system. Data is thus made readily accessible to the people who need it, especially organizational decision makers, without interrupting online operational workloads.

The key value of a data warehouse is that it provides a single, more quickly accessible, and more accurately consolidated image of business reality. It lets organizational decision makers monitor and compare current and past operations, rationally forecast future operations, and devise new business processes. These benefits are driving data warehousing's popularity and have led some advocates to call the data warehouse the center of IS architecture in the years ahead.

THE BASICS OF DATA WAREHOUSING TECHNOLOGY

According to Bill Inmon, author of *Building the Data Warehouse* (NY: John Wiley, 1993), a data warehouse has four distinguishing characteristics:

1. Subject-orientation.
2. Integration.
3. Time-variance.
4. Nonvolatility.

As depicted in Exhibit II-3-1, the subject-oriented data base characteristic of the data warehouse organizes data according to subject, unlike the application-based data base. The alignment around subject areas affects the design and implementation of the data found in the data warehouse. For this reason, the major subject areas influence the most important part of the key structure. Data warehouse data entries also differ from applications-oriented data in the relationships. Although operational data has relationships among tables based on the business rules that are in effect, the data warehouse encompasses a spectrum of time.

A data warehouse is also integrated in that data is moved there from many different applications (see Exhibit II-3-2). This integration is noticeable in several ways, such as the implementation of consistent naming conventions, consistent measurement of variables, consistent encoding structures, and consistent physical attributes of data. In comparison, operational data is often inconsistent across applications. The preprocessing of information aids in reducing access time at the point of inquiry.

Exhibit II-3-3 shows the time-variant feature of the data warehouse. The data stored is about five to ten years old and used for making consistent comparisons, viewing trends, and providing a forecasting tool. Operational environment data reflects only accurate values as of the moment of access. The data in such a system may change at a later point in time through updates or inserts. On the contrary, data in the data warehouse is accurate as of some moment in time and will produce the same results every time for the same query.

The time-variant feature of the data warehouse is observed in different ways. In addition to the lengthier time horizon as compared to the operational environment, time-variance is also apparent in the key structure of a data warehouse. Every key structure contains—implicitly or explicitly—an element of time, such as day, week, or month. Time-variance is also evidenced by the fact that data warehouse data is never updated. Operational data is updated as the need arises.

The nonvolatility of the warehouse means that there is no inserting, deleting, replacing, or changing of data on a record-by-record basis, as is the case in the operational environment (see Exhibit II-3-4). This difference has tremendous consequences. At the design level, for example, there is no need to be cautious about update anomaly. It follows that normalization of the physical data base

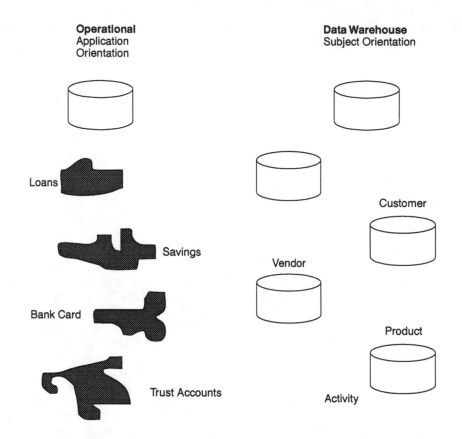

Operational
Application
Orientation

Data Warehouse
Subject Orientation

Loans

Savings

Bank Card

Trust Accounts

Customer

Vendor

Product

Activity

SOURCE: W.H. Inmon, *Building the Data Warehouse* (New York: John Wiley, 1993).

Exhibit II-3-1. The Data Warehouse Is Subject-Oriented

design loses its importance, because the design focuses on optimized access of data. Other issues that simplify data warehouse design involve the nonpresence of transaction and data integrity as well as detection and remedy of deadlock, which is found in every operational data base environment.

Effective and efficient use of the data warehouse necessitates that the data warehouse run on a separate platform. If it does not, it will slow down the operations data base and reduce response time by a large factor.

DESIGN AND CONSTRUCTION OF A DATA WAREHOUSE

Preliminary Considerations

Like any other large undertaking, a data warehousing project should demonstrate success early and often to upper management. This ensures high visibility

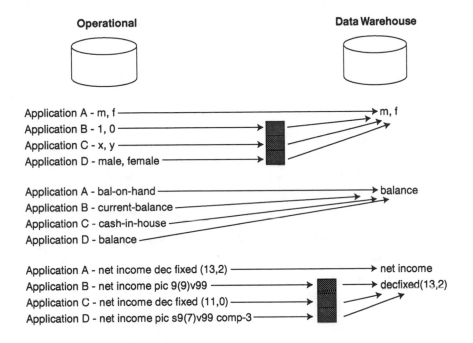

Source: W.H. Inmon, *Building the Data Warehouse* (New York: John Wiley, 1993).

Exhibit II-3-2. Integration of Data in the Data Warehouse

and justification of the immense resource commitments and costs associated with the project. Before undertaking the design of the data warehouse, however, it is wise to remember that a data warehousing project is not as easy as copying data from one data base to another and handing it over to users, who then simply extract the data with PC-based queries and reporting tools.

Developers should not underestimate the many complex issues involved in data warehousing. These include architectural considerations, security, data integrity, and network issues. According to one estimate, about 80% of the time that is spent constructing a data warehouse is devoted to extracting, cleaning, and loading data. In addition, problems that may have been undetected for years can surface during the design phase. The discovery of data that has never been captured as well as data that was captured but then altered and stored are examples of these types of problems. A solid understanding of the business and all the processes that have to be modeled are also extremely important.

Another major consideration important to up-front planning is the difference between the data warehouse and most other client/server applications. First, there is the issue of batch orientation for much of the processing. The complexity of processes (which may be executed on multiple platforms), data volumes, and resulting data synchronization issues must be correctly analyzed and resolved.

Operational

Data Warehouse

Current value data:

- Time horizon—60 to 90 days
- Key may or may not have an element of time
- Data can be updated

Snapshot data:

- Time horizon—5 to 10 years
- Key contains an element of time
- Once snapshot is made, record cannot be updated

SOURCE: W.H. Inmon, *Building the Data Warehouse* (New York: John Wiley, 1993).

Exhibit II-3-3. The Data Warehouse Is Time-Variant

Operational

change　　　replace

replace　　　insert

change

Data Warehouse

load

access

Data is regularly updated on a record-by-record basis.

Data is loaded into the warehouse and is accessed there, but once the snapshot of data is made, the data in the warehouse does not change.

SOURCE: W.H. Inmon, *Building the Data Warehouse* (New York: John Wiley, 1993).

Exhibit II-3-4. The Data Warehouse Is Nonvolatile

Next, the data volume in a data warehouse, which can be in the terabyte range, has to be considered. New purchases of large amounts of disk storage space and magnetic tape for backup should be expected.

It is also vital to plan and provide for the transport of large amounts of data over the network. The ability of data warehousing to support a wide range of queries, from simple ones that return only limited amounts of information to complex ones that might access several million rows, can cause complications. It is also necessary to incorporate the availability of corporate metadata into this thought process. The designers of the data warehouse have to remember that metadata is likely to be replicated at multiple sites. This points to the need for synchronization across the different platforms to avoid inconsistencies.

Finally, security must be considered. In terms of location and security, data warehouse and non-data warehouse applications must appear seamless. Users should not need different IDs to sign on to the different systems, but the application should be smart enough to provide users the correct access with only one password.

Designing the Warehouse

After having addressed all the preliminary issues, the design task begins. There are two approaches to designing a data warehouse: the top-down approach and the bottom-up approach. In the top-down approach, all of an organization's business processes are analyzed to build an enterprisewide data warehouse in one step. This approach requires an immense commitment of planning, resources, and time and results in a new information structure from which the entire organization benefits.

The bottom-up approach, on the other hand, breaks the task down and delivers only a small subset of the data warehouse. New pieces are then phased in until the entire organization is modeled. The bottom-up approach lets data warehouse technology be quickly delivered to a part of the organization. This approach is recommended because its time demands are not as rigorous. It also allows development team members to learn as they implement the system, identify bottlenecks and shortfalls, and find out how to avoid them as additional parts of the data warehouse are delivered.

Because a data warehouse is subject-oriented, the first design step involves choosing a business subject area to be modeled and eliciting information about the following:

- The business process that needs to be modeled.
- The facts that need to be extracted from the operational data base.
- The level of detail required.
- Characteristics of the facts (e.g., dimension, attribute, and cardinality).

After each of these areas has been thoroughly investigated and more detailed information about facts, dimensions, attributes, and sparsity has been gathered, still another decision must be made. The question now becomes which schema to use for the design of the data warehouse data base. There are two major options: the classic star schema and the snowflake schema.

The Star Schema. In the star design schema, a separate table is used for each dimension, and a single large table is used for the facts (see Exhibit II-3-5). The fact table's indexed key comprises the keys of the different dimension tables.

With this schema, the problem of sparsity, or the creation of empty rows, is avoided by not creating records where combinations are invalid. Users are able to follow paths for detailed drilldowns and summary rollups. Because the dimension tables are also relatively small, precalculated aggregation can be imbed-

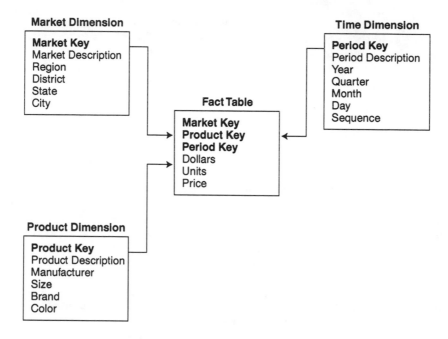

SOURCE: N. Raden, "Modeling a Data Warehouse," *InformationWeek* (January 29, 1996), pp. 60-62.

Exhibit II-3-5. The Star Design Schema

ded within the fact table, providing extremely fast response times. It is also possible to apply multiple hierarchies against the same fact table, which leads to the development of a flexible and useful set of data.

The Snowflake Schema. The snowflake schema as depicted in Exhibit II-3-6 is best used when there are large dimensions such as time. The dimension tables are split at the attribute level to provide a greater variety of combinations. The breakup of the time dimension into a quarter entity and a month entity provides more detailed aggregation and also more exact information.

DECISION SUPPORT SYSTEMS AND DATA WAREHOUSING

Because many vendors offer decision support system (DSS) products and information on how to implement them abounds, insight into the different technologies available is helpful. Three concepts should be evaluated in terms of their usability for decision support and relationship to the so-called real data ware-

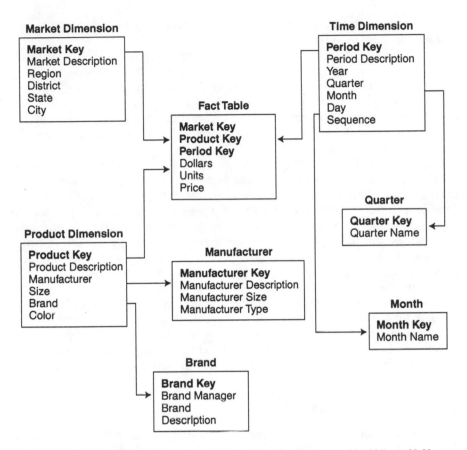

SOURCE: N. Raden, "Modeling a Data Warehouse," *InformationWeek* (January 29, 1996), pp. 60-62.

Exhibit II-3-6. The Snowflake Design Schema

house. They are virtual data warehouses, multidimensional online analytical processing (OLAP), and relational OLAP.

The Virtual Data Warehouse

The virtual data warehouse promises to deliver the same benefits as a real data warehouse but without the associated amount of work and difficulty. The virtual data warehouse concept can be subdivided into the surround data warehouse and the OLAP/data mart data warehouse. In a surround data warehouse, legacy systems are merely surrounded with methods to access data without a fundamental change of the operational data. The surround concept thus negates a key feature of the real data warehouse, which integrates operational data in a way that allows users to make sense of it.

In addition, the data structure of a virtual data warehouse does not lend itself to DSS processing. Legacy operational systems were built to ease updating, writing, and deleting and not with simple data extraction in mind. Another deficiency with this technology is the minimal amount of historical data that is kept, usually only 60 to 90 days worth of information. A real data warehouse, on the other hand, with its two-to-five years of information, provides a far superior means of analyzing trends.

In the case of direct OLAP/data marts, legacy data is transferred directly to the OLAP/data mart environment. Although this approach recognizes the need to remove data from the operational environment, it too falls short of being a real data warehouse. If only a few, small applications were feeding a data mart, the approach would be acceptable. The reality is, however, that there are many applications and thus many OLAP/data mart environments, each requiring a customized interface. This raises the problem of building and managing such interfaces, especially as the number of OLAP/data marts increases.

Because the different OLAP/data marts are not effectively integrated, different users arrive at different conclusions when analyzing the data. It is thus possible for the marketing department to report that business is doing fine and another department to report just the opposite. This drawback does not exist with the real data warehouse, where all data is integrated. Users who examine the data at a certain point in time would all reach the same conclusions.

Multidimensional OLAP

Multidimensional data base technology is a definite step up from the virtual data warehouse. It is designed for executives and analysts who want to look at data from different perspectives and have the ability to examine summarized and detailed data. When implemented together with a data warehouse, multidimensional data base technology provides more efficient and faster access to corporate data. Proprietary multidimensional data bases facilitate the organization of data hierarchically in multiple dimensions, allowing users to make advanced analyses of small portions of data from the data warehouse. The technology is understandably embraced by many in the industry because of its increased usability and superior analytical functionality.

As a standalone technology, multidimensional OLAP is inferior to a real data warehouse for a variety of reasons. The main drawback is that the technology is not able to handle more than 20 to 30 gigabytes of data, which is unacceptable for most larger corporations, whose needs range in the 100 gigabyte to several terabyte range. Furthermore, multidimensional data bases do not have the flexibility and measurability required of today's decision support systems because they do not support the necessary ad hoc creation of multidimensional views of products and customers. Multidimensional data bases should be considered for use in smaller organizations or on a department level only.

Relational OLAP

Relational OLAP is also used with many decision support systems and provides sophisticated analytical capability in conjunction with a data warehouse. Unlike multidimensional data base technology, relational OLAP lets end users define complex multidimensional views and analyze them. These advantages are only possible if certain functionalities are incorporated into relational OLAP.

Users must be removed from the process of generating their own structured query language (SQL). Multiple SQL statements should be generated by the system for every analysis request to the data warehouse. In this way, a set of business measurements (e.g., comparison and ranking measurements) is established, which is essential to the appropriate use of the technology.

The shortcoming of relational OLAP technology involves the nonintegration of corporate data. Thus although relational OLAP works well in conjunction with a data warehouse, by itself, the technology is somewhat limited.

Examination of the three preceding decision support technologies leads to the only correct deduction—that the data warehouse is still the most suitable technology for larger firms. The benefit of having integrated, cleansed data from legacy systems together with historical information about the business makes a properly implemented data warehouse the primary choice for decision support.

BENEFITS OF WAREHOUSING FOR DATA MINING

The technology of data mining is closely related to that of data warehousing. It involves the process of extracting large amounts of previously unknown data and then using the data to make important business decisions. The key phrase here is *unknown information buried* in the huge mounds of operational data that, if analyzed, provides relevant information to organizational decision makers.

Significant data is sometimes undetected because most data is captured and maintained by a particular department. What may seem irrelevant or uninteresting at the department level may yield insights and indicate patterns important at the organizational level. These patterns include market trends, such as customer buying patterns. They aid in such areas as determining the effectiveness of sales promotions, detecting fraud, evaluating risk and assessing quality, or analyzing insurance claims. The possibilities are limitless and yield a variety of benefits ultimately leading to improved customer service and business performance.

Data that is needed but often located on several different systems, in different formats and structures, and somewhat redundant provides no real value to business users. This is where the data warehouse comes into play. As a source of consolidated and cleansed business data, it provides a better means of discovering hidden data and facilitating analysis than do regular flat files or operational data bases.

Three steps are thus needed to identify and use hidden information:

1. The captured data must be incorporated into a view of the entire organization instead of only one department.

165

2. The data must be analyzed or mined for valuable information.

3. The information must be specially organized to simplify decision making.

Data Mining Tasks

In data mining, data warehouses, query generators, and data interpretation systems are combined with discovery-driven systems to provide the ability to automatically reveal important yet hidden data. The following tasks need to be completed to make full use of data mining:

- Creating prediction and classification models.
- Analyzing links.
- Segmenting data bases.
- Detecting deviations.

Creating Models. The first task makes use of the data warehouse's contents to automatically generate a model that predicts desired behavior. In comparison to traditional models that use statistical techniques and linear and logical regression, discovery-driven models generate accurate models that are also more comprehensible, because of their sets of if-then rules. The performance of a particular stock, for example, can be predicted to assess its suitability for an investment portfolio.

Analyzing Links. The goal of the link analysis is to establish relevant connections between data base records. An example here is the analysis of items that are usually purchased together, like a washer and dryer. Such analysis can lead to a more effective pricing and selling strategy.

Segmenting Data Bases. When segmenting data bases, collections of records with common characteristics or behaviors are identified. One example is the analysis of sales for a certain time period, such as President's Day or Thanksgiving weekend, to detect patterns in customer purchase behavior. For the reasons discussed earlier, this is an ideal task for a data warehouse.

Detecting Deviations. The fourth and final task involves detection of deviations, which is the opposite of data segmentation. Here, the goal is to identify records that vary from the norm, or lie outside of any particular cluster with similar characteristics. This discovery of deviance from the clusters is then explained as normal or as a hint of a previously unknown behavior or attribute.

Data Mining Techniques

At this point, it is important to present several techniques that aid data mining efforts. These techniques include the creation of predictive models, and the performing of supervised induction, association, and sequence discovery.

Creating Predictive Models. The creation of a so-called predictive model is facilitated through numerous statistical techniques and various forms of visualization that ease the user's recognition of patterns.

Supervised Induction. With supervised induction, classification models are created from a set of records, which is referred to as the training set. This method makes it possible to infer from a set of descriptors of the training set to the general. In this way, a rule might be produced that states that a customer who is male, lives in a certain zip code area, earns between $25,000 and $30,000, is between 40 and 45 years of age, and listens more to the radio than watches TV might be a possible buyer for a new camcorder. The advantage of this technique is that the patterns are based on local phenomena, whereas statistical measures check for conditions that are valid for an entire population.

Association Discovery. Association discovery allows for the prediction of the occurrence of some items in a set of records if other items are also present. For example, it is possible to identify relationships among different medical procedures by analyzing claim forms submitted to an insurance company. With this information the prediction could be made, within a certain margin of error, that for a certain treatment usually the same five medicines are required.

Sequence Discovery. Sequence discovery aids the data miner by providing information on a customer's behavior over time. If a certain person buys a VCR this week, he or she usually buys videotapes on the next purchasing occasion. The detection of such a pattern is especially important to catalog companies, because it helps them better target their potential customer base with specialized advertising or catalogs.

Tools

The main tools used in data mining are neural networks, decision trees, rule induction, and data visualization.

Neural Networks. A neural network consists of three interconnected layers: an input and an output layer with a hidden layer in between (see Exhibit II-3-7). The hidden processing layer is like the brain of the neural network because it stores or learns rules about input patterns and then produces a known set of outputs. Because the process of neural networks is not transparent, it leaves the user without a clear interpretation of the resulting model, which, nevertheless, is applied.

Decision Trees. Decision trees divide data into groups based on the values that the different variables take on (see Exhibit II-3-8). The result is often a complex hierarchy of classifying data, which enables the user to deduct possible future behavior. For instance, it might be deducted that for a person who only uses a credit card occasionally, there is a 20% probability that an offer for

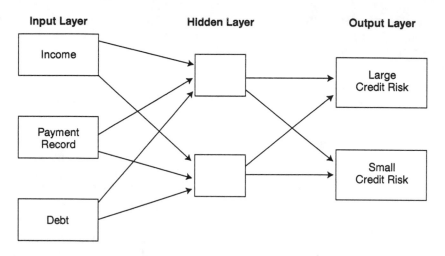

SOURCE: H. Edelstein, "Technology How To: Mining Data Warehouses," *InformationWeek*
(January 8, 1996), pp. 48-51.

Exhibit II-3-7. Neural Network

---

another credit card would be accepted. Although decision trees are faster than
neural networks in many cases, they have drawbacks. One of these is the handling of data ranges as in age groups, which can inadvertently hide patterns.

Rule Induction. The method of rule induction is applied by creating nonhierarchical sets of possibly overlapping conditions. This is accomplished by first
generating partial decision trees. Statistical techniques are then used to determine
which decision trees to apply to the input data. This method is especially useful
in cases where there are long and complex condition lists.

Data Visualization. Data visualization is not really a data mining tool. However, because it provides a picture for the user with as many as four graphically
represented variables, it is a powerful tool for providing concise information. The
graphics products available make the detection of patterns much easier than is
the case when mere numbers are analyzed.

Because of the pros and cons of the varied data mining tools, software vendors
today incorporate all or some of them in their data mining packages. Each tool
is essentially a matter of looking at data with different means and from different
angles.

One of the potential problems in data mining is performance-related. To speed
up processing, it might be necessary to subset the data either by the number of
rows accessed or by the number of variables that examined. This can lead to

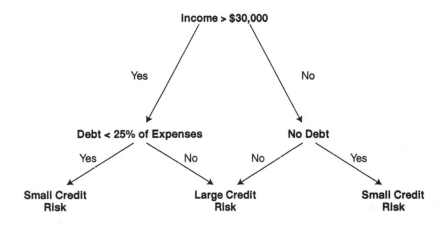

Source: H. Edelstein, "Technology How To: Mining Data Warehouses," *InformationWeek*
(January 8, 1996), pp. 48-51.

Exhibit II-3-8. Decision Tree

slightly different conclusions about the data set. Consequently, in most cases it is better to wait for the correct answer using a large sample.

MANAGERIAL AND ORGANIZATIONAL IMPACTS OF DATA WAREHOUSING

Although organizational managers eagerly await the completion of a data warehouse, many issues must be dealt with before the fruits of this new technology are harvested. This is especially true in today's fast changing enterprise with its quick reaction times.

The subject of economic benefit also deserves mentioning when dealing with data warehousing because some projects have already acquired the reputation of providing little payback on the huge investments involved. Data warehouses are sometimes accused of being pits into which data disappears never to be seen again.

Managers have to understand at the outset that the quality of the data is of extreme importance in a data warehousing project. The sometimes difficult challenge for management is to make the data entering the data warehouse consistent. In some organizations, data is stored in flat, VSAM, IMS, IDMS, or SAS files and a variety of relational data bases. In addition, different systems that were designed for different functions contain the same terms, used with different meanings.

If care is not taken to clean up this terminology during data warehouse construction, misleading management information results. The logical consequence of this requirement is that management has to agree on the data definition for elements in the warehouse. This is yet another challenging task. People who use the data in the short term and the long term must have input into the process and know what the data means.

The manager in charge of loading the data warehouse has four ways to handle erroneous data. If the data is inaccurate, it must be completely rejected and corrected in the source system. Data may also be accepted as is, if it is within a certain tolerance level and if it is marked as such. An example here is a nine-digit zip code field in the data warehouse that is to be populated with a five-digit zip code from the operational data base.

A third option for handling erroneous data is the capture and correcting of the data before it enters the warehouse. Capture and correction are handled programmatically in the process of transforming the data from one system to the data warehouse. An example might be a field that was in lowercase and needs to be stored in uppercase. A final means of handling errors is to replace erroneous data with a default value. If, for example, the date of February 29 of a non-leap year is defaulted to February 28, there is no loss in data integrity.

Another way that data warehousing affects management and organizations in general concerns today's business motto of working smarter, not harder. Today's data warehouse users can become more productive, because they will have the tools to analyze the huge amounts of data that they store instead of just collecting it.

Organizations are also affected by the invalid notion that implementing data warehousing technology simply consists of integrating all pertinent existing company data in one place. Managers need to be aware that data warehousing implies changes in the job duties of many people. For example, in an organization implementing a data warehouse, data analysis and modeling become much more prevalent than just requirements analysis. The data base administrator position does not merely involve the critical aspect of efficiently storing data but takes on the central role in the development of the application. The individual filling this position requires an in-depth understanding of the data, its usage, and how it can be transformed into usable information. Furthermore, because of its data model-oriented methodology, data warehouse design requires a development life cycle that reverses traditional development approaches. The development of a data warehouse virtually begins with a data model, from which the warehouse is built.

In summary, it must be noted that data warehouses are high-maintenance systems that require their own staffs. In this way, future changes are implemented in a timely manner by experienced personnel. It is also important to remember that a technically advanced and fast warehouse that adds little value will probably be abandoned by users from the start, reiterating the immense importance of clean data.

CONCLUSION

The value of data warehousing to an organization is multidimensional. An enterprisewide data warehouse serves as a central repository for all data names used in an organization and therefore simplifies business relationships among departments by using one standard. Users of the data warehouse get consistent results when querying this data base and understand the data in the same way without ambiguity. By its nature, the data warehouse also allows quicker access to summarized data about products, customers, and other business items of interest. In addition, the historical aspect of such a data base (i.e., information is kept for two to five years) allows users to detect and analyze patterns in the business items.

Organizations beginning to build a data warehouse should not undertake the task lightly. It does not simply involve the moving of data from the operational data base to the data warehouse but rather the cleaning of data for future usefulness. It is also important to distinguish the different types of warehouse technologies (i.e., relational OLTP, multidimensional OLTP, and virtual data warehouse) and understand their fundamental differences.

Other issues that need to be addressed and resolved range from creating a team solely dedicated to the design, implementation, and maintenance of a data warehouse to the need for top-level support from the outset and management education on the concept and benefits of corporate sharing of data.

A further benefit of data warehousing results from the ability to mine the data using a variety of tools. Data mining aids corporate analysts in detecting customer behavior patterns, finding fraud within the organization, developing marketing strategies, and detecting inefficiencies in internal business processes.

Because the subject of data warehousing is immensely complex, outside assistance is often beneficial. It provides organizational members with training in the technology and exposure, both theoretical and hands-on, that enables them to continue with later phases of the project.

The data warehouse is without doubt one of the most exciting technologies of our time. Organizations that make use of it increase their chances of improving customer service and developing more effective marketing strategies.

II-4

A Practical Approach to Information Engineering

Martin Kansky

INFORMATION ENGINEERING (IE) METHODOLOGY is extremely popular in the information resource management field. It is designed to improve data management and integrated systems and to develop a top-down road map for an organization's information systems. Information engineering differs from other methodologies because it considers the organization's information to be a significant asset. The methodology also assumes that when an organization understands and manages its data more effectively, it can increase its competitive advantage.

IMPLEMENTATION SUCCESS

A general set of organizational characteristics is key to explaining how a high level of IE success may be achieved. An important characteristic is excellent communication between senior management and line management as well as between each department. Organizations that contain levels of management that are knowledgeable and committed to the organization's goals are more likely to succeed. In this type of organization, senior management effectively communicates the organization's goals to the line managers. They closely monitor the line managers' activities and hold line managers accountable for successful IE implementation.

Another characteristic for success is that the organization currently uses a formal system development methodology. Organizations that have experience with a methodology understand the magnitude of methodology changes. Organizations can often modify the standards and procedures of the existing methodology to support an IE methodology. Organizations that use a formal structure to develop data processing or information systems are also likely to succeed. The organization must form a new IE department if one does not exist before implementation, and the staff's responsibilities must be clearly defined.

Successful organizations should constantly evaluate the latest technical developments. IE technology is changing rapidly and integrated computer-aided engineering tools are evolving, as are programming languages and hardware. A successful IE implementer must research and use the latest technology. In addition, at least one senior manager should initiate the organization's migration to IE. This manager should be able to allocate both financial resources and staff to the IE project and ensure senior management's ongoing support.

The final characteristic that the organization must possess is flexibility. Managers must encourage experimentation and learn from their own mistakes. Most IE users conduct pilot projects with the methodology to determine how to tailor it to the organization's needs. Most organizations experience a trial-and-error period from which they gain useful experience.

STAFF RAMIFICATIONS

The introduction of IE raises expectations by management, users, and the staff. Being aware of expectations ensures that the needs of these groups are addressed. If the expectations are ignored, they eventually surface as the IE project commences.

The organization must review and address these expectations to prevent potential project delays. In addition, senior management may expect IE to quickly resolve many existing systems problems (e.g., duplicated data bases and systems). It may also expect IE to immediately reduce systems development costs. Although IE can resolve some problems in the short term, its real benefits are not realized for approximately five years. In fact, several short-term losses may occur before any benefits are realized.

From the user's perspective, the expectation is that IE should create systems at a faster pace. This may be true in some cases if the organization effectively uses fourth-generation languages and code generators. In most cases, there is no significant difference; however, as an organization matures with its use of IE, this expectation may be realized to some extent.

Users also expect that their level of involvement in the project will remain the same; they should be told at the onset that they will be heavily involved in the project from the beginning. Users who do not have a clear understanding of the IE methodology may become disturbed by the level of detail and involvement expected from them.

Impact of IS Department

The introduction of IE into the IS department usually results in mixed responses. If the IS department encourages change and is willing to experiment with methods and tools, IE should gain immediate acceptance. The members of the department often are eager to learn and apply methodology and excited about their new roles, which include facilitator, documentarian, data modeler, and activity modeler. IS departments that respond ineffectively to change have different responses. They are usually upset that their current work may be interrupted and become frustrated when they attempt to learn IE methodology. There may also be concern about the stability of their current jobs if code generators are introduced.

STAGES OF INFORMATION ENGINEERING

Information engineering methodology consists of four stages: the information strategic plan, business area analysis, business system design, and construction. Each stage complements and feeds into the next.

Stage 1. The Plan. The information strategic plan (ISP) concerns the organization's overall strategic objectives, including senior management goals and critical success factors. The ISP deals with how technology, both current and future, is used to create new opportunities and competitive advantages. Management creates a high-level overview of the organization that describes its functions, data, and information needs for the present as well as the future.

Architectures. An ISP's goal is to develop a blueprint or framework that supports successive IE activities. This blueprint consist of three architectures—the information architecture, the business systems architecture, and the technical architecture.

The information architecture defines the organization's activities and the information that is necessary to support them. These high-level activities are logically grouped together to form business areas. The business system architecture describes possible business systems and the data that supports the information architecture. This represents the first attempt to project the number of systems that the organization will need. The technical architecture describes the hardware and software environment that support the business systems architecture.

Stage 2. Analysis. Business area analysis (BAA) concerns the processes that run a selected business area, how these processes interrelate, and their required data. This stage defines and documents the business functions. These functions are broken down into processes that describe how the functions perform. To document the data requirements, management must define the entities that group common business information, define the relationships between these entities, and determine how the entities interact with the various functions and processes.

Stage 3. Design. Business system design is concerned with how management integrates the business processes into procedures and how these procedures work. Users must be extensively involved in this stage to assist in the design. The objective is to define the human-computer interactions necessary to perform the business activities identified during the BAA.

Stage 4. Construction. The construction stage involves implementing the procedures. This includes the use of code generators as well as manual programming. It uses the fourth-generation languages and other technical resources that the technical architecture defines.

PROJECT MANAGEMENT CONCERNS

All organizations that implement IE experience problems. One common problem occurs when the organization initiates a BAA project before it implements an ISP project. Most organizations move immediately to the BAA project because problems currently exist in this area. Although this is often unavoidable, the risk is that the BAA and its associated systems may not support the organization's overall objectives.

The BAA project scope may also be too large to complete within a reasonable time frame. Because the BAA project may span more than one business area, management may have to redefine the project's boundaries and modify the project schedule. If an information strategy plan develops after the BAA has begun, the BAA must be modified to support the plan. If a BAA takes precedence, the project manager should limit the scope of the BAA as much as possible.

Another common concern is that an organization may initiate several BAAs without successfully completing the first one. Instead of gaining experience from the pilot project's methodology, the organization compounds mistakes that occur during the initial BAA. If an ISP is not available, IE benefits are significantly reduced. The project team may then become frustrated when activities and data overlap with other BAAs.

Management can avoid this problem by picking a small project in a well-known area that the users can contribute to. It is more worthwhile to complete a small project and learn from it than to begin multiple BAAs and duplicate the problems. The BAA team must also be aware of other BAA projects and must be sensitive to the constraints on users' time.

Using Core Project Teams. Problems may develop also when inexperienced or part-time personnel work on the project because of the organization's staff shortage. Users may become frustrated if the project proceeds at a slow pace or if they must repeat tasks because the team did not complete them correctly the first time. If this is the case, management must develop a core project team and assign a minimum of additional part-time members. The core team should consist of the data modeler, the activity modeler, and at least one full-time user representative. Additional staff, both part-time and full-time, may be added. This ensures continuity among the team, and the users are pleased to have a full-time team in their business area.

Pilot Projects. An organization cannot conduct an IE project without standards and procedures, because these are necessary to manage the information from multiple projects. For example, if management performs multiple BAAs without standards, the entity-relationship diagram documentation often differs from one model to another. Procedures must be defined so that the model's information is not omitted. During a BAA, some modelers prefer to create process dependency diagrams, whereas others may produce data flow diagrams. Users notice these inconsistencies and begin to question the methodology's validity. Management must conduct a pilot IE project to create and document both standards and procedures.

Accountability. A final concern may arise if line managers are not held accountable for the project. Senior management initiates the project but often does not develop controls to ensure that line managers achieve the project's goal. The line manager is frequently too busy to work on an IE project or does not believe it is a worthwhile effort and therefore does not allocate the necessary staff support. This results in significant delays and can ultimately hinder the project's success. To avoid this scenario, a management steering team should be developed to help the project team obtain the resources and cooperation needed to succeed as well as to monitor line manager participation.

STANDARDS AND PROCEDURES DEVELOPMENT

Standards and procedures are critical to IE development. Standards ensure that management documents systems requirements in a consistent manner on all projects. If no standards exist, project delays result from document reinterpretation. Standards also provide the basis for communication between projects. With standards, individuals newly assigned to the project become productive in a shorter period of time. For example, during a BAA, all entities are recorded in uppercase. Data stores that are defined during current systems analysis are recorded in lowercase. This method greatly assists the project team members during the confirmation step by allowing them to distinguish between the two entities.

Procedures provide guidelines to help management conduct all IE stages. They define the initial project plan and ensure that the same level of detail is applied to all projects. For example, a data model review should be performed by the quality assurance team before the plan is distributed to the users and to management.

INTEGRATED CASE TOOLS

The selection of the integrated computer-aided software engineering (CASE) tool ensures that all the information from a preceding IE stage can be accessed and used in the current stage. For example, a tool that uses the entity-relationship diagram information from the BAA stage helps develop the logical data base design during the business system design stage.

Organizations often believe that if they acquire an integrated CASE tool they can perform successful IE implementation; however, the tool exists to support methodology. Several integrated CASE tools do not fully support the information engineering methodology. In selecting an integrated CASE tool, the organization must determine which IE aspects the tool does not support. For example, the methodology uses problem definitions to describe the problem diagram. Many integrated CASE tools cannot produce this graphic. The user must then determine how to define a problem with the tool. The manner in which it is defined becomes a standard.

JOINT APPLICATION DEVELOPMENT

A joint application development (JAD) session is the most efficient method for obtaining IS requirements because it optimizes the use of both the project team's and the user's time. It creates an environment in which issues and problems are resolved quickly. Participants in a JAD session include business functions experts, executive sponsors, systems analysts, a facilitator, and a documentarian. A meeting is often called by the executive sponsor, who must obtain the cooperation of the functional experts. During the session, the sponsor must resolve open issues.

The facilitator leads the group through the IE process and must possess excellent communication skills, have IE experience, and be able to use such presentation aids as flip charts and overhead slides. The facilitator must also have patience with the group, maintain a calm composure, and work well under pressure. Success depends on the facilitator's effectiveness. Many organizations feel that anyone can facilitate a session and often assign programmers or analysts to do so, even though many do not understand the organization and cannot conceptualize a high-level view of the organization's activities.

The Two-Minute Rule. The facilitator should be able to handle problems and enforce a two-minute rule, when appropriate, to curtail open-ended discussions that present no immediate solutions. When a discussion appears to be going in circles, the facilitator should stop it and ask for a two-minute recap of the issue. If a solution is not reached the issue or problem is added to a list that is continuously maintained and displayed. In addition, a documentarian records the business requirements and issues that are defined during the session. This individual must ensure that the facilitator stays on track. At the end of the session, the documentarian inputs the information into the integrated CASE tool and prints the documentation.

Functional Experts' Role. Although that JAD approach is the most effective method for collecting IS requirements, it does require significant time to prepare for each session. Locating and obtaining the participation of appropriate functional experts is often the most difficult task when scheduling a session. As a project moves from the higher stages of IE to the lower-level stages of design and construction, the functional experts change. Functional experts that participate at the ISP and BAA stages often do not fully understand the design and construction stages. A JAD session usually reaches a level at which the participants cannot provide accurate information. If the appropriate person cannot be found to participate in the session, the session should end. The facilitator then schedules another session with the appropriate functional experts.

Individual Interviews. When it is impossible to schedule a JAD session because of the functional experts' conflicts schedules, an individual interview often takes place. Although it is preferable to conduct a JAD session, individual interviews are commonly used during the ISP project when it is difficult to schedule time with senior management. Because there are fewer people to interview at this level than at the BAA level, the drawbacks to individual interviews are minimized. During a BAA, individual interviews drastically impact the overall project schedule. This may cause project delays, and extended completion dates often result. Individual interviews are an insufficient use of the project team's time. Although the quality of information obtained is high, it is difficult to resolve the users' issues concerning the project. Users often do not see the overall picture and remain focused on their own area. If individual interviews take place, the project team must keep the users informed of the project's progress.

A JAD session must still take place at the end of each stage of the IE project. The session's purpose is to walk both management and users through the models to ensure the project's correctness and completeness and to obtain approval. If no intervening JAD session is held before the final JAD session, the facilitator may have a difficult time because the participants have never seen the entire model. From the users' perception the project team owns the model and must defend it. During this session, the facilitator must transfer project ownership to the users. If this transfer does not occur, the project is likely to be unsuccessful.

ISP GUIDELINES

The organization may conduct the information strategic plan using three different approaches—a mini-ISP, multiple ISPs, and a full ISP.

A mini-ISP is completed in three to six months. It obtains a general understanding of the organization's overall activities. It is not concerned with detail; it provides an architecture from which information systems can be built. Although this approach does provide some guidance for future BAA projects, it requires modification as each of these projects uncovers additional detail. A mini-ISP is recommended if a full ISP cannot be performed.

Multiple ISPs are often performed when the organization has an immediate need in a particular business area and has severe time constraints. For example, the organization can conduct an ISP for the financial services division and another for the manufacturing division. Performing multiple ISPs reduces the benefits of a high-level plan for the organization. It fosters an atmosphere of individual efforts that often do not support the organization's overall objectives. Multiple ISPs are usually performed for a particular division or department. The goal of the methodology is to develop an ISP that is independent of the organization's formal structure. In addition, multiple ISPs often create a fragmented and nonintegrated plan for IS development. They increase the probability of redundant information systems, redundant data, and overlapping business functions and decrease the level of integrated systems.

The ISP Project Team. The ISP project team consists of a minimum of three people. Two project team members must conduct the interviews and the third member is the documentarian. The two interviewers have IE training and practical experience. They must also understand the organization and possess excellent communication and presentation skills. The documentarian records the information during the meetings and inputs the information into the integrated CASE tool.

The Management Steering Team. The formation of a management steering team is essential to the success of an IE project. It ensures that senior management is committed to IE and reviews the ISP documentation. The senior manager who supports and encourages the use of IE is on this team. The manager ensures that the other steering team members remain committed to the project. An ISP that does not get the appropriate support from management usually results in marginal results for the organization. The members of the steering team must undergo a two-day introduction to the IE methodology. The steering team should also meet once a month to review the project's progress and the findings of the ISP project team.

A completed ISP project does not mean that the overall plan is complete. The ISP must be updated periodically to reflect the organization's current activities. The ISP is a living document that constantly changes and grows with the completion of each BAA project.

A CUSTOMIZED APPROACH TO BUSINESS AREA ANALYSIS

Business areas are defined in the ISP. The BAA project should take 6 to 12 months; however, many organizations find that BAA projects take longer. The BAA stage is divided into two phases.

The first phase concerns defining the overall activities of the business area and decomposing these functions into elementary processes. The phase usually takes six months to complete. Management creates an entity-relationship diagram that describes the data that is needed to support the functions and processes. This phase should define 100% of the attributes and 80% of the relationship between entries. A process dependency diagram illustrates the elementary processes.

The second phase continues with the analysis performed in the first and gathers additional information to prepare for the business systems design stage. It may take three to six months to complete. The second phase defines 100% of the attributes, the entity life cycles, process logic diagrams, and process action diagrams.

Management and Technical Steering Teams. The organization should form two steering teams at the beginning of a BAA project. The management steering team sets the overall priorities of the project, resolves any project-related issues, and allocates resources to the project team. The team also ensures that line managers support the effort and are aware of the project's objectives. The project team should meet with the management steering team at least once every other month.

The technical steering team consists of a group of functional experts from the user community. This group reviews the activity and data models for content and correctness. It also clarifies the information contained in the models and assists the project team when selecting participants for JAD sessions. During a review, the project team's user representative presents the model.

The BAA Project Team. The project team usually consists of six members—two IS staff members and four user representatives. The IS staff creates the data and activity models. The user representatives set the meetings with the users and present the model and objectives. When the users present the model, it instills in them a feeling of ownership. The project team must not take ownership of the models.

Conducting a BAA without an ISP. Many organizations perform BAAs without first performing an ISP. To conduct a BAA in this fashion, it is necessary to add an additional task in the BAA project plan. This task obtains the base information usually available to the BAA from the ISP: a subset of the information found in the ISP. This subset of information includes the organization's structure, the high-level function decomposition diagram, the high-level entity-relationship diagram, and the inventory of current systems and data stores.

Encyclopedia Management. An encyclopedia is a central repository for all information systems requirements and is usually part of the integrated CASE tool. Because many integrated CASE tools reside on microcomputers, management of the encyclopedia may become difficult. The use of one microcomputer per project team resolves the encyclopedia issue but creates other problems. When only one machine is used, the BAA's productivity is greatly reduced because of machine contention. For example, when the data modeler enters information into the encyclopedia the activity modeler must wait.

It is more common for team members to have their own encyclopedia on their own microcomputers. If more than one encyclopedia is used during a BAA, one contains the data model and the other contains the activity model. Many integrated CASE tools have utilities that can consolidate encyclopedias. The project manager must maintain the core encyclopedia. Usually, during a BAA project the encyclopedias are merged toward the end of the effort. When the BAA has been completed, the core encyclopedia goes to the data administration department so that it can be consolidated into the enterprise model.

Current Systems Analysis. Current systems analysis identifies existing applications, procedures, and data. This activity usually parallels the modeling activities and a project team subset is assigned this task. The data model created by the BAA team documents both immediate and future data needs. Toward the end of the BAA effort, the BAA team performs the confirmation of the BAA models. During this activity, the BAA data models are compared with the organization's current information. The objective is to determine what information exists to support the BAA, what information no longer needs to be collected, and what information must be created.

Selecting systems to be reviewed can be difficult because the project team can look at only a few of the organization's systems. Systems that are likely to be reviewed are those for which money has been allocated and used recently.

During the first phase of the BAA, current system analysis activities usually identify and collect documentation. During the second phase, several major systems are selected and reviewed in detail. If a system has good documentation, there is no need to redocument it for confirmation. Systems that do not have documentation should not be reviewed.

ACTION PLAN

Several critical success factors can be taken from organizations that have already set up and actively use the IE methodology. To ensure successful IE implementation, these factors must exist in the organization.

Executive Sponsorship. Often, middle management introduces and supports IE methodology. These managers are close to the day-to-day business operations and are aware of the IE benefits. The support of middle management is not enough; a senior manager can greatly influence IE success. This manager often provides support at a high level to ensure that the proper resources and facilities are available to the project team.

IE Group. The information engineering group is the focal point of all IE projects. This group introduces the methodology to the organization and acts as a resource center for all IE activities. The groups assist new project teams in the application of IE, researches integrated CASE tools, and organizes initial training sessions. It also helps data administrators to develop IE standards and procedures, conducts the ISP, and facilitates JAD sessions.

Functional Experts. IE is based on the assumption that the business-function experts participate in the project. Most project managers discover that the functional experts are usually working on other important projects and have limited time. Many times these experts assign subordinates to stand in; however, these subordinates do not have the knowledge or the authority to make project decisions. This often results in project delays and incorrect information systems requirements.

Project Scope. An organization that does not have an ISP increases its chances of conducting IE projects that are too large in scope; a lack of timely systems delivery to management and users may result. Projects should be completed within a six-month period or segmented into smaller projects. The IE methodology supports the organization's goals by conducting smaller projects that ensure better control and provide more frequent deliverables.

Standards and Procedures. Standards and procedures are essential to successful communication, documentation, and analysis. The project teams must maintain a common method to discuss IE with the users. If the users hear conflicting information, they immediately dismiss any good that IE can provide. Standards also reduce the effort required to maintain the organization's models by eliminating the need to modify the models that the project teams produce.

CONCLUSION

Communication is vital to the success of the project. This communication should flow both up and down the organization. Many times, IE projects do not achieve their objectives because of the lack of communication throughout the organization. Senior management must communicate IE goals and objectives to line managers and explain why their participation is vital. Line management also must communicate upward any problems that they encounter as a result of IE implementation. The project team should always make the project visible to the organization. The more people who learn about the effort and the purpose, the more input the project team receives. In addition, the team should report the information on a timely basis, and it should review the IE project documentation frequently with management and the users.

II-5

Managing Advanced Information Technology

Louis Fried

THE DRIVE TO SEEK IMPROVEMENT has resulted in a burgeoning interest in technology management ranging from acquisition through exploitation of new technologies. Although many technologies are of primary interest to one or two industries, all industries have a common interest in information technology.

Most major corporations would not be able to operate their businesses without computer systems. Executives have awakened to the fact that technology management is as important for their information systems as it is for their manufacturing facilities and their products. As a result, IS departments are under pressure to meet the same goals of efficiency, cost reduction, and responsiveness as other departments within their companies. This chapter details the objectives of technology management and the link between the introduction of new information technology and business process redesign.

TECHNOLOGY MANAGEMENT

Two major issues, inexorably linked together, trouble IS directors of major corporations. These issues are:

- Supporting the redesign of company business processes.
- Replacing legacy systems that are impeding the IS function's ability to respond flexibly to business needs.

The link between these two issues is the ability of the IS function to plan for, acquire, and deploy new information technology for the development and operation of new applications.

The internal needs of IS—better price/performance ratios, faster software development paradigms, specialized application capabilities, smaller increments of capacity increases, and improved architectural flexibility—generally require a substantial short-term increase in new technology costs before the long-term benefits can be realized. Although business needs have spurred the adoption of new information technology, in many cases, IS departments are poorly equipped to deal with the entire scope of managing the introduction of new technology to the organization. Technology management is not only a problem for IS. In fast-paced industries such as bio-technology or electronics, technology management is critical to continued survival.

The scope of technology management includes a broad range of activities such as:

- Strategic technology positioning.
- Tracking technology trends.
- Aligning technology needs with business needs.
- Identifying appropriate new technology.
- Identifying the technology rendezvous (i.e., the relative importance of technologies to the business compared with the time at which the technology should be adopted by the company).
- Justify technology acquisition.
- Acquiring new technology.
- Introducing new technology.
- Adapting technology to the business needs.
- Deploying technology.

This chapter offers a strategic perspective on some of these issues.

STRATEGIC TECHNOLOGY POSITIONING

Although many aspects of technology management could be assumed under this activity, this chapter takes a narrower view. Strategic technology positioning consists of adopting policies and procedures that set forth the management position regarding technology. An organization may determine that its competitive position is best served by being an adopter, an adapter, or an inventor of technology. Adopters frequently use off-the-shelf products and thus trail others in technology acquisition. Adapters make technology acquisitions. Adapters make technology an essential element of their value-based planning and use new technology in innovative ways. Inventors seek opportunities through creating new or innovative uses of technology to stay far ahead of competition.

Furthermore, producers must align the feedback of technology opportunities with the business plans. Strategic business plans developed without regard to the competitive threats and opportunities supported by technology advances can be blind sided by more aggressive users of technology.

TRACKING TECHNOLOGY TRENDS

Large IS groups frequently have specific positions created to track technology. Technology tracking activities are often part of a technology planning or systems architecture group within the IS division. Although some managers feel that technology tracking is a part of every systems analyst's job, more successful results are obtained when the effort is not so diffused. Because successful technology planning must be continuous, specific assignment of responsibility is necessary. In smaller companies that cannot dedicate full-time personnel to this task, the responsibility should be made explicit for one or two individuals as a part-time function with defined results expected.

Technology tracking can only work properly in the framework of strategic technology positioning. It is futile to track emerging technologies if the company's position is that of an adopter. However, for technology adapters and inventors, a vision and understanding of the future and of technology life cycles is imperative. This understanding of the technology life cycle serves as a means to determine areas where skills need to be developed, to identify new projects, to improve productivity and quality, and to anticipate potential competitive advantages and disadvantages. A technology's life cycle consists of six stages:

1. *Breakthrough and basic research.* The technology is invented and advanced to a stage at which product development is feasible.
2. *Research and development.* Initial products are developed.
3. *Emergence.* Products are introduced and the market is educated to accept the products.
4. *Growth.* New products using the technology continue to be offered at a rapid pace.
5. *Maturity.* The market for the technology stabilizes and products become commonly applied.
6. *Decline.* The technology is superseded by newer technologies that have functional cost, or performance advantages.

Usually, businesses must consider technology applications rather than simply individual technologies. Technology applications are generally constructed by blending individual technologies; personal computing, for example, was made possible by technology trends in miniaturization, local area networking, and graphical user interfaces, among other technologies. By monitoring the stages of information technologies, which develop at different paces, a company can observe the applications to which technologies are applied.

ALIGNING TECHNOLOGY STRATEGY WITH BUSINESS NEEDS

Tracking technology without regard to the needs of the business can waste a lot of time and money. It is absolutely critical that those tracking the technology be aware of its potential uses in the processes and products or services of the company. This knowledge enables technology planners to:

- Appraise those technologies that may be used immediately for competitive advantage.
- Appraise changes or new developments that may be used for competitive advantage over the next three to five years.

- Identify potential applications of current and future technology and how such applications may affect the competition.
- Identify potential changes in current applications driven by market demands or technology developments.

Aligning technology strategy with business needs is one of the most frequently identified problems facing corporate IS executives. This alignment requires knowledge of the business operations of each strategic business unit in the corporation, their competitive business strategies, and the best available information on the technology and business strategies of their competitors. In addition, it requires the active participation of both information technologists and users to develop an understanding of the potential for use of information technology and a consistent vision of the future.

Research in technology management and product development has shown the success of triad management. Triad management techniques create teams of marketing, technology suppliers (e.g., engineering or R&D), and manufacturing representatives to rapidly introduce new technologies or bring new products to market. Similarly, most successful IS implementations are those that were required and driven by the users. To achieve alignment, IS must form alliances and occasional task forces or teams with user organizations. Technology planning is no exception if the alignment of technology and business strategy is the goal.

Alignment of technology strategy with business strategy requires two modes of operations:

- The technology strategy must be able to respond quickly to changing business needs.
- The technology potential and vision must be able to influence the development of business strategy.

Technology Planning Specialists

Even though the profusion of personal computers has forced employees and managers to become more computer-literate, there are still noncomputing professionals who do not have insights into the full potential of computing and communications technology.

Technology planners should make a point of meeting with user managers, not only to educate them informally, but to learn about their business operations and needs. Most people are flattered to be asked about their job, and most line managers will readily respond to requests from IS personnel to learn more about their business operations.

Technology planners also need to discover how competing companies are using information technology. This does not imply industrial espionage, but simply tracking the trade press, attending industry or information technology conferences, and talking with prospective vendors and suppliers. Innovative applications can arise outside the company's industry and be applicable to company needs, so this intelligence effort should not be confined to the company's industry.

Increasingly, close cooperation with both suppliers and customers is needed to be competitive and to respond to market conditions. It is now necessary to view the organization and its business processes as part of an extended enterprise composed of the organization, its allies, suppliers, and major customers. Technology planners must either create relationships within the businesses that will provide the perspective of the extended enterprise or they must initiate relationships with key suppliers and customers to understand their uses of information technology and how the company's processes need to interface to those of its business partners.

Building the Business Case

As technology planners acquire a knowledge of the industry and the business, they must document this knowledge so it may be used to build a business case for new applications. They also need to find a way to translate the needs of strategic business units into a projection of when and how new or emerging technologies will influence the company's industry.

First, with a knowledge of the industry's business processes, technology planners can construct a value chain. For example, in a manufacturing company, the value chain may contain the major elements of R&D, engineering, logistics, operations and manufacturing, marketing, distribution, sales, and service. New technology can affect any aspect of this chain through such areas as product design tools, materials or components procurement, inventory management, manufacturing methods and controls, maintenance of plant and equipment, packaging processes, and sales and service tracking. The potential applications that can provide leverage and maximize the contribution to the corporation from technology investment can be recognized. In addition, the technologies capable of supporting those applications can be identified.

Working with user managers, technology planners can gain an understanding of both the leverage points and the perceived priorities of managers. These factors determine the relative importance of various technologies to the organization compared to the time at which the technology should be adopted by the company. Exhibit II-5-1 illustrates how the technology rendezvous may be presented to management.

From the example of Exhibit II-5-1, it is possible to identify clusters of technology that must be introduced over time to keep a company competitive. In this example, the cluster of technologies in the upper-left corner of the chart indicates those technologies that will be most important in the shortest period of time. Technologies of immediate and short-term importance must be considered in the early stages of implementing a technology plan. The plan must consider other technologies in an evolutionary fashion.

The technology rendezvous is important for meeting strategic alignment objectives. When constructed on the basis of knowledge of the industry and the competition, it provides a senior executive view of the needs for introducing new technology or extending the use of currently available technology to remain competitive, thus influencing corporate business strategy. Simultaneously, it provides IS with a view of when and how it should anticipate training, experimentation, and application development using these technologies. This knowledge

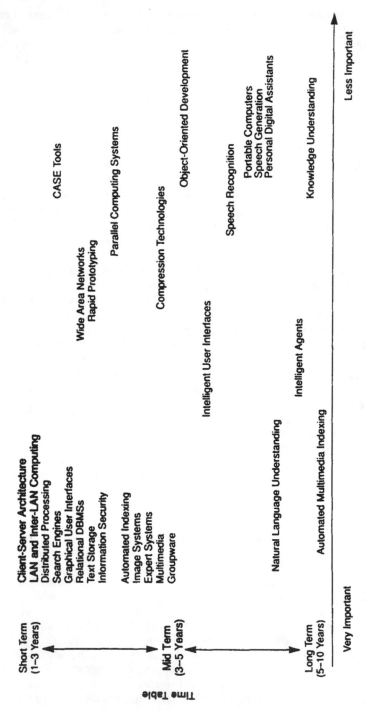

Exhibit II-5-1. Changing the Relative Importance of Information Technologies (Technology Rendezvous)

Note: The technologies being tracked and the technology rendezvous will be different for each industry and, to some extent, different for each company, based on the technology position maintained by the company and the relative level of technology already in place.

directly influences long-range application planning, IS budgeting, and hardware and software selection.

INFORMATION TECHNOLOGY DEVELOPMENT

Professionals involved in technology management have discovered that technology per se is rarely deployable. Instead, technology is deployed through application.

For example, a new technology for the manufacture of integrated circuits is deployed through the change of processes and equipment on the production line; a new materials technology for manufacturing automobile fenders will be deployed through changes in the molding equipment and the manufacturing and assembly processes. Similarly, new information technology is deployed through its application for the benefit of the company and its processes and users.

Information technology deployment through applications means changes and requisite training for the application designers and developers as well as for the users. The processes that will change may include requirements definition, systems design specification, programming, user documentation development, application testing, system installation, system support, and end-user business processes. These changes may result in an organizational change to the systems development and maintenance function to permit more rapid applications development and better field support to users. Substantive changes that allow companies to take full advantage of enabling information systems technology almost invariably require changes in the business processes and in the organizations that perform these processes.

Up to this point, this chapter has dealt primarily with technology planning, including identification of business needs and identification of appropriate new technologies to support those needs. The actual acquisition of new technology requires justification in terms that meet the approval of senior management. Even if the planning stages have been carried out with appropriate approval by senior management, the introduction of specific elements of the plan requires detailed planning and justification based on the benefits of applications to the business.

Two types of technology introduction predominate: those that do not have an impact visible to the users and those that have a direct effect on the user. In the first case, for example, an IS department may introduce a subsystem for massively parallel processing that adds capacity to process additional transactions of the same nature and, in the same apparent manner, as current transactions. Such a technology introduction may be invisible to the users of the systems (other than the impact from system testing before implementation). However, it would be highly visible to the IS department as new design and programming techniques are introduced. Many other introductions of new technology inevitably change the way employees who use the applications perform their tasks.

Reengineering

The goal of many organizations is to replace old systems with new ones that are easier to maintain and modify, support the types of user interfaces to which users of personal computers have become accustomed, and support new modes of business operation. For many applications, downsizing of equipment and the introduction of client/server architectures will provide significantly lower operating costs.

One reason that apparently successful systems development projects have not achieved expected benefits is that the applications were designed to support existing business processes. When existing business processes form the foundation for the requirements definition, even anticipated gains in efficiency may be unrealized. The result: companies have begun using a collection of methods and tools under the general name of business process redesign (BPR), or reengineering.

Many consulting firms have adopted variations of the BPR methods. SRI International defines BPR as a methodology for transforming the business process of an enterprise to achieve breakthroughs in the quality, responsiveness, flexibility, and cost of those processes in order to compete more effectively and efficiently in the enterprise's chosen market. BPR uses a combination of industrial engineering, operations research, management theory, performance measurement, quality management, and systems analysis techniques and tools simultaneously to redesign business processes and to harness the power of information technology to support these restructured business processes more effectively.

BPR projects are designed to take a fresh look at a major business process from a customer perspective. The customer of a process may be the external customer of its products or services or may be the internal recipient of the process output.

Redesigning business processes using new technology benefits the company by improving efficiency and making business processes more responsive to customer needs. In the end, it represents a clear manner in which technology influences the strategy of the company.

SRI's seven-step BPR methodology and some of the associated elements, tools, and techniques used are outlined in Exhibit II-5-2. By using this or similar BPR or reengineering approaches, organizations can overcome resistance to change and the problems of maintaining the effectiveness of employees during the period of transition to the new systems and business processes that affect the way employees perform their jobs.

Guidelines for Successful Technology Deployment

Project Framework. Few companies succeed with projects that massively change the entire company's operating structure. New technology introduced within such a project framework runs a high risk of adding to the company's problems rather than solving them. Generally, projects that address a single process (e.g., materials procurement, loan approval processing, or claims processing) have a greater likelihood of success. Success in the initial project that introduces new technology is crucial to expanding the use of the technology in the organization. In fact, it may be critical to the entire process of acceptance of new technology by management and employees.

Keep Employees Well Informed. All affected or potentially affected company personnel should be provided with regular information about the BPR project's

1. *Taking the Customer's View.* Identifying Factors that lead to customer satisfaction, from the customer perspective.
   - Customer focus groups.
   - Customer surveys.
   - Customer interviews.
   - Customer evaluation.
   - Transaction analysis.

2. *Taking Management's View.* Indentifying management's perspective of factors that lead to customer satisfaction and the differences between the two perspectives.
   - Management interviews.
   - Management focus groups.

3. *Current Process Defintition and Measurement.* Documenting the current business process; measuring ouput of the process and subprocess to develop a foundation for the redesign.
   - Work flow observation.
   - Structured interviews.
   - Process flow diagramming.
   - Data and Control flow diagramming.
   - CASE tools.
   - PERT tools.
   - Process cycle time analysis.
   - Intermediate and output measurement.
   - Statistical analysis of ouput production.
   - Problem root-cause analysis.

4. *Working Group Education.* Educating client's BPR participants about the potentials of process restructuring and IT as applicable to the problem domain.
   - Presentations on:
     — Process redesign methods.
     — Process definition and tools.
     — IT capabilities.
     — Business cases.
   - Workbook of articles and papers.

Exhibit II-5-2. Seven-Step BPR Methodology

---

purpose, status, effect on existing processes and employees, training schedules, and interface needs. This information should be delivered within the context of the expected benefits to the company's competitive positions so that everyone involved retains a focus on the value of the change.

Acknowledge Downsizing Effects. If staff reductions are anticipated, employees should receive a statement from senior management at the beginning of

5. *Indentifying Change Opportunities.* Brainstorming with BPR team members to create a new business process.
   - Process and data/control flow diagrams.
   - CASE tools.
   - Measurement results interpretation.
   - Process walkthrough methods.
   - Alternatives evaluation methods
   - Redesign requirements analysis and deployment.
   - Measurement and metrics design.
   - PERT tools

6. *Analysis of Recommended Action.* Conducting detailed planning of the new process; identifying IT implications, new technology trends, new system architectures; developing cost/benefit estimates; and preparing a proposal to management for the business process change.
   - Management and organizational change analysis.
   - Business case for redesign.
   - System architecture.
   - Technology forecasting.
   - Project cost estimating (including training).
   - Hardware/software cost estimating.
   - Cost/benefit analysis.
   - Project planning.
   - Risk analysis.
   - Presentation preparation tools.

7. *Approval, Commitment, and Implementation.* Obtaining management approval and detailed project planning.
   - Presentation preparation tools (implementation is not covered by the BPR methodology).

SOURCE: SRI International

Exhibit II-5-2. (*continued*)

---

the project about how such reductions will be managed. For example, a statement that all staff reduction will be managed through normal attrition can make a major difference in employee cooperation. Management positions on how employees will be retrained, what options exist for transfer or early retirement, or what types of outplacement support will be provided should be publicized. Some companies have offered bonuses to employees facing displacement so that they will remain at their positions until new systems or processes are completely installed.

Continuity. The implementation time for redesigning a major business process and introducing new technology that enables the new process may be two or more years. It is vital to the success of the project that the senior management sponsors of the project be committed to this time frame and that they be kept informed of progress throughout the implementation.

Coordinated Assistance. Advice or assistance should be sought and encouraged from both management and the employees involved in the project during implementation. For example, it may be appropriate to seek the assistance of the human resources or personnel functions to deal with changes that affect employees. Furthermore, coordination with the managers of functions that use the output of the redesigned process must be maintained to ensure that new interfaces operate smoothly. If the process involves external suppliers or distributors, it may be necessary to set up a help desk or hotline during the implementation phase or train field representatives to assist such external participants.

Evaluate and Monitor the New Process. Where possible, it is essential to set up and evaluate a pilot operation of the redesigned business process. Debugging is always easier when only a limited part of the company's activities have been committed.

The results of the process change should be monitored in terms of the measures and goals established during the analysis phase, and these results should be periodically reported for at least the first year or more of the implementation. The measurement devices should be built into the new process to provide continuous measurement and to form a basis for further evolution in the future.

CONCLUSION

Acquiring appropriate technology that can advance an organization's competitive position requires a dedicated and continuing effort. Acquisition of appropriate technology first requires that the technology planners understand the needs of the business and the strategic technology position that senior management has adopted.

Second, technology planners must maintain an awareness of technology trends to ensure the organization's ability to support its business needs and technology position. Technology planners must position IS not only to support the organization's business directions but to influence company or business strategy. Next, IS and technology planners must justify the adoption of new technology in terms of the new or improved business processes that the new systems will support. The key to successful introduction and deployment of new information technology is business process redesign that combines the business strategy, the process improvement justification, and the technology into a coherent approach that can be readily conveyed to senior management.

II-6

Shifting to Distributed Computing

Richard Ross

MANY OF THE TOP CONCERNS of senior IS managers relate directly to the issues of distributing information technology to end users. The explosive rate at which information technology has found its way into the front office, combined with the lack of control by the IS organization (ostensibly the group chartered with managing the corporation's IT investment), has left many IS managers at a loss as to how they should best respond. The following issues are of special concern:

- Where should increasingly scarce people and monetary resources be invested?
- What skills will be required to implement and support the new environment?
- How fast should the transition from a centralized computing environment to a distributed computing environment occur?
- What will be the long-term impact of actions taken today to meet short-term needs?
- What will be the overall ability of the central IS group to deliver to new standards of service created by changing user expectations in a distributed computer environment?

The inability to resolve these issues is causing a conflict in many organizations. Particularly in large companies during the past decade, the rule of thumb for technology investment has been that the cost of not being able to respond to market needs will always outweigh the savings accruing from constraining technology deployment. This has resulted in a plethora of diverse and incompatible systems, often supported by independent IS organizations. In turn, these developments have brought to light another, even greater risk—that the opportunity cost to the corporation of not being able to act as a single entity will always outweigh the benefit of local flexibility.

This conflict was demonstrated by a global retailer with sales and marketing organizations in many countries. To meet local market needs, each country had its own management structure with independent manufacturing, distribution, and systems organizations. The result was that the company's supply chain became clogged—raw materials sat in warehouses in one country while factories in another went idle; finished goods piled up in one country while store shelves were empty in others; costs rose as the number of basic patterns proliferated. Perhaps most important, the incompatibility of the systems prevented management from gaining an understanding of the problem and from being able to pull it all together at the points of maximum leverage while leaving the marketing and sale functions a degree of freedom.

Another example comes from a financial service firm. The rush to place technology into the hands of traders has resulted in a total inability to effectively manage risk across the firm or to perform single-point client service or multi-product portfolio management.

WANTED—A NEW FRAMEWORK FOR MANAGING

The problem for IS managers is that a distributed computing environment cannot be managed according to the lessons learned during the last 20 years of centralized computing. First and foremost, the distributed computing environment is largely a result of the loss of control by the central IS group because of its inability to deliver appropriate levels of service to the business units. Arguments about the ever-declining cost of desktop technology are all well and good, but the fact of the matter is that managing and digesting technology is not the job function of users. If central IS could have met their needs, it is possible users would have been more inclined to forego managing their own systems.

Central IS's inability to meet those needs while stubbornly trying to deliver with centralized computing has caused users to go their own way. It is not just the technology that is at fault. The centralized computing skills themselves are not fully applicable to a distributed computing environment. For example, the underlying factors governing risk, cost, and quality of service have changed. IS managers need a new framework, one that helps them to balance the opportunity cost to the business unit against that to the company while optimizing overall service delivery.

DEFINING THE PROBLEM: A MODEL FOR DCE SERVICE DELIVERY

To help IS managers get a grip on the problem, this chapter proposes a model of service delivery for the distributed computing environment (DCE). This model focuses on three factors that have the most important influence on service as well as on the needs of the business units versus the corporation—risk, cost, and quality (see Exhibit II-6-1). Each factor is analyzed to understand its cause and then to determine how best to reduce it (in the case of risk and cost) or increase it (as in quality).

Risk in any systems architecture is due primarily to the number of independent elements in the architecture (see Exhibit II-6-2). Each element carries its own

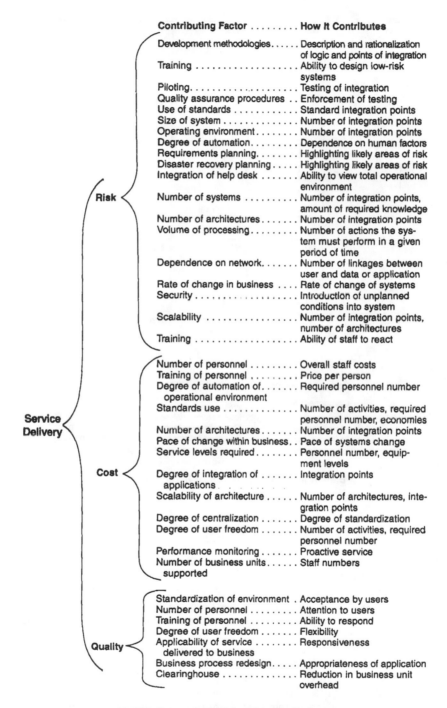

Contributing Factor	How It Contributes

Risk

Contributing Factor	How It Contributes
Development methodologies	Description and rationalization of logic and points of integration
Training	Ability to design low-risk systems
Piloting	Testing of integration
Quality assurance procedures	Enforcement of testing
Use of standards	Standard integration points
Size of system	Number of integration points
Operating environment	Number of integration points
Degree of automation	Dependence on human factors
Requirements planning	Highlighting likely areas of risk
Disaster recovery planning	Highlighting likely areas of risk
Integration of help desk	Ability to view total operational environment
Number of systems	Number of integration points, amount of required knowledge
Number of architectures	Number of integration points
Volume of processing	Number of actions the system must perform in a given period of time
Dependence on network	Number of linkages between user and data or application
Rate of change in business	Rate of change of systems
Security	Introduction of unplanned conditions into system
Scalability	Number of integration points, number of architectures
Training	Ability of staff to react

Cost

Contributing Factor	How It Contributes
Number of personnel	Overall staff costs
Training of personnel	Price per person
Degree of automation of operational environment	Required personnel number
Standards use	Number of activities, required personnel number, economies
Number of architectures	Number of integration points
Pace of change within business	Pace of systems change
Service levels required	Personnel number, equipment levels
Degree of integration of applications	Integration points
Scalability of architecture	Number of architectures, integration points
Degree of centralization	Degree of standardization
Degree of user freedom	Number of activities, required personnel number
Performance monitoring	Proactive service
Number of business units supported	Staff numbers

Quality

Contributing Factor	How It Contributes
Standardization of environment	Acceptance by users
Number of personnel	Attention to users
Training of personnel	Ability to respond
Degree of user freedom	Flexibility
Applicability of service delivered to business	Responsiveness
Business process redesign	Appropriateness of application
Clearinghouse	Reduction in business unit overhead

Service Delivery

Exhibit II-6-1. A Model of Service Delivery

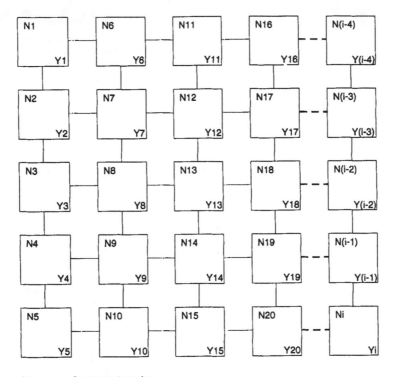

N   = Component number

Y   = Component risk

Given fully independent components, total network risk is equivalent to the sum of
the individual component risks, 1 to i. Thus, the way to minimize risk is either to minimize i (i.e., to have a centralized computing environment) or to minimize Y for each
component by standardizing on components with minimum risk profiles.

Exhibit II-6-2. Optimization of Risk in a Network

---

risk, say for failure, and this is compounded by the risk associated with the
interface between each element.

This is the reason that a distributed computing environment will have a greater
operational risk than a centralized one—there are more independent elements in
a DCE. However, because each element tends to be smaller and simpler to
construct, a DCE tends to have a much lower project risk than a centralized
environment. Thus, one point to consider in rightsizing should be how soon a
system is needed. For example, a Wall Street system that is needed right away
and has a useful competitive life of only a few years would be best built in a
distributed computing environment to ensure that it gets online quickly. Conversely, a manufacturing system that is not needed right away but will remain in
service for years is probably better suited for centralization.

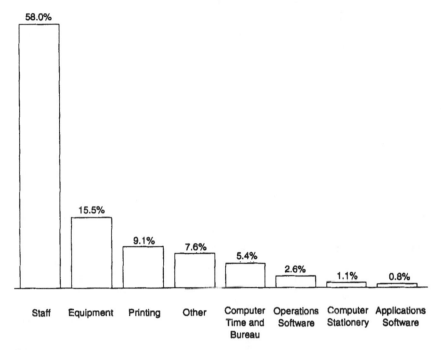

Note:

*Average of organizations studied; total IS costs

SOURCE: Decision Strategies Group, Greenwich CT

Exhibit II-6-3. Cost Profile of the IS Function

One other difference between a distributed environment and a centralized environment is the impact of a particular risk. Even though a DCE is much more likely to have a system component failure, each component controls such a small portion of the overall system that the potential impact of any one failure is greatly reduced. This is important to take into account when performing disaster planning for the new environment.

Cost is largely a function of staff levels (see Exhibit II-6-3). As the need for service increases, the number of staff members invariably increases as well. People are flexible and can provide a level of service far beyond that of automation. Particularly in a dynamic environment, in which the needs for response are ill-defined and can change from moment to moment, people are the only solution.

Unfortunately, staff is usually viewed as a variable cost, to be cut when the need for budget reductions arises. This results in a decrease in service delivered that is often disproportionately larger than the savings incurred through staff reductions.

Finally, quality is a subjective judgment, impossible to quantify, but the factor most directly related to the user's perception of service where information technology is concerned. In essence, the perception of quality is proportional to the user's response to three questions:

- Can I accomplish my task?
- Am I able to try new things to get the job done?
- Am I being paid the attention I deserve?

One of the most important factors in the perceived quality of service delivery is the ability of the support technology to work unnoticed. Because of the similarities between this need and the way in which the US telephone network operates (you pick up the phone and the service is invariably there), the term *dialtone* is used to describe such a background level of operation.

One problem with highly functional IS environments is that users must think about them to use them. This is not the case with the telephone system, which operates so dependably that we have integrated it into our routine working practices and use it without much conscious effort. The phone companies maintain this level of usefulness by clearly separating additional features from basic service and letting the customer add each new feature as the customer desires.

Contrast this with the typical business system that represents an attempt to deliver a package of functions on day one and to continually increase its delivered functionality. The impact on users is that they are forced to continually adapt to changes, are not allowed to merge the use of the system into the background, and must continually stop delivering on their jobs just to cope with the technology. This coping might be as simple as looking something up in a manual or changing a printer cartridge, or it may mean not working at all while the system is rebooted.

Does anyone ever call up AT&T and congratulate it for keeping the nation's phones working that day? Of course not. Yet IS organizations are continually disappointed when users do not seem to appreciate that they have delivered 99.99% availability and a 24-hour help desk.

Complexity: The Barrier to Service Delivery

In general, the basic driver to each of the three service factors is complexity. Complexity increases risk by increasing the number of interfaces between system elements as well as the number of elements themselves. It increases cost by increasing the need for staff as the only way to deal with ill-defined environments. Finally, it affects quality by making it harder to provide those services that users base their perception of quality on (i.e., dialtone and personal attention), in response to which even more staff are added.

This, then, is the paradoxical environment in which IS managers operate. To improve the quality of service, they find themselves increasing the risk and cost of the operation. Improved application delivery cycles result in more systems to manage. End-user development tools and business unit-led development increases the number of architectures and data formats. Increasing access to corporate data through networks increases the number of interfaces. Conversely, trying to improve the risk and cost aspects, typically through standardization of the environment, usually results in decreased levels of service delivered because of the constraints placed on the user freedom. This paradox did not exist in the good old days of centralized computing, when the IS organization dictated the service level.

ADVICE FOR MANAGING DISTRIBUTED COMPUTING

The measure of success in a distributed computing environment is the ability to deliver service through optimizing for the factors of risk, cost, and quality while meeting the needs of both the business units and the corporation. It sounds like a tall order but it is not impossible. There are five key practices involved in corporate information processing:

- Manage tightly, but control loosely.
- Organize to provide on three levels.
- Choose one standard—even a single bad one is better than none or many good ones.
- Integrate data at the front end—don't homogenize on the back end.
- Minimize the use of predetermined architectures.

Manage Tightly, Control Loosely

The situation for the Allied paratroopers at the Bulge was grim. Vastly outnumbered, outgunned, and in a logistically poor location, they faced a greater likelihood of total annihilation than of any sort of victory. Yet they managed to hold out for days, waiting for reinforcements and beating an orderly retreat when they finally came.

In Korea, the First Marine Division at Chosin Reservoir and the Second Infantry Division at Kanu-ri faced the Chinese backlash from the UN decision to cross the 38th parallel. The marines retreated in good order, bringing their dead and wounded and all their equipment with them and disabling between a quarter and a third of all Chinese troops along the way. The army in Korea, in contrast, suffered many casualties, lost most of its equipment, and escaped as a scattered bunch of desperate men.

What do these battle stories signify for the manager of a distributed computing environment? They highlight the need for flexible independence at the front lines, based on a solid foundation of rules and training and backed up with timely and appropriate levels of support. The army at the Battle of the Bulge reacted flexibly to the situation at hand; in addition, they were backed by rigorous training that reinforced standards of action as well as by a supply chain that made action possible. In contrast, the army in Korea suffered from a surfeit of central command, which clogged supply lines and rendered the front line troops incapable of independent action.

In the distributed computing environment, the users are in the thick of battle, reacting with the best of their abilities to events moment by moment. IS can support its troops in a way that allows them to react appropriately or can make them stop and call for a different type of service while the customers get more and more frustrated.

BPR and Metrics. Two tools that are key to enabling distributed management are business process redesign (BPR) and metrics. BPR gets the business system working first, highlights the critical areas requiring support, builds consensus between the users and the IS organization as to the required level of support, and reduces the sheer number of variables that must be managed at any one time. In essence, applying BPR first allows a company to step back and get used to the new environment.

Without a good set of metrics, there is no way to tell how effective IS management has been or where effort needs to be applied moment to moment. The metrics required to manage a distributed computing environment are different from those IS is used to. With central computing, IS basically accepted that it would be unable to determine the actual support delivered to any one business. Because centralized computing environments are so large and take so long to implement, their cost and performance are spread over many functions. For this reason, indirect measurements were adopted when speaking of central systems, measures such as availability and throughput.

But these indirect measurements do not tell the real story of how much benefit a business might derive from its investment in a system. With distributed computing, it is possible to allocate expenses and effort not only to a given business unit but to an individual business function as well. IS must take advantage of this capability by moving away from the old measurements of computing performance and refocusing on business metrics, such as return on investment.

Pricing should be used as a tool to encourage users to indulge in behavior that supports the strategic direction of the company. For example, an organization used to allow any word processing package that the users desired. It then reduced the number of packages it would support to two, but still allowed the use of any package. This resulted in an incurred cost to the IS organization due to help desk calls, training problems, and system hangs. The organization eventually settled on one package as a standard, gave it free to all users, and eliminated support for any other package. The acceptance of this standard package by users was high, reducing help calls and the need for human intervention. Moreover, the company was able to negotiate an 80% discount over the street price from the vendor, further reducing the cost.

In addition to achieving a significant cost savings, the company was able to drastically reduce the complexity of its office automation environment, thus allowing it to deliver better levels of service.

Organize to Provide Service on Three Levels

The historical IS shop exists as a single organization to provide service to all users. Very large or progressive companies have developed a two-dimensional delivery system: part of the organization delivers business-focused service (particularly applications development), and the rest acts as a generic utility. Distributed computing environments require a three-dimensional service delivery organization. In this emerging organization model, one dimension of service is for dialtone, overseeing the technology infrastructure. A second dimension is for business-focused or value-added service, ensuring that the available technology resources are delivered and used in a way that maximizes the benefit to the business unit. The third dimension involves overseeing synergy, which means ensuring that there is maximum leverage between each business unit and the corporation.

Dialtone IS services lend themselves to automation and outsourcing. They are complex, to a degree that cannot be well managed or maintained by human activity alone. They must be stable, as this is the need of users of these services. In addition, they are nonstrategic to the business and lend themselves to economies of scale and hence are susceptible to outsourcing (Exhibits II-6-4 and II-6-5).

Value-added services should occur at the operations as well as at the development level. For example, business unit managers are responsible for overseeing the development of applications and really understanding the business. This concept should be extended to operational areas, such as training, maintenance, and the help desk. When these resources are placed in the business unit, they will be better positioned to work with the users to support their business instead of making the users take time out to deal with the technology.

The third level of service—providing maximum leverage between the business unit and the corporation—is perhaps the most difficult to maintain and represents the greatest change in the way IS does business today. Currently, the staff members in charge of the activities that leverage across all business units are the most removed from those businesses. Functions such as strategic planning, test beds, low-level coding, and code library development tend to be staffed by technically excellent people with little or no business knowledge. IS managers must turn this situation around and recruit senior staff with knowledge of the business functions, business process redesign, and corporate training. These skills are needed to take the best of each business unit, combine it into a central core, and deliver it back to the business.

Choose One Standard

In the immortal words of the sneaker manufacturer, "Just do it." If the key to managing a distributed computing environment is to reduce complexity, then implementing a standard is the thing to do. Moreover, the benefits to be achieved from even a bad standard, if it helps to reduce complexity, will outweigh the risks incurred from possibly picking the wrong standard. The message is clear: there is more to be gained from taking inappropriate action now than from waiting to take perfect action later.

It should be clear that IT is moving more and more toward commodity status. The differences between one platform and another will disappear over time. Even if IS picks a truly bad standard, it will likely merge with the winner in the next few years, with little loss of investment. More important, the users are able to get on with their work. In addition, it is easier to move from one standard to the eventual winner than from many.

Common Operational Problem	Responsiveness to Automation
Equipment hangs	●
Network contention	◑
Software upgrades	●
Equipment upgrades	○
Disaster recovery	●
Backups	●
Quality assurance of new applications	●
Equipment faults (e.g., print cartridge replacement, disk crash)	○
Operator error (e.g., forgotten password, kick out plug)	○
Operator error (e.g., not understanding how to work application)	●

Responsiveness
High   ●
Medium   ◑
Low   ○

SOURCE: Interviews and Decision Strategies Group analysis

Exhibit II-6-4. Responsiveness of Operations to Automation

Even if you pick the winner, there is no guarantee that you will not suffer a discontinuity. IBM made its customers migrate from the 360 to 370 architecture. Microsoft is moving from DOS to Windows to Windows NT. UNIX is still trying to decide which version it wants to be. The only thing certain about information technology is the pace of change, so there is little use in waiting for things to quiet down before making a move.

Integrate Data at the Front End

At the very core of a company's survival is the ability to access data as needed. Companies have been trying for decades to find some way to create a single data model that standardizes the way it stores data and thus allows for access by any system.

Dialtone Function	Applicability
Equipment maintenance	●
Trouble calls	●
Help desk	●
Installations	●
Moves and changes	●
Billing	◑
Accounting	◑
Service level contracting	○
Procurement	○
Management	○

Applicability
High   ●
Medium ◑
Low   ○

SOURCE: Interviews and Decision Strategies Group analysis

Exhibit II-6-5. Applicability of Outsourcing to Dialtone

The truth of the matter is that for any sufficiently large company (i.e., one with more than one product in one market), data standardization is unrealistic. Different market centers track the same data in different ways. Different systems require different data formats. New technologies require data to be stated in new ways. To try to standardize the storage of data means ignoring these facts of life to an unreasonable extent.

The standardization approach also ignores the fact that businesses have 20 to 30 years' worth of data already. Are they to go back and recreate all this to satisfy future needs? Probably not. Such a project would immobilize the business and the creation of future systems for years to come.

Systems designed to integrate and reconcile data from multiple sources, presenting a single image to the front end, intrinsically support the client/server model of distributed computing and build flexibility into future applications. They allow data to be stored in many forms, each optimized for the application at hand. More important, they allow a company to access its data on an as-

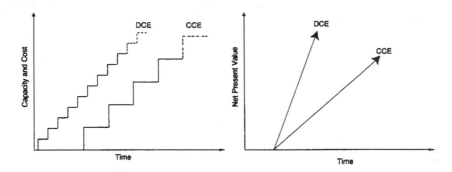

Distributed computing environment (DCE) has a higher net present value because its capacity can be used sooner relative to its marginal costs when compared with the centralized computing environment (CCE).

SOURCE: Decision Strategies Group, Greenwich CT

Exhibit II-6-6. Net Present Value of Distributed Versus Centralized Computing

---

needed basis. These integration systems are an important component to successfully managing future growth.

Less Architecture Is More

To overdesign a systems architecture is to overly constrain the organization. Most architecture arises as a function of rightsizing of applications on the basis of where the data must be stored and used. Understanding this helps the IS manager size the network and associated support infrastructure.

The management of risk and impact also drives architecture by forcing redundancy of systems and, in some cases, mandating the placement of data repositories regardless of user preferences. Assessing project versus operational risk helps to determine whether a system is built for central or distributed use.

This view is one in which the business needs drive the shape of the architecture. It results in a dynamic interconnection of systems that respond flexibly to business needs. Under a centralized computing environment, it was impractical to employ such an approach. It took so long and cost so much to implement a system that investment had to come before business need. This necessitated preplanning of an architecture as an investment guide.

The economics of distributed computing are different. Systems cost much less and can be quickly implemented. This means that their use can be responsive to business needs instead of anticipative. It also results in a greater net present value, for even though their operational costs might be higher, distributed computing environments are more immediately useful for a given level of investment (see Exhibit II-6-6).

CONCLUSION

For IS managers, there indeed exists a framework for managing the new computing environment, one that in fact more directly relates to the business than their old way of managing. If you are able to master it, you enhance your opportunities to become a member of the corporate business management team instead of simply a supplier of computing services.

Success in a distributed computing environment requires a serious culture shift for IS managers. They must loosen up their management styles, learning to decentralize daily control of operations. They must provide direction to staff members so that they can recognize synergies among business units. Some jobs that were viewed as low-level support activities (e.g., value-added services such as help desk and printer maintenance) must be recognized as key to user productivity and distributed. Others, viewed as senior technical positions (e.g., dialtone functions such as network management and installations), might be outsourced, freeing scarce IS resources.

Most important, IS managers must understand the shift in power away from themselves and toward users. The IS organization is no longer the main provider of services; it now must find a role for itself as a manager of synergy, becoming a facilitator to the business units as they learn to manage their own newfound capabilities.

II-7

Leveraging Developed Software

Hal H. Green
Ray Walker

THE RESULT OF LEVERAGING , successfully done, is a continual reduction of both the upfront costs associated with acquiring and installing software applications as well as the long-term support costs. These cost reductions are achieved through economies of scale realized as common elements of a software application are repetitively applied and costs are prorated across a larger set of installations.

This chapter examines the issues of leveraging application software in a manufacturing context. Leveraging, or the reduction in application life cycle costs (illustrated in Exhibit II-7-1), can be discussed in either a context of spanning multiple locations within a single business unit or spanning multiple businesses within a single product type.

Although the example used comes from the manufacturing sector, the principles can be applied to nearly any IS environment.

MAINTENANCE AND SUPPORT COSTS

The process of acquiring automation, process control, and plant information systems has traditionally involved one-of-a-kind development projects. The needs of each site were individually assessed, software was selected and purchased, and local resources (often in the form of system integrators) were contracted to provide services. This business paradigm is easily executed, but it has resulted in a profusion of unique, site-specific systems. By multiplying atypical applications, manufacturers discovered that the costs of supporting these unique sets of applications increased because of the lack of commonality or no economy of scale. Compounding the problem is the plethora of applications serving different needs or functions of the manufacturing user communities.

Business structure and organization are also issues to consider. Manufacturing sites are usually autonomous, with at least cost-center responsibility if not profit-

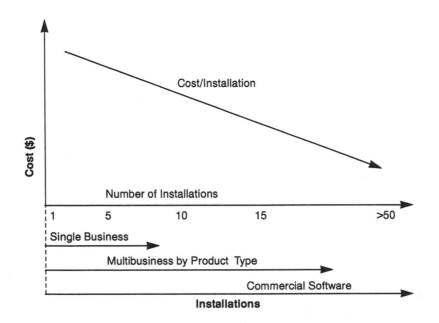

Exhibit II-7-1. Application Life Cycle Cost and Leverage Opportunity

and-loss responsibility. A large manufacturing site may make multiple products or product grades. Production areas within a single site are usually product defined. A single product may be made at multiple sites for sourcing, distribution, and marketing reasons. Business units are formed around one or more product lines. A business unit may span multiple products on multiple sites. Hence, any given site may support many business units.

Information and control systems (e.g., process control, product control, and quality control) are usually required to be in-plant systems. Enterprise systems (e.g., material requirements planning or MRP, warehousing and distribution, and order management) are most often shared across multiple sites as supply-chain functions.

Manufacturing applications that directly affect the manufacturing process have historically been contained on site. This situation has resulted in a physical architecture different from a central mainframe with dumb terminals at remote sites. Whereas manufacturing applications are often computer-intensive, few simultaneous users are served.

Unique site manufacturing applications crowd the IS landscape, owing their existence to differences in product or area requirements. Factors that have contributed to this condition include:

- The need to exchange process control data with other manufacturing systems.
- The autonomy of manufacturing sites to make IT investment decisions.
- The lack of a vision for future integration of direct manufacturing systems with other plant IS applications and systems.

Historically, manufacturing applications were constructed in a purely vertical sense with automation, not integration, in mind. Shared applications, while functionally isolated, were often interfaced with other applications through a variety of means. The resulting set of disparate and unique legacy systems has driven support costs higher, even as support resources are shrinking.

One textile fibers manufacturing business estimates that for every dollar invested in development, $0.25 per year is incurred for maintenance and support, including both direct and indirect costs. Taken over a 10-year anticipated life of an application, this amounts to a present value for support of about 2.5 times the total costs of the initial development.

ECONOMICS OF LEVERAGING

Because leveraging is foremost a business objective, it is important to note the economic effects of leveraging as a capital decision process. As leveraging occurs, the costs of application software go down per site. Assuming benefits from the software are constant, the net present value (NPV) of the per-site investment increases. Exhibit II-7-2 shows a sample discounted cash flow (DCF) curve from an initial investment in an application.

The word "benefits" is frequently used to describe enhancements to the manufacturing operation that result from using an application. For an investment to yield a net present value, the sum of the present value of the future cash flows resulting from the initial investment must be greater than the present value of the costs associated with realizing the benefits.

The investor (plant site) expects benefits (future cash flows) from the investment (initial costs) at an appropriate discount rate. One method of organizing and analyzing benefits in manufacturing seeks to maximize net present value of a set of information system projects identified through a strategic planning activity.

Exhibit II-7-2 illustrates the shape of the curve of cash flows over time when costs are assigned as negative cash flows for an application project and benefits as positive cash flows. Development costs (cash outflow) initially cause the curve to go down. After commissioning and allowing some time for use of the application to reach maximum effectiveness, cash flows become positive as benefits (cash inflow) begin to be realized. Support costs (cash outflow) continue but should be small compared to the benefits accruing per period. Discounted cash flow causes the net cash flow over time to steadily decline, assuming constant support cost for the application. A break-even point occurs when cumulative DCF is equal to zero. That is, the present value of application benefits are equal to the present value of costs.

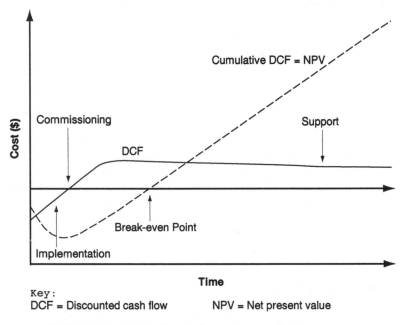

Exhibit II-7-2. Application Software Investment Performance

The principal business drivers for leveraging are economic, not technical. A successfully applied program of leveraging an application or capability across multiple manufacturing sites reduces the installed costs per site while minimizing the ongoing support and maintenance costs of the delivered applications. Assuming manufacturing benefits result from the application, the result of leveraging is a maximum net present value of the investment across one or more manufacturing sites.

Exhibit II-7-3 illustrates the economy-of-scale effect as a measure of the resources required per installation. Leveraging has the effect of driving the total costs per site to some base level that is set by the costs of off-the-shelf components plus resources required to install the system at each respective site and make it operational.

Leveragability, then, can be economically measured as a function of the costs associated with planning and implementing each site's respective requirements. Exhibit II-7-4 demonstrates the effect of leveraging as the number of sites to receive the application increases. Leveraging is therefore an economy-of-scale effect. The greater number of sites in the leveraged effort, the greater the net present value of the investment across the collection of target sites or installations.

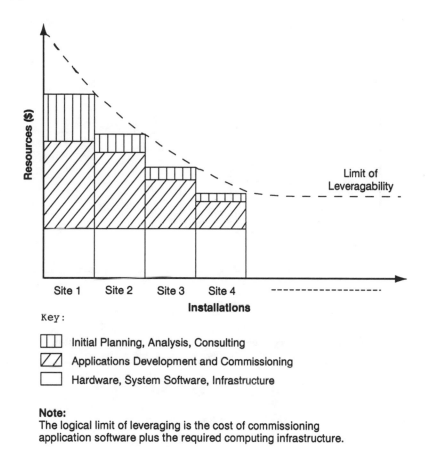

Key:

▯▯▯ Initial Planning, Analysis, Consulting

▱▱▱ Applications Development and Commissioning

▭ Hardware, System Software, Infrastructure

Note:
The logical limit of leveraging is the cost of commissioning
application software plus the required computing infrastructure.

Exhibit II-7-3. Leveragability as a Measure of Resources per Installation

CONSISTENT DATA

Because the business drivers for leveraging are clear, it is reasonable to ask why leveraging is not a pervasive business practice. Some of the barriers to achieving leveragable software include:

- Misperceptions of leveraging.
- Absence of a long-term manufacturing applications migration plan.
- Lack of a consistent architectural framework.

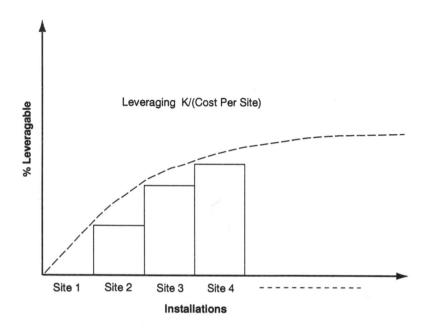

Exhibit II-7-4. Leveragability as Affected by Number of Sites

- Corporate culture ("not invented here" thinking).
- Ad hoc approaches to applications development.
- Conflict between corporate IS/engineering and the manufacturing sites' objectives.

Whereas application leveraging does not have to mean "one size fits all," some consistent framework for applications must exist. Although the data content of applications varies between businesses or sites, applications should fit into a common architecture across the domain of sites or businesses over which the application is implemented. Without such a common framework, leveraging does not occur.

The common architecture must not stop at the point of defining hardware, communication protocols, or even data base types. A consistent way of understanding the types of data to be stored and the type of repository for those data should be planned. If a real-time data base is to be employed for process control and monitoring, for example, what are the valid types of data to be managed by this portion of the plant manufacturing application architecture? What are the valid functions to be addressed by this part of the architecture?

Often, the answers to these questions are blurred by misconceptions. Sometimes companies select a particular data base vendor and perceive they have accomplished leveraging. On the contrary, the applications must be able to use the target data base existing at the sites if they are to be leveragable. This degree of leveraging implies that ideally, standards for the development of applications that use the selected data base are determined before development and acquisition of applications.

Data Standards. Data standards go beyond a textual specification of functionality, however. Textual specifications usually emphasize desired functionality of the application and fall short of defining a model of the data to be employed. Information engineering, which is now rapidly becoming a standard for data base design, is an improvement over narrative specifications. This methodology incorporates data modeling and functional modeling using the entity-relationship diagram (ERD) and activity hierarchy diagram (AHD), respectively.

To prepare for leveraging applications, IS management must answer several questions:

- What are the standards of technology to support applications?
- What are the standard tools for building and maintaining manufacturing applications?
- What are the standards for modeling and describing requirements?
- What types of data will be stored and operated on by these systems?
- What are the standards for screen design and user interaction?

TECHNICAL ARCHITECTURE ISSUES

It is often impractical to know every aspect of each plant's or site's technical IS architecture before commencing development. It is important, however, to have a common way of viewing existing and proposed systems in the context of the types of data they are to manage.

A data-centered architectural framework can be used for information and control systems in manufacturing. Exhibits II-7-5, II-7-6, and II-7-7 provide an overview of the framework. Data is viewed in three distinctive categories: "in-area" for process control, "manufacturing operations and control" for plantwide product control, and "production history" for plantwide and businesswide decision support.

This architecture places such functions as MRP, order management, and inventory management as business-level or supply-chain functions. Exhibit II-7-6 decomposes these three categories of data stores into further detail. Exhibit II-7-7 maps a specific example set of data concerning the subject of quality to the architecture of Exhibit II-7-6.

These diagrams portray a topological way of viewing manufacturing data. They also reveal a fundamental obstacle to leveraging. Without some type of agreed-on taxonomy of data, leveraging becomes difficult. An application framework should recognize and position the desired suite of application software during the planning or analysis phase of the project.

Many workable approaches can be applied for modeling manufacturing applications, but both data types and functions must be central to any discussion of

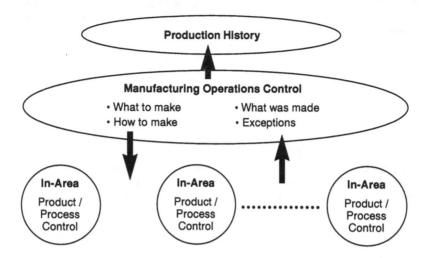

Exhibit II-7-5. Typical Manufacturing Information Processing Requirements

modern plant IS architecture. In the end, it is the types of data to be managed that must be understood between applications if cost-effective integration is to be obtained.

Given optimum conditions of similar or like plant logical and physical architectures, the ability to leverage an application is, nevertheless, still influenced by a variety of factors. First, it is important to note that the closer the application is to the manufacturing process equipment, the less the application will lend itself to leveraging. Process control applications are inherently tied to the site process control systems, which often vary greatly. It is rare for a process control application to be totally independent from the field/equipment instrumentation that is measuring and controlling the process.

A valid approach to designing leveragable applications is to place only real-time applications at the process control level of the architecture. These applications must, by the nature of the data they manage, be positioned such that they can access the real-time process control data base.

In contrast, applications that can be logically positioned farther from the front-end process control data base tend to offer greater leveragability. Such applications are usually data base applications related to the product rather than to the manufacturing process. The integrated operations or manufacturing execution level of type 3 applications in Exhibit II-7-6 lists some of these product or subject-based applications.

Leveragability, then, is limited to the extent that it is specific to the site's manufacturing processes. It is therefore incumbent on the applications designer or purchaser that applications be properly positioned within an IS architecture

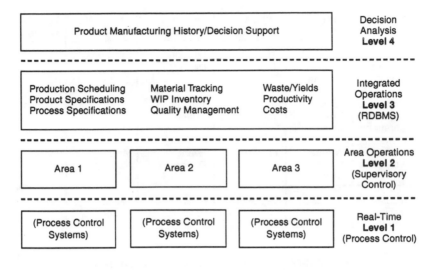

Exhibit II-7-6. Direct Manufacturing Application Framework

and, further, be constructed to be as generic as practical with respect to the process control system.

MEASURING LEVERAGABILITY

The ultimate measure of leveraging is the resulting business benefit—the reduced cost of delivering a working capability from site-to-site across an enterprise. Can leveraging be quantitatively analyzed before commencing a rollout to forecast the required resources/costs per incremental site? The following paragraphs illustrate one approach.

It is helpful to map the various functional elements of the application against an expected quantity of its leveragability. A spreadsheet can be used to view the various components according to three categories:

- Core functionality.
- Auxiliary functions.
- Site-specific functions.

The data types and functions that constitute the application set can be decomposed into these three categories. Exhibit II-7-8 depicts such a decomposition.

Because the data model underpinning the application is foundational and can be understood independently of user screens and reports, it is likely the purest representation of core capability. Functions of one or more applications can be

	Example Data	Typical Orientation	Typical Usage	Integration Scope	Typical Volume
Multisite Decision Support	Lot/Batch Quality Summary	Subject / Table	Multisite Read Only	Business	Low
Cross-Area Integrated Operations	Lot/Batch Quality Detail	Subject/ Table	Transaction Driven	Site	Medium
In-Area Operations	In-Area Quality Result	File/Field	Event Driven	Area	Medium
Process/ Machine Control	Process/ Quality Parameter	Tag or I/O	Real-Time	Machine/ Process Step	High

Exhibit II-7-7. Manufacturing Data Types and Usage

Category	Examples	Leveraging
Core Capability	User/Client Screens, Data Model, Data Base Implementation	High
Auxiliary Functions	Reports, Specialized Functions, Particular Screens	Medium-High
Site-Specific Functions	I/O Processing, Data Occurrences, Quantity of Users, Hardware, System	None-Low

Exhibit II-7-8. Taxonomy of Leveraging

separated by some rationale into independent subfunctions. Subfunctions can be assigned to one of the three leveraging categories.

Site-specific functions are those that must be customized to a particular site and therefore offer the least leveragability. Auxiliary functions may be items like reports, specialized functions, or perhaps particular user screens. Such functional items may be leveragable across a subset of sites but perhaps not across the whole of the target domain. Core functions are those that can be leveraged across every site in the target domain without modification.

Analysis of Target Domain. There are two possible sources for meaningful information on which to base estimates for assessing leveraging. The best information source results from an exhaustive analysis across the target domain. An analysis of the subject areas across multiple sites is able to calibrate the data/functional model according to the extent of changes likely as the application is moved from one site to the other.

Pilots and Prototypes. Another valid source of information comes from completed projects, pilots, or prototypes. These assets are excellent, because they provide a test bed for quantifying leveragability. In any case, usually more than two sites must be sampled to have a meaningful representation of the whole.

Weighting Factors. Assuming one of these two possible sources of information, it is possible to assign leveraging weighting factors to each function or process of the application. Weighting factors should reflect, more or less, the extent of variability in each subfunction or capability of the application with respect to target sites. Weighting factors can also be used to assess resource requirements as measured in development resources or costs. This discipline reinforces an effort to maintain as much of the application as practical within what may be called the core, thus driving greater leveragability.

Breakpoint values for leveragability can be defined, quantitatively, by what is in the core, auxiliary, or site-specific categories. One hundred percent leveraging means that no modification is required in the application to make the application operational across all sites in the target domain. Zero percent leveragability implies unique tailoring of the applications to each site. The end state of the leveraging analysis should portray the incremental costs of moving the complete application across all sites in the target domain.

CONCLUSION

Leveraging is driven by management's recognition of the inherent costs of unique site-specific solutions. An effective business practice explicitly declares leveraging to be part of the IS strategy. The goal of leveraging is to maximize the net present value of the IS investment.

As companies continue to invest in information technology, significant assets are being created. The management of these assets should be performed with the same decision process that is applied to other important capital assets.

II-8

Client/Server as an Enabler of Change

Donald Saelens
Stuart Nelson

THE BUSINESS PRESS is full of talk of reinventing business to be more responsive to customers. The Minnesota Department of Revenue (DOR) has put words into action. For Minnesota businesses, a redesigned sales tax system will mean less heartburn at tax time and fewer dollars spent on lawyers and accountants to help them comply with the state's sales and use tax laws. For the state, the project is expected to enable the DOR to resolve delinquent accounts in less than 90 days (instead of up to two years) and reduce the number of paper tax returns it receives.

This chapter describes the reengineering project and its targeted changes in the department's mainframe-based systems and in the DOR's functional orientation. The IS principles guiding the client/server development effort are also shared.

FAILINGS OF A RIGID SYSTEM

As the state's tax- and revenue-collecting arm, the DOR has frequent and direct contact with Minnesota businesses and citizens. Under the old system, DOR employees operated within a compliance process that offered no systematic capability for accommodating exceptions or being responsive to changing business practices and demographics.

Continuous changes in the state's complex sales and use-tax laws often left businesses confused about how to comply with the law. They simply received their tax forms in the mail with directions to file and pay. When they called the DOR for help, they found employees there hard to reach or unable to answer their questions because employees did not have the appropriate information readily available. As a result, businesses ended up sending in the wrong returns, the incorrect payment, or nothing at all.

The state sales tax system is one of three major tax systems under the DOR's jurisdiction (the others are income and property) serving 151,000 businesses and administering more than $2 billion in sales and use taxes annually. The sales tax was originally created in the late 1960s as a temporary measure. Over the years, the tax became permanent; businesses also sprang up in the state and many companies evolved into new lines of business. The tax system and the DOR fell out of sync with the times and with business growth, reaching the breaking point in the late 1980s. Employees could no longer trust the taxpayer data because maintaining updated information on all businesses was virtually impossible. In addition, it took months to properly process returns. The agency had trouble effectively dealing with problems and questions taxpayers had about their returns and payment.

The chief contributor to these problems was the DOR's outdated computer system built 25 years ago. Although the business environment changed over the years, the agency was constrained by the computer. Employees could not apply common sense to case management because the computer set the parameters. As a result, many good suggestions from employees on how to improve the process could not be accommodated because all cases had to be made to conform to the computer.

Another problem that hamstrung the agency was its functional orientation. Employees were charged with narrowly focused tasks—auditing, data entry, and payment processing, for example. This job design, which contributed to employees' inability to address taxpayer concerns outside of their area of expertise, was reinforced by the DOR's criteria for judging and rewarding employees. Instead of measuring employee performance with an eye toward the entire sales tax process, the agency emphasized the accomplishment of individual tasks and activities.

These inefficiencies took a toll on the state's bank account. Because the DOR could not swiftly identify taxpayer problems, many businesses—more than 13,000 in 1992—paid either less than they should have or nothing at all. Problems in collecting this revenue were further compounded because more than half of the delinquent companies were no longer in business. Inaccurate information in the agency's computer made it difficult for employees to know who was in business and who was not. Because every delinquent case represents potentially lost revenue to the state, the DOR needed to make it easier for companies to file (and file correctly), and it had to give employees the tools and means to reduce the time it takes to track down and collect taxes from nonfilers.

THE REENGINEERING PLAN

The agency began by redefining its mission and its image among taxpayers. In addition to holding regular meetings of top officials, the DOR solicited input from employees and taxpayers. The result of the two-year strategic planning process, the agency's first, was a comprehensive business plan and a succinct mission statement: to win compliance with Minnesota's revenue system.

What is unique about this mission statement is how it positions the agency in relation to taxpayers. Embodied in the new mission statement is a philosophical shift from demanding compliance to winning it. This new philosophy places equal emphasis on making it easy for people to file and pay the right amount of taxes, helping them understand what is expected of them, and enabling them to choose the right option for compliance that best fits their needs.

To reach this goal, changes were targeted in the department's mainframe-based systems and the DOR's functional orientation. Among the other constraints the agency faced were increasing workload, a declining cost/benefit ratio, and the end of several years of budget growth.

The agency had already received funding to replace the 1960s computer system, and DOR executives wanted to ensure that it was invested intelligently. After reviewing the challenges, the agency decided that rather than reautomating the current inefficient work processes, it should apply its funding to a complete reengineering of the sales tax system.

The overall reengineering project involved three phases:

- *Phase 1: Vision and strategy.* The DOR reviewed the sales tax system business processes and established targets that would bring significant improvements in performance.
- *Phase 2: Business process redesign.* Ten months later, teams used the targets established during phase 1 as the basis for creating and developing redesigned business processes.
- *Phase 3: Development and implementation.* Another 10 months later, the final phase began involving the development, testing, and implementation of appropriate information systems, job designs, and management systems.

PROCESS REDESIGN FOR CUSTOMER SERVICE

With the support of the highest levels of DOR leadership, the project team—which comprised representatives from all areas of the agency, as well as several external consultants—redesigned the sales tax system's six key processes:

- Taxpayer registration and profiling.
- Sales and use tax filing.
- Filing processing.
- Ensuring compliance accuracy.
- Ensuring payment.
- Performance information dissemination. The IT support for each of the redesigned processes will be discussed later.

All processes cross divisional lines within the organization.

The result of this process redesign effort is a significantly streamlined sales tax system, the heart of which is the taxpayer registration and profiling process.

A New Registration and Profiling Process

Because many of the problems the DOR experienced in the past stemmed from its lack of timely and accurate taxpayer information, the new process was designed to handle greatly expanded registration options available to taxpayers and more detailed information about companies' operations.

Today, when a taxpayer registers, a unique, customized profile is created in the system's data base. This profile—which can be updated by any DOR employee or even by the taxpayer—collects and stores pertinent company information, including the type of business the company is involved in, how long it has been operating, its location, and its relationship with the DOR. In addition, taxpayers can register in several different ways (e.g., phone, fax, or mail) and can be accommodated easily regardless of whether they are filing permanently, seasonally, or just once.

DOR employees are now able to send customized return forms to taxpayers instead of generic forms. Another advantage of the customized taxpayer profile is that it allows the agency to be proactive in supplying businesses with tailored education and service options. Using information from the profile, the DOR can inform taxpayers of industry-specific educational offerings or potential law changes.

Process Improvements

Before the process redesign, a taxpayer with payment problems was continually handed off to different employees, each of whom handled a small portion of the case. Now, most delinquent taxpayers can be assigned to one DOR employee who follows the case through to completion or resolution. Employees are empowered to make decisions throughout the process to resolve cases more quickly and to take accelerated enforcement action to ensure that every case is resolved within 90 days. The future work load of cases for businesses still operating is expected to improve significantly. Today's volume is 5,700 cases with an average age of 20 months and a balance of $3,500. By interceding early, DOR employees will be able to ensure that the state receives the revenue due more quickly and that the process does not become a burden on the taxpayer.

The greatest opportunity for improved efficiency is in the area of filing processing. Each year, the agency receives approximately 900,000 paper returns filed by Minnesota businesses. Every one of these returns must be received in the mailroom, opened, sorted, and routed. The information on the returns is entered manually into the computer by a team of employees, after which the team microfilms the returns for future reference. The DOR expects to reduce the number of paper returns it receives to 600,000 because many more taxpayers will qualify for quarterly or annual filing, and state and local option tax returns have been combined. Initially, high-dollar businesses will be required to file and pay electronically; the agency plans to further trim the number of paper returns it handles because 95% of businesses, regardless of dollar volume, will have the option to file electronically.

INFORMATION SYSTEMS DECISIONS

Crucial to the success of the redesigned processes is the information technology developed to support them. The agency's 25-year-old mainframe system, patched numerous times throughout the years, could no longer deliver the performance necessary for DOR employees to be effective. Therefore, the team designed and built a completely new system, based on a client/server architecture that would be flexible enough to meet the needs of the redesigned environment.

Technological underpinnings of the system are Macintoshes and Windows-based PCs that are linked by local area networks to both a miniserver and the agency's existing legacy mainframe. This technology provides users with a much more friendly environment in which to operate. The client/server architecture brings processing power and critical information to employees' desktops. It can also easily accommodate the multitude of exceptions and differences that DOR employees encounter from taxpayers as large as Northwest Airlines to those as small as traditional mom-and-pop businesses.

In addition to the technology, several other factors contributed to the project's success. First was the fact that the agency's IS department was involved from the beginning of the project, not brought in at the tail end to build supporting systems in a vacuum. IS was also represented on the executive steering committee. This provided the opportunity for a better understanding among the IS staff of what the users needed to do their jobs more effectively.

Another factor was the set of IS management principles relating to, among other things, technology infrastructure, applications development, and data that the project team devised to guide itself through the project. These principles helped to keep team members focused on their mission. They are:

- Infrastructure:
  - Networked intelligent workstations will be available on every knowledge worker's desk.
  - Cooperative processing will be implemented in a client/server architecture, with each platform handling what it does best.
  - An open systems architecture is favored over closed proprietary architectures.
  - Mainframe systems will not be modified beyond what is necessary to support existing data.
- Applications development:
  - Pilots and prototyping are to be used to help us learn and apply new ways to deliver information products.
  - The underlying information technologies will be transparent to the user.
  - Applications will be written to operate on or as close to the workstation as possible.
  - Development will occur on IBM and compatible equipment, not on Macintoshes.
- Data:
  - Information will be located at the level that will best facilitate its shared access.
  - Existing data will continue to be stored and maintained on the mainframe. New data will be stored and maintained on the client/server platform.

— Data redundancy will be avoided except in cases of performance.

— Data will be transmitted electronically whenever feasible.

Finally, the team selected a proprietary Accelerated Applications Development (X/AD) methodology that makes extensive use of pilots and prototyping. The team thus avoided spending two years defining requirements and then pushing the system onto users all at once. The resulting system more closely reflects users' true requirements (and was more readily accepted by employees). The methodology promotes the use of time-boxing, which limits the amount of time available for a task, a design iteration, a work session, or even an entire development project. The project team was able to define, design, and implement applications in 12 months.

THE INFORMATION TECHNOLOGY CHANGES

The redesign effort resulted in significant changes to the way the DOR conducts business. Without the appropriate enabling technology, however, the improved operations would exist only on paper. The following sections discuss the major changes enacted in the agency's six key processes and how information technology was applied to make these changes a reality.

Taxpayer Registration and Profiling. The redesigned registration and profiling process expands the registration options available to taxpayers as well as the information about their operations. To support this new process, the project team developed the following:

- New registration screens, expanded taxpayer profile data base structure, and follow-up structure.
- A process to synchronize the new taxpayer data base with existing mainframe systems.
- A system to distribute and manage the disbursement of available taxpayer identification numbers.
- A system to accommodate taxpayers' electronic access to and update capability of their profile information.

Sales and Use-Tax Filing. New filing cycles have been established; returns can be customized; and taxpayers can have options concerning the method they use to file and remit sales and use taxes. To address these changes, the team developed:

- A new flexible sales tax system to create customized, computer-readable sales tax returns based on a taxpayer's profile.
- A 24-hour touchtone computer response and personal computer bulletin.
- Software for customers' use in filing, paying, and requesting information on sales and use taxes.

In addition, the team upgraded the existing workstations to accommodate responses to and interactions with taxpayers, and it modified the core tax system on the mainframe to handle individual use-tax reporting.

Filing Processing. The actual processing of the tax returns received by the agency presented several challenges to the project team. To satisfy new mandates—for example, that critical tax return information be available online on the day of arrival (with the balance of that information being available within seven calendar days), that filing information must be 98% accurate, and that tax remittances be deposited immediately—the team implemented an imaging and scanning system to capture filing and payment information and developed interfaces between the core sales tax system and other existing systems.

Ensuring Compliance Accuracy. The profile information obtained during the registration process is used by the agency to customize its services to taxpayers on the basis of their needs and compliance history. To provide access to such information, the team developed.

- Software for electronic links to taxpayers.
- A compliance screening data base with an automatic update from transaction data bases.
- Statistical applications and expert systems to identify candidates for specific compliance actions.
- A work flow management and case preparation system to automatically assign and track compliance activities.

Ensuring Payment. This process is divided into accounts receivable resolution and registered nonfiler resolution. In the new process, delinquent taxpayers are to be assigned to one DOR employee who has discretion in choosing how to deal with them. The goal is to resolve all cases within 90 days of identification. The process is supported by PC-LANs that provide access to the work flow management system as well as other information systems and offer a place to record case activity for future reference in assistance with case resolution.

Performance Information Dissemination. As part of the process redesign, new measures have been instituted to monitor the department's progress in winning compliance. To help management in applying these measures, the project team developed automated performance measurement capabilities and links in all new information systems. The team also created and installed a technological infrastructure to support the development and operation of the online systems containing the sales tax performance measures.

EFFECTS ON THE ORGANIZATION

An effort of this magnitude has profound implications for the organization. The project generated uneasiness among the IS staff, especially certain technology specialists who were faced with a completely new technological environment and new job responsibilities.

One of the biggest challenges the project team faced was approaching the redesign of jobs with the same openness of mind used in redesigning the processes. When it comes to the human perspective, friends and colleagues are involved. It does not help the organization or its employees to adapt to the changes if these human issues are avoided.

Project leaders initially thought they could gain employee acceptance of the new system if they just communicated to employees what was happening. But the team quickly discovered that communication is not enough; people actually have to start doing the new work and using the new technology as soon as possible.

This was accomplished in part by instituting a technology lab in which employees could feel safe to experiment and make mistakes. Although communication about what the organization is trying to accomplish alleviates some concerns employees have about new technology, there's no substitute for hands-on experience.

All DOR employees experienced firsthand how IT could be used to break down organizational barriers and draw people together. One of the earliest deliverables of the project was an integrated infrastructure of 500 workstations equipped with groupware products. Within four to five days of installing the application, employees who were formerly not allowed to communicate with each other were in touch by electronic mail, and DOR leadership could communicate directly and immediately with all employees.

Lotus Notes was installed as a departmentwide tool for team members in the sales tax reengineering project. Discussion data bases were established for the individual teams that focused on the separate business processes that made up the overall sales tax system. These data bases were used to share ideas and concepts and to coordinate information across teams. In addition, data bases were established to track project issues and their resolution, as well as to document decisions that could be referred to as needed by any member of the project.

CONCLUSION

The project demanded a lot of energy, time, and brainpower from the entire DOR staff. The outcome, however, benefits both taxpayers and the agency.

For taxpayers, the new system not only makes it easier to comply with the tax laws, but it will also reduce their cost of compliance. (The Minnesota Retail Merchants Association has estimated that Minnesota businesses spend $55 million a year to comply, most of which is spent on professional expertise to help them navigate through changes in tax law.) If they have questions about taxes, businesses now know where to go for help, and they can get that help much more quickly and completely. In addition to being given instructions on what they must do to comply, they are asked what changes would help them comply more easily. Businesses are encouraged to participate in the policy-making process to ensure that sound, realistic policies are created.

For DOR employees, the reengineered environment and new technology give them the flexibility to be creative in how they approach their work. Instead of single-function jobs, employees have multifunctional responsibilities. Rather than focusing inward on information flow, employees emphasize an external customer view. They are no longer handcuffed to a rigid method of dealing with taxpayers, largely because the new client/server system supports the new processes and provides everyone with instant access to important information.

II-9

Client/Server Computing: Politics and Solutions

Steve Guynes
John Windsor

A POLITICAL SITUATION is developing in corporate America that will have a major impact on corporate information systems (IS) departments. The IS department's role within the corporation is changing—especially in its interaction with corporate end users, who are demanding increased control over corporate data. IS departments must voluntarily begin to move in a new direction, or they risk encountering serious problems with top management. For example, the increased interest in outsourcing is one indication of the many problems that now must be addressed.

Client/server computing is another such issue. Client/server is a growing presence because top management believes it saves money, and end-users believe it solves all their computing needs. Unfortunately, neither of the above is completely true.

The major problem is neither top management nor end users fully understand all that is involved in corporate computing. Top management easily understands that the hardware for a client/server system costs less than a mainframe but has a more difficult time understanding the costs involved in client/server multiplatform software or controls to protect data integrity. In addition, end users may have little understanding of why they cannot upload their local data to the corporate data base.

The IS department is committed to protecting the systems and data that are the lifeblood of the organization. For their part, IS managers are concerned with where corporate data is being downloaded and with the stability of the networking systems that are installed throughout the organization. Historically, data centers have been charged with maintaining the integrity of the corporate data base and have established extensive controls to protect the data. They view client/server systems as presenting additional problems to an already complex situation. So, it seems that IS management, top management, and corporate end users are working at cross purposes.

This chapter suggests some ways to implement client/server systems that will satisfy all three. Terms associated with client/server computing are defined. The advantages attainable from a successful implementation are identified, and some of the problems are addressed. Finally, some suggestions for implementation are presented.

CLIENT/SERVER DEFINITIONS

Client/server computing is still being defined by the computer industry. Even though client/server is still evolving, however, there are some fundamental concepts in place. The main force behind client/server computing is the empowerment of end users through delivery of data, programs, and processing power to their desktops.

The distinguishing feature of client/server is it contains cooperative processing capabilities. This means client (i.e., workstation or desktop PC) processing is split from server processing, while presenting a single logical picture to the user.

Clients have processing capabilities and make requests to servers. The server is the machine on which a process or set of processes resides; it is made available to the clients via a local area network (LAN). Software applications tie the three components—client, server, and network—together to form the client/server architecture.

At the workstation, the user can manipulate data and execute the programs received from the server. The client addresses the server by way of queries or commands and waits for the appropriate response from the server. The server, for its part, never initiates contact with clients; rather, it monitors the network, waiting for requests from clients. Depending on the specific implementation, the queries can be performed synchronously or asynchronously.

When it receives data from the server, the client performs data analysis and presentation locally. In a heterogeneous client/server environment, the server can run under several operating systems that the client can run under. These include, for example, DOS, OS/2, Windows, and UNIX.

The Server's Varied Tasks

The nature of the server depends on the goal of the client/server system. The server could be:

- A print server, which requires low machine power.
- A file server, which requires medium machine resources.
- A data base server, which requires a high amount of machine resources—usually DASD and I/O.
- An application server, which also requires a high amount of machine resources—usually processing speed and memory.

These types of servers are not mutually exclusive. It is possible, for example, for a server to be a print server for one client, a file server for another, and a data base server running the data base application for yet another.

Ideally, servers hide the composite client/server system from clients and users. As long as the server is transparent, its configuration (e.g., the underlying hardware platform and operating system) disappears from the client view of the system. If a data base server uses an SQL interface, for example, the clients do not need to know if the server is running UNIX, OS/2, Windows NT, or any other type of operating system. The client is just concerned about the SQL interface.

In many large corporations that have data residing on a mainframe computer, the server provides a gateway from the clients to the mainframe. The server must have adequate disk space to serve the clients, since in many client/server systems, the server is the central location for data. In other large computer installations, the mainframe operates as the server.

In addition, the client/server relationship can be hierarchical. For example, a client to one server may be a server to another client. However, to qualify as true client/server computing, a system must have the ability to install the client and server on two separate platforms. Either the client or the server hardware platform can be upgraded without having to upgrade the other platform. The user must be able to change one of the nodes in the client/server system without having to change the other nodes. Each node is independent of the others, with the server able to service multiple clients concurrently.

BENEFITS FROM CLIENT/SERVER

In client/server environments, the user begins to feel direct involvement with the software, data, hardware, and the system performance. This sense of ownership has long been an objective of data centers. It allows users to define the critical needs of the information system and allocate the resources they are willing to spend on those needs. One implication of this benefit is that user involvement with the process of developing their information system is considered key to the success of any new development project. There are other benefits that appeal to top management, however, including cost savings, boosts in efficiency, and extended life span of existing equipment. These are discussed in the sections that follow.

Keeping Costs Down

If installed properly, client/server reduces the operating costs of information systems departments; one reason is that the hardware typically uses replicated configurations, thereby allowing a greater coverage of the sophisticated support environment.

Client/server systems also permit applications to be put on PCs or workstation servers, which are less expensive than mainframe and midrange systems. Additionally, client/server data base management systems are less expensive than mainframe DBMSs.

Server performance is cheaper than equivalent mainframe performance. The client/server model provides faster performance because the processing is done locally and does not have to compete for mainframe central processing unit (CPU) time. PC MIPS (millions of instructions per second) can provide a cost advantage of several hundred to one, compared with mainframe MIPS. The response time to complex queries on a mainframe takes from three to five seconds. With a client/server system, those same responses are sub-second. This allows users to customize complex queries that were difficult in the mainframe environment.

In addition, client/server provides a better return on technology investment because it allows niche or specialized technology to be configured as common resources, widely available within the computing environment. If the proper controls are in place, client/server computing allows greater access to corporate data and information while providing for appropriate data security.

Increasing Efficiency

The workstation controls the user interface in the client/server environment. Most commonly, user-interface commands are processed on the client. Because the server is free of user-interface and other types of computations performed by the client, the server is able to devote more resources to specific computing tasks, such as intensive number crunching or large data base searches. Only answers to requests from the clients are sent. Full data files do not need to be sent to another workstation for processing, as in networked, PC-based environments. Also, because the user interface is controlled at the client, not every keystroke is sent through the network. Both of these cut down on network traffic dramatically.

Client/server systems can speed up the application development process, as the developers do not have to compete for mainframe resources. This provides faster response time due to fewer bottlenecks and may reduce the systems development backlogs found in most companies. Because the client receives only the data requested, network traffic is reduced and performance is improved. Access is also easier because resources are transparent to users.

Powerful PCs and workstations can run applications that are impractical to run on the mainframe. Central processing unit-intensive applications, such as graphics and data analysis, require client/server's distributed processing and shared data to be efficient. By running these applications on PCs or workstations, the server is freed up to process other applications.

Data base servers centralize the data that allows remote access to the data. Client/server computing allows multiuser access to shared data bases. With client/server computing, users can access the data that was stored in their departments in the past and any other corporate data that they need to access. Besides access to more data, users have broader access to expensive resources, use data and applications on systems purchased from different vendors, and tap the power of larger systems. Client/server end users also have access to large data bases, printers, and high-speed processors, all of which improve user productivity and quality.

Reallocating Resources

Client/server computing allows organizations to extend the life span of existing computer equipment. The existing mainframes can be kept to perform as servers and to process some of the existing applications that cannot be converted to client/server applications. Mainframes can also be used as enterprise data management and storage systems after most daily activity is moved to lower-cost servers. The users' PCs can be kept to act as front-end processors.

Client/server computing is flexible in that either the client or the server platform can be independently upgraded at any time. As processing needs change, servers can be upgraded or downgraded without having to develop new front-end applications. As the number of users increases, client machines can be added without affecting the other clients or the servers.

PROBLEMS ASSOCIATED WITH CLIENT/SERVER

Companies either moving from a mainframe environment to client/server computing or adding client/server to an existing environment face several problems. Client/server technology requires more diverse technical support, as it allows a company to use hardware and software from many different vendors. Support contracts are required with each of these vendors. To minimize the difficulties of having multiple vendors, a company must establish internal hardware and software standards. The full-scale implementation of client/server computing also requires either reorganizing existing departments or creating new ones.

Staffing Changes

Client/server systems require more staff experts because of the diversity of technologies that must be brought together to create an effective system. For example, most developers understand either the mainframe system or the PC, but not both. Because there is a shortage of skilled and experienced developers, there is more trial and error in developing client/server applications than for older, well-understood mainframe applications. Developers need to go through extensive training to learn the new technology.

In addition to retraining current employees, a company may need to hire specialists in LAN administration, data base administration, application development, project management, and technical support for users.

Lack of Standardization

As with most developing technology, client/server computing does not have agreed-on industry standards. Currently no standard exists for retrieving, manipulating, and maintaining such complex data as graphics, text, and images.

Standards must be established for client/server to allow the use of products from different vendors. SQL is one such standardized data-access language; however, each DBMS vendor has its own SQL dialect. This adds to the complexity of building transparent links between front-end tools and back-end data base servers.

Communications Barriers. If the company is to take advantage of its investment in existing computing resources—typically a mainframe—during the move into a client/server environment, a set of communications problems is introduced. The ability to communicate with the mainframe as a server is available through several protocols, most typically, TCP/IP. However, the client software is most generally based on other protocols.

The network hardware needed to include the mainframe in the system is an expenditure generally not included in the prices quoted by the software vendors. Additionally, most client/server software that is compatible with the mainframe links only to specific software on the mainframe. The company must either have current versions of that software or maintain both platforms at compatible levels.

Security Issues

There are serious security and access control issues to be considered. Since the server is usually the central location for critical data, adequate physical security and operational security measures need to be taken to ensure data safety. Backup and recovery procedures are improving, but logging procedures are still lacking.

A large number of tools exist that perform security and control functions on mainframe systems. All of them can help with the client/server effort, but they are not designed specifically for client/server computing. The lack of automatic backup and recovery tools is a problem. Until these tools are developed, companies should not place mission-critical applications on a client/server system.

Performance monitoring and capacity planning tools are a low priority for many client/server administrators. Buying a new server and adding it to the system is much less expensive than spending much time on capacity and performance analysis.

Tools

The tools necessary to support client/servers are not yet fully developed. There is a lack of client/server-oriented communications, diagnostic, and applications tools. These troubleshooting tools are more powerful than they were three years ago, but they are less robust than those readily available for mainframes. There is also a lack of tools for converting existing applications to client/server routines. This forces client/server users to either write new applications or use the existing applications on the mainframe. But because most companies have invested in millions of lines of mainframe code that cannot be converted to client/server use, mainframes will not be turned off for many years.

Ownership Problems

The sense of ownership that can be such a positive for user involvement in application development may also create major problems. Systems may be viewed as "belonging to" the user departments, creating situations in which hardware and software expenditures introduce products that are incompatible with the existing system. Users may well develop the attitude that the system must be modified to meet their design decisions, rather than the users adhering to company standards.

The ownership problem can also extend to user-developed software and data. Data and data bases must be viewed as a corporatewide resource to be used effectively. As the data becomes associated with a department, the development of private data bases can be a major problem. In addition, the rest of an organization depends on timely data. As a department develops private data bases, the updates to the corporate data base become less critical, in their eyes, and the data used for decision making by the rest of the organization becomes out of date. The ownership of software creates the same type of problem. As a department develops software that provides enhanced decision-making capabilities, department members' willingness to share the software may well decline. In addition, the private software bank creates situations in which other users are forced to develop the same tools on their own. This increases the potential cost of new applications.

IMPLEMENTATION CONSIDERATIONS

The successful implementation of client/server systems requires involvement by top management, representatives from user groups, and key IS representatives. It is possible for the IS group to become so involved in the selection and decision process that it ignores the business's strategic concerns. As well, the business users may not see the value of careful and reasoned selection and end up with a solution that at the beginning may seem to meet their requirements but that in reality may not solve their business needs.

Therefore, the major issues to be considered include ways that the system would improve efficiency, reduce costs, provide a competitive advantage, and reduce cycle time. Users should build a model of the organization's work flow and data flow. This helps in designing networks and will help determine how data should be distributed.

The front-end tools that are selected (i.e., SQL based, user interface builders, and integrated development tools) are determined by the data and systems requirements and require a great deal of investigation to decide which application suits the system's particular needs.

The users must be made aware of the client/server technology and the benefits it can provide, and a knowledge of the proposed client/server solution should be disseminated to all concerned groups. This knowledge is key to the proper use of the technology, as it puts users in a better position to evaluate the technology. If the users can make use of the system, and it solves the business problem, the chance for acceptance is high.

Training

Training is another key to the successful implementation of a client/server solution. Without proper training, the risk of failure is high. If the information systems personnel are not aware of the fundamental limits of current client/server technology, they can either exceed the system's limits or underuse the system, which would seriously damage the success of the project.

Checklist

There are several key decisions that must be made before implementing a client/server system. Technical questions include:

1. What operating systems will be used?
2. What graphical user interface will be used?
3. What hardware platforms will be used?
4. What on-line transaction processing monitors will be used?

Political questions include:

1. Who owns the data?
2. Who owns the applications?
3. Who is responsible for maintenance?
4. How are the standards established and enforced?

CONCLUSION

The changes occurring in the IS environment are the result of the increasing desire to move the technology beyond its present state. With the availability of new and cheaper personal technologies, the drive for direct exploitation of available information is increasing. IS managers should approach client/server with a clear view of the lasting business benefits that it can bring rather than opposing or resisting the migration. However, to accommodate this change, IS managers must also recognize that in client/server computing, no answer is perfect. Technology that is selected now may not be the solution in the future, and IS managers should realize that some custom components may have a limited lifetime.

IS managers' responsibilities have shifted to the establishment and management of infrastructures, services, and data quality. It is no longer appropriate for them to position themselves between the end user and the technology. Most of the issues presented by client/server systems are similar to those presented by mainframe computing. For this reason, IS managers are uniquely positioned to lead in the correct implementation of client/server systems. Top management, end-users, and the IS group must work together to solve the types of business problems for which client/server is best suited.

The leadership in implementing client/server computing should be placed with the user. However, implementation must be guided and supported by IS. All the knowledge gained from years of dealing with problems of standardization, documentation, maintenance, and performance should be shared with the rest of the organization.

Section III

IT Infrastructure

As SYSTEMS BECOME MORE INTEGRATED and users assume more responsibility for information processing, the responsibilities of the IS organization are changing. There is less centralized development of applications and fewer large computer operations centers. The enterprise's IS manager is less responsible for doing information processing and more responsible for providing the infrastructure within which IT is put to use. This includes both the intellectual and physical infrastructure.

This section provides some of the guidelines the IS management team needs to carry out its mission of providing the appropriate environment for users to engage in constructive information usage.

The idea of information sharing is central to the concept of reducing organizational complexity. As organizations flatten and functions merge into processes, information must be perceived as more than the property of single line departments. However, not much is known about information sharing—why sharing happens, why it does not, how much sharing is desirable, and how to manage it. Chapter III-1, "Providing Access: Corporate Information Sharing," discusses these important issues and the behavioral norms that make up the rules of the game.

With impressive improvements in the mobility of IS technology, the constraints of a fixed work place are fast disappearing. The virtual office is now a reality. When attempting to ensure the integrity of the virtual office, however, new approaches to management and control are needed. Providing confidentiality and availability for mobile users may require modified business rules and technology. Chapter III-2, "Providing Virtual Office Integrity and Security," suggests a framework for achieving these objectives and addresses the issues and options for improved security in the mobile environment.

"Distributed Network Support and Control" describes the need to properly size and maintain networks to support the internal customer. As distributed or client/server networks increase in number, size, and importance, managers must prepare to deal with complex security requirements. Chapter III-3 considers the exposures in today's distributed networks and the role of network management in building cost-effective security measures.

To manage an increasing work load, hold costs down, and keep services levels high, IS departments must adopt a consultative approach to supporting internal customers. This means shifting the primary focus of internal client support from happy clients to productive clients. Chapter III-4, "The Consultative Approach to Client Support," provides guidelines for moving to such a customer-focused approach, which allows support groups to explore alternative ways of providing service to internal customers. The chapter discusses steps for conducting client feedback sessions and outlines the skills needed to implement a successful consultative approach to client-based computing. This is a workable solution to otherwise unmanageable increases in technology support requests.

"User Involvement in Project Success," Chapter III-5, offers ideas on how to improve user involvement in projects designed for more effective and productive use of information technology. Often a systems project is never given the chance to succeed, because the expectations placed on it conflict and can never be met. The chapter describes an approach to project leadership called Theory W, which is designed to prevent such a situation. This theory is used to manage expectations and risk by having all parties involved in a project—users, systems developers, and senior management—negotiate a set of win-win conditions that guide the project throughout. A case study is used to illustrate win-win conditions and show why they are necessary for a successful project of any size.

Executive information systems (EIS) are among the fastest-growing applications in corporate America. Today's executives should be able to access information concerning stock prices, competitors, customers, key performance indicators, and internal operations at a moment's notice. Chapter III-6 describes the most crucial element of the EIS, the EIS interface. "Guidelines for Designing EIS Interfaces" examines studies of actual EIS interfaces and discusses methods of designing the interfaces to best meet the unique information needs of executives.

As end users begin to develop applications that have organizationwide implications, the end-user computing support function becomes more involved in providing strategic direction for the organization. In this mature environment, reviewing end-user applications is a valuable tool for improving the quality of end-user computing, protecting the investment in end-user computing, and pointing out potential problems with user-developed applications that could damage business. Chapter III-7, "Reviewing End-User Applications," suggests guidelines for preparing for such a review.

Providing effective security is an important building block of the IT infrastructure. In the end-user computing environment, security poses even more of a challenge. First, what is meant by security must be defined, and then the services required to meet management's expectations concerning security must be established. Chapter III-8, "End-User Computing Control Guidelines," examines security within the context of an architecture based on quality and offers specific suggestions on steps that can be taken to protect the corporation's assets in this departure from the traditional IS environment.

III-1

Providing Access: Corporate Information Sharing

Randall H. Russell

NEW TECHNOLOGY is making ever-mounting volumes of information more readily available, through more types of media, to more recipients, than ever before. However, the broadening array of communication channels that has eased information transmission has coincidentally reduced the certainty that the message a sender tries to transmit will be received by the intended party.

These changes have complicated the jobs of those who manage their companies' information behavior, that is, the way their companies' personnel acquire and use information. This chapter explores the factors that affect information behavior and contribute to information sharing.

A DEFINITION OF INFORMATION SHARING

Information behavior is the way people act regarding the information they need, hold, or manage. It includes creating, retrieving, or modifying information, storing information, and providing access to information.

Information sharing is the voluntary act of making information available to others. It is not the routine reporting of information (e.g., the submission of a time and expense form at the end of a pay period), nor is it the routine exchange of information between intelligent devices. Exhibit III-1-1 highlights some of the differences between information sharing and reporting.

Information sharing represents one end of a continuum of information access. It describes voluntary information access that takes place on an ad hoc basis. Together, information sharing and information reporting represent the two ends of the information access continuum.

The information access continuum is dynamic because information behavior changes over time. Consequently, when organizations have positive experiences with information sharing, they try to systematize this sharing into a more formal exchange process (i.e., information reporting).

For example, a large domestic oil company recently developed an operational model for the production of gas in one of its major fields. This new model, which was created only after its reservoir and facility engineers repeatedly shared information, was more accurate than the organization's past model. It also demonstrated to the engineers' business partner, the pipeline company, that it could ship more gas for the organization without reducing the volume of gas it shipped for its other partners. Now that the new model has been accepted, the oil company has institutionalized the information sharing that led to it as formal information reporting. Information sharing has been acknowledged as a beneficial practice and is encouraged elsewhere in the organization.

THE IMPORTANCE OF INFORMATION SHARING

Information sharing has been seen to increase employee commitment, improve decision making, quicken an organization's response to potential shortages or delays in production, and predict shifts in organizational structure. However, some employees use it to advance their personal goals. They leak sensitive information to outsiders and withhold critical information from their managers.

A study of how information sharing affected the results of collective bargaining processes found that, in Japan, sharing led to shorter and easier negotiation processes, accompanied by the acceptance of lower wage increases by unions. In contrast, a similar study in the US concluded that information sharing increased labor's bargaining power. Thus, sharing can result in undesirable consequences, depending on which side of the relationship the observer sits.

Experience shows that information sharing does not occur easily. Because individuals try to maximize their gain, they restrict their information sharing when they believe that their unique value to the firm is reflected in the information they control and selectively share.

THE IMPORTANCE OF CONTEXT

The context and content of information determines a lot about how and when it will be shared. Over time, relationships between humans and technology inevitably change. Consequently, human access to, use of, and sharing of information over the life cycle of a business process can also change.

Organizations must try to ensure that truly useful information is made available to those who will use it to serve the interests of the firm. Organizations can achieve this by using appropriate incentives for encouraging the sharing of this information. However, incentives, in and of themselves, are not enough. Other contextual matters play a role—namely, the implicit and explicit organizational structures (i.e., norms and rules) that help determine information behavior. These three factors: incentives, rules and norms are discussed in the following sections.

WHAT DETERMINES INFORMATION BEHAVIOR?

The following discussion of the factors that determine information behavior demonstrates that it is impossible to treat any of these factors in isolation. In fact, two of the factors, behavioral norms and incentives, are always in evidence, regardless of whether formal or written rules are in place. Oddly, people often act as if they must attend to only the written rules of an organization when they attempt to change information behavior within it.

Explicit Rules

All large and complex organizations use explicit rules to define their intended information flows. For information reporting, they use standard formats that specify information types and the frequency of reporting required to systematically support decisions, trigger events, and in other ways drive organizational outcomes. For example, personnel in most large organizations must report time and expenses, evaluate themselves and others, and report on a project's status or sales activity. Organizational units prepare budgets, develop tactical and strategic plans, and report on projects and initiatives. Increasingly, organizations report on quality achievements; and measurement, in general, is becoming increasingly popular. Most of this reporting is in response to explicit rules that identify the information required to run the business.

Personnel, however, tend to adhere only selectively to such explicit reporting requirements. People comply more often when adherence is linked to incentives. Directly linking payment to the timely submission of time and expense reports, for example, is tremendously successful in increasing the percentage of reports delivered on time. (An inverse relationship exists between the number of reporting rules and the level of compliance.) People conform to rules selectively; those linked to outcomes of interest are observed first.

Explicit rules need not always be written. Official policy manuals often lag behind operational reality. For example, someone with senior-level authority can quickly change the rules that govern information access in a company, simply by leading through example.

In one company, a senior manager of a corporate IT group wanted to impose a new software product as a communication standard. To enforce this rule, he refused to communicate through any other mechanism. He tied following the communications rule to the incentive of communicating directly with him. Thus, are new rules written.

As the IT manager introduced his official communication mechanism, he also encouraged staff members to use a formal rumor mill data base for posting rumors or responding to posted rumors. Rumors included facts, beliefs, or mere suspicions. Intended to promote an environment of trust and openness, this mechanism provided a very informal opportunity for information sharing.

Following this experimental phase in the data base's introduction, the bank merged with a financial institution that did not encourage as much communication among its personnel. The potential value of the rumor mill conflicted with the underlying norms of the new organization. In short order, the data base was discontinued, though it continued to be used for other reporting purposes.

The executives of the newly created financial institution were concerned about how the automation of informal information flows would affect the regulation of the organization. Information sharing can encourage organizational changes that may have little to do with an organization's formal structure. Informal systems of information access (i.e., rumor mills) must support an organization's formal goals, or chaos will ensue.

Because information sharing tends to flatten hierarchies and further democratization, it is strenuously resisted by those who expect to lose through such a transformation. Those who see an advantage in it support it. Organizations considering information sharing must decide whether organizational flattening, improved morale, and democratization are the goal of the transformation they are considering or an unintended result.

BEHAVIORAL NORMS—THE UNWRITTEN RULES OF THE GAME

Although the terms *organizational culture* and *behavioral norms* are often used interchangeably, they do not mean the same thing. *Culture* refers to the decision styles and customary forms of interaction that characterize an organization's work environment and significantly determine behavioral norms. Culture is ever present and can encourage or discourage desired changes in information behaviors.

Many people describe culture as "the way things work around here." This can include being polite to one's boss, not revealing personal information to staff members one does not know well, and leading a discussion if one has the highest status in the room.

As Exhibit III-1-1 suggests, explicit rules of access are typically associated with information reporting, but the implicit norms of organizations relate more closely to information sharing. Reporting, which is more formal and systematic, is defined by rules. Information sharing, which is more informal and ad hoc, occurs through the tacit agreement of people acting according to behavioral norms.

One commonly observed behavioral norm is the fact that people prefer to share information associated with positive outcomes. In a recently reengineered R&D function within a major chemical company, for example, cross-functional teams were observed to be much more comfortable sharing information about the progress they were making, rather than negative results teams or team members had received.

Another behavioral norm is the preference for sharing when it is possible. Seeding, a mechanism for encouraging the use of shared discussion data bases, works because people like to share. When an organization seeds a data base, it makes the data base freely available in the expectation that users will, in turn, contribute to it. The more useful, interesting, or rewarding the information in a seeded data base, the more people want to contribute information to it, to reseed it. Without such reseeding, a data base declines in value, and sharing diminishes in a downward spiral of use and usefulness. Therefore, systems in which users receive information but do not contribute to it tend to be unstable and in decline.

Information Sharing		Information Reporting
Informal	◀--▶	Formal
Ad hoc		Periodic
Unstructured		Structured
Voluntary		Mandatory
Nonsystematic		Systematic
Implicit exchange value		Explicit exchange value

Exhibit III-1-1. Forms of Information Access

The more useful, interesting, or intrinsically rewarding information is, the easier it is to share. The easier it is to share, the more an organization can rely on behavioral norms to ensure its communication. However, when information supports formally structured processes or when it can have a negative impact (e.g., the reporting of negative project results), organizations must systematize its communication by establishing formal reporting relationships and mechanisms. Information reporting is thus associated with increased formalization of the information access process.

INCENTIVES

When information sharing is not seen as risky and when people depend on each other to accomplish work, the intrinsic value of information is often a sufficiently attractive incentive to support an adequate level of information sharing. However, when traditionally conservative information behavior is entrenched or when the information required is negative and is mandated, organizations desiring to encourage information sharing should examine the value of the incentives they offer for it.

The evolution of customer support at Lotus Development Corp. provides a useful example of the important role that incentives can play in transforming an organizational culture and enforcing formal rules of behavior. Early customer service at Lotus was a cumbersome process in which a support representative would take a call, document the caller's problem, attempt to reproduce it, and develop a solution. The solution would then be documented and catalogued so that the next support representative who faced that problem would not have to recreate the solution.

One challenge in managing this support function was cataloguing its solutions in such a way that representatives could retrieve them; another challenge was changing the way representatives shared their knowledge. At the time, Lotus was actually discouraging representatives from sharing information. Representatives

Degree of interdependence	Nature of tasks	
	Routine	**Nonroutine**
Pooled efforts	Reporting	Semistructured sharing

Exhibit III-1-2. Optimal Information Access According to the Nature of Tasks and Interdependence of Workers

could get promoted by carving out a problem area and becoming such an expert in it that other representatives with questions in it would come to that representative for help. Representatives learned that by hoarding information, by making other representatives come to them, they could seem valuable to the company. Doing so, however, did not help the department do its job of providing answers to customer problems as quickly as possible. To best support the customer, the department had to document all known solutions and make this material as readily available as possible.

Lotus now takes a two-prong approach to customer service. Representatives are now supported by a combined groupware and data base system that provides full text search and retrieval. This system helps them to determine rapidly whether a solution has been developed for any specific problem. However, Lotus recognizes that a technical solution by itself can not ensure optimal customer support. Therefore, it has instituted a formal performance evaluation standard whereby 10% of each support representative's annual appraisal involves a peer evaluation of how well he or she shares information. This is an appropriate incentive when information sharing is an important goal.

WHEN DOES SHARING PAY OFF?

Information sharing is most appropriate when information behavior is unformalized or when individuals need to adjust their behavior to coordinate their activities or work with others. As Exhibit III-1-2 illustrates, information sharing is most useful to people who are reciprocally interdependent (i.e., who work together) performing nonroutine tasks. They need to communicate frequently, making mutual adjustments to complete tasks.

This situation often occurs after an organizational transformation or when the external environment is changing rapidly. In many organizations, business process reengineering is disrupting information flow; information sharing can improve cohesion within and across processes.

The importance of information access can also depend on the type of work being done. For example, concurrent product development can benefit from information sharing practices because it involves people working together. Other activities that can be improved through information sharing include:

- Conceptual design.
- Technology demonstration.
- Feasibility demonstration.
- Process capability demonstration.
- Design review.
- Production readiness.

Product development is one of the most promising places to implement work-group computing meant to support cooperative work. Here, idea sharing and parallel development promise dramatic reductions in the development cycle. Within product development, the less structured components usually provide the best opportunities for sharing. Information reporting is likely to be more useful within routine components of the process.

III-2

Providing Virtual Office Integrity and Security

Ralph R. Stahl, Jr.

USERS TODAY EXPECT to be able to connect their notebooks by modem from any location to the server at headquarters. Tomorrow, mobile users may expect connectivity for their notebooks in the air as they fly and on the ground as they drive to their next destination. The business traveler may anticipate that all applications will behave in exactly the same manner on the road as in the office.

Large companies are encouraging employees to telecommute for three major reasons:

- Telecommuting benefits employees by allowing them more flexibility in managing their professional and personal lives.
- State and local governments are requiring companies to take action to reduce air pollution and traffic congestion.
- Office sharing allows companies to reduce their real estate expenses.

LINK Resources Corp., a New York City consulting firm, reports that Americans bought for home use a record 5.85 million microcomputers last year. One out of three American households already has a microcomputer. BIS Strategic Decisions estimates that 45 million workers in the United States are considered part of the mobile work force. Other surveys estimate that, in addition to the time spent in the office, the average white collar worker spends six hours a week working at home.

Against this backdrop, the challenge for the application developer is to develop systems that may be used in any environment. The information architecture for the enterprise must also accommodate many methods of remote connectivity (i.e., dial-up, Integrated Services Digital Network [ISDN], Cellular Digital Packet Data [CDPD], Internet, wireless, video, and image transmission) in addition to the traditional local area network and wide area network connectivity.

This chapter is divided into four major sections: availability and continuity; integrity; confidentiality; and new technology considerations, which briefly reviews the security implications for some of the emerging technologies. The ar-

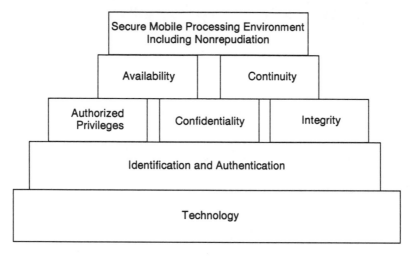

Exhibit III-2-1. Model of Security Services

chitectural model of the security services in Exhibit III-2-1 provides a high-level view of the interdependence of identification and authentication, authorized privileges, availability, continuity, integrity, and confidentiality in providing a trustworthy environment that supports nonrepudiation and mobile power user security.

AVAILABILITY

In this chapter, availability is defined as the assurance that an authorized user's access to an organization's resources will not be improperly impaired. Achieving such assurance involves properly categorizing information privilege keys and ensuring that the mobile user's authorized privileges are properly associated with these privilege keys. Availability also involves physical considerations (e.g., theft prevention, device identification, mobile uninterrupted power supply), notebook connectivity (e.g., a power source, telephone communications tools), and miscellaneous toolkit necessities.

Scheduling Considerations

Information availability is an operations scheduling issue, although some organizations believe that all availability needs are covered by their business resumption practices. Security practitioners must be aware of the need to maintain operational schedules. If the backup and batch processing is scheduled to end at a precise time so that the online or remote transaction processing may start, then the credibility of the central staff to meet their commitments to the field are tested every day. Although capacity planning is not a security issue, the complete information protection plan will make sure that the topic is adequately addressed by the appropriate operational staff members.

Physical Considerations

Concerns associated with the desktop microcomputers in the corporate office also apply to notebooks for the mobile user. However, with respect to mobile computing, security practitioners may need to be more creative to achieve the desired results.

Theft-Prevention Devices. Such theft-prevention devices as cabling and bolting plates can be used to minimize the potential of notebook theft by opportunity. The cables are designed so that they may be looped through an opening in a stationary object to tie the laptop down while the user is traveling. Resistance to these devices exists because many users feel that having these devices gives the impression of not trusting coworkers or business associates. However, security administrators who use theft-prevention devices in their companies indicate that they have experienced a significant decrease in loss. Although the products are effective, corporate procedures with strong enforcement practices are usually required before these products are put into use.

Device Identification. Device identification is critical to the ability to identify a misplaced or stolen notebook. In addition to traditional identification methods (e.g., serial number registers, tags, labels, and engraving), microcomputers can be marked by using invisible ink to record the company's name and the notebook's serial number on the inside of the lid just under the monitor display area. The invisibly inked number must match the serial number recorded in the corporation's asset inventory register. This practice can also be used to resolve disputes associated with ownership of the microcomputer.

Mobile Uninterrupted Power Supply. Mobile uninterrupted power supply implies that each mobile user should have a portable surge protector with sufficient electrical outlets for each device that is connected to the microcomputer or notebook. Electricity follows the path of least resistance, and it will reach the microcomputer through any device cable if the power source for the device is not protected. Surge protector plugs are available at most hardware and electronics retailers. It is also recommended that the user carry a fully charged spare battery pack for the notebook. Usually the battery can be purchased from the dealer that sells the notebook.

Notebook Connectivity

To ensure the notebook's power source, the electrical wall connection should be used so that the notebook's battery can be conserved or recharged. For proper grounding, the electrical code requires that all computer male plugs be three-pronged. In some facilities, the female electrical wall receptacles may accept only two-pronged male plugs. In such cases, the problem can be averted by carrying a female/male converter plug that converts three-pronged plugs to two-pronged plugs and has a grounding wire that may be attached to the wall receptacle's holding screw. This type of converter is available in hardware stores. If the user travels internationally, the toolkit must also include an international voltage adapter that eliminates the need to carry different converter plugs.

Telephone Communications Tools

Offices and hotels are updating their telephone PBXs to digital service, but modems and the majority of the PBXs are still analog systems. To ensure connectivity in either environment, a converter should be purchased that can covert the phone line to analog at the modem connection. The complete converter kit should include alligator clips for phones that do not have RJ-11 jacks. The phone line converter requires an AC adapter as the power source; the full-functioning kit will have the capability to use a 9-volt battery when an electrical wall outlet is not available.

Data transmission may be interrupted if a phone system with call-waiting capability is used. The feature can be suspended during the data transmission call by adding *70 (occasionally #70) at the beginning of the dial string. As a general rule, this is probably not required. However, if a transmission session is interrupted for an unknown reason, call waiting may be responsible.

A miscellaneous toolkit should include the following items:

- A small screwdriver with both a flat and a phillips head.
- An extra-long telephone cord with male RJ-11 connectors on both ends.
- A connector with two female RJ-11 receptacles.

CONTINUITY

Continuity is defined in this section as the processes of preventing, mitigating, and recovering from service disruption. The terms business resumption planning, disaster recovery planning, and contingency planning may also be used in this context; they concentrate on the recovery aspects of continuity that ensure availability of the computing platform and information when needed.

Recovery diskettes will reduce the user's lost time when access to information and remote computing resources is lost due to a major breakdown of the notebook. Recovery diskettes should be customized to the exact configuration of the notebook and should contain the following files:

- DOS system files (COMMAND.COM and the two hidden files), which make the recovery diskettes bootable.
- PARTNSAV.FIL, which contains the hard drive partition table, boot sector, and CMOS information.
- CONFIG.SYS, which contains the appropriate values for files, buffers, stacks, and hard disk drivers.
- REBUILD.COM, which restores the CMOS.

- SYS.COM, which enables the transfer of the operating system to another disk.
- Copies of any special drivers that are necessary to meet the standard operational configuration of the organization's protocol stack.
- Communication programs that support emergency downloading of files and programs.

Creating Mobile Backup

The fastest and easiest backup medium is a cassette tape; however, this represents an additional cost and bulky devices to transport. Most (if not all) mobile users want to travel as lightly as possible. The diskette then becomes the most acceptable medium, but care must be taken to minimize the number of diskettes needed.

The proper organization of the notebook's hard drive will minimize the time and number of diskettes required to create a backup. A directory called *data* can be created with all of the subdirectories necessary to easily organize, track, and access data unique to the user. Application software, data bases, or operating software should not be included in the data directory. These files should be stored in other appropriately established directories. When necessary, these software and data base files may be downloaded from the office server.

The backup diskettes should be kept in a different place than the notebook. If the notebook is lost, the backup diskettes will not be lost.

Loss of Computing Resources

Loss of computing resources as a result of the loss or theft of a notebook or its mechanical failure is the most difficult problem to deal with. The user is away from the office yet may need to repair or replace the notebook immediately, and canceling the next few days' appointments is not an acceptable option.

The recovery process requires planning and discipline on the part of the notebook's user. The data files on the notebook should be backed up regularly. If any ingredient of the business resumption plan is missing or incomplete, lengthy delays are unavoidable.

Mechanical Failure. When the problem is a mechanical failure, the existence and awareness (by the user) of a national maintenance agreement with a rapid response clause can ensure fast repair. The remote user should be able to easily obtain the location of the nearest repair location. This can be handled by the organization's help desk service, which should provide 24-hour accessibility. After the notebook is repaired, the user must determine if any data or application programs were lost or damaged. Lost or damaged software may be replaced by using the emergency recovery diskettes to download the needed data from the office server. Although this may be a long transmission session, it is preferable to getting on an airplane and flying back to the office. After this is accomplished, the notebook owner can restore individual files from the backup diskettes. This process may consume a full day and may require the active participation of the remote user, but after it is done, the user's machine has been restored with the least amount of lost time.

Lost or Stolen Notebook. When the notebook is lost or stolen, the plan must provide rapid delivery of a new notebook to the user in the field. Spare notebooks with the standard operating software, application software, master files, and current data bases should be available at the data center. After a call is received to ship a backup machine, the only step necessary is to find the quickest method. Airlines and bus depots should be called to determine whether they provide shipment that is faster than that provided by the standard 24-hour service providers. After the shipment arrives, the next step is for the user to restore personal data from the separately stored backup diskettes.

If backup procedures for mobile users are to be effective, they should be tested and adjusted frequently. The recovery process may never be needed, and these procedures may be regarded as taking a lot of time. However, without such procedures, it may take the better part of a week just to obtain a replacement microcomputer. After that, it must be determined what applications and data bases must be loaded. Finally, if backup procedures are not enforced, how will the user's personal information be restored to a new notebook?

INTEGRITY

In this section, integrity is defined as the process of ensuring that the intended meaning of information is maintained. Information integrity is provided by allowing only authorized persons and processes to perform only those tasks that they are authorized to perform. Everything else is prohibited.

Software Considerations

Virus Protection. Remote users are no more or no less susceptible to viruses than their office-based counterparts. Therefore, they are likely to experience a virus infection of their computer within the next few years. Proper procedures can prevent the virus from attaching itself to any of the hard drive's files. A corporate contract should be purchased for one of the leading virus detection and eradication software products. The cost can be suprisingly reasonable, in some cases less than $15.00 per user per year, although some very good products may be even cheaper. The secret of success is to have the detection software active in memory at all times as a terminate and stay resident (TSR) program. Some vendors can relocate the TSR so that base memory is not used at all, while other TSR programs may take as little as 5K bytes of base memory. With the detection software active in memory, the computer's user may remain passive. Many organization's detection programs have limited success because they require the user to execute a scan program to check diskettes or hard drives after disconnecting from a bulletin board. During a normal day's workload, users may often overlook this program. If TSR software is used, however, the microcomputer should lock up when it detects a virus, and it should not allow the user to proceed. The eradication or cleaning program should be run immediately to remove the virus. The remote user should be provided with a system-bootable diskette that contains a virus detection and eradication program. If additional directions are required to remove the virus, the mobile user should contact the organization's help desk for assistance.

Notebook Configuration Integrity. Although the name *personal computer* may have been appropriate at one time, today it is a misnomer because the microcomputer has become such an integral part of information processing for the business community. By personalizing the notebook to the configuration of their choice, users may cause incompatibility with their organizations' requirements.

As part of the processing infrastructure, it is important that the IS manager control the protocol layers to ensure proper connectivity, memory management, and execution of the company's processes. This is not to say that mobile users are not allowed to install some software of their choice. An example of coexistence on the microcomputer would be to control the autoexec.bat by not allowing anyone except the systems administrator to modify the file. However, the last line of the autoexec.bat is a call to an autouser.bat that gives the user the ability to add activity to the booting process. Today, a number of client-based products exist that are designed to establish administrative control over the configuration and accessibility of directories and files on the hard drive of notebooks. These products support multiple-user confidentiality and several levels of administrators.

The content of certain directories on the hard drive should also be under the control of the systems administrator. Again, policies and guidelines should allow a section of the hard drive to be used at the discretion of the mobile user. Procedures and guidelines should clearly state the areas that may not be altered and the latitude the mobile user has to customize the microcomputer.

These same robust access control products provide complete control over the DOS computing environment. Access to printers, ports, disk drives, modems, files, and directories may be constructed so that several users of the same microcomputer may not have the same authorized use of these facilities. In addition, passive DES (or proprietary) encryption and decryption of files or directories may be established. The major vendors provide the ability to configure directories on the hard drive to be encrypted or decrypted on the fly. When a file is written to the designated directory, it is automatically encrypted. Conversely, when a file is read into memory by the authorized owner of the directory, it is automatically decrypted. This prevents unauthorized access to information even though several users may share the same microcomputer.

Nonrepudiation. In this section, nonrepudiation is defined as the process of ensuring that a user—either the originator or the recipient—can be identified as having engaged in a particular transaction. This facility may also be used during the normal course of business activity to identify the originator of information. Nonrepudiation involves the following procedures:

- Electronic identification of the sender is accomplished by digital signaturing. Depending on the nature of an organization's business, a choice of standards can be followed.

- Message hashing ensures that the content of the message is not altered. This is accomplished by cycle redundancy checks, which sum up a total value of the message's bits and stores that hash with the digital signature in an encrypted envelope for the message.

- A copy of the message (the hash) with the digital signature of the originator is sent to the message archive. Each message must have an established retention date that will vary according to the message content. The records retention policy should serve as the guide for establishing the date. The message should be automatically removed from the archive data base when this date is reached.

- Proof that the message was delivered requires an electronic acknowledgment containing the date stamp of the activity to be sent to the archive and matched to the message.

- Proof that the message was opened requires an electronic acknowledgment containing the date stamp of the activity to be sent to the archive and matched to the message.

- When the importance of a transaction dictates that nonrepudiation is required, a utility should monitor the message activity to ensure that the message was received and opened. Electronic message status should be returned to the sender for appropriate follow-up as required (e.g., to determine why the message was not received or opened).

- The trustworthy information processing infrastructure must provide assurances that the message and audit details cannot be altered.

Many business uses exist for nonrepudiation in the mobile world (e.g., purchase orders, expense statements, strategic management directives, conflict of interest forms, and other important documents) that can provide technology an opportunity to reduce today's manual administrative efforts. However, nonrepudiation does not address confidentiality of the message; this is accomplished through encryption.

Remote Access Authentication. After the decision is made to allow modem, Internet, CDPD, or ISDN access to the infrastructure, the risk of unauthorized access to the infrastructure increases. Call forwarding and other advances in technology have eliminated the security effectiveness of dial-back modems. The Gartner Group considers dial-back modems to have limited effectiveness.

The most effective way known to authenticate a remote user is through two-phase authentication: with something that the remote user knows and something that the remote user possesses. Another means of authentication is through the use of biometrics, which includes voiceprint, fingerprint, or retinal scan. Currently, the cost and technical problems associated with remote biometrics scans render them impractical for common use.

When selecting one of the myriad products that support two-phase authentication, the following items should be added to the functional requirements list:

- It must provide the capability for centralized administration of access system controls (e.g., personal identification numbers, passwords, alarms, use analysis), while the actual authentication platforms may be decentrally deployed.
- It must function independently from the network infrastructure. The product should be independent of modem type, BAUD rate, or any other characteristics related to transmission of data. When changes are made to the network infrastructure, the product should not require modification.
- It must function independently from the hardware infrastructure. The product must function with all hardware platforms and operating systems. When changes are made to the hardware infrastructure, the product should not require modification.
- The product must function independently from all application software. Changes to the application should not dictate changes to the product. A one-time modification to the application software may be required to request the user's identification to the product.
- It must provide a random-number challenge (algorithm) to the product making the call that is in the possession of the caller. The challenge response (one-time password) must be unique for each authentication session. This ensures that the caller has the product in his or her possession each time a call is placed.
- The product must allow encrypted data to be processed. It is not a requirement for this product to perform the encryption.
- The product must accommodate caller mobility. The caller may need to call different processors or locations that are not part of the infrastructure. In addition, the caller may want to place calls from different devices (e.g., a different microcomputer in a different location); therefore, the authentication process must be capable of being relocated.
- The product must provide magnetic and printed reports of audit trail activity. The following data should be included in the audit log: date and time for all access attempts, the line on which the call entered, entry time, disconnect time, reason for disconnect, caller associated with the call, and system violations or other unusual occurrences.
- If the product in the caller's possession fails, a backup capability should exist that will grant the requester access to the infrastructure. The backup process must be available at all times. The most economical way would be to place a call to the network help desk; the help desk will then grant one-time access on verbal authentication of the requester. This requires a process that will mitigate social engineering.
- The process device must provide controlled one-time access for some individuals (e.g., vendors or customers) that is granted by a remote authority. An example of this feature would be a one-time password generator that would relay the challenge and response over the phone. The central unit issues a random challenge number, the hand-held password generator calculates a response through an algorithm using the personal identification number of the requestor, and this unique response is compared to the central

1. Be aware of the surroundings.
2. Make portable devices inconspicuous.
3. Shred confidential documents before discarding them.
4. Lock portable devices out of sight when leaving them unattended.
5. Hide physical security tokens; do not carry them in the same case with the notebook.
6. Establish a regular schedule for performing appropriate backup practices.
7. Select nontrivial passwords.
8. Change passwords frequently.
9. Follow the common-sense rule to question whether you are in an appropriate place and time to be working with the company's information assets.
10. Most important of all: treat sensitive company information as if it were the combination to your personal safe.

Exhibit III-2-2. Safe Practices for Mobile Users

unit's response for that user. If the two responses agree, then access is granted.

- It must support the establishment of alternate dial authentication hot sites when the primary site goes offline for any reason.

CONFIDENTIALITY

Confidentiality in this section refers to the facilities by which information is protected against unauthorized reading. To facilitate establishing adequate levels of protection for information, data trustees (owners) must provide a classification to all information. This classification is based on the level of damage to the enterprise that may result from allowing individuals to gain access to information that they do not need.

Mobile Employee Information Security Recommendations

Ensuring compliance with appropriate security procedures and practices depends not on the security tools that are provided but on the effectiveness of the security awareness program. Awareness contributes to the success of the effective information protection program. Exhibit III-2-2 provides a baseline of awareness requirements for the mobile user. Posters in the organization's facilities are very effective and are an important part of the overall awareness program. However, use of E-mail and articles in the organization's internal communication media are more effective than posters with mobile users.

Software Considerations

Version Management. If the application software or data base on the notebook is not the current version, the mobile user may create and transmit incorrect data into the organization's record of reference data bases. If this happens, the integrity of the data bases may be damaged and inaccurate information may be given to a business partner.

Each time the mobile user connects into the organization's network servers, a process should be performed to ensure that all of the software (both application and operating system) on the mobile client is current. The same is true for all data base subsets that are mirrored on the notebook. The organization's change control process should notify the synchronization process when production environment changes are made and should have the updates available when a remote connection is made. If the notebook does not have all the current software and data base data, the infrastructure servers must not accept information uploads until the software and data base are synchronized. The mobile user must recreate the information before attempting to upload using the current versions of the processing environment.

When the mobile user connects into the infrastructure and determines that a download of updates is required, the user should have an option to delay the download. The user is not allowed to upload information to the server, but queries may be made. This is important if the user is with a customer and wants to obtain status information; if a potentially lengthy download takes place, the user would waste the customer's time.

Encryption and Decryption. To date, encryption is the most effective security measure to ensure information confidentiality. One type of technology uses a two-part key in which the private key is kept by the owner and the public key is published. The recipient's public key is used to encrypt the data, which can only be decrypted by the recipient's private key. To reduce the computational overhead, encryption is often used to create a digital envelope that holds a DES encryption (symmetric) key and DES-encrypted data. Message nonrepudiation uses document hashing and digital signature as a means of verifying the message sender. This is accomplished by encrypting a message with the sender's private key and letting others decrypt the message with the sender's public key.

The major security concern is maintaining integrity and confidentiality of the keys. Each organization must devise a process to distribute the public keys to everyone who is involved in the encrypted messaging process, including customers and other business partners. The recommendation is to establish a comprehensive public-key data base on a central server that may be accessed by everyone (this means that it is located outside the security firewall) and to have each mobile user keep a subset of public keys on his or her notebook for major business partners.

Another concern is protecting corporate equity. Consideration must be given to the necessity for the corporation to decrypt messages when the owner of the private key is not available. One method may be to include the symmetric DES key (discussed in the previous encryption section) in an extractable format in the message archiving facility (discussed in the nonrepudiation segment). A tightly controlled process to extract the DES key would allow the message to be decrypted without compromising the private key of the originator. Right-to-privacy concerns are outweighed by corporate equity considerations, because company resources were used to create the messages.

NEW TECHNOLOGY CONSIDERATIONS

It is important to have an appreciation for new connectivity technologies so that users may determine their potential threats and vulnerabilities. By looking at what is coming, users should be able to develop the mitigating security measures before deploying the technology. Many security concerns exist, but very few proven answers are associated with emerging technologies. However, the technologist (and to some degree the mobile user) often wants to implement the technology quickly, before the technology itself has reached commercial strength.

PC Card. The unified standard that combines Personal Computer Memory Card International Association (PCMCIA) standard and the Japan Electronic Industry Development Association (JEIDA) standard is called PC Card. The credit-card-sized devices take the form of memory cards, modems, and disk drives that can be plugged into slots in computers. The card's security measures should be the same as those for the hard drive. Experience indicates that many users do not take the card out of the drive when it is not in use; therefore, passive encryption is recommended. For this reason, cards may be ineffective if used as a removable security lock. The card is effectively used in several applications, most notably as a removable modem.

Smart Cards. Although they are the size of an ordinary credit card, smart cards use an embedded processor that gives both the system designer and the system user a powerful authentication tool. Smart cards are a subset of the rapidly growing integrated circuit card industry.

Two types of smart cards exist: contact cards and contactless cards. The contact-type interface uses an eight-position contact located at one corner of the card. A contact card reader also uses a matching set of contact points to transfer information between the card and the reader. The contactless card does not come in direct contact with the card reader, but uses an inductive power coil and transmit and receive capacitor plates to transfer information to the contactless reader. AT contactless card product is essentially an 8-bit computer with a proprietary operating system and either 3K bytes or 8K bytes of user-accessible, nonvolatile memory inside the smart card.

Many applications take advantage of the strong authentication capabilities of smart cards. The most common application is electronic money. Another example of smart card flexibility is in a building security system protected by card access that otherwise requires a large network of cables connecting the door reader, door controllers, and host computer. By converting systems to a smart card system and using the onboard data base and encryption capabilities, the miles of network cabling and host computer may be eliminated. However, the need for a special card reader coupled with the mobile user's desire to travel light all but eliminates the smart card as a practical security mechanism in today's mobile arena.

Cellular Digital Packet Data. Cellular digital packet data (CDPD) technology is rapidly building its infrastructure right alongside the traditional analog cellular infrastructure. In fact, much of the existing analog base stations are being used. Similar to most public communications efforts, a coalition of common carriers have cooperated to ensure interoperability. CDPD was developed by a group of major cellular communications companies. This prestigious group makes it clear that by leveraging existing technologies and infrastructures, they will have a nationwide network available within a very short time.

As the caller moves from one analog cell to the next, the ability to transmit data digitally means faster transmission speeds and a solution to the current problem of lost or repeated transmission. The user will be able to send and receive data while in a moving car.

Technologists believe that because existing common protocols are used, today's applications can use CDPD without modification. The one drawback that may impede CDPD deployment is the development of effective modems that support CDPD.

Three types of CDPD services are provided. CDPD Network allows subscribers to transfer data through their applications as an extension of their internal network. In CDPD Networked Applications Services, the cellular provider provides specific application services like E-mail, directory services, and virtual terminal services to subscribers. CDPD Network Support Services provides network management, use accounting, and network security. The security practitioner should not assume that cellular carrier's security interpretation or objectives are the same as those of the organization. The carrier's direction is intended to protect its investment in the CDPD Network and principally to ensure that only authorized paying subscribers use the network. A secondary concern is providing data privacy. It is not practical for the carrier to comply with each subscriber's security policies.

The user should already be aware of the security risk associated with cellular transmissions. Cellular (analog or digital) transmission is a miniature radio station broadcasting to everyone who has the receiving equipment. Digital transmission will be able to scramble the transmission by channel hopping, which makes interception more difficult but not impossible for the motivated eavesdropper. Therefore, the best solution to maintain data integrity and confidentiality is through digital signaturing and encryption. Because of the administrative overhead associated with key management of symmetric keys (the same passphrase is used to encrypt and decrypt the message), public and private key encryption is recommended.

Wireless Communication. Although cellular communication is a desirable tool for mobile users, wireless communication may be a valuable capability for those roving from place to place within the confines of their own building.

Wireless networking employs a number of different methods. One of the most popular of these methods is spread-spectrum radio. Wireless adapters connect into a computer either internally or through the parallel port, and they then communicate to a base station or what could be called a wireless hub. The communication receiving area of the base station is usually several hundred feet, and the microcomputer and adapter may be placed anywhere within that radius. Other methods use a line-of-sight transmission technology.

Because the transmission takes place within the confines of the company's buildings, the security requirements may be fewer. However, depending on the organization's type of business and the need for confidentiality, consideration might be required to neutralize the motivated eavesdroppers who may be within the company or stationed outside the facility.

Other technologies use radio frequencies to forward messages to a central station, which in turn (when requested) sends the message to the recipient. This method is known as store and forward, and it is not normally used for interactive messaging. The best answer to ensure confidentiality in all wireless applications is encryption.

The World Wide Web. The World Wide Web (known as the Web or WWW) provides an infrastructure for accessing information. The Web provides a simple means of attaining almost any type of information that is available on a Web server that is attached to the Internet. All information on the Web is stored in pages, using a standardized hypertext language. Many companies use the Web to provide timely information to their customers. They typically provide information about products, upgrades and patches, and feedback areas. The Web is considered a companion to E-mail because it provides information by using an interface. However, the Web is designed to accommodate limited two-way communication.

Security considerations must be the same as with all other Internet connections. The Web server should be placed outside of the firewall. If confidential information is placed on the Web, then encryption should be used. In general, the Web architecture does not provide for the integrity and confidentiality of information. Access to the infrastructure should not be allowed through the Web server.

CONCLUSIONS

Security procedures, guidelines, and practices accepted by end users must enable them to do their jobs. If end users interpret security to be a roadblock, they will often find ways to circumvent security requirements. To ensure that this does not happen, the security practitioner should spend time learning the problems and security concerns of users. The practitioner should consider scheduling one day per month to stay at home and telecommute in addition to dialing in while on business trips. This practice enhances understanding of the remote access conditions, and assists the security practitioner in developing more effective security practices.

Information today is stored not only in the data center but also on desktops, notebooks, and home computers; it is stored wherever mobile users have taken the data. By understanding the implications of this fact of business life, practices can be better established to secure assets while supporting employees' requirements to perform optimally and competitively through having access to the most current information and computing power.

III-3

Distributed Network Support and Control

Ira Hertzoff

DURING THE 1980s, distributed processing was performed by a client using files and programs stored on a server. Today, distributed processing is split between the server and client to offer desktop power, more efficient network resource use, and better control of network traffic. As these open systems grow in importance, supporting them becomes a networkwide process as opposed to a node-based operation.

Clients on networks using structured query language (SQL) data base servers can access a common data base. Transparent access to data shared between applications and to data bases on multiple servers is provided by widely used front ends (e.g., Lotus 1-2-3, Paradox, and Excel).

In the layered networks of the 1990s, mission-critical data is stored on corporate hosts; local data is stored on multifunctional servers providing back-end SQL data base services; and dissimilar clients run multiple front-end applications. Numerous paths into the network permit access to data from both internal and external sources. Distributed networks can be compromised from remote locations and their security seriously jeopardized if a single node is accessible. Distributed network security now depends on how the desktop computer is used and how desktop power is managed. No operation is safe unless all connected operations are secure—protecting only the host is no longer sufficient.

THE NEED FOR PROTECTION

Appropriate procedures and techniques are necessary to protect the network from external or internal manipulation of its programs, data, and resources. An effective program protects a network from physical destruction, unauthorized modification, or disclosure of software and records.

High-technology network penetration is a glamorous subject and certainly a possible threat, but low-tech penetration is much more common and much more dangerous. Intruders who take advantage of procedural weaknesses represent a greater threat to distributed networks than do industrial spies with, for example, emissions-sensing devices. Protection costs should be balanced against risk reality. The IS manager can offer cost-effective protection against real risks by carefully rating the problems that can occur. If the maximum potential risk is from a disgruntled employee, the IS manager should not specify distributed network security measures to protect against a resourceful and determined adversary (e.g., an unfriendly government).

The Protection Decision

There are three ways to protect a network against risk: self-insurance or risk retention, insurance against economic loss, and installation of security measures.

Insurance. Protection measures should be selected with the assistance of the corporate risk or insurance manager. Large enterprises commonly self-insure for minor risks, obtain outside insurance against larger risks, and use security measures when they are less expensive than insurance. Insurance is a proven method of managing risk and should always be part of a network security plan. A corporation can elect to risk an event and make no insurance or security provisions; if this is a planned rather than accidental decision, it is a valid approach.

SECURITY MEASURES

Distributed networks require three types of protection: disaster recovery, physical security, and data security. Because these areas overlap, a single well-selected security measure can offer protection in more than one. Distributed networks, because their elements are geographically separated, are inherently protected against physical destruction. Distributed networks do not require the complex backup procedures for disasters that single-site operations must have, because the distributed architecture is based on distributed servers and the IS staff is trained to automatically back up files.

Geographically distributed systems pose security risks because if processing does not take place in physically secure computer rooms, desktop computer users can create openings into the network, and software can be introduced into the network from many workstations. Preventing viruses, worms, and the unauthorized copying of software and files requires proper administrative procedures.

The distributed network needs in-depth protection involving multiple layers of defenses. Any single security measure can be improperly installed, be bypassed, or have flaws; the use of multiple security measures protects against weaknesses in any individual measure (see Exhibit III-3-1).

Unplanned shutdowns and restarts of any programmable network device must be considered a breach of security and investigated. In particular, operating system restarts should not be allowed or accepted as normal operating practice.

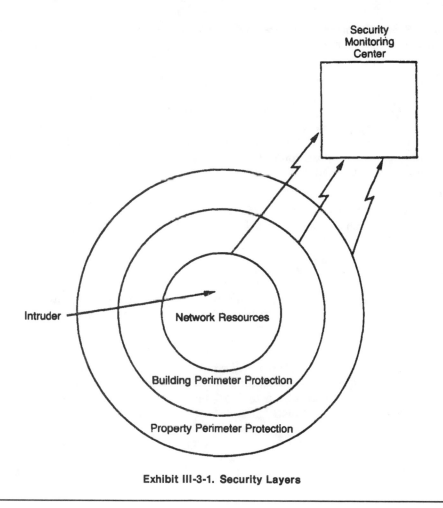

Exhibit III-3-1. Security Layers

The Auditor's Role. The effectiveness of any security plan must be proved by ongoing reviews conducted by technically competent auditors. To ensure their independence, these auditors must not have any other operating responsibilities. If there is no existing program for conducting audits and training auditors, or if the existing program is inadequate, an effective program should be established.

Multilevel Processing Security. Because various classes of work are usually performed on the same network, processes requiring different levels of security interact. (When all work on a network is at the same security level, processes do not need to be compartmentalized to prevent a user from accessing another user's data.) Security requires that these processes interact only in a known, predictable, and authorized manner.

Secure Processing Over an Insecure Network. The public telephone network and many private communications facilities are inherently insecure. The telephone network is universal (with many points of access and exit) and has well-known security weaknesses. Any secure process that operates over such links must use handshaking techniques to verify that only the desired parties are in communication. Cryptographic techniques applied to the data ensure that data cannot be interpreted if it is intercepted.

Risks of LANs as Gateways to Hosts. When a LAN is used as a front end to a secure mainframe, a serious LAN security breach will compromise the mainframe's security. Such breaches can occur because both secure and insecure operating systems run on the same network and because of the nature of workstation intelligence. A user can program a personal computer to set up a connection through the LAN to secure host data bases. Anyone with physical, logical, or remote control access to that personal computer can log into the host as a user. Gateways should be evaluated by how they support network management and security.

LAN Security. Many LAN security systems are inadequate because they were designed for work group automation and are the product of a single-user design mentality. One indication of a poor LAN operating system is the availability of many add-on security products. Access rights tools should be an integral part of the operating system.

The use of LANs as host gateways is economical, but many LAN operating systems expose host passwords by letting them reside in buffers or making them available for decoding by protocol analyzers. In both LAN token-ring and Ethernet networks, each station sees all LAN traffic. A true star topology is therefore superior. Personal computer-based LANs suffer from the weaknesses of their DOS operating system and are prone to virus and worm attacks. Diskless workstations prevent users from loading unauthorized software. Requiring users to load local operating system software from the network reduces the risk of infection.

DISASTER RECOVERY

A network must function under all normal and anticipated abnormal conditions and must be robust enough to survive the unexpected. The IS manager must provide the systems, staff, and procedures necessary to ensure that a disaster—intentional or otherwise—does not cripple the organization. As mission-critical applications become more time-dependent, the acceptable time for restoration of service is decreasing.

Accurate recordkeeping in a configuration control data base with connectivity information is fundamental. This type of network management can be the core of a disaster plan generated as the network control staff performs daily operations. Problem management tools can simulate outages, and change control tools can track recovery operations. IS managers responsible for maintaining distributed networks must integrate disaster planning into their daily operations—not as a separate project but as a large problem that requires immediate action and resolution.

Reserve Capacity

To protect data against disaster, the IS manager can, at different locations, place multiple servers with reverse capacities so that each server can absorb additional work loads as necessary. This dual-operating capability should be part of installation planning so that critical applications operate at two or more sites.

Disaster recovery of a distributed network is a network process. Under both normal and abnormal conditions, the network should use its interconnections and facilities to support recovery of failing network elements. SQL data bases, mirror-imaged at different points on distributed client-server networks, can be designated to allow automatic recovery.

THE NETWORK AS A BACKUP SYSTEM

With enough network bandwidth, critical files can be stored on several servers. Provisions to switch network control should be made if an event disables the primary control point. In large networks, management domains must be established and plans developed for switching control if a domain management point is unavailable. The swapping of network control should be a normal and periodic procedure familiar to the staff at all sites.

The process of restoring an operation includes recognizing that a disaster has occurred. This function is best handled by the network management center. Incomplete information generated by inadequate tools can delay action until the problem becomes uncontainable. Disaster recovery resources should enhance the network management system rather than create disaster recovery systems that operate in parallel.

Backup Types

Backups are file copies used to protect against errors; they are recycled on a regular schedule. (Archival backups may be retained permanently.) The process of recycling media is called rotation. Backup media can be recycled on a generation basis or on a longer cycle. A distributed network with enough bandwidth can be backed up to a server in a central location. Central backup ensures that remote devices are backed up according to plan and taken off site.

PHYSICAL SECURITY

A network's physical security can be penetrated and seriously compromised in many ways, for example, through the destruction of or damage to buildings, cables, switches, power sources, computer hardware, computer files, and programs.

Seals

Security seals are attached to equipment to detect tampering or entry. A damaged seal indicates that a unit has been opened. Managers of large networks should use prenumbered, bar-coded seals. Each seal number should be recorded in a data base and assigned to a technician. The number on the seal should be then linked to a specific machine or component.

When a seal is broken during an authorized repair, a new seal should be installed and the number linked to the old seal number, the repair technician, and the actions taken. This approach keeps the configuration data base current. It allows the IS manager to positively identify each component, which technician worked on it, and whether it was altered by anyone—authorized or unauthorized—from the state recorded in the configuration data base.

Hot Work: Flame, Electrical, and Plumbing Service Control

Effective security practice requires any employee or contractor working on critical systems to obtain proper clearance—a hot work permit or license. The license ensures that responsible managers know when work that can cause a fire or cut power or water is being performed. Permits should be required for all work on communications lines and power sources and for construction that can affect lines. In some cases, a prudent manager may assign a staff member to follow the worker with a fire extinguisher. Avoiding extended downtime makes this precaution a wise investment.

CONTROL OF EXECUTABLE CODE

The IS manager should control the loading onto the network of all executable code to reduce the risk of infection from computer viruses and worms and from unauthorized modification of programs. On a distributed network, all paths that can result in code execution must be blocked, including unintended paths. On a DOS machine, for example, using the Type command on a text file with embedded ANSI.sys commands results in execution of the commands. In this case, a text editor is safer than using Type when checking suspect files. Other unintended paths include service diagnostic disks that can carry viruses and worms from machine to machine; problems can also be caused by a failure to maintain a network operating system software at current revision levels.

Remote Library Control

Effective control practice requires valuable assets to be concentrated and placed in a protected environment. To protect executable code and programs, data center managers must ensure that all network programs are kept in program libraries, which are stored in directories with controlled access.

Authorization to make changes or add programs to the library and the ability to authorize such changes should be assigned to different individuals. Activity journals on each file must be kept to verify accountability. Library data structures should be maintained to improve management insight and control.

ACCESS CONTROL

Access controls limit a particular user's access to specific network resources. One method is to request information that only one person knows (e.g., a password). User IDs, authorization tables, and access rights lists can also be used to control access. Such controls (as audit trails and alarms) are essential to protect against authorized users and should be implemented in the software that controls the program library as well as in the software that prevents unauthorized access to production data. Various access control methods are discussed in the following sections.

Handshaking

Network users should be informed of the last use of their ID and password when they log on the network. They should report any suspicious activity to management or be held responsible for the results of this activity.

Lockwords

Lockwords permit users with knowledge of the file code to access that file without verifying their identity. Multiple resources require multiple lockwords, a problem that, on some LANs, is counteracted by writing batch files to issue lockwords. This practice results in very poor control, and it is difficult to change lockwords. The use of lockword security is unacceptable and is particularly dangerous when LAN gateways to hosts are used.

Passwords

Passwords use access control as their primary security mechanism. Password secrecy is needed to prevent system access, but passwords can be compromised in many ways, the easiest being user carelessness. Procedural controls on the contents of passwords—restrictions on using common words or names—and forcing users to periodically change their passwords are essential. Real-time monitoring of log-ons to discern the use of password detection programs should be continuous, as should logging of unsuccessful attempts into an audit trail.

COMMUNICATIONS SECURITY

Data interception is a concern in organizations with valuable or sensitive data. Data can be intercepted by wiretapping or emission sensing, by unauthorized access to data, and by authorized or unauthorized terminals. Communications security is the key to protecting against data interception. To be effective, communications security must be enforced in five areas: line, transmission, cryptography, emissions, and technical. These areas are discussed in the following sections.

Line Security

Line security involves protecting telephone lines against wiretapping or other types of interception and employs the same techniques used for protecting voice conversations against interception. Often, PBXs and other premises-based equipment are not properly protected. Intruders can enter them in the supervisor mode and, with the right resources, capture and analyze their data stream for user IDs and passwords. The intruders can then use these identifications to gain access to the network.

The IS manager should double-check that PBX security is adequate. Because protocol analyzers and other diagnostic devices can discover passwords and cryptographic keys, their use should be restricted. It is preferable to employ high-level diagnostic programs to identify failing components.

Transmission Security

Transmission security is the protection of network communications against their interception by workstations authorized to access the network but not particular data. It provides protection against the use of authorized devices for unauthorized purposes.

Cryptographic Security

Cryptographic security protects the keys, encryption software, and hardware used for encoding transmissions. The cryptographic system must operate at a level higher than the highest-level information the system protects. Just as a lock does not protect against a stolen key, cryptography does not protect against a lost or intercepted cryptographic key.

Emissions Security

Emissions security protects against the interception of computer emissions (e.g., electrical fluctuations, radiation, or other modulations), by an adversary who can determine what is being processed. This risk is posed primarily by an adversary who is both resourceful and determined (e.g., a foreign government or determined competitor). Emissions-suppressed equipment and cables are available to protect against it.

Optical fiber does not radiate much and is not prone to emission leakage and interception. Light does leak at cable bends, however, and emissions can be read by interception equipment that is very close to the cable. However, the skill required for tapping optical fiber is very high and the risk of discovery is great. Fiber is useful in areas prone to electrical storms because it is nonconductive and can isolate equipment from voltage surges.

Technical Security

Technical security is a broad phrase used to describe protection against non-computer intrusion devices. Some known devices include:

- Microphones or free-space transmitters (e.g., radio taps).
- Devices built into equipment (e.g., modulators).
- Carrier-current bugging equipment (e.g., power-line modulators).
- Visual or optical surveillance equipment (e.g., video cameras).
- Telephone bugging devices (e.g., infinity transmitters).

DANGERS OF PRIVATE SECURITY

Coordinated networkwide security is necessary to protect against the conversion of corporate resources to private use. It is difficult to determine whether user-developed security measures are adequate or well implemented, and networkwide measures are needed to ensure that the network is not vulnerable to its users.

Control of Client/Server Processes

Host-to-host links over the public network require multilevel security, which must be provided with access control. Network services needing multilevel security are mail, file transfer, and remote log-on. Network applications should be usable without modification.

In UNIX, for example, users should have access to any network application usually provided on the UNIX host. The objective is to create a user-transparent wall around network applications that permits initial setup of the connection to another host but does not allow application-to-host communication until security is ensured for both systems. Users must be allowed to enter passwords, but password interception must be prevented. A verified trusted path mechanism, with network accesses that are audited at the called host, is desirable.

NETWORK MANAGEMENT INTEGRATION

Network management control centers are similar to central station burglar alarm control centers. Alarms in both are forwarded to central monitoring points where computer programs either take action on them or display recommendations for action. Network management control centers monitor line conditions and check for device failures. Message displays indicate if, for example, there is an outage, an intruder is testing passwords, or unauthorized access has been detected.

New Risks

As network management interfaces are standardized according to internationally accepted models, the possibility of an attack on the management system increases. Reconfiguring routers can effectively destroy communications. In assessing the security of the network management system, the IS manager should ask:

- What are its message authentication procedures?
- Should network control messages be encrypted?
- How are trusted users controlled within the system?
- How are external messages from linked networks validated?

Protecting Against Trusted Users. Perhaps the most difficult security problem is caused by authorized users doing unauthorized work. Every authorized user should sign a standard statement indicating that access to the network is granted for approved business purposes only, that they will limit their use to those purposes, and that they will adhere to network security practices.

The audit trail and trend analysis of audit logs are used to detect use changes of authorized users. This is a task for independent, technically competent auditors who should periodically review use patterns. However, this is primarily an after-the-event check. To turn it into a deterrent, these audits must be publicized, along with the corporation's intention to prosecute when users convert network resources to their own use.

The Super-User Problem. Most security systems have a super-user who can start and stop the system. The ID and password of the super-user should be used only when absolutely essential—in other words, the person or persons having super-user authority should also have less powerful rights under a different ID and password. Super-user IDs and passwords should never be used when a less dangerous method of completing a task exists.

Network Administration. The policy of separating IDs for security management from IDs for daily use also applies to network administrators. On LANs, a log-on from a machine with a corrupted hard disk can propagate through the network if the user has management privileges. The risk can be minimized by restricting where and how the management ID is used. It is safest to execute management tasks from a console, not a workstation.

Viruses, Worms, and Contaminated Hard Disks. Conditions that do not seriously affect mainframes (e.g., viruses and worms) are hazardous to distributed networks if proper procedures are not followed. Total protection is impossible, but the risks can be controlled. The key is to control how executable code gets on the network. This can be done by restricting the ports of program entry to the minimum, installing quality-control mechanisms, and educating end users.

Control of unintended portals is more difficult. Software must be kept at current release levels, and new software must be audited before installation. Users with administrative rights must be instructed to protect their boot diskettes and to avoid logging on from unknown machines when using their administrative rights. Then practical, administrative users should work only at their terminals.

CONCLUSION

The prerequisite for distributed network control is a stable, well-managed network. To achieve this, IS managers must ensure comprehensive data security as hardware and software implementation decisions are made and as network processes migrate from host-to-terminal environments to peer-to-peer environments, in which computer resources are often located in open user areas under the control of nontechnical managers.

III-4

The Consultative Approach to Client Support

Kate Nasser

IN THIS AGE of technology cost containment, fluid information-based organizations, and cuts in training budgets, productivity is key; embracing a productive customer philosophy is well worth the effort. The IS department's biggest hurdle in implementing this philosophy is a lack of customer trust. To earn it, support staff members must understand their end-user customers' business needs and expectations and, just as important, exhibit that understanding. Awareness is not enough; trust is built first by listening and then by visibly and audibly illustrating understanding. Without first building a base of trust, end-user computing (EUC) support groups cannot implement a productive customer philosophy.

This philosophy does not sanction discourteous or patronizing behavior that is sure to make the customer unhappy. The manner in which the problem is solved or the request is filled remains important with the productive customer philosophy. The difference is that the problem solvers using a happy-customer philosophy do not always use their expertise to find the solution that will make the customer productive. Instead, they yield to the customers' definition of what will make the customer happy. The staff members using a productive-customer philosophy start with the customers' trust in them and their respect for the customers' business needs. From there, customers and staff work in partnership to find the best solution.

To bring customer service excellence to end-user computing, the IS manager must exemplify to the support staff the customer service philosophy they need to communicate to and use with end users. Defining a customer service philosophy is the first essential move toward service excellence. If technology support groups overlook this phase, miscommunication with end users is likely to follow, resulting in an antagonistic and uncooperative working relationship.

Defining the philosophy involves knowing whether the support group's primary customer service goal is to have happy end-user customers or productive end-user customers. Although having both would certainly be ideal, it is not always feasible. For example, an end user might call and request a word processor that is different from the standard everyone else in the organization is using.

The happy-customer philosophy that says the customer is always right would lead the support group to automatically submit a purchase order for a new word processor. The productive-customer philosophy would lead the support staff manager to ask the customer a series of questions about his or her technology needs. (In requesting a new word processor, the customer has not articulated a need but has described his or her proposed solution to it.) This needs analysis must be done in a timely, action-oriented manner so the end-user customer knows that the questions are not just another IS bureaucratic stall.

HISTORY OF CLIENT SUPPORT

The prevalence of a happy-customer philosophy has its roots in the history of end-user computing. The introduction of microcomputers came at a time when non-IS staff (i.e., end users) had reached their limit on long IS backlogs and slow responses to requests. IS at that point was focusing neither on happiness nor on productivity.

Even as IS and end-user customers were both approving large numbers of microcomputer purchases, IS did not analyze the total support needs and infrastructure demands of using those computers. IS allowed end users to do whatever they wanted until they had a problem. The stage was set for a reactive troubleshooting approach to end-user computing and a happy-customer philosophy.

Because IS did not take microcomputers seriously or see the financial and business impact of computing without standards, procedures, or expert input, it established a precedent of nonpartnership with end users. In fact, end-user computing support groups were not even part of IS in the early years. The EUC philosophy was to be responsive by giving customers what they wanted—as quickly as possible. Their goal was to distinguish themselves from the slow-moving, unresponsive IS departments by focusing primarily on customer happiness.

MAKING THE TRANSITION

Organizations have changed dramatically since the introduction of personal computing. Networks alone have brought legitimacy to standards for the desktop. Client servers now hold and run critical applications. Access to information highways is becoming a strong factor to competitive advantage. Compatibility and accessibility among applications, functions, and departments are tacit if not explicit end-user customer expectations. Customers expect their end-user support groups to provide expert guidance—even though these same customers may complain if the answer is not what they initially wanted to hear. The stage is now set for a consultative approach to end-user computing, a shift from the historically reactive setup.

Making the transition to a consultative approach to client support requires the IS manager to learn and understand the differences among needs, expectations, problems, and solutions and then train the support staff to make the same distinctions. The following case study illustrates these differences.

An end-user customer, Pat, is a marketing manager with responsibility for key product launches. Her department is a fast-moving unit that needs a lot of support. Pat is dissatisfied with IS service and is meeting with senior IS staff members who support marketing. She complains, "My department has problems using the company's systems and technology. When a crunch hits, the technology seems either to break down or my department cannot make it do what is needed."

IS responds that its training courses explain how to use the technology and systems. Pat answers that her department does not always have time to refer to the training manuals and follow them step by step, especially right before a product launch, and these launches are central to the company's business. IS replies that it is understaffed and cannot always respond immediately to Pat's department's requests. The discussion continues with no resolution, and the marketing department eventually goes outside the company for technical guidance and support.

This case study did not have to have an unhappy ending, however. IS could have adopted a consultative approach. If, during discussions with Pat, IS managers had asked themselves the following questions, the encounter would have gone much differently:

- What are Pat's needs?
- What are the marketing department's needs regarding technology?
- What problems is marketing facing?
- What are marketing's expectations, and how can IS best address those expectations?
- What are the solutions to marketing's needs and problems?

In the previous case study, the managers were looking for the end user to define the problem and maybe even the solution. Because IS was exploring very little and listening even less, customer service during this discussion was very poor. Without effective questioning and objective listening, IS cannot solve Pat's problems, nor will Pat feel that the department has provided excellent customer service.

By using a productive-customer philosophy, the IS staff could be much more consultative in discussions with end users. A consultative dialogue would enable support people to determine needs, problems, expectations, and solutions. In the previous case study, IS established a position based on what it thought marketing was requesting. If the staff had listened effectively, the conversation would have gone something like this. Pat would have said, "We have problems in using the systems and technology we already have. When the crunch hits, the technology either breaks down or we cannot figure out how to get it to do what we need."

IS would have responded, "We have some logs with us from the calls your group has made for help. Could we discuss a few of these calls?"

Pat might have replied, "Good. This first one noted here was two days before our launch. We were gearing up and producing slides for our key customer campaign. The graphics package would not let us change style backgrounds from one slide to the next. The technology would not do what we needed, and we could not stop at that point to take a training course."

IS might have answered, "We understand your frustration. Certainly, a training course at that point is not appropriate. One of your people did take the training course last year. Evidently you are still experiencing some difficulties?"

At this point, Pat would probably say, "Yes, but can't your staff give us help when we need it?" IS would reply that it certainly wants to help marketing and it has a few ideas. Does marketing have a schedule for its launches—those crunch periods? If IS could plan for these periods, it might be able to put a staff member on call for marketing's after-hours requests.

In addition, IS might offer to meet with marketing before the launch to understand the functions it would be trying on the system. A staff member could coach marketing staff on how to perform those functions. If marketing needs hands-on support, IS could explain these needs to a contractor and provide figures on how much a contractor would cost.

The differences between the two versions of this meeting are striking. Both happen every day in organizations throughout the country. The first scenario is plagued with assumptions, disclaimers, and traditional claims about being too busy to provide the service requested. The second version is exploratory and searches to define needs and expectations before deciding on solutions. The IS group neither refuses Pat's solution to just send someone down when they need help nor accedes to it without comment. The IS manager negotiates for a proactive solution to gain information about launch support needs. In addition, the manager mentions that if hands-on support is essential, the support group can coordinate and oversee contract help. The staff members are providing value because they understand the marketing department's real need.

The answer to the first question (What are Pat's needs during the discussion?) is simple. Pat needs to see the technical staff involved and committed to making things better. Furthermore, Pat will be more likely to explore solution alternatives if IS shows more interest in finding workable outcomes.

The answer to the second question (What are marketing's needs regarding technology?) is that they need technology that helps them have a successful launch or at the very least does not hinder a successful launch. The problem marketing is facing is using the technology during crunch times—when it is critical that they be able to use the technology. Marketing's expectations are that the IS support group will help them use the technology they procured for them.

The best way for the IS department to address marketing's expectations and find a solution to marketing's problems and needs is to realize that help does not have to be hands-on during a crisis. In fact, the most effective help prevents the crisis in the first place. The IS manager can offer a proactive solution during a tense meeting in which Pat, the customer, is focusing on the next expected crisis. The happy-customer philosophy would give Pat whatever Pat suggested. The productive-customer philosophy allows the experts to offer other solutions to meet the customer's needs and prevent crises. Knowing the difference between a need, a problem, an expectation, and a solution is key to excellence in customer service.

Addressing all aspects of marketing's needs is another key factor in service excellence. The meeting just outlined could have lasted much longer and investigated issues of business challenges, technology training, and other outside solutions. In this case, IS will bring focus to marketing's business challenges during the next meeting and beyond.

IS will build rapport, credibility, and trust by using a consultative approach in such meetings and by delivering expert, timely solutions.

To achieve a consistent reputation of customer service excellence, IS must address other business units in the same manner. At no time should IS act as if its purpose for existence is technology. Customer-focused end-user support, the productive-customer philosophy, is the future of IS.

MARKETING AND IMPLEMENTING CUSTOMER-FOCUSED SUPPORT

Step 1: Understanding Customer Needs and Expectations

Customer-focused support begins with IS working to understand customer needs and expectations. End-user feedback sessions, which are a special type of focus group, help do this in two ways. First, by their very design, these in-person sessions focus on customers and communicate the IS manager's interest in providing excellent service. Second, the sessions allow for much clearer understanding of end-user customers' needs and expectations than paper or electronic surveys. For example, an end-user customer statement such as "We need immediate support for all the in-house application systems" is vague. What does the term *immediate* mean? Does the word *all* mean all the ones they use, all the ones ever coded, or all the ones they see as critical to the business? In-person feedback sessions allow for immediate clarification of end-user requests. Such clarifying questions eventually help dispel the myths that end-user customers are unreasonable and want the moon for no cost. Consultation, partnership, influence, and negotiation seem real and feasible during and after successful feedback sessions. Planning, structure, and expert facilitation are critical to the success of these sessions.

The planning starts with the IS manager and key staff brainstorming about how they believe end-user customer departments view current service and support. This step generates possible caution signs to consider in running the sessions. It also allows the IS manager and staff to sense which end-user customer departments are their allies and which oppose them.

At this point in the planning, the IS manager should formulate three open-ended questions to pose in every feedback session. Standard questions from which to work include:

- What support services does the end user find valuable?
- What services are not valuable?
- What services does the department need that IS is not providing or procuring?

The important components of the session are timing, notetaking, and communication. A 75-minute session with 25 participants usually works well, and assigning two scribes to take notes on the proceedings ensures more accurate capture of the feedback. One notetaker should work from a flip chart to make the notetaking visible to the participants so they will know IS is listening. The IS support group can review these flip charts later. The second notetaker should be typing directly into a computer. Flip charts are for summarizing an idea, and word processing is good for noting details IS can use after the sessions for developing its new customer-focused support processes.

The session needs rules about communication; otherwise it can deteriorate into an endless voicing of complaints that serves only as a therapy session for venting dissatisfactions. The meeting leader, which may be a specialized facilitator or the IS manager, opens the meeting with purpose, goals, exit outcomes, and suggested communication formats. A wish-list format requires participants to pose their opinions in action-oriented suggestions for IS rather than in broad accusations.

IS support staff members should attend the session to listen and to understand end-user customers, and they must follow these guidelines. The guidelines should be covered in a separate preparation session held for IS members before the feedback sessions. The main rule is that IS support group members should not participate verbally except to ask clarifying questions or respond to nonrhetorical questions from participants.

These guidelines prevent IS staff from becoming defensive during the feedback sessions and taking time with attempts to explain why they cannot fulfill individual requests.

When running multiple feedback sessions, the IS manager and support staff—as well as the outside facilitator if one is used—should debrief each other between sessions. Staff need a chance to vent any frustrations they felt but did not voice during the previous session. Having done this, they are more prepared to listen during the next session. In such feedback sessions, IS support staff who are used to hearing mostly negative feedback are typically amazed at the amount of positive reactions they receive. These sessions give end users a chance to verbalize their needs and hear what other end-user customers need. Fears that hearing other end users' needs would escalate all end users' expectations are typically unrealized. Instead, the sessions show users the true scope of demands the IS support department is juggling and attempting to meet. More realistic expectations are likely to follow.

Step 2: Analyzing the Data

Whether these feedback sessions or some other method is used to help support groups understand current and near-term customer expectations, analyzing and acting on the data is crucial. End users assume and expect changes to follow such inquiries about their expectations. If the IS department does not intend to explore new strategies for delivering service and support, it should not perform any kind of inquiry. Surveying needs and expectations with no visible attempts at change will simply solidify end users' views of IS as a nonresponsive bureaucracy. It may also reduce the customer response rate on future inquiries and surveys.

As the support group analyzes the data collected, it should consider the following points:

- There are many contributing factors to each response.
- Cause-and-effect deductions cannot be made from response data alone.

- Low-frequency responses are not automatically unimportant. They may be critical to the business of the organization.
- The responses should be scrutinized for misdefinitions and the differences between needs and solutions.
- The customer trends that indicate high dissatisfaction and high satisfaction should be carefully charted. Emotional components to these responses provide opportunities for early change and marketing of strengths.
- The organizational factors that contributed to these points of high dissatisfaction and high satisfaction should be listed. A short, targeted meeting to create this list produces the best results. After the list has been created, it should be put aside for two to three days and then revisited in the first of IS's changing strategy meetings (which are discussed in the following section).

Step 3: Thinking Creatively in Client Support Strategy Sessions

Many IS professionals argue that if they could change the way they are providing services, they already would have. It is difficult to transform the way people think about technology service and support; however, it is not impossible. Changes in attitudes and thinking almost always occur when something or someone helps people see the particular situation from a different perspective.

Feedback sessions are one mechanism for reaching this new insight. Customers' discussion, examples, and frustrations may help the support group envision that different way of providing service. Assigning IS staff members to rotating stints in end-user customer departments frequently helps as well. Yet, a more concentrated effort may be necessary after the feedback sessions are over. Within six months of the feedback sessions, participants will want to see some change—however small.

Creative thinking and problem-solving techniques in a series of strategy meetings can bring IS staff to the new insight that is required. Many exercises have been designed to help support personnel break through mental blocks and status quo thinking. The IS manager may find that an expert facilitator can help create momentum and initiate creative thought in the first few meetings. Expert facilitators can be found in an organization's human resources department or outside the organization. Facilitators need to be impartial and unaffected by the decisions that are made in the meeting. End-user customers and IS support staff members are often too close to the issues to be objective facilitators.

The following practices are useful in running a strategy session:

- Questioning what has always been assumed to be fact.
- If IS support staff total six or more, breaking into pairs or trios for five-minute creative bursts and then sharing the results as a total group.
- As a whole group, expanding on the ideas from the creative bursts.
- Noting any comments on roadblocks to each idea.
- Having pairs argue in favor of ideas they believe will not work. This forces people to see the ideas from a different angle.
- Structuring each strategy session to allow time for introduction, creative

bursts, discussion, and closure on at least two ideas. Each session should be a step closer to building the new customer-focused strategy.

- Resisting the tendency to analyze and discuss without coming to closure. IS divisions often make this mistake. It may be the analytical personalities, the belief that a perfect solution exists, or a basic resistance to change that produces this detrimental behavior. In any case, the objective facilitator can help IS move past endless discussions and toward action.

- Visibly posting the reasons for changing or transforming the IS support group during every strategy session along with the question: What about the customer? Support staff members can look to these visual cues to avoid slipping into status quo thinking.

Communicating the New Strategy

At the end of the strategy sessions, IS will have a new mission statement that reflects its primary purpose for existence and its customer service philosophy. From that statement should flow new or revised processes, procedures, and service-level agreements. Once these changes have been drafted, a pilot to test the new strategy should be undertaken to determine how well it works.

If the new mission statement accurately reflects the overall business technology needs of the end-user departments, a trial implementation of the new service strategy is all that is needed to determine necessary modifications. Endless analysis, discussion, and planning do not test the efficacy of new procedures and processes.

For a pilot program to be effective, the support department must make sure end users are aware that the new strategy is in a pilot phase. This does not mean, however, that the program should be implemented in a lackluster manner. IS must be confident about its new strategy and simultaneously open to end-user feedback on the new strategy. In fact, the IS manager must solicit feedback at regular intervals to determine needed changes along the way.

The timing of a pilot is also important. Many departments want to stress test their new service strategy during peak periods of customer use. Although stress testing is an excellent idea, it should not be the first thing IS does in a pilot. End users need time to adjust to changes, and IS support needs an initial test of its understanding of business technology needs before it conducts a stress test.

An initial test requires introducing a new service strategy six weeks before an expected peak period. This provides time to modify processes and an adequate window to motivate end users to work with the support group in the short term to minimize the potential for crises during the peak period. If new technology platforms or systems for customer use are part of the new strategy, the timing of the initial test must be adjusted according to technical demands.

Communicating the changes to end users is critical to a new strategy's success. IS managers must be creative to catch the attention of jaded end users and to signal that changes really are in the works. The manager should improve the standard communication mechanism in place in the organization.

An option is to conduct kickoff sessions that end users attend during lunch or before work. Attending this event enables customers to meet the IS support staff, see the new technology they have been requesting, or attend a software package clinic. During the kickoff, the support staff can distribute information highlighting the new service strategy and processes. These highlights can be outlined in repeated 10- to 15-minute briefings. This event can also be customized for each end-user department and presented in each unit's area. Above all, the support group must reach out and market changes in the best possible light by showing confidence and enthusiasm.

Step 5: Phasing in the Final Plan

The pilot is a short, concentrated period to test the new strategy and changes in processes and procedures. During the pilot, the support group should collect end-user feedback, monitor its own assessments, and work to remain open to potential changes to the initial plan. IS support staff members may be tempted to rationalize the design of the new strategy and resist modifications because they experience creative ownership. The consultative approach, however, requires that support personnel create a strategy in partnership with end-user customer knowledge and input. The pilot phase is a partnership opportunity.

After revising the service strategy and processes as needed, the IS manager must finalize the plan and phase it in throughout the organization. Often a pilot period tests a plan with key end-user representatives. The final plan is phased in to eventually service all end users with the new processes and procedures. If IS does not phase in the services, end-user customers will expect 100% availability of all promised services the first time they make a request.

SKILLS FOR THE CONSULTATIVE APPROACH

The consultative approach to client support service, based on the productive customer philosophy, involves much more than technical acumen. The IS department's specialized technical analysts may not have the consulting skills needed for such an approach. The IS manager must therefore assess staff skills and provide appropriate training where necessary. The IS manager may also need to implement training to help support staff members shift from a focus on solving technical problems to a focus on addressing business technology challenges.

In assessing staff skills, the IS manager should look for:

- *Telephone skills.* These are required for gathering customer feedback and diagnosing and solving problems, especially at remote sites. This is not purely the domain of a front-line help desk.
- *Consultative communication skills.* Techniques include asking open-ended questions, assessing customer priorities, and opening exploratory discussions.
- *Interpersonal skills.* The key interpersonal skills are assessing personal space requirements, reading body language, estimating personality types and social styles, and adjusting behavior as needed.
- *Time management and organizational skills.* IS staff must juggle many priorities, keep communication flowing on status, market as they go, and of course, solve technical problems. This presents quite a challenge to IS man-

agement and staff, especially to IS members who were previously assigned only to long-term projects.

- *Negotiation skills.* Win-win negotiation skills do not come naturally to everyone. Yet all IS staff members can and should learn this valuable skill set. The productive customer philosophy requires IS staff to negotiate with end users whenever various options to solve service problems and meet service needs are explored.

- *Listening skills.* Listening is the most important skill for consultative service. Success in this endeavor is not possible without hearing end users' viewpoints.

This list is meant to guide an overall staff development effort. The organization's human resources department may be able to help assess these skills and then search for specific training and mechanisms to teach support personnel in unskilled areas. Customized courses are the most valuable because staff can spend the training time applying skills to the IS department's environment. Generalized customer service training courses provide overall principles and leave IS to translate them to their environment after the course.

IS managers need the same skills as the support staff. Moreover, managers must exemplify to the staff the attitudes and philosophy they want the staff to exhibit to end-user customers. For example an IS manager who wants the support staff to listen to end users can teach by example by listening carefully to staff members. The manager must outline the vision and strategy and then exemplify it to staff.

If IS managers want to encourage teamwork and participation with end users—both key aspects of a consultative service environment—they must develop the support group into its own team. A group rarely begins as an empowered team that is able to make strategic decisions without guidance.

CONCLUSION

The consultative approach to customer service is the basis for IS department service excellence. In everyday terms, it means anticipating and understanding the end users' immediate needs and expectations, exploring options to meet their broader strategic needs, following through on details, communicating throughout the process, and delivering what is promised.

When dining out at a restaurant, for example, patrons evaluate the service according to these same criteria. Excellent service includes all of these elements. People are drawn to go back to similar experiences because they meet their needs and expectations without hassle. Even when mix-ups in service occur, the diner trusts that the service provider will aptly handle the mix-up.

IS department customers respond in the same way if the support group has built that base of trust. Small steps that show the group is changing for the better help gain that trust. Thinking and planning alone do not.

III-5

User Involvement in Project Success

Stanley H. Stahl

A BASIC PRINCIPLE for implementing a sustainable software productivity improvement program is to make everyone a winner. A systems project that comes in on budget and schedule and meets user requirements makes winners out of users, senior management, and the IS department. Making winners of people helps get their commitment and involvement as well as helps them overcome their natural resistance to change. It is a critical component of quality improvement in pioneer W. Edwards Deming's precept of putting everyone to work to accomplish quality improvement.

Theory W is a way to make everyone a winner. This theory was developed by Barry Boehm in the late 1980s. At the time, Boehm was chief scientist in TRW's defense systems group and a professor of computer science at the University of California, Los Angeles. In their paper, "Theory-W Software Project Management: Principles and Examples," Boehm and Rony Ross presented a unifying theory of software project management that is simultaneously simple, general, and specific. In the introduction to this paper, Boehm and Ross wrote:

"The software project manager's primary problem is that a software project needs to simultaneously satisfy a variety of constituencies: the users, the customers, the development team, and management. ... Each of these constituencies has its own desires with respect to the software project. ... These desires create fundamental conflicts when taken together. ... These conflicts are at the root of most software project management difficulties—both at the strategic level (e.g., setting goals, establishing major milestones, and responsibilities) and at the tactical level (e.g., resolving day-to-day conflicts, prioritizing assignments, and adapting to changes)."

Theory W is a way to help project managers cope with the difficulty of simultaneously satisfying different constituencies. Theory W has one simple but very far-reaching principle: Make everyone a winner by setting up win-win conditions for everyone.

THEORY W: BACKGROUND AND BASICS

Theory W contrasts with such theories on management as Theory X, Theory Y, and Theory Z. The Theory-X approach to management originated in the work of Frederick Taylor, who was active at the beginning of this century. Taylor contended that the most efficient way to accomplish work was to organize jobs into a well-orchestrated sequence of efficient and predictable tasks. Management's responsibility was to keep the system running smoothly; this task was often accomplished by coercing and intimidating workers.

For obvious reasons, Taylor's Theory X is inappropriate for managing software projects. Theories Y and Z, dating from approximately 1960 and 1980, respectively, were intended as alternatives to Theory X. Theory Y's perspective is that management must stimulate creativity and initiative, which are both important qualities for a quality software project. The difficulty with Theory Y, however, is that it provides inadequate mechanisms for identifying and resolving conflicts.

Theory Z seeks to improve on Theory Y by emphasizing the development of shared values and building consensus. The problem with Theory Z is that consensus may not always be possible or desirable; this can be the case with different constituencies that have their own unique set of individual constraints and requirements.

If the Theory-X manager is an autocrat, the Theory-Y manager a coach, and the Theory-Z manager a facilitator, then the Theory-W manager is a negotiator. The manager in the Theory-W model must proactively seek out win-lose and lose-lose conflicts and negotiate them into win-win situations. Delivering software systems while making winners of all stakeholders seems, at first glance, to be hopelessly naive.

Users want systems delivered immediately and they want them with all the bells and whistles imaginable. Management not only wants systems delivered on schedule and within budget, they also want a short schedule and a low budget. Developers want technical challenges and opportunities for professional growth, and they often do not want to document their work. Maintainers want well-documented systems with few bugs and the opportunity for a promotion out of maintenance. How can a project manager expect to successfully negotiate the conflicting needs of all constituents?

Importance of Negotiating

Although it may seem like a naive theory, there is an accumulation of evidence that Theory W works. In fact, Theory W is coming to be seen as fundamental to project success. The reason lies in the character of a win-win negotiation. The objective in win-win negotiating is for all parties to recognize each other's specific needs and to craft a resolution that allows all participants to share in getting their needs met. This is very different from traditional styles of negotiation, which are too often win-lose.

In the absence of an explicit commitment to foster win-win relationships, software projects have the capability of becoming win-lose. For example, building a quick but bug-laden product may represent a low-cost win for an over-pressured development organization but it is a loss for the users. Alternatively, when management and users force developers to add extra features without giving the development organization the time and resources needed to develop the extra features, the result may be a win for users and a loss for developers. Software maintenance personnel often lose as management, developers, and users fail to ensure that software is well-documented and easily maintainable.

At their worst, software projects can become lose-lose situations where no one wins. It is common for management to set unreasonable schedule expectations, and as a result, the development department tries to catch up by adding more and more people to the project. The result is, all too often, a poor product that comes in over cost and over schedule. In this case, everyone loses.

THE COSTS OF NOT NEGOTIATING

The following example illustrates how ignoring Theory W affects a project. Although this example is fictional, it is based on actual experience.

A Lose-Lose Project

A growing specialty retailer had just hired a new chief information officer. The new CIO was given the charter to modernize the company's antiquated information systems but had been explicitly told by the CEO that budgets were extremely limited and that the new systems would have to be implemented as the size of the IS staff was decreased.

The first task was to conduct a user-needs survey, which indicated that, except for the payroll systems, all of the company's existing systems were inadequate. The inventory control system was barely usable, there was no integration between different systems, and each system served, at best, only the limited needs of the department for which it had been designed. The survey also indicated that most users were unaware of the potential productivity boost that up-to-date information systems could give to the company's business.

After the survey had been analyzed, the following four recommendations were made:

- All existing systems should be replaced by a client/server system capable of supplying timely and accurate information to both operational personnel and senior management.
- The changeover should be implemented in stages. The first system should be a relatively simple, low-risk, standalone application.
- The system should be procured from an outside contractor, preferably a software house that has a package that could be used with minimal modifications.
- Training should be provided to middle managers to enable them to better guide the IS department in implementing the new systems.

Basics Were Ignored. These recommendations were enthusiastically approved. A request for proposal (RFP) for a new inventory distribution management control system, everyone's favorite candidate to be implemented first, was requested. However, nothing was said abut budgets, schedules, or personnel needs, and in their enthusiasm, everyone seemed to forget about training.

Lack of User Input. The RFP was developed with little input from distribution personnel, so it was rather open-ended and not very explicit. The result was that the IS group received from outside contractors eight responses ranging in price from $70,000 to $625,000. After lengthy negotiations, a contract was awarded to a single company that would provide both hardware and software for the new inventory control system.

Unclear RFP. The contractor was a leader in inventory management systems, though its largest account was only half the size of the retail company's. To land the account, the contractor promised to make any necessary modifications to the system free-of-charge. The contractor's reading of the RFP led its developers to believe that there was little technical risk in this promise.

The contractor's interpretation of the RFP was wrong. Although the contract called for the system to be up and running in six months at a cost of $240,000, a year later the contractor was still working on changing the system to meet the client's needs.

Neither was the retailer's IS department experiencing a good year. It was continually at odds with both distribution personnel and the contractor over the capabilities of the new inventory system. The users kept claiming that the system was not powerful enough to meet their needs, while the contractor argued that the system was in use by more than 200 satisfied companies.

Costly Failure. At the end of an acrimonious year, the contractor and retailer agreed to cancel the project. In the course of the year, the contractor was paid more than $150,000, and the contractor estimated that its programming staff had spent more than 10 worker-months modifying the system to meet the client's needs. The retailer estimated that it had invested the time of one senior analyst as well as several hundreds of hours of distribution personnel.

HOW THEORY W WOULD HAVE HELPED

Losers and the Consequences

The most apparent source of difficulty on the project was the explicit win-lose contract established between the retailer and the contractor. By requiring the contractor to cover any expenses incurred in modifying the system, the retailer set up a situation in which the contractor would only make changes reluctantly. This reduced the likelihood that distributors would get the modifications they needed and increased the likelihood that changes would be made in a slap-dash way, with too little attention paid to quality.

The IS department's relationship with senior management was also win-lose. There was little likelihood that the CIO could emerge victorious. The system had to be brought in on time and within budget, though the user community was inadequately trained to help properly identify its needs and requirements. The result was that neither the retailer nor the contractor had an adequate handle on the inventory system's requirements and were consequently unable to adequately budget or schedule the system's implementation. The IS department's situation was made worse because senior management wanted to decrease the size of the IS staff.

The users lost the most. Not only did they not receive the system they needed and had been promised, but they wasted time and money in diverting attention from their primary jobs to help develop the new system. Both the retailer's and the contractor's developers lost the time they invested in a failed project, the ability to grow professionally, and the opportunity to work on a successful project.

From a Process Perspective

Fault lies both in the contracting process and in the systems requirements management process by which the retailer and the contractor defined and managed systems requirements. These front-end processes are often the source of project management difficulties, but problems were aggravated in this case by the IS department's inability to identify the real needs of stakeholders and to negotiate an appropriate win-win package. Although the IS group was neglectful, it is not to blame—the problem lies with the process.

Steps to Improve the Process

To improve these processes, Theory-W principles of software management can be used. The following three steps, which are adapted from "Theory-W Software Project Management," can be used to implement Theory-W software management:

1. Establishing a set of win-win preconditions by performing the following:
   — Understanding what it is that people want to win.
   — Establishing an explicit set of win-win objectives based on reasonable expectations that match participants' objectives to their win conditions.
   — Providing an environment that supports win-win negotiations.
2. Structuring a win-win development process by accomplishing the following:
   — Establishing a realistic plan that highlights potential win-lose and lose-lose risk items.
   — Involving all affected parties.
   — Resolving win-lose and lose-lose situations.
3. Structuring a win-win software product that matches the following:
   — The users' and maintainers' win conditions.
   — Management's and supplier's financial and scheduling win conditions.

Application of Theory W

There are several actions that the retailer's IS department could have taken to increase the project's probability of success. It could have trained distribution personnel on the key role they have in properly identifying and articulating their needs. Following this training, IS staff could have worked with these users to draft a more thorough RFP. After receiving RFP responses, the IS department could have involved senior management in identifying limits on resources and schedules. It could have foreseen the difficulties the contractor would have if significant program modifications were needed.

By identifying constituent win conditions, the IS department would then have been in a position to negotiate a fair contract that would have explicitly taken into account all win conditions. Having done these critical up-front tasks, the IS department would then have been in a position to structure both a win-win development process and software product. Unfortunately, the IS department never set win-win preconditions.

In the absence of an explicit philosophy to make everyone a winner and an explicit process for accomplishing this, the IS department lacked the necessary support to identify stakeholders' needs and negotiate a reasonable set of win-win objectives. Thus, it was only a matter of time until incompatible and unobtainable win conditions destroyed the project.

CONCLUSION

Boehm's work on Theory-W software project management continues. Currently, a professor of computer science at the University of Southern California and chairperson of its center for software engineering, he has a research program to develop a tool for computer-aided process engineering (CAPE) that has Theory W built in. As part of this research, he is prototyping an interactive system in which constituents can enter their win conditions and all stakeholders can then simultaneously analyze and negotiate a combined set of win conditions.

Theory W has been shown to be successful—both in terms of the paradigm it offers and in terms of its ability to help managers explicate and simultaneously manage the win conditions of all constituents and stakeholders. Theory W is even more important for organizations embarking on a systematic program to improve productivity. Productivity improvement programs, and such similar programs as total quality management (TQM) or process reengineering, require the full and complete support of all stakeholders. Theory W offers both a theory and a process for getting and keeping this needed support.

III-6

Guidelines for Designing EIS Interfaces

Hugh J. Watson
John Satzinger

EXECUTIVE INFORMATION SYSTEMS are among the faster-growing applications in corporate America. They are designed to supply senior executives with needed information, such as news and stock prices and information about competitors, customers, key performance indicators, and internal operations. Most large firms have an EIS in place or are planning to develop one, and even smaller firms are implementing them.

Several factors have led to the development of EISs. One major factor is executives' need for better information in today's competitive business environment. Another is the availability of special-purpose EIS software that facilitates the development of an EIS. Pilot Software, Inc. (Boston MA) and Comshare, Inc. (Ann Arbor MI) led the way when they first offered their products in the mid 1980s, and since then a host of other products have become available. As successes with EISs at such corporations as Northwest Industries, Ltd., Lockheed Aeronautical Systems Co., Xerox Corp., Quaker Oats Co., and Beneficial Corp. have become widely known, executives and IS managers have recognized the potential of EISs and championed their development.

As part of the research program on executive information systems at the University of Georgia, EIS practices at leading-edge firms have been studied. Investigators have talked with many EIS developers, worked with most available EIS products, and consulted on the development of EISs. This research has led to guidelines for developing EIS user interfaces. These guidelines can also be helpful to executives in knowing what to ask for and to expect in their EISs. The guidelines are illustrated with specific examples, many of which are from Lockheed Aeronautical Systems, which developed one of the first EISs and has been successful with their system.

THE DEVELOPMENT OF EISS

Building a successful EIS is challenging. A myriad of technical, organizational, and managerial issues must be addressed. Of utmost importance is creating an EIS that is easy to use. Executives have little time or patience with systems that are difficult to use. Consequently, system designers should pay careful attention to the design of the user interface for the system.

Definition of EIS User Interface. The term *user interface* refers to how the user directs the operation of the system (e.g., keyboard, mouse, or touch-screen; question/answer, command language, or menus) and how the output is given to the user (e.g., graphical, tabular, or textual; color or monochrome; paper or online). For the system to be easy, the user must know how to make it work and what the output means. A user interface must be designed to make operating the system and interpreting the output as easy as possible.

Designing an EIS user interface is different from designing other information systems. Because of the nature of executive users, the system must be more than user friendly; it must be user intuitive, even user seductive. Another difference is the flexibility the system must have, because it is difficult to determine how a particular executive will use an EIS. Also, because of advances in hardware and software, system designers have many new options to choose from when implementing an EIS.

A successful EIS often benefits other users in addition to executives. For this reason, it has been argued that EIS also stands for "everybody's information system." These users are more likely to accept complex user interfaces than senior executives and may be willing to trade off simplicity for flexibility. In many instances, however, the more complex applications created for lower-organizational-level users are not given to executives. For example, an executive may not need an application that provides advanced query capabilities to analyze sales data. The focus of this chapter is the design of user interfaces for executives rather than for lower-level organizational personnel.

DESIGN GUIDELINES

Developers of an EIS are typically building their first system of this type. EIS users, information content, and software are often different from previous systems development projects. Although developers will learn how to better build EISs, poor initial choices can undermine or even eliminate the chances for successful implementation. Some EISs have not been as successful as they might be or have even failed because of poor user interface designs.

The following guidelines on designing an EIS interface should help developers successfully implement an EIS:

- Involve executives in the design of the user interface.
- Set standards for screen layout, format, and color.
- Make the system intuitive.
- Use standard definitions of terms.

- Design the main menu as a gateway to all computer use.
- Design the system for ease of navigation.
- Strive to make response time as fast as possible.
- Expect preferences in user interfaces to change.

In the following sections these guidelines are examined and illustrated with examples from successful EIS implementations.

Involving Executives in the Design of the User Interface

Although user involvement in the systems development process is critical for all types of information systems, executive involvement in the design of an EIS user interface is especially important. Executives might have limited experience working directly with a computer, and if they do have some computer experience, the EIS will look and feel quite different from any organizational information systems the executives might know. Designers should be prepared to show a variety of prototype screens and navigation approaches because the executive might have limited knowledge of what an EIS can actually do. Evaluating these prototypes is also likely to get apprehensive executives more committed to the EIS as they begin to see the system's potential. For this reason, it is important to involve all executive users in the process, not just the executive sponsor.

Prototyping Approaches. Early prototyping should be used to help decide on the basic look and feel of the system. Two fundamental approaches should be presented:

- A full-screen interface with large buttons and icons.
- A multiple window interface with pull-down menus and dialog boxes.

The first approach might be less intimidating but the second approach conforms to the popular interface design standards. The preferred look and feel should be used to finalize the development environment that will be used (e.g., Windows), as some development environments might more easily accommodate one or the other type of system. Additionally, differences in preference for the look and feel reveal early in the development process the amount of individual tailoring that might be required for each executive.

Although rapid prototyping and extensive user feedback are important, the prototypes do not have to be computer-based. Paper screen mock-ups (i.e., storyboards) can be quite effective because the executive can review the screens as time permits and consider the alternatives before providing feedback to the designer. Computer-based prototypes, however, are useful when showing the executive the potential of the technology and when exploring navigation approaches the executive might prefer.

Executives also must be involved in the design of the interface, because preferences for screen prototypes can provide clues about the importance of screen content and design. This aids a designer in uncovering additional information requirements. The relationships among importance of data, the level of detail desired, and the frequency of need for the information can help a designer to understand the way an executive will actually use the EIS.

Because of the almost endless number of possible screens that can be provided, it is important for the designer to narrow the number down to the most important screens for each executive. This not only reduces development time and system overhead, but also makes it possible to provide a system that makes it easy for the executive to find the information that is actually needed.

Any later changes to the interface of an EIS should be discussed with its users. This is specially true when a designer considers deleting seldom-used screens. It is not easy to tell the value of a particular screen just by tracking usage. An executive may have looked at a particular screen only once, but that screen could have provided critical insight that day. Months later, the same screen might be needed once again when the same critical need arises.

Setting Standards for Screen Layout, Format, and Color

Currently available EIS software offers an array of screen design alternatives. Screens can display graphs, tables, and text in hundreds of formats and colors. Unfortunately, this cornucopia of choices can be detrimental. There is a temptation to use many of these alternatives to add sizzle to the screens, but yielding to this temptation can create displays that are confusing. IS managers should carefully develop screen design standards that use only a few layouts, formats, and colors.

The EIS at Lockheed illustrates the use of screen design standards; a sample screen is presented in Exhibit III-6-1. The top of the screen presents the screen number, a title for the screen, and the date of the last update. The right-hand corner gives the names of those who are knowledgeable about the information and their work telephone number. This information makes it easier for users to go directly to the person who is best able to answer any questions about the information. Some EISs allow users to click on the person's name to have the telephone number dialed automatically.

Layout Standards. Lockheed's standard layout is to present graphical information at the top of the screen, more detailed tabular data below it, and textual information at the bottom of the screen. The graph provides a quick visual presentation of a situation, the table gives specific numbers, and the text provides explanations, assessments, and describes actions being taken.

Graphs Standards. Graphs of historical and current data always use bar charts. When actuals are compared against plans or budgets, paired bar charts such as those shown in Exhibit III-6-1 are used. The bars are of different widths to allow users with color perception problems to correctly identify each bar. Projections into the future use line graphs. Pie or stacked bar charts are used to depict parts of a whole. On all charts, vertical wording is avoided and abbreviations and acronyms are limited to those on an authorized list. All bar charts are set to zero at the origin to avoid distortions, scales are set in prescribed incre-

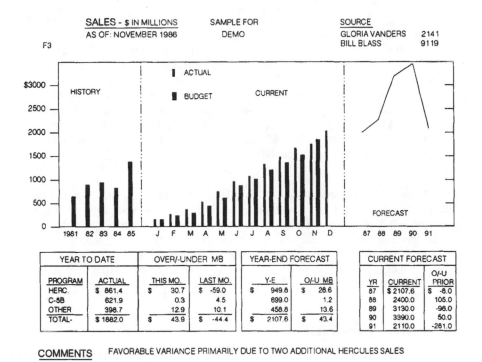

SALES - $ IN MILLIONS
AS OF: NOVEMBER 1986
F3

SAMPLE FOR
DEMO

SOURCE
GLORIA VANDERS   2141
BILL BLASS   9119

YEAR TO DATE		OVER/-UNDER MB		YEAR-END FORECAST		CURRENT FORECAST		
PROGRAM	ACTUAL	THIS MO.	LAST MO.	Y-E	O/-U MB	YR	CURRENT	O/-U PRIOR
HERC.	$ 861.4	$ 30.7	$ -59.0	$ 949.8	$ 28.6	87	$ 2107.6	$ -8.0
C-5B	621.9	0.3	4.5	699.0	1.2	88	2400.0	105.0
OTHER	398.7	12.9	10.1	458.8	13.6	89	3130.0	-98.0
TOTAL-	$ 1882.0	$ 43.9	$ -44.4	$ 2107.6	$ 43.4	90	3390.0	50.0
						91	2110.0	-281.0

COMMENTS   FAVORABLE VARIANCE PRIMARILY DUE TO TWO ADDITIONAL HERCULES SALES

Exhibit III-6-1. Sample Screen from the Lockheed Aeronautical Systems EIS

ments and are identical within a subject series, and bars that exceed the scale have numerical values shown.

Color Standards. Lockheed's EIS uses only a few carefully selected colors. Yellow is used to show actual performance, cyan (i.e., light blue) is used for company goals and commitments to the corporate office, and magenta represents internal goals and objectives. A traffic-light pattern is used to highlight comparisons: green is good, yellow is marginal, and red is unfavorable. For example, under budget or ahead of schedule is in green, on budget or on schedule is in yellow, and over budget or behind schedule is in red. Organization charts use different colors for the various levels of management. Colors have been selected to minimize color differentiation problems—about 6% of all men and less than 1% of all women have color perception problems—and all displays are designed to be effective with black-and-white hard copy output.

Standard layout, formats, and colors offer many advantages. They provide a consistent look and feel for the system. Users are less likely to misinterpret or misunderstand the information presented. Standard displays require less cognitive effort on the part of the user and take less time to understand.

Use of Text. Textual material is entered by the EIS support staff to make the information displayed more useful. The information itself may not reveal the full story, but the purpose of the commentary is to add value to the information displayed. Although Lockheed's EIS presents commentary information on the same screen to which it applies, other EISs place it on a separate screen.

The power of a textual commentary is illustrated by the following example from Lockheed: Both a graph and tabular data indicated that actual cash flow was below budget by $20 million. A commentary revealed, however, that payment for a plane in the amount of $20 million was in route from a foreign country and would be in a Lockheed account by the end of the day.

Advanced Capabilities. A few EISs allow voice commentaries to be associated with screens. This is an appealing feature because executives are used to receiving information verbally and voice is richer for communications than printed words. Voice annotations to screens is currently the best accepted of the multimedia enhancements to EISs. Other possibilities such as video and personal teleconferencing have good potential, but the business case for them has yet to be made.

Making the System Intuitive

Ideally, an executive should be able to use an EIS without training. At the most, no more than 15 minutes of instruction should be required to teach executives how to use the basic information-retrieval capabilities. Systems more complex than this are unlikely to be used.

Most successful EISs are operated by point-and-click technology. By picking from among menus, icons, or buttons, an executive navigates through the system to a desired capability (e.g., E-mail or information). Experience with decision support systems has shown that most executives will not use a command language with a verb-noun syntax because it is too time-consuming to use and difficult to learn and remember.

In one easy-to-use EIS, 35-inch monitors were installed in executives' offices; each was able to simultaneously display up to 10 windows of information. Screens were customized for each executive and varied with the day of the month. This EIS was essentially a ticker tape of relevant information.

A recent development is the use of World Wide Web technology for building and operating EISs. This takes one of two possible forms. In the first, Web rather than traditional EIS software is used. The second approach takes advantage of vendors' integration of Web technology into their product lines, such as an EIS client that functions as a browser. In either case, users operate the system with a browser, which for most people is a familiar and easy-to-use interface.

User Documentation. Systems developers are typically expected to write user documentation for new applications. However, this is usually unnecessary or inappropriate for EISs. The system should be sufficiently intuitive that instruction manuals are not needed. Even more so than with other types of users, executives do not read documentation. If an executive is having a problem using the system, it is best if the user call the EIS support staff to correct the difficulty.

Users may request documentation, and in this case, it should be provided, either within the system and/or as hard copy. Ideally, the instructions should fit on a single page or a few screens.

Standard Definitions of Terms

Most organizations have data dictionaries that include definitions for the data elements used in transaction processing application. There are other terms that are widely used throughout organizations and are very important to EISs that are not as precisely defined. Everyone in a company uses these words and has a general understanding of their meaning, but slight differences exist and can cause misunderstandings.

For example, the term *sign-up* at Lockheed had different shades of meaning. A sign-up involves a company interested in purchasing an aircraft. To marketing personnel, a sign-up occurred when customers said they were going to make a purchase. The legal department, however, recorded a sign-up only when a signed contract was received. Finance waited until a down payment was received. Each group generally knew what the term meant, but slight differences based on their organizational perspective led to timing differences as to when a sign-up was recorded. Because an aircraft can cost between $20 to $30 million, such differences can result in considerably different impressions as to how the organization is doing. A sign-up has now been defined as a signed contract with a down payment.

A Dictionary of Terms. Lockheed has an executive data dictionary that contains definitions for all of the terms used in its EIS. The definitions can be accessed through the EIS and is available to all users. Creating an executive data dictionary is useful because it makes executives consider what terms are being used inconsistently and to develop definitions that reflect an organizationwide rather than functional-area perspective.

The Main Menu as a Gateway

Most organizations have a variety of applications designed to support executives: E-mail, electronic filing, decision support, and access to external news and stock prices. It is common for many of them to require their own access procedures and passwords. This requirement, and the resulting difficulty and inconvenience, discourages hands-on computer use. The development of an EIS provides an excellent opportunity to deliver all of these capabilities in a single, integrated system. An EIS provides the logical and physical umbrella under which all of the executives' computer applications are placed.

A number of EISs use their main menus to display all information and applications available through picks (i.e., menus, icons, or buttons). The kinds of information usually provide one set of options. For example, there may be screen picks for financial, production, marketing, and human resources information.

Separate picks may exist on the basis of products, geographical location, and organizational units (e.g., corporate, division). The choices reflect the information contained in the EIS and how it is organized. Lower-level menus let users move to specific information desired within a general category.

Access to these applications should be transparent to the user and not require any additional log-on procedures or passwords; these activities should be handled automatically by the system.

Designing the System for Ease of Navigation

Vendors' demos often show executives moving easily through a system, looking at current status information and drilling down to more detailed information when a problem or item of interest is identified. This scenario is possible in practice, but only if careful attention is given to navigation issues early in the system's design.

Navigation problems may be marked when there are few screens in the system. As the number of screens grow, as they inevitably do, users find it more difficult and time-consuming to move through a system. For example, an executive is looking at financial information and wants to move to operational production data. In a poorly designed system, the user will have to back out of the financial application, screen by screen, until the main menu is reached, and then enter the production application, and move through screens to the desired information.

The starting point in designing navigation for an EIS is understanding the mental models that executives have of the organization. If the structure of the information does not match their mental models, users will have a difficult time finding the information they want. For example, do executives look at the firm in terms of geographical location, products, functional areas, or divisions? Each view of the organization may call for a pick on the main menu and set of related screens.

A complicating factor is when one or a few executives have unique mental models. During the development of one EIS at a hospital, designers found that the director of nursing wanted information structured much differently than other users. Her view of the hospital could be accommodated but required custom designing the system for her use. The decision of whether to do this was a business rather than a technical one.

Navigation Features. There are features that can be included in an EIS to make navigation easier. Some systems have a screen that shows where the user is in the system. Often, users get lost and are uncertain about how to move elsewhere, short of turning off the system and starting over. Another feature is to have a home key or pick that takes the user directly back to the main menu. Some systems provide a retrace capability that allows users to easily backtrack to screens viewed previously. Another helpful feature is to include a pick on the main menu that takes the user to a screen that lists the user's most popular screens. From this screen, a user can go directly to any screen on a personalized menu.

Also, a single menu can be created that provides direct access to a large number of screens. For example, a company has five plants; each produces 20 products and there is work-in-process and finished goods inventory. The various combinations result in 200 screens (i.e., $5 \times 20 \times 2 = 200$). A single menu where the user picks the plant, the product, and the type of inventory provides direct access to the desired screen.

Lockheed switched recently from custom developed to commercial EIS software. Before Lockheed signed the contract, however, the vendor had to agree to support keyboards as an input device to the system, largely for navigation reasons. Lockheed's executives were accustomed to point-to-point navigation in the system. Each screen could be accessed from any place in the system by simply entering its screen number. Most executives remembered or kept a list of the screen numbers of their favorite screens.

Fast Response Time

When incorporating text and graphics, internal and external data, hundreds of individually tailored screens and views, and multiple navigational paths through the system, EIS developers must continually monitor the response time of the system. Executives are intolerant of slow response times. A recent survey of EIS development practices found that response times for EISs had actually degraded from an average of 2.8 seconds in 1988 to 5.3 seconds in 1991 despite the increased use of powerful desktop computers and local area networks. Although the same survey found satisfaction with ease of use and the effectiveness of the EISs to be relatively high, satisfaction with response times was extremely low.

Response time problems can be anticipated when the EIS must dynamically build a screen each time it is requested by searching corporate data bases. Response time can be much faster if the screens are static and updated each night, though designers must evaluate the trade-off between timeliness of data on the screens and response time. Response time can also be affected by the narrow bandwidth of today's network.

When Speed Counts. Generally, executives expect very fast response times when flipping through their usual set of screens each morning. One EIS developer suggested thinking of the maximum acceptable time to move from screen to screen as the time it takes the executive to turn a page of *The Wall Street Journal*.

Executives can usually tolerate a slow response to ad-hoc queries. When an executive is used to waiting several days for the staff to gather information for a specific question, several minutes may be an acceptable wait for directly retrieving the same information through the EIS. The differences between predefined screens and ad-hoc query screens should be made clear to the executive, however. In either case, when any system function takes more than a few seconds, a message should always provide feedback that the system is processing the executive's request.

USER PREFERENCES

Almost all aspects of an EIS, including the user interface, change in time. Several examples illustrate this point. So much information is displayed on a screen in the Lockheed EIS, that a first-time viewer may be confused. However, this is what Lockheed's executives prefer. They want information on a single screen rather than having to page through several screens. This approach also better supports making comparisons, such as when an executive wants to check graphical against numerical presentations of data.

Lockheed's screens were not originally designed this way; rather, they have evolved in response to executives' requests. This same phenomenon has been noted in other organizations as their EISs have matured.

Often, organizations developing an EIS order touchscreens for technophobic executives. These users quickly discover the disadvantages of touch-screens and also find that using a mouse is easy after a little practice. Although touchscreens may help sell the idea of an EIS to some executives, these executives will probably prefer mouses eventually.

As an EIS evolves, the number of its users usually increases. Quite possibly, training given to first-time users will have to change. For example, more time may have to be spent discussing how to interpret the information presented on the most complex screens. Another approach is to include less complex screens in the system. This was done in one manufacturing firm where the new CEO had a strong background in engineering and production but was relatively weak in finance. Recognizing this fact, the EIS staff developed a number of simple screens that displayed key financial information. Within a few months, after the CEO has become experienced in finance, the special screens were phased out of the system.

CONCLUSION

From a user's perspective, the user interface is the EIS. Most users care little about which hardware or software is used, where data resides, or which communications protocols are used. EIS users focus on what they must know in order to use the system, how the system's actions are directed, and how the system's output is presented. If executives have to spend much time learning to use the EIS or finding the information they need from it, they will not use this system. To make sure that executives will use an EIS, developers must pay close attention to the user interface. The eight guidelines discussed in this chapter can aid in the design of a successful EIS.

III-7

Reviewing End-User Applications

Steven M. Williford

IN MOST ORGANIZATIONS, sophisticated end users are building their own applications. This trend began with computer-literate end users developing simple applications to increase their personal productivity. End-user applications development has since evolved to include complex applications developed by groups of users and shared across departmental boundaries throughout the organization. Data from these applications is used by decision makers at all levels of the company.

It is obvious to most end-user computing and IS managers that applications with such organizationwide implications deserve careful scrutiny. However, they are not always familiar with an effective mechanism for evaluating these applications. The method for reviewing end-user-developed applications discussed in this chapter can provide information not only for improving application quality but for determining the effectiveness of end-user computing (and end-user computing support) in general.

A review of user-developed applications can indicate the need for changes in the end-user computing support department and its services as well as in its system of controls. In addition, such a review can provide direction for strategic planning within the organization and is a valuable and helpful step in measuring the effectiveness of end-user computing support department policies and procedures. Auditors might also initiate an audit of end-user applications as part of a continuous improvement or total quality management program being undertaken in their organization.

End-user computing or IS managers should also consider performing a review of end-user applications as a proactive step toward being able to justify the existence of the end-user computing support department. For example, a review may reveal that some end-user applications contribute heavily to increased productivity in the workplace. In an era marked by budget cuts and downsizing, it is always wise to be able to point out such triumphs.

DEFINITIONS AND CHARACTERISTICS

Each organization may use a different set of definitions to describe various aspects of end-user computing support, and it is important to have a common understanding of the terms to be used. The following definitions are used in this chapter.

- *Application.* An application is a set of computer programs, data, and procedures that is used to resolve a business-specific problem. For example, the accounting department may develop and implement an application that generates profit-and-loss statements.
- *Product.* Products are the software used to develop or assist in the development of computer systems. Examples are spreadsheets, word processors, fourth-generation languages (4GLs), CASE tools, or graphics packages. Tools is another common term for product.
- *System.* A system is a combination of computer applications, processes, and deliverables. During a review of end-user applications, it is important to determine the fit of the application within the system. An example is a budget system in which individual managers collect data from individuals using a manual process (e.g., paper forms), transfer the data to spreadsheets, and then electronically transmit the information to the accounting department, which consolidates the information and uses a budget forecasting application to create reports for senior management.
- *Work group.* This is a group that performs a common business function, independent of organizational boundaries, and is tied together by a system or process. The managers who collect data for the budget system make up a work group; they may all report to different managers in different departments, and each person is probably in more than one work group. Other work groups include project development and training.
- *Work unit.* This term is used for such organizational units as departments, divisions, or sections (e.g., accounting, human resources, and engineering). An application typically resides in a work unit (i.e., is run by that unit) but affects other units throughout the organization.

It is also important to review the unique characteristics of the end-user computing environment as they relate to user-developed applications:

- *Point of control.* In an end-user computing environment, the person using the application has either developed the application or is typically closer to the developer organizationally.
- *The critical nature of applications.* End-user applications tend to be valued less and are often not developed under the strict guidelines of traditional IS applications. Because of this, the impact of the application's being in error or not working at all is often not considered until it is too late.
- *Range of measuring criticality and value.* End-user applications may range from trivial to mission critical. Applications created by IS have a much narrower range but are concentrated toward the critical end of the scale.
- *Development.* In an end-user computing environment, the people who

handle any one application may be scattered organizationally; the applications may also be scattered over time and across products. For example, an application may originally be developed on a word processing package. If the math requirements for the application become too complicated, the application would be transferred to a spreadsheet product. Finally, the application may be converted to a data base product to handle complex reporting requirements.

- *Quantity of applications.* There are more applications developed by end users than by the IS department, but they are usually smaller in scope and more tuned to individual productivity.

- *Type of products.* End-user development products usually provide a group of standard functions (e.g., Lotus 1-2-3 provides built-in functions). Creating a complex application using these products may require a high degree of knowledge about the development product or may necessitate using several development products to create a single application.

OVERCOMING MISCONCEPTIONS

In some organizations senior IS management initiates a review on behalf of user managers who may be concerned that their applications are getting away from them. In these cases, the end-user computing support department managers may be asked to help sell the idea to corporate managers. In most organizations, gaining any management commitment to reviewing end-user applications requires overcoming several obstacles. The following sections discuss common management objections to reviewing end-user applications and ways to overcome this mind-set.

End-User Applications Are Not Significant. This is a typical misconception on the part of either corporate managers or senior IS managers. End-user applications may be perceived as transient, disposable, and not production oriented—therefore not significant. Traditional applications that are run by the IS department and cannot be tampered with in any way by anyone other than a technical expert are viewed as much more substantial, stable, and worthwhile. Senior management may be unwilling to approve an investment in reviewing what they perceive to be insignificant applications. To change this viewpoint and bring them up to date, end-user computing managers should make the effort to point out particular end-user-developed applications that are currently providing critical data or contributing more concretely to improved productivity and increased bottom-line benefits.

Ease of Use Results in Effective Applications. This is another common misconception of senior management and IS management. They may believe that the ease of use of end-user applications development products would prevent users from creating anything but the most effective applications. Again, managers would be reluctant to spend resources on reviewing applications that they feel are typically well created. This misconception has been amplified by sales promotions that vigorously emphasize the ease of use of these products. IS managers should point out to senior managers that development products have limitations and that ease of use not only cannot guarantee that applications do what they were intended to do but can contribute to end users creating unnecessary applications and duplicating effort.

The End Users Will Not Cooperate with the Review. This is a common objection of user management and their employees. IS managers should promote the concept of an informal reviewing method (e.g., an inventory or statistics review, both of which are discussed in a later section) that would be less of an imposition on end users and therefore less of a threat to those users who are very protective of their current work processes.

In some organizations, end users react to a review of their applications in much the same way they would react to an audit of their personal finances by the IRS—that is, they view it as a hassle and something they would like to avoid at all costs, regardless of whether they feel they have anything to hide (e.g., pirated software). If this is the case, the IS department might want to consider setting up self-audit guidelines with the cooperation of the users, or have them participate in the first central review. Review guidelines explain what the review team will be looking for. When end users know what to expect and have a chance to evaluate their own applications using the same criteria the reviewers will be using, they are typically far more willing to cooperate with the actual review. In addition, involving them in the review can alleviate an us-versus-them attitude.

PREPARING FOR THE REVIEW

The reviewing process follows a life cycle similar to that of any other project. The steps discussed in this chapter cover preparation for a review; they provide the background necessary to begin a review. These steps are designed as a general guideline. Not all companies may need all the steps, and early reviews (undertaken when end-user development is still relatively new to the organization) will usually not follow all the steps. Preparing for a review requires:

- Defining the review objectives.
- Defining the review method.
- Defining the scope and content of the review.

Each of these is discussed in the following sections and summarized in Exhibit III-7-1.

Defining Review Objectives

Review objectives help determine the results and essentially guide the process by defining the intent of the review. IS and user managers should define and agree to the objectives before proceeding. In general, reviews are more successful if they focus on a particular objective. For example:

Once it has been established that a review of end-user applications would be helpful or even necessary in a particular organization, careful preparation for conducting the review should begin. Although some more informal reviews may not require all the steps discussed in this chapter, for the most part, each step is an important and necessary component of a succesful review (i.e., one that provides useful and valuable information). The following is a checklist of these steps:

Define Review Objectives
The audit may be designed to:

- Determine, indentify, or resolve end-user applications problems.
- Evaluate end-user computing support group services.
- Respond to financial issues.
- Collect specific information.
- Provide input to strategic or long-range planning.

Define the Review Method
Four of the most effective methods are:

- Formal audit.
- Inventory.
- Statistical review.
- Best guess review.

Define the Scope and Content of the Review
Determining the scope and content helps:

- Define what the end-user computing department will consider as end-user computing.
- Define which environments a particular review will evaluate.

Exhibit III-7-1. Steps to Prepare for End-User Computing Audit

---

- *Determine, identify, or resolve end-user applications problems.* This common objective ensures that the review will provide answers to such questions as:
  — Is there a problem with end-user applications (e.g., are particular applications proving to be error-prone, duplicating effort, or providing inaccurate data)?
  — What is the exact problem with a particular application (e.g., why is the application providing inaccurate data)?
  — How can this problem be solved? For example, what can be done to make this application more effective, or should a better set of checks and balances be implemented to validate end-user applications? A better set of checks and balances might involve comparing the results of an end-

user application that reports sales volume by region to the results of a traditional IS application that tracks the same information.

— What are the consequences of ignoring the flaws in this application?

— Who should fix this application?

— Is it worth the cost to fix the application, or should use of the application be discontinued? For example, end users might create an application that automates the extraction and compilation of sales data from a larger system. The cost of maintaining or repairing such a system could be prohibitive if the data from the larger system could just as easily be compiled using a calculator.

● *Evaluate end-user computing support group services.* When there are complaints from the user areas (e.g., users may feel that they are not getting enough support to develop effective applications) or when the end-user computing department takes on new levels of support, it may consider a review of end-user applications to help them evaluate current services. For example, such a review can reveal a large number of error-prone or ineffective applications, which would indicate a need for more development support. The review might reveal that a number of users are duplicating applications development effort or are sharing inaccurate data from one application. Users may have developed applications that are inappropriate or inadequate for solving the problems they were designed to address. Any of these scenarios would indicate an increased need for support of end-user applications development. Typical questions to be answered with this objective are:

— Can the services be improved?

— Should new services be added?

— Should services be moved to or from another group of end users?

— Are resources being allocated effectively (e.g., is the marketing department the only user group without any productivity-increasing applications)?

● *Respond to financial issues.* This objective can provide pertinent information if budget cuts or competition within IS for resources threatens the end-user computing support department. A review of user-developed applications may lend credence to the need for end-user computing support of the development and implementation of valuable computer applications by pointing out an application that may be saving a great deal of time and money in a particular user area. A review with this objective provides information similar to the answers provided when evaluating services in the objective; however, the information is then used to answer such questions as:

— Can the end-user computing support group be reduced or eliminated?

— Can the services to user-developers be reduced?

— Can some budgetary efficiencies be gained in supporting end-user applications development?

● *Collect specific information.* Corporate or IS management may request information about end-user applications, especially if they receive data from them on a regular basis or if (as in applications run in the payroll department) many people would be affected by an inaccurate application. It is also not unlikely that user management would request an investigation of end-user applications in their area. Both of these cases are more common in companies that are committed to a continuous improvement or total quality program.

- *Provide input to strategic or long-range planning.* A review with this objective would highlight much of the same information found in a review to evaluate services or respond to financial issues but would add a more strategic element to the process. For example, this objective would answer such questions as:
  - Do end-user applications contribute to accomplishing corporate goals?
  - Are there end-user applications that might create strategic opportunities if implemented on a broader scale?
  - Are resources adequate to initiate or foster development of end-user applications that might eventually contribute to achieving strategic goals?

Defining the Review Method

The methods of collecting data should be determined by the political climate, the people who will act on the results of the audit, and the resources available to perform the work. The following sections discuss five of the most common and effective methods for reviewing end-user applications and examine the most appropriate instances for using each of them.

Formal Audit. This method for auditing end-user applications is usually selected if the audit is requested by corporate management. They may be concerned about applications that are built to provide financial information or about the possibility of misconduct associated with user applications. Because most organizations are audited in a financial sense, corporate and user management are familiar with the process and the results of a less formal method. However, a formal audit is more expensive and often more upsetting to the participants (i.e., the end users).

Inventory. Taking an inventory of end-user applications involves gathering information about the products and applications on each workstation. Although an inventory is a less formal variation of an audit and may be perceived by corporate and senior IS management as less significant than a formal audit, it provides much of the same information as a formal audit. An inventory can be useful when the information will be used for improving the user environment, preparing for later, more formal audits, or providing feedback to management. The end-user computing support department may initiate this type of review for purely informational purposes to increase support staff awareness of end-user applications development (e.g., the objective may simply be to determine the number of end-user applications or to evaluate their sophistication). Inventories can be done in less time and are less expensive than formal audits. In addition, they can easily be done by the IS department without the consultation of a professional auditor. An inventory is more low-key than a formal audit, and taking an inventory of applications is far less threatening to end users.

Statistical Review. Statistical reviewing involves collecting raw data from the help desk, support logs, computer transactions, or similar sources. This method of auditing is useful only if the support department generates a statistically significant amount of readily available data. This implies a large number of applications, a large number of users, and centralized support or centrally controlled computing resources (e.g., mainframes and local area networks). A statistical review is most appropriate when minor tuning of end-user computing services is the objective. This is an extensive process that provides enough information to confirm or deny perceptions or indicate the need to change; it has a product focus, which can tell how many people are using Lotus 1-2-3 or how many are using WordPerfect. These statistics often come from LANs as users go through the network to access the product. This product focus does not provide much useful information for deciding how to change.

Best-Guess Review. This is the most informal type of review. When time is a critical element, a best-guess review can be performed on the basis of the existing knowledge of the end-user computing support staff. This can even be classified as a review of the IS department's impression of end-user applications. Corporate or senior IS management may request a report on end-user applications within the organization. Such a review can be useful if support people and users are centralized and the support people are familiar with the users and their applications. The IS staff can also use the results to make changes within their limits of authority. Although a best-guess review does not gather significant unbiased data, it can be surprisingly useful just to get end-user computing staff impressions down on paper.

Defining the Scope and Content of the Review

The scope defines the extent of the review and should also state specific limits of the review—that is, what is and is not to be accomplished. In most organizations and with most types of review, it may be helpful to involve users from a broad range of areas to participate in defining the scope. Knowledge of the review and involvement in the definition of the review scope by the users can be valuable in promoting their buy-in to the results.

The review may be limited to particular products, environments, a type of user, a work unit, or a specific application or system. Determining the scope and content focuses on the appropriate applications. As part of defining the scope and content of the application it is necessary to determine the types of end-user environments to be audited. This definition of environment is used to:

- Define what the IS department consider end-user computing, that is, to determine whether a particular application will actually be considered a user-developed application. For example, in some organizations a programmer's use of an end-user product to create an application would be considered end-user computing and the application would be included in a review of end-user applications. In most companies, however, applications that should be included in a review of end-user applications come from the point-of-origin, shared work unit, and work group environments. Applications in a turnover environment can also be included because, although development may be done by another group, end users work with the application on a daily basis. Each of these environment classifications is discussed at the end of this section.

● Define which environments a particular review will evaluate. For example, the application developed by the programmer using an end-user development tool would fit in the distributed environment, which is also discussed at the end of this section. However, the review might be designed to investigate only applications developed in a point-of-origin environment.

In each organization, end-user computing may consist of several environments. Each environment is defined by products, support, and resources. Although there may be a few exceptions or hybrids, an end-user computing environment can usually fit into one of the general categories discussed in the following sections.

Point-of-Origin Environment. In this environment, all functions are performed by the person who needs the application. This is how end-user computing began and is often the image management still has of it. These applications are generally developed to improve personal productivity. They are typically considered to be disposable—that is, instead of performing any significant maintenance, the applications are simply redeveloped. Redevelopment makes sense because new techniques or products can often make the applications more useful.

Shared Environment. In a shared environment, original development of the application is performed by a person who needs the application. However, the application is then shared with other people within a work unit or work group. If any maintenance is done to the application, the new version is also distributed to the other users in the unit.

Work Unit Environment. In this environment, applications development and maintenance are performed by people within an organizational unit to meet a need of or increase the productivity of the work unit as a whole (unlike point-of-origin and shared applications, which are developed for the individual). The applications are usually more sophisticated and designed to be more easily maintained. They may also be developed by someone whose job responsibilities include applications development.

Work Group Environment. In a work group environment, applications development and maintenance are performed by people within a work group for use by others in the work group. The developer is someone who has the time and ability to create an application that fulfills an informally identified need of the work group. In most cases, the application solves problems of duration (e.g., expediting the process) not effort (i.e., productivity).

Turnover Environment. In this type of environment, applications are developed by one group and turned over to another group for maintenance and maybe to a third group for actual use. There are many combinations, but some common examples are:

- The application is developed by the end-user computing support group and turned over to a work unit for maintenance and use. This combination is popular during end-user computing start-up phases and during the implementation of new product or technology.
- The application is developed and used by the end user but turned over to end-user computing support for maintenance.

Distributed Environment. In a distributed environment, applications are developed and maintained by a work unit for use by others. The developing work unit is responsible for development and maintenance only. They may report to a user group or indirectly to central IS. The development products may be traditional programming products or end-user computing products. Although this is not typically considered an end-user computing environment, in some organizations the work unit is the end-user computing support group.

Centralized Development and Support Environment. In this environment, a programming group under the direct control of central information systems develops and maintains applications for use by end users. Although centralized programming groups in some organizations may use end-user computing products for applications development, in general, applications developed in this environment are not reviewed with other end-user applications.

Reseeded Environment. A common hybrid of environments occurs when a point-of-origin application becomes shared. In these instances, the people receiving the application typically fine tune it for their particular jobs using their own product knowledge. This causes several versions of the original application to exist, tuned to each user's expertise and needs. Maintenance of the original application is driven by having the expertise and time available to make alterations rather than by the need for such alterations. This reseeded application grows into another application that should be grouped and reviewed with point-of-origin applications. However, these applications should be reviewed to ensure that they are not duplicating effort. Fifteen to twenty applications may grow out of a single application. In many cases, one application customized for each user would suffice.

Determining the Application Environment

During preparation for the review, the scope and content phase helps determine which end-user environments should be included. For example, the IS staff may decide that only point-of-origin applications will be included in a particular review. The first step in actually performing the review is to identify the environment to which particular applications belong. To do this, it is necessary to isolate who performed the functions associated with the life cycle of the individual application. These functions are:

- *Needs identification.* Who decided something needed to be done, and what were the basic objectives of the application created to do that something?
- *Design.* Who designed the processes, procedures, and appearance of the application?
- *Creation.* Who created the technical parts of the application (e.g., spreadsheets, macros, programs, or data)?
- *Implementation.* Who decided when and how the implementations would proceed?
- *Use.* Who actually uses the application?
- *Training.* Who developed and implemented the training and education for the application? Typically, this is an informal and undocumented process—tutoring is the most common training method.
- *Maintenance.* Who maintains the application? Who handles problem resolution, tunes the application, makes improvements to the application, connects the application to other applications, rewrites the application using different products, or clones the application into new applications?
- *Ongoing decision making.* Who makes decisions about enhancements or replacements?

The matrix in Exhibit III-7-2 matches answers to these questions with the different environments.

Evaluating Applications Development Controls

This step in performing the review provides information about the controls in effect concerning end-user applications. It should address the following questions:

- Who controls the development of the application?
- Are there controls in place to decide what types of applications users can develop?
- Are the controls enforced? Can they be enforced?

Determining Application Criticality

This checklist helps determine the critical level of specific applications:

- Does the application create reports for anyone at or above the vice-presidential level?
- Does the application handle money? Issue an invoice? Issue refunds? Collect or record payments? Transfer bank funds?

FUNCTION / ENVIRONMENT	Needs Identification	Design	Creation	Implementation	Use	Training	Maintenance	Ongoing Decision Making
Point of Origin	User	User	User	User	User only	User	User	User
Shared	Original User	Original user	Original user	Original user	Original user and users with similar needs	Original user or subsequent users	Original user and subsequent users	Original user
Work Unit	Work unit expert or manager	Work unit analyst and product expert	Work unit expert	Work unit expert	Someone other than the developer	Initially by work unit expert—later by user	Work unit expert	Work unit management
Work Group	Work group	Analyst and product expert	Work group expert	Work group	Work group	Initially by work group expert—later by user	Work group expert	Work group users
Turnover	User or work unit	Developing group	Developing group	Developing group and user	User	Developer	Developing group	Work unit management
Distributed	Work unit	Developing group	Developing group	Developing group and work unit	Portion of the work unit	Developing group, work unit expert	Developing group	Work unit management
Centralized	Work unit	Developing group	Developing group	Developing group and work unit	Portion of the work unit	Developing group, work unit expert	Developing group	Work unit management

Exhibit III-7-2. Environment-Function Matrix

- Does the application make financial decisions about stock investments or the timing of deposits or withdrawals?
- Does the application participate in a production process; that is, does it:
  — Issue a policy, loan, or prescription?
  — Update inventory information?
  — Control distribution channels?
- What is the size of the application? The larger the application (or group of applications that form a system), the more difficult it is to manage.

Determining the Level of Security

This set of questions can help determine not only the level of security that already exists concerning end-user applications but the level of security that is most appropriate to the particular applications being reviewed. The following questions pertain to physical security:

- Are devices, work areas, and data media locked?
- Is there public access to these areas during the day?
- Is the room locked at night?
- Is access to the area monitored?
- Is there a policy or some way to determine the level of security necessary?

The following questions relate to the security of data, programs, and input/output and to general security:

- How is data secured? By user? By work unit? By work area? By device?
- Is the data secured within the application?
- Who has access to the data?
- Is use of the programs controlled?
- Are data entry forms, reports, or graphs controlled, filed, or shredded?
- Is there some way to identify sensitive items?

Reviewing the Use and Availability of the Product

Creating complex applications using an end-user development product often requires more product knowledge than creating them using a comparable programming language. The end-user developer may go to great lengths to get around end-user product limitations when the application could probably be created more easily using traditional programming or a different tool. The questions in the following checklist help evaluate the appropriateness of the development products in use to create specific applications:

- Are products being used appropriately? To answer this question, it is necessary to match user application needs to the tool used to create the application. This can help indicate the inappropriate use of development products (e.g., use of a spreadsheet as a word processor or a word processor as a data base).

- Is the user applying the product functions appropriately for the applications being developed or used? For example, a row and column function would not be the most effective function for an application designed to generate 10 or 15 reports using the same data but different layouts.

As part of this step, the availability of end-user applications development products should be assessed. The following questions address this issue:

- Which products are available to this user?
- Which of the available products are employed by this user?
- Are these products targeted to this user?

Reviewing User Capabilities and Development Product Knowledge

This step in conducting a review of end-user applications focuses on the end user's ability to develop applications using a particular product and to select an appropriate development product for the application being created. The questions to answer are:

- Is the user adequately trained in the use of the development product he or she is currently creating applications with? Is additional training necessary or available?
- Does the end user understand the development aspects of the product?
- Is the end user familiar with the process for developing applications? With development methodologies? With applications testing and maintenance guidelines?
- Has the end user determined and initiated or requested an appropriate level of support and backup for this application?
- Is the end user aware of the potential impact of failure of the application?
- Are the development products being used by this end user appropriate for the applications being developed?
- If the user is maintaining the application, does that user possess sufficient knowledge of the product to perform maintenance?

Reviewing User Management of Data

Because the data collected using end-user applications is increasingly used to make high-level decisions within the organization, careful scrutiny of end-user management of that data is essential. The following questions address this important issue:

- Is redundant data controlled?
- Is data sharing possible with this application?
- Who creates or alters the data from this application?
- Is data from traditional IS or mainframe systems—often called production data—updated by end-user applications or processes? If so, is the data controlled or verified?

- If data is transformed from product to product (e.g., from spreadsheet to data base), from type to type (e.g., HEX to ASCII), or from paper to electronic media, is it verified by a balancing procedure?
- Are data dictionaries, common field names, data lengths, field descriptions, and definitions used?
- Are numeric fields of different lengths passed from one application to another?

Reviewing the Applications

This is obviously an important step in a review of end-user applications. The following questions focus on an evaluation of the applications themselves and assess problem resolution, backup, documentation, links, and audit trails associated with these applications:

- Problem resolution:
  - Is there a mechanism in place to recognize whether an application has a problem?
  - Is there an established procedure for reporting application problems?
  - Is there a formal process in place to determine what that problem may be or to resolve or correct the problem?
  - Are these procedures being followed?
- Backup:
  - Is the application backed up?
  - Is the data backed up?
  - Are the reports backed up?
  - Is there a backup person capable of performing the activities on the application?
  - Are backup procedures in effect for support, development, and maintenance of the application?
- Documentation:
  - What documentation is required for the application? Is the application critical enough to require extensive documentation? Is the application somewhat critical and therefore deserving of at least some documentation? Is the application a personal productivity enhancer for a small task and therefore deserving of only informal or no documentation?
  - If documentation guidelines are in place, are they being followed?
  - How is the documentation maintained, stored and updated?
- Links:
  - How are the data, programs, processes, input, output, and people associated with this application connected?
  - What is received by the application?
  - Where does the application send data, information, knowledge, and decisions?
  - Are these links documented?

- Audit trail:
- Are the results of this application verified or cross-checked with other results?
- Who is notified if the results of the application cannot be verified by other results?

GUIDELINES FOR IMPROVING END-USER APPLICATIONS DEVELOPMENT

Reviewing end-user applications requires that some resources (i.e., time and money) be spent. In most companies, these resources are scarce; what resources are available are often sought after by more than one group. A review of end-user applications is often low on senior management's priority list. Reducing the time it takes to collect information can greatly improve the IS department's chances of gaining approval for the review. However, reducing the need to collect information can decrease the need to conduct a review at all. This can be done by setting up and enforcing adherence to end-user applications development guidelines. It is a cost-effective way to improve the end-user applications development environment and help conserve limited resources.

To begin, general guidelines should be created and distributed before a planned review is started. The effectiveness of the guidelines can then be evaluated. The following checklist outlines some areas in which guidelines established by the IS department can improve end-user applications development:

- *Use of end-user development products.* Users should be provided with a set of hypothetical examples of appropriate and inappropriate uses of development products (i.e., which products should be used to develop which types of applications).
- *Documentation.* A checklist or matrix of situations and the appropriate documentation for each should be developed. This could also include who will review an application and whether review of the application is required or optional.
- *Support for design and development.* A quick-reference card of functions or types of problems supported by various groups can be distributed to end-user developers.
- *Responsibility and authority.* A list of responsibilities should be distributed that clearly states who owns the application, who owns the data, and who owns problem resolution.
- *Corporate computing policy.* Corporate policies regarding illegal software, freeware or shareware, and security issues should be made available to end-user developers.

One tactic to improve the quality of end user applications development while avoiding some of the costs in time and money of a full-fledged review is to set up work group auditors. These people may report to a corporate auditing group or to the IS department on a regular basis concerning end-user applications development. This is particularly effective with remote users.

CONCLUSION

The increase in the number of end users developing complex applications and the corresponding reliance of decision makers at all levels of the organization on the data produced by these applications make a careful evaluation of the applications a necessary endeavor. In the current end-user environment, a failed application can seriously damage the business of the organization. To ensure that a review meets the objectives set out for it, IS managers must carefully plan the details of each aspect of the review. This chapter outlines the steps that should be taken before and during an actual review of end-user applications.

III-8

End-User Computing Control Guidelines

Ron Hale

THIS CHAPTER EXAMINES end-user computing control within the context of an architecture based on quality. As end-user computing systems have advanced, many of the security and management issues have been addressed. A central administration capability and an effective level of access authorization and authentication generally exist for current systems that are connected to networks. In prior architectures, the network was only a transport mechanism. In many of the systems that are being designed and implemented today, however, the network is the system and provides many of the controls that had been available on the mainframe. For example, many workstations now provide power-on passwords; storage capacity has expanded sufficiently so that workers are not required to maintain diskette files; and control over access to system functions and to data is protected not only through physical means but also through logical security, encryption, and other techniques.

ARCHITECTURAL APPROACHES TO INFORMATION PROTECTION

Although tools are becoming available (e.g., from hardware providers, security product developers, and network vendors) that can be used to solve many of the confidentiality and integrity problems common in end-user computing, the approach to implementing control is often not as straightforward as is common in centralized processing environments. The goals of worker empowerment, increased functionality and utility, and the ability of end-users to control their environment must be guarded. In many organizations, end-users have the political strength and independence to resist efforts that are seen as restrictive or costly. In addition, networks, remote access, distributed data servers, Internet tools, and the other components that have become part of the end-user environment have made control a difficult task.

To address the complexity of end-user computing, an architectural approach is required. It helps to ensure that an organization's control strategy and technical strategy are mutually supportive. The components of an information protection architecture include management, confidentiality and integrity controls, and continuity controls.

MANAGEMENT STRUCTURE

Perhaps the best and most expedient means of bringing stability to the end-user platform is to develop an effective management structure.

Distributed Management

Because end-user computing is highly distributed, and because local personnel and managers are responsible for controlling the business environment where end-user solutions are implemented, it is appropriate that control responsibilities are also distributed. Centralized administration and management of security in a highly decentralized environment cannot work without a great deal of effort and a large staff. When authority for managing control is distributed within the organization, management can expect a higher degree of voluntary compliance; in particular where adherence to policies and procedures is included in personnel evaluation criteria. If distributed responsibility is properly implemented, ensuring that the goals of the program are consistent with the requirements and goals of the business unit is more likely to be successful.

Distributing security responsibilities may mean that traditional information protection roles need to be redefined. In many centralized organizations, security specialists are responsible for implementing and managing access control. In a distributed end-user environment, this is not practical. There are too many systems and users for the security organization to manage access control. Even with the availability of network and other tools, it may not be appropriate for security personnel to be responsible for access administration. In many distributed environments where advanced networks have been implemented, access controls may best be managed by network administrators. In a similar manner, server, UNIX, and any other system security may best be managed by personnel responsible for that environment.

With many technologies that are used in distributed and end-user computing environments, no special classes of administration are defined for security. Administrators have access to root or operate at the operating system level with all rights and privileges. In such cases, it is not appropriate for security personnel to take an active role in managing access security. Their role should be more consultative in nature. They could also be involved with monitoring and risk management planning, which are potentially more beneficial to the organization and more in line with management responsibilities.

Security Management Committee

Because security in end-user computing environments is distributive, greater acceptance of policies and procedures can be expected if the organization as a whole is involved with defining the environment. To achieve this, a security management committee can be created that represents some of the largest or most influential information technology users and technology groups. This committee should be responsible for recommending the security policy and for developing the procedures and standards that will be in force throughout the enterprise.

Representation on the committee by the internal audit department is often beneficial, and their support and insight can be important in developing an effective security management structure. However, consideration must be given to the control responsibilities of audit and the need to separate their responsibility for monitoring compliance with controls and for developing controls as part of the security committee. In some enterprises, this is not a major issue because internal audit takes a more consultative position. If maintaining the independence of audit is important, then audit can participate as an observer.

Senior Executive Support

The internal audit department traditionally had an advantage over the security organization because of its reporting relationship. Internal auditors in most organizations report to senior executives, which enables them to discuss significant control concerns and to get management acceptance of actions that need to be taken to resolve issues. Security has traditionally reported to IS management and has not had the executive exposure unless there has been a security compromise or other incident. In a distributed environment, it may be beneficial to have the security department and the security management committee report to a senior executive who will be a champion and who has sufficient authority within the enterprise to promote information protection as an important and necessary part of managing the business. Such a reporting relationship will also remove security from the purely technical environment of information systems and place it in a more business-focused environment.

POLICY AND STRATEGY

The ability to communicate strategy and requirements is essential in an end-user computing environment. This communication generally takes the form of enterprisewide policy statements and is supported by procedures, standards, and guidelines that can be targeted to specific business functions, technology platforms, or information sources.

The Information Protection Policy Statement

An information protection policy statement should define management expectations for information protection, the responsibilities of individuals and groups for protecting information, and the organizational structure that will assist management in implementing protection approaches that are consistent with the business strategy. Because the statement will be widely distributed and is meant to clearly communicate management's and users' responsibilities, it should not take the form of a legal document. The effectiveness of the information protection policy depends in large part on its effective communication.

Classification of Information

To protect information, users and managers need to have a consistent definition of what information is important and what protective measures are appropriate. In any organization, local management will be inclined to feel that their information is more sensitive and critical than other information within the organization. From an organizational standpoint, this may not be the case. To ensure that the organization protects only to the appropriate level the information that has the highest value or is the most sensitive, a classification method must be in place.

In the mainframe environment, all information was protected essentially to the same level by default. In a distributed and end-user computing environment, such levels of protection are not practical and represent a significant cost in terms of organizational efficiency. The information protection policy should clearly identify the criteria that should be used in classification, the labels that are to be used to communicate classification decisions, and the nature of controls that are appropriate for each class of information.

Classifying information is a difficult task. There is a tendency to view variations in the nature of information or in its use as separate information classes. However, the fewer the classes of information that an enterprise defines, the easier it is to classify the information and to understand what needs to be done to protect it. In many organizations, information is classified only according to its sensitivity and criticality. Classes of sensitivity can be highly sensitive, sensitive, proprietary, and public. Classes of criticality can be defined in terms of the period within which information needs to be made available following a business disruption.

Monitoring and Control

A method of monitoring the control system and correcting disruptive variances must be established. Such monitoring can include traditional audit and system reports, but because the system is distributed and addresses all information, total reliance on traditional approaches may not be effective.

In an end-user computing environment, relying on business management to call security personnel when they need help is unrealistic. Security needs to be proactive. By periodically meeting with business managers or their representatives and discussing their security issues and concerns, security personnel can determine the difficulties that are being experienced and can detect changes in risk due to new technology, the application of technology, or business processes. By increasing dialogue and promoting the awareness that security wants to improve performance, not to block progress, these meetings can help ensure that business management will seek security assistance when a problem arises.

Standards, Procedures, and Guidelines

The other elements of effective management—standards, procedures, and guidelines—define in terms of technology and business processes precisely how controls are to be implemented. Standards could be developed for documenting end-user applications and spreadsheets, access controls and access paths, system implementation and design specifications, and other elements that need to be consistent across an enterprise. Procedures define how something is done, such as testing applications, managing change in end-user environments, and gaining approval for access to information and systems. Guidelines provide a suggested approach to security when differences in organizations make consistency difficult or when local processes need to be defined. Policies, procedures, standards, and guidelines are each a significant component in the information protection architecture.

CONFIDENTIALITY AND INTEGRITY CONTROLS

Confidentiality and integrity controls are intended to operate on physical, logical, and procedural levels. Because end-user computing is primarily business and user focused, security solutions need to be tightly integrated into the way the business is managed and how work is done.

Physical Controls

In early end-user computing solutions, physical security was the only available control to ensure the protection of the hardware, software, and information. This control helped to ensure the availability of the system as well as to prevent unauthorized access to information and functions.

With the spread of distributed computing and local networks, physical controls still maintain a certain significance. Devices such as data, application, and security servers need to be protected from unauthorized access; and continuity of service needs to be ensured. For example, the integrity of the system must be protected in cases where local users have been given access to servers and have installed programs or made modifications that resulted in service interruptions. Contract maintenance personnel should be prevented from running diagnostics or performing other procedures unless they are escorted and supervised. System code should be protected from unauthorized modifications. Vendor personnel should be monitored to ensure that any modifications or diagnostic routines will not compromise system integrity or provide unknown or unauthorized access paths.

The network represents a critical element of end-user computing solutions. Network devices, including the transmission path, need to be protected from unauthorized access. Protection of the path is important to ensure the continuity of network traffic and to prevent unauthorized monitoring of the traffic.

Lastly, media used with end-user systems need to be protected. As with mainframe systems, files on user workstations and servers need to be backed up regularly. Backup copies need to be taken off-site to ensure that they will be available in the event of a disaster. During transit and in storage, media need to be protected from unauthorized access or modification. Media that are used with the local workstation may also need to be protected. Users may produce magnetic output to store intermediate work products, to provide local backup of strategic files, or to take home to work with. These media, and all media associated with end-user systems, need to be protected to the highest level of classification of the information contained therein.

System Controls

System security in end-user computing solutions is as significant as mainframe security is in centralized architectures. The difference lies in the tools and techniques that are available in the distributed world, which are often not as all-encompassing or as effective as are mainframe tools, and in the types of vulnerabilities.

Tools. In the mainframe world, one tool can be used to identify and to protect all data as well as system resources. For each device in the distributed environment, there may be an associated internal security capability and tool. Tools are often not consistent across platforms and are not complementary. They do not allow for a single point of administration and provide little efficiency from an enterprise standpoint. To gain this efficiency, additional security products need to be installed.

Even when a multiplicity of security tools is used, a decision needs to be made about where to place the locus of control. In some central management solutions, the mainframe becomes the center of access control and authentication. However, this may not be appropriate in organizations that have made the decision to move away from a mainframe. Distributed security management solutions may be practical for some of the many systems used in an environment, but may not address security in all environments.

Another approach to implementing a consistent access control system across all environments is to use tools such as Kerberos. Because considerable effort may be required to link existing mainframe applications and users with end users, such an approach requires a strong commitment from the organization.

Vulnerabilities. In some systems used to support end-user computing, problems in the design of the operating system or with the tools and functions that are bundled with the system have resulted in security vulnerabilities. For example, UNIX administrators have reported compromises of system integrity due to bugs in system software such as editors and main programs. These compromises have been well publicized and exploited by system crackers.

The lack of experience in effective system management has introduced other vulnerabilities. Distributed, open systems may be easier to break because they are open. UNIX source code is available, and high schools and universities teach classes on how to work with UNIX. Persons intent on breaking UNIX systems have these systems available to practice on, the ready documentation to learn a great deal about the system, and a cracker underground that can mentor their activities and provide additional insights. MVS systems, on the contrary, are not open, available, or easy to break. Although UNIX is frequently pointed to as a security problem, similar vulnerabilities can be found in many systems typically used in the end-user or distributed system environment.

The task of security then is to identify areas of risk or technical compromise and to find ways to mitigate the risk or to detect attempts to compromise the integrity of the system. The risk of outsiders penetrating system security should not be management's only concern. Insiders represent a substantial risk, because they not only have all of the knowledge that is available to the cracker community but also understand the security environment, have increased availability to systems, and have potentially more time to attempt to break the system. Thus, an internal compromise may be more significant than an attack from outside of the organization.

Data Base Controls

Access to the data base represents another area of risk in a distributed environment. In mainframe systems, access paths to data are limited, and the security system can be used to control both the data and the paths. In distributed and end-user systems, data can be distributed across an enterprise. In many instances, the path to the data is expected to be controlled through the application. However, users frequently are given other software tools that can provide access through an alternate, unprotected path. For example, user access to data may be defined within client software provided on their systems. Controls may be menu driven or table driven. At the same time, users may be provided with interactive SQL products that can be used to define SELECTS and other data base operations. If the data base is implemented on top of the system level, and if access is provided through a listener port that will acquiesce to any request, users may have the ability to access, modify, and write anything to the data base.

Network and Communications Controls

Access path controls also need to be implemented at the network level. In many environments, various access paths are used, each with different security characteristics and levels of control. One path may be intended for after-hour employee access. Another may be developed to provide system manager access for trouble shooting and testing. Vendors and support personnel may have an entirely different path. In addition, individual users may implement their own access path through internal modems using remote communication software such as PC Anywhere.

Multiple and inconsistent paths can create an opportunity for system compromise. Some paths might not be effectively monitored, so if a compromise were to occur, security and system management might not be aware of the condition.

Access Path Controls. To help ensure the integrity of the network, it is best to provide only limited access points. This helps both in detecting unauthorized access attempts and in correcting problems. Different levels of access may require different levels of security. Because system support personnel will be operating at the system level, they may require the use of one-time passwords. Individual users may be given multiple-use passwords if their access is not considered to be a significant security risk and if monitoring and detection controls are effective.

Access Path Schematic. If multiple access paths are provided, the cost of security may not be consistent with the risks or the risks may not be effectively controlled to support business protection requirements. To identify where control reliance is placed and the consistency of controls across an environment, an access path schematic should be used. This schematic depicts users; the path that they take to system resources, including data; control points; and the extent of reliance on controls. Often, it shows control points where major reliance is placed or where the control is inappropriate and the level of reliance is inconsistent with the general security architecture.

Some users (e.g., network support) may employ several diverse access paths, including dial-in access, Internet access, or private network access, depending on the type of maintenance or diagnostic activities that are required. For system and application access, reliance is placed at the control level on the use of shared identifiers and passwords. Routers that are accessed by network support may have two levels of access provided, one that permits modification of router tables and another that permits only read access to this information. This could represent a significant security vulnerability, in particular when certain routers are used as firewalls between the Internet and the internal network.

External users are provided with dial-back access to the network. At each level of access through the application, they are required to enter individual user identifiers and passwords, user authentication is performed, and an access decision is made. This may be a burden for the users and could provide ineffective security, in particular when password format and change interval requirements are not consistent. An excessive number of passwords, frequent changes, and a perception on the part of users that security is too restrictive may lead to writing passwords down, selecting trivial passwords, or using other measures that weaken the level of security.

Application and Process Controls

The last component of confidentiality and integrity controls is involved with applications and processes. Because end-user computing is generally highly integrated into the management of a business function, security solutions need to address not only the technology but also the process. Application development controls need to be consistent with the type and extent of development activity within the end-user area. At a minimum, spreadsheets and other business tools should be documented and the master preserved in a secure location. It may be appropriate to take master copies off-site to ensure their integrity and recovery.

Work flow management software can be used to protect the integrity of processes. This software is generally a middleware system that allows management to develop rules that define what is expected or the limits imposed on a process as well as to create graphical images that define the process flow. For example, work flow rules can be developed that establish the organization's purchase authorization limits. If a purchase order exceeds the defined limit, the process flow will control what happens to the transaction and will automatically route it to the user with the appropriate signature authority. Through work flow management software, processes can be controlled, end-user solutions can be tightly integrated into business functions, and effective integrity controls can be ensured throughout the process.

CONTINUITY CONTROLS

Because much of the data and processing capability is distributed in end-user computing environments, continuity controls need to be distributed across an organization if systems are to be adequately protected. Centralized solutions for continuity may not be acceptable. Servers may be backed up by a centralized administration group, but this may not adequately protect work in progress or work that is completed on the user's workstation.

In some instances, network backup strategies have been developed to periodically back up the user's workstation. This can be a costly undertaking given the number of workstations and the size of local disk drives. However, with the availability of higher bandwidth networks, compression algorithms, and a strategy of periodically backing up only modified files, a centrally controlled process may be effective in such cases.

In many organizations, the risk of business disruption is not in the mainframe environment but in the systems that have been distributed across an enterprise. End-user computing systems need to be considered when the recovery and continuity strategy for an enterprise, and in particular the business function, is developed. Plans need to be developed to address the criticality of end-user systems to each business function and to determine the best approach to recovering these systems as defined by their importance to the overall enterprise.

CONCLUSION

End-user computing represents a significant departure from traditional data processing. It also represents a unique opportunity to integrate confidentiality, integrity, and continuity with business processes and with the use of information within business units. The following are the confidentiality, integrity, and continuity efforts that should be considered in the context of end-user computing.

- *Establish an enterprisewide information protection policy.* Because information and technology are distributed, responsibility for protecting information also needs to be distributed. A policy should define individual and organizational responsibility for protecting information, the classes of information that need to be protected, and the nature of the protection controls that are required. In addition, the policy should express management's concern for information protection and should provide the basic structure for achieving its goals.

- *Develop a management structure for information protection.* The role of traditional security organizations needs to change to support end-user computing environments. Security needs to be less involved with directly administering access control and more involved with designing controls. Protection management may need to be supported by an enterprisewide committee to represent technical groups as well as users' organizations. The security committee should be chaired by the security manager and should be responsible for managing changes to the protection policy and its implementation throughout the enterprise.

- *Develop appropriate technical components.* An appropriate technical architecture needs to be developed to support the distinct protection requirements of end-user computing. The use of new technologies, increased dependence on networks, easy access to data, and the challenge of protecting end-user-developed applications must be addressed. From a network standpoint, external access points need to be consolidated for better manageability and increased control. Authentication and monitoring controls need to be implemented at the boundary point between the external and internal networks. Access paths to data need to be identified, and all access paths should be secured to the same level. Application development and change control processes need to be adjusted to reduce integrity and continuity risks. Within the end-user environment, controls must be implemented to ensure that access is authorized, that users can be authenticated, and that responsibility for individual actions can be assigned. Auditability controls, to help ensure that unauthorized actions can be detected, also need to be in place. New software solutions, including workforce management middleware solutions, may be used to help ensure that sensitive business processes are effectively controlled.

- *Provide an execution and feedback mechanism.* The end-user computing environment is characterized by rapid and frequent change. The systems that users have available, the software that can be used, and the utilities that can be purchased change daily. To manage change and to provide consistency and control, a means needs to be developed to detect changes either in business processes or requirements or in the technology or its use within an enterprise. To be effective, confidentiality, integrity, and continuity need to be considered in advance of change and throughout the life cycle.

For more detailed suggestions for controlling the use of computing technology by the fully mobile end user, refer to Chapter III-2.

Section IV

Data, Development, and Support

THE ROLE THAT IS has traditionally played in systems development is changing from that of builder to consultant, contract manager, and to some extent, mechanic. In the latter capacity, IS professionals often have to fix something that someone else built—and sometimes not all that well. The IS team is also the keeper of the data, a role that is growing in importance, and almost always responsible for extending the life of legacy systems.

In those cases where the IS group is responsible for building application systems, the approaches used are changing rapidly in response to pressures to shorten development life cycles and reduce the cost of systems. One newer approach is rapid application development (RAD).

Even though RAD promises to avoid such problems as losing upper management support during a long development cycle, information engineering (IE) has more to offer as a development methodology for complex, integrated systems. Modifications to traditional IE, such as defining the business in terms of events that require action rather than functions that are performed, enhance IE's effectiveness in today's business environment. In "A Comparison of RAD and Information Engineering," Chapter IV-1 evaluates the two methodologies and describes approaches to systems development that are responsive to rapidly changing business conditions.

For an application to meet user needs, the human dimensions of systems design must be combined with the technical considerations of systems development. Usability engineering, or "User-Centered Design," can enhance the creation of business information systems. Chapter IV-2 describes the benefits of these techniques, including greater user satisfaction and productivity and reduced development cycle costs and risk.

To take full advantage of the data that has been accumulated in both newer and legacy systems, some companies are also moving toward the data warehouse concept. A dedicated plan and the activities of coordinated data management teams—the data team, the technical infrastructure team, and the direct-use tools teams—are required to establish a data warehouse and directory. Chapter IV-3, "Building and Using a Data Warehouse," presents an approach to data access and availability that is both appropriate for most companies and vital for advancing a company's information architecture into the future.

Decentralization of systems development is becoming common. Today, users are relying on IS staff to supply them with the tools and support that allow them to develop their own systems. For many companies, this kind of empowerment has increased overall efficiency and fostered a stronger connection between the users and their work. Chapter IV-4, "The Decentralization of Systems Development," describes the factors that contributed to this trend and offers IS managers suggestions on how to inaugurate such an effort in their own organizations.

Client/server technology is projected to be the de facto computing standard by the end of this decade. It is therefore imperative that companies planning to capitalize on the benefits of faster and greater access to corporate data understand the architectural and implementation issues they will encounter. Chapter IV-5 reviews the issues in "Client/Server Architecture and Implementation" and provides IS managers with the information they need to lead their companies' use of technology into the twenty-first century.

In today's fast-paced technological environment, organizations are continually seeking ways to do business more effectively, efficiently and, most of all, economically. The move to a client/server environment is one way to achieve those goals. To be successful in the move, however, IS managers need to take specific action to transform IS staff from one development methodology to another. Transforming a mainframe development and operations staff to a client/server environment involves four major tasks with 18 distinct steps. Chapter IV-6, "Moving to a Client/Server Environment," presents a detailed description of each of these steps. Managers who try to take shortcuts on the road to client/server run the risks of failing in the implementation of this new and promising technology.

Object-oriented programming is another methodology that can help in systems development. Chapter IV-7 discusses this relatively new technique, which promises to take the reuse of code out of the realm of hope and into implementation. "Object-Oriented Programming: Problems and Solutions," explains potential problems as well as various methods that can be taken to correct them.

At the same time that organizations move toward newer technologies such as object orientation and client/server, legacy systems support remains a major chore for most IS managers. Even organizations committed to downsizing have business applications that need support and maintenance during the transition. By ensuring that IS personnel keep a reasoned, objective perspective on legacy systems and their remaining business value, IS managers can help bolster the morale of their legacy maintenance staff. "Supporting Legacy Systems," Chapter IV-8, shows IS managers how to minimize apathy toward legacy systems maintenance.

Open systems integration in today's environment challenges IS managers to address issues of performance, communication, and compatibility in what is often a new development arena. Chapter IV-9, "Open Systems Integration: Issues in Design and Implementation," presents seven guidelines for integration that help

IS managers and project teams produce an open system that is fluid, stable, and intuitively usable.

Incorporating hypertext into group decision support systems (GDSSs) is an effective way of improving group decision making and enhancing communication among group members. Chapter IV-10, "Using Hypertext for Group Decision Support Systems," describes the development of HyperGroup, a hypertext-based GDSS designed to be a communication and recording medium for supporting small, face-to-face group decision making. By providing an environment appropriate for GDSSs, hypertext presents an additional way IS managers can deliver rapid and valuable solutions to user needs.

IV-1

A Comparison of RAD and Information Engineering

Edward A. Forbes
Il-Yeol Song

OVER TIME, SEVERAL INFORMATION SYSTEMS development methodologies have been proposed and tried in practice. Only a few, however, have become widely recognized as valuable in effective systems implementation. One of these few is the structured analysis and design technique (Olle, et al 1991), which evolved during the 1970s and early 1980s as a rigorous specification methodology. This structured method emphasized process modeling and functional decomposition and is characterized as a process-centered method. Many systems have been developed successfully using the structured method, but there is a perception that it is too rigid and does not work for large-scale projects.

In the early 1980s, a data-centered methodology known as Information Engineering (IE) was proposed and adopted by a number of IS practitioners (Martin 1991). The techniques employed in IE were similar to those used in structured methods (e.g., functional decomposition, data-flow diagramming, and entity-relationship diagramming). However, IE stresses data over process modeling in the initial stages. More significantly, IE takes a broad, top-down view of the overall enterprise in the initial planning and analysis stages to provide an integrating framework for all succeeding computer applications.

At first, the concept of an enterprisewide integrated information system, as espoused by IE, was very attractive. In practice, however, several problems caused IE to fall into disfavor among all but its most ardent supporters. The complaints generally fell into one of two areas:

- To be successful, IE requires the commitment and support of the very highest levels of management over a long period, while the total system is being developed.
- The initial stages of planning and high-level analysis can be very open-ended, drag on for long periods of time, and produce nothing tangible to the sponsors of the effort.

These are both valid areas of concern. The first, however, is true of the development of any large, complex system requiring an engineering discipline (e.g., construction of an office building, superhighway, or manufacturing facility). If it is necessary to undertake a project of this complexity, the issue of management support must be dealt with as a political problem rather than a technical problem.

The second area, on the other hand, can be addressed with techniques and approaches other than those used in the past. Recently, the rapid application development (RAD) methodology received much attention (Gane 1989). Its intent is to allay the concerns over structured methods by speeding up the development process and removing some of the tedium. This is accomplished primarily through Computer Aided Software Engineering (CASE) tools and more powerful development languages, such as 4GL, but RAD has not changed the basic approach of structured methods. All structured approaches focus on the processes involved in performing localized functions within the business. RAD's goal is to implement each process as quickly as possible; it is important to note that an integrated system is not the primary goal of a development effort using RAD.

Because of the appeal of quicker systems development and the perceived problems with IE, RAD is gaining favor in many business IS groups as the methodology of choice. This chapter proposes that RAD and other structured methods, though superficially attractive to business and IS managers, are fundamentally flawed when producing business information systems that provide either competitive advantage or lower cost of operations.

In fact, IE techniques could be modified to address many of the objections to IE. With these modifications, IE offers the greater promise as a methodology for development of integrated business information systems in the immediate future.

No single methodology is applicable to the development of all possible types of computer systems. Before any meaningful comparison of RAD and IE can be made, it is necessary to define the business problem area to be addressed by these methods.

PROBLEM AREA DEFINITION

For the purposes of this chapter, the problem area is limited to business information systems, primarily for the support of management information systems (MIS), decision support systems (DSS), and business performance analysis and planning.

Viewed in abstract, a business organization is, in itself, a system designed to manage and control, as efficiently as possible, the events involved in the transformation of labor, capital, and materials into goods and services. Earlier standard business organizations and their business functions have been organized in a hierarchical structure.

Generally speaking, normal events are handled and disposed of at the lowest level of the hierarchy. The remaining layers are primarily organized to handle and dispose of unusual events that vary in degree of importance to the business. Each successive layer of the hierarchy, from the bottom up, is involved in the collection and abstraction of information about these events, and their impact, for the next higher level of management. Information flows up the hierarchy to the lowest level at which an appropriate response can be formulated and implemented to deal with an exceptional condition. As the business environment changes, causing previously uncommon events to become more prevalent, the organization of the hierarchical functions is adapted to handle these events at lower levels.

What has been described is an integrated information system that is essential for the operation of any large business organization. Considerable effort is involved in the collection of raw data and its abstraction into usable information by a diverse group of clients. Typically, each computer application collects data and produces information for only one, or a few, nodes in the hierarchy. The personal and nonautomated communication links established within the hierarchy still provide for integration of the total business information system.

At this time, there are two significant trends that affect the course of business information systems over the next few years:

- Most simple systems that automate the manual processes have been built.
- Many large organizational hierarchies are being dismantled (i.e., downsized, rightsized, outsourced, or decentralized).

Looking for new areas that involve the automation of manual processes does not give the business many opportunities for savings. New business information systems require far more complex human interactions than traditional application systems do. These realizations have a tremendous effect on the level of discipline required for information systems engineering in the near future.

Therefore, an effective methodology for the development of these new systems must provide for a global perspective of the organization as well as the inherent complexity of such a large system. Whereas everyone involved would like to believe that systems enhancements will come about through small, relatively simple, incremental steps, indications are that this is a fallacy. Research supports the premise that additional effort to improve the current independent systems randomly has no beneficial effect for the overall business. Research also suggests that the impact on businesses using this approach may even be negative.

For automated information systems to significantly augment the current business hierarchy function, they must be enterprisewide and adaptable to rapidly changing business conditions. Additionally, the number of different events these systems must process will expand as the system takes on more of the burden that used to be handled by business staff.

For the development of complex new information systems, an effective methodology should not only provide for a global perspective of the organization but be systematic and adaptable to rapidly changing business conditions.

TARGET ENVIRONMENT

The effectiveness of any methodology is heavily dependent on the target computer system architecture for the implementation of the information system solution. Whereas current technology provides a wide variety of hardware and software tools from which to choose, the individual organization must define the particular architecture to be used for its own information system infrastructure (Andriole 1990). Determining the architecture is based on the problem area definition; the following is an information system architecture that could be used to realize the new information system requirements described in the previous section.

Businesses, by their nature, are event driven. Sensors, whether human or automated, are established to react to significant events that have been determined by management. These sensors trigger activities to deal with particular events and process them to a conclusion that is most beneficial, or least harmful, to the enterprise. All the processes are governed by business policies that are intended to ensure business continuity and stability. In a stable economy among mature businesses, the business policies must differentiate one business from another.

To be effective as an integrating medium for the business, the new information system must be a working model of the business. This model can be defined in terms of data objects that interact through events within the bounds of defined business policies. In such a model, the data objects represent capital assets and human resources available to the business. These objects interact and change state in response to significant business events, which are represented by the relationships among the data objects. The business policies are represented by the constraints placed on the relationships and the states that the data objects are allowed to assume.

Certain functional requirements can be established for a computer system based on the preceding description. To be beneficial, an information system modeling the business must be easily adaptable to changing conditions in the business world. Additionally, the model must be guaranteed to accurately reflect the state of the business at any point in time, current or historical. External to the underlying business model, provision must be made for management to do projections and what-if analysis based on data from the business model. The computer system must be designed to be extensible with minimal rework, because it will be developed and installed incrementally over a lengthy period.

Based on these functional requirements, the architecture and tools that best support the implementation of this type of computer system can be defined. Fundamentally, the architecture should be data centered, meaning the data base acts as the integrating medium for all processes and activities associated with the business. With this perspective, the rest of the architecture can be viewed from the center out.

Given a central repository of data, discriminating between event processing applications (i.e., those responding to events that cause a change in state of one or more objects) and business applications such as MIS/DSS/EIS (i.e., functions relying on a base of available data, for retrieval purposes only, to prepare decision-making information) is essential. These discrete types of applications must be isolated from one another to ensure the integration of the information system through the underlying data. In such an environment, every relevant fact is recorded in the data base, with validity and integrity, before it is used as a basis for any information requirement within the business. This separation of applications is illustrated in Exhibit IV-1-1.

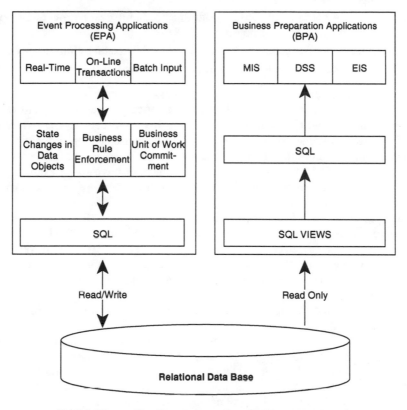

Exhibit IV-1-1. The New Information System Architecture

The separation of these two fundamental types of processes is predicated on their inherent differences. The processes that are permitted to change the data representing the state of the business must be constrained to ensure compliance with policy. Care must be taken that the changes recorded are accurate and that the integrity of the data is not compromised. Access to data by any particular event activity is limited to just what is needed to complete the process. On the other hand, the data required for support of strategic and tactical planning and decision making is unpredictable. Therefore, it must be readily available in a multitude of views and aggregations, rather than bounded by predefined access paths.

With the characteristics of event processing applications (EPA) in mind, it would be reasonable to structure the implementation modules using object-oriented concepts. These would include:

1. Encapsulation of event processing, object state changes, and integrity rules within business objects.

2. Inheritance of data attributes, relationships, and event-processing operations from more general objects.

3. Polymorphism for generic add, change, and delete state change operations.

4. Internal identity of objects for entity and referential integrity.

In this environment, most of the event processing is incorporated as rules and integrity constraints in the form of operations encapsulated in object-oriented data base management systems (ODBMS). In this architecture, any external or user interface to the data base, which is used to trigger common processing logic to handle any predictable event defined by the business, is encapsulated as an operation in the ODBMS module. Access paths also have to be established to allow for exception processing by authorized personnel when any unexpected event or condition occurs.

The characteristics of business preparation applications (BPA) require unencumbered access to all relevant data in various combinations. The best available technologies for this purpose are relational data base and 4GL development languages coupled with the Structured Query Language (SQL). The applications have direct access to the underlying relational database for querying or extraction to local data bases for additional manipulation. Direct data retrieval can be supported through fully normalized SQL views and meaningful aggregate SQL views of the underlying data. Data warehouse concepts are useful to support what-if processing and repetitive access of the same data. By extracting copies of data to separate physical files, end users can modify the data at will without affecting the primary data. For extensibility and flexibility, the data base design must be stable and robust against business changes. By maintaining the intersection tables for 1:1 and 1:M relationships in the data model, a change in policy that changes a cardinality to M:N would have no adverse effect. To ensure adequate performance, the physical data base tables can be denormalized. As long as the specific representation of the data in the physical tables conforms to the normalized data model, the data can be mapped to the interface through the SQL views.

This information system architecture contains two separate front-ends to a relational data base system. The event-triggered applications are processed using a standard SQL interface.

EVALUATION OF METHODOLOGIES

After the problem area and the target environment have been defined, it is feasible to compare and evaluate RAD and IE methodologies. This section discusses techniques that address IE's technical problems, but it is not intended to be an objective evaluation of RAD and IE for the purpose of deciding which is better. This chapter takes the position that IE clearly is better for the problem area and target environment previously described. The rest of the chapter is intended to support this position by describing and evaluating the significant contrasting aspects of RAD and IE, pointing out the flaws in RAD, and identifying alternative techniques for those areas of IE that are currently believed deficient. Quotes critiquing RAD are from Gane (1989) as an expert source on the subject.

The IE methodology is applied in a top-down fashion. The initial phase is strategic planning, which involves a high-level analysis of the entire enterprise to produce a functional decomposition and a global data model. This provides the framework within which all subsequent application development occurs. All local views of data are ultimately mapped back to the global data model to ensure total system integration.

To achieve an integrated business information system, the RAD requires synthesis of many independently developed component systems. This means that relatively small functional areas of a business are analyzed independently when defining system development projects. There is no integrating medium; if integration is desired, it must be accomplished through a synthesis technique.

Synthesis is effective if applied at the logical level before design. However, this requires design and development of the entire system as a single unit. Both RAD and IE advocate incremental development of large systems, as massive development projects have never proved successful. This means that synthesis techniques must be applied to physical models and active systems, requiring large amounts of rework and disruption to users.

The focus of analysis in IE is on the data, which will ultimately be the integrating factor of the total business information system. In RAD, the focus is process oriented. Step one, according to Gane, "is to develop a systemwide data flow diagram."

If the scope of the undertaking is small enough (i.e., one individual can conceptualize an abstract image of the problem area at one time), it makes little difference which model is undertaken first. Process-oriented people will gravitate toward data flow diagramming (DFD), whereas data-oriented people prefer entity relationship diagramming (ERD). Given a much larger scope (i.e., the abstract model must be built incrementally and refined iteratively until a full understanding of the problem evolves), it is essential to model those aspects of the model that:

1. Are most stable over time.
2. Involve the fewest abstract object instances.
3. Can be assumed to encompass the entire problem area.

Based on the most abstract model of reality (i.e., objects interact through events over time, resulting in changes in the state of those objects), the choice is of modeling how the objects and events interact (i.e., the processes) or what is being acted on (i.e., data representing objects and events). It is generally agreed that modeling data is preferred for stability and minimalism over processes.

Although data analysis is used as a technique within RAD, it is applied independently within the scope of each application or subsystem. Gane says that programmers must limit their view of the data to the immediate needs of the application under consideration. "Sometimes it helps us to discover entities which we have not detected when drawing a data flow diagram. Sometimes, however, it tends to include entities which are not actually needed. The entity-relationship model tends to 'model the world,' to be more comprehensive than is needed."

This points out the fundamental difference between RAD and IE. In IE, it is precisely because ERD can model the world accurately and precisely that it is used in the global analysis of the enterprise. In RAD, however, the assumption is that as the individual computer applications are constructed, the incremental data requirements can be added to the model in a seamless fashion.

Once the data model for an application is complete, the technique of normalization should be applied to the resulting logical data base design. Step four, according to Gane, is to normalize the tables. The problem is that the normalized state is dependent on the universal relation in which normalization occurs. Because each application has a different universal relation, what is normalized in the context of one application may not be normalized in another. For example, if the scope of a university application is confined to enrollment, the attribute Instructor___name could very well normalize to the Course entity. In the larger context, Instructor___name is an attribute of the Instructor entity, which may, in turn, generalize to Employee___name, or even Person___name. Ignoring this, assuming that these tables are implemented incrementally as each application is developed, the resulting data base requires significant rework to keep it in normal form as it evolves. Rework is also required in the application programs that are dependent on the original data base structure.

Performing isolated data analysis within the scope of a single application produces, by definition, a local view of data. The premise of RAD is that, once normalized, these local views can ultimately be combined into a global view through canonical synthesis. The problem is that local view analysis, by ignoring the global view, almost always misses the need for generalized entities. Once potential subtype entities have been implemented independently, it is extremely difficult to combine them within a more general entity. This is because the general entity identity has been lost by implementing the subtypes with independent identity. Rectifying this condition requires significant rework of the data base structure and dependent process logic in application programs.

Rather than prescribing a formal design process, RAD applies the concept of modularization, based on the classical Input-Process-Output model of programs, to the process model resulting from analysis. According to Gane, the sixth step is partitioning the logical model of process and data into "procedure units."

The logical model is a DFD, in which the data stores have been correlated to entities in a local ERD. Each procedure unit maps to the Input-Process-Output model. No distinction is made between event processing and information processing. It is left to the designer's discretion to determine what should be recorded in the data base and when that should occur. There is no proscription of direct transformation of input data into output information. The problem is this bypasses the data base as the integrating medium.

IE stresses the central importance of the data base. One result of strategic planning and analysis is a matrix detailing which functions ultimately create, read, update, and delete instances of the entity classes in the data model. Although IE uses DFD as a technique in the later stages of analysis to define detailed process flow, the requirement to record changes in the data as a function of the processes has already been established. The modifications to IE suggested later in this chapter make the difference between event processes and information processes even more distinct.

The cornerstone of RAD, and the source of its name, is rapid application development. Toward this end, the use of high-level, nonprocedural languages (i.e., 4GL) are advocated for application development. Gane allocates a complete chapter to applying 4GL to the development and implementation of information systems. Whereas the 4GL technology advocated by Gane is valuable in the implementation of MIS/DSS applications, it is not currently able to deal with OO event-driven processes. This is a minor point, as more powerful OO languages are imminent, as well as CASE-generated code.

IE, however, does not preclude the use of any particular implementation mechanism, as long as the basic precepts of the methodology are not violated. It is RAD's emphasis on speed of development over effectiveness of the system in business terms that is objectionable.

Finally, there is an area in which the proponents of RAD and IE both agree: A tremendous amount of understanding, support, and commitment on the part of executive management is required for the development of large, complex systems. Gane says that having one senior business executive, known as the executive sponsor, take responsibility for the project's success is a key to that success. The executive sponsor "convenes . . . the 'Steering Body,'" which is to reach a consensus on project scope, system objectives, policy, and organizational impact.

The problem is that the level of commitment implied is rarely present in the organization. If it were, it would be just as easy to make a case for IE, or its results, as it would be for RAD. In reality, one reason that RAD is more popular than IE is that it offers IS a path of least resistance. This phenomenon is epitomized by the goals typically stated by IS (e.g., incremental development, small successes, or quick implementation). These goals can be accomplished through RAD without high-level executive commitment. The problem is all of this activity is taking place in a strategic vacuum. Incremental development of what? Small successes measured against what criteria? Quick implementation of indiscriminate pieces of code with no logical connection to any other code? If IS were to try to use RAD for the development of true integrated information systems, RAD's deficiencies would soon become apparent.

PROBLEMS WITH IE

The primary objection to IE, as previously stated, is that it provides the potential for open-ended analysis during the strategic planning stage. There is no doubt that this potential exists, and as a result there are more than a few instances of IE projects never getting out of the strategic planning phase.

Currently, strategic planning is conducted using the techniques of functional decomposition (FD) and ERD to record and structure information gathered through individual or group interviews. The interviewees always include the executives and managers of the enterprise and may also include key lower-level personnel as deemed necessary. The emphasis during the interviews is on the functions performed in conducting the business. Business entities and their relationships for ERD are deduced from the functional descriptions by looking for nouns (i.e., entities) and action verbs (i.e., relationships).

The problem is that there is no objective way to determine when strategic planning is done. Whereas IE suggests that there is a cutoff point at which functions become processes, this is a very subjective concept and not easily discerned. As a result, analysts working on projects are reluctant to declare the strategic phase complete for fear that some significant aspect has been overlooked. Because functional responsibilities lie with personnel below the management level, no one in a position of authority has the knowledge necessary to make a determination of completeness. The only viable alternative has been to set a predetermined amount of time for strategic planning and analysis and declare that phase complete when the time is up. This is not satisfactory for anyone involved. The analysts are sure they need more time, and management is just as certain that time is being wasted.

This situation can be alleviated by changing the approach taken during strategic planning and analysis without sacrificing the goals or quality of the results. The fundamental difference suggested is to define the business in terms of events that require action, rather than functions that are performed. Events trigger functions; they represent why things happen in the business rather than how they are handled. Because all functions occur as a result of events, strategic planning can be limited to events without affecting the scope of the analysis. Using this approach, no functional analysis needs to, or should, occur during strategic planning. There are several advantages to this approach. Events:

1. Are more discrete than functions or processes.
2. Are the purview of management.
3. Map directly to the ERD as relationships or associative entities.
4. Directly determine the existence and state of object entities.

Because there are fewer events than functions or processes designed to handle them, it takes less time to perform a complete analysis. The interviews are limited to executive and management personnel, because all necessary knowledge about events that are important to the business exists at that level. This limits the amount of time necessary for the interview process. Additionally, the managers can determine when they have discussed all significant events, thus terminating the information gathering. This is the critical event that is missing when performing a functional decomposition that limits the amount of time spent on strategic planning.

The process of event analysis includes event decomposition (ED) as opposed to FD. For the purpose of information system specification, events must occur at a point in time from the perspective of the system. Many events, as perceived by management, occur over a period of time. For example, management may define a sale as an event, but it occurs over a period of time. Depending on the policies of the business, a sale could decompose into the following sequence of point-in-time events:

1. A customer places an order.
2. An order is picked for delivery.
3. A picked order is loaded for delivery.
4. A delivery arrives at a customer location.
5. Title transfers to a customer when a delivery is accepted.

These point-in-time events are also referred to as business rules. Business-rule analysis has been proposed elsewhere for use within the context of bottom-up analysis. It is the incorporation of this technique within IE strategic planning that is different.

Data entries can be of two types, objects or events. Event entities are represented as associative entities in ERD and are mapped directly from an equivalent business event. In the absence of an associative entity, a relationship in an ERD represents an implied business event. Object entities are directly dependent on business events. For these reasons, the ERD resulting from ED is ensured to be more accurate than one based on FD. No intuition is required on the part of the data analyst, as the process of producing the ERD should be a transcription rather than a creative endeavor.

Once the ED and ERD are complete, the strategic planning process should continue normally. Using a CRUD matrix (Martin 1991), the ED and ERD are grouped into application projects of reasonable granularity to be the subject of succeeding business area analysis (BAA) and development. The events and information needs most critical to the business are identified by the business managers. BAA projects are then scheduled based on ranked function and information requirements.

During BAA functional analysis, a modification of the rules for data flow diagram (DFD) can be adopted to support the discrimination of event processes from information processes. Rather than allow individual data stores to be represented in a DFD, all data stores are combined into a single data repository representing the integrated data base. Additionally, data is allowed to flow in only one direction between the data repository and any single function. The direction of flow indicates whether that function was associated with event or information processing. Correlation of the DFD to the ERD occurs at the attribute level based on the content of individual data flows entering and leaving the data repository.

With these suggestions, IE is a better methodology than RAD for the development of the new, complex business information systems that are both imminent and inevitable.

CONCLUSION

In the last few years, RAD methods have been proposed as an alternative to IE methods. The clear implication of the RAD proponents is that IE is too complex, too time-consuming, and therefore unworkable. Whereas acknowledging that IE has deficiencies, this chapter shows that those deficiencies can be addressed to produce a methodology that is effective for the development of business information systems. Given the necessary management support and commitment, IE seems to be a better available approach than RAD to broad scale, integrated system development. Alternatively, RAD is inappropriate as a development methodology for these complex, integrated information systems. It would be short-sighted to abandon IE in favor of RAD just as the need for more complex information systems is imminent.

IV-2

User-Centered Design

James J. Kubie
L.A. Melkus
R.C. Johnson, Jr.
G.A. Flanagan

EVERY YEAR, COMPANIES WORLDWIDE spend billions of dollars developing information technology (IT) solutions to meet their business needs. One increasingly popular approach is client/server computing. Like other technical solutions that have emerged over the years, client/server architectures are complicated to implement because of complex interconnectivity, data base, and performance issues. Consequently, the most important element in an effective system—the end user—often gets relegated to the back burner. This is a self-defeating mistake, however, because for IT to support business, it must produce successful end products: applications that users can and will use to achieve the needs of the business. Even the most ingeniously designed application and user interface will not produce the desired results if end users do not, or cannot, use it.

GOAL: USEFUL, EFFECTIVE SYSTEM SOLUTIONS

Usability engineering, or user-centered design, is a framework of principles, methodologies, and techniques that maximize the likelihood that users will find an application solution useful and effective. Frequently, though, usability engineering is labeled as something that would be nice to do but is too expensive. Misleading impressions and the false sense of security created while pursuing other IT initiatives may also undermine efforts to implement highly usable systems. For example:

- *Reengineering.* Those believing that new streamlined processes somehow translate into usable, effective system solutions when computerized are frequently disappointed.
- *Object-oriented technology.* Techniques for object creation appear to provide adequate task and end-user focus by creating business objects that are linked

349

to perform tasks. The risk is that usability engineering principles and adequate testing may not be undertaken to ensure that the user can easily and productively use these objects to accomplish key tasks.

- *Rapid application development (RAD).* The ease of producing alternative interfaces can blur project focus. A practical question is, when should developers stop prototyping? The answer is, when the requirements and performance criteria specified by the users have been met.
- *Graphical user interfaces (GUIs).* It is a myth that adhering to general GUI standards will produce an acceptable interface, much less the most productive solution.

Why is it important to create effective solutions? It is a simple matter of economics. A client might have incurred a $7 million increase in yearly operations costs just to match the service levels of an old command-line system. Why? Because task performance by average users on critical tasks was not established as a system objective. Instead, the company spent time and money working with exceptional end users during iterative prototyping. Although users had ample time to acclimate to the interface and build-up their performance times, the developers failed to focus adequately on specific, critical tasks. They overlooked the average user and did not recognize the importance of continual testing to confirm the achievement of performance objectives.

FITTING USABILITY TECHNIQUES INTO THE DEVELOPMENT CYCLE

Most design efforts focus primarily on technical, interface, and design validation issues. Effective end-user involvement is, however, one of the most critical factors in creating successful application solutions. The cognitive dimension of interface design deserves greater focus than the presentation, look, and feel of an application. In fact, it is estimated that 50% to 75% of the interface design effort should deal with the murky, difficult issues such as:

- Better task structuring to make the interface intuitive to most users.
- Improving data availability to minimize cognitive load.
- Enhancing screen topologies and screen flow to minimize errors.

This is also the time in the development cycle that the question of usability evaluation should be addressed. Determining what, how, and when evaluations should be conducted, as well as the number and type of subjects, can help maintain a user-centered focus.

Concentrating on the cognitive area of the design challenge produces solutions that are easier to learn and less error prone, which reduces user training and support costs. It also ensures that new processes are performed as designed and can potentially reduce project cost and cycle time.

A user-centered focus can be significantly enhanced by employing usability engineering techniques during all phases of the development cycle (see Exhibit IV-2-1). The next section discusses the requirements process and some useful techniques for identifying, validating, and ranking requirements quickly and

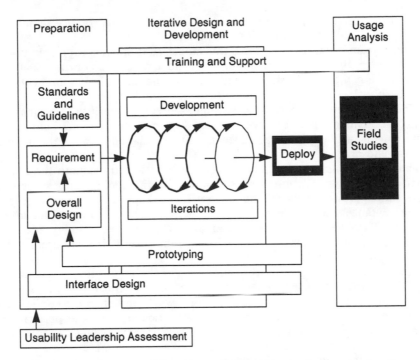

**Exhibit IV-2-1. Incorporating Usability Engineering Techniques During the
Requirements and the Design Phases**

cost-effectively. This work is critical to the definition of the user's conceptual
model, which is needed to establish key tasks, quantitative objectives, and acceptance criteria.

THE USER-CENTERED REQUIREMENTS PROCESS

Understanding the users' perspective is integral to producing a system that
users will accept. The user-centered requirements process provides a way to
describe an application that is likely to be accepted and usable in tangible terms
and that preserves the user's viewpoint during ongoing design and development.
This tangible description allows the entire development team to share the users'
model, which is a critical success factor.

Understanding How Users Work

The key to designing a successful user interface is understanding the user's
conceptual model of the work domain being supported. The user's model has as

much, if not more, to do with how users integrate work in practice than it does with the look and feel of a particular interface.

Project team members must share this model, so it must be expressed in some tangible form. By working with hundreds of end users on many projects, efficient methods for gathering and sharing end-user information with development teams have been refined over the past five years.

Model requirements come from multiple sources, primarily the business and technical points of view. IT is often faced with making tradeoffs among these requirements or filling in missing pieces in the translation between the intention and the end product.

The users' model is an additional framework used to evaluate the requirements. Given a team's commitment to maximize the application's success, the users' model provides a "lens" that focuses IT work on the key requirements for user acceptance and user performance. This lens acts as a guideline for making the tradeoffs in the design. As the design takes shape, the lens also provides a way to determine if an application's design has met the criteria that end users will use to judge the final product.

User-centered requirements keep the scope of the application in perspective. An IT team that is focused on developing one application can easily lose sight of the fact that users have many tools to perform work, and often the application is not the primary determinant of how easy a particular task is to perform. The amount of time required to perform a set of work tasks may be affected more by nonapplication factors (e.g., management policies or complexity of customer materials) than by an application's ease-of-use. User-centered requirements highlight the fact that any application is part of a larger whole from the users' perspectives.

Describing the Work Task and Context

There are two primary types of information required to develop a user model. The first type is information about the work that users perform. The second type of information is the context in which users form opinions about an application. The context can be a set of expectations users have about a new application or a set of attributes users describe as being important. User-centered requirements are a combination of both types of information.

Task Information. Task information includes both procedural and evaluative components. Procedural information consists of descriptions of the process, the work objects, and the expected outcome of a task. It provides a basis for developing the set of objects with which users work. Evaluative information consists of judgments about the tasks, such as the importance, the difficulty, and the frequency of the task. It provides a way to rank the tasks according to their importance and to identify most-probable paths and exception conditions.

Context. User-centered requirements address users' expectations. Users describe attributes of an application they would find satisfying. Through structured techniques, it is possible to refine this description to a set of measurement dimensions that are used to determine how well an application meets users' expectations. By using measurement dimensions, it is possible to measure existing applications and processes to determine a baseline on which the new application should improve.

Taken together, expectations, measurement dimensions, and baseline measures define measurable usability objectives for the new application. Measurable objectives provide the IT team with a way to share the users' perspectives, make decisions about which requirements are likely to be perceived positively by the end users, and agree when and how to address usability problems discovered later in the development cycle.

Identifying Representative End Users

The key difference between user-centered requirements and the more familiar business and technical requirements is that user-centered requirements come directly from representative end users who possess the same attributes as the general user population. They are not necessarily the most experienced in the population or the most technically capable.

Although the application must support very experienced and technical users, it is the average user's skills that need to be apparent in the design. It must meet their needs and provide the support they require to perform work. It does not matter that an expert can perform a task 20% faster with a new application if the average user requires 10% more time to do it. To be valid, user-centered requirements must take into account the user population's demographics.

User-centered requirements are only as valid as the sample used to develop them. For example, job titles and years of experience often do not provide the required information to define a sample adequately. Geographic differences in experience with a new operating environment, in the types of customers the user supports, and in how the user is motivated to perform work can all affect the facets of the application the user attends to. It is important to create representative samples and to collect and treat the data appropriately to develop valid user-centered requirements.

Techniques for Developing User-Centered Requirements

Although most useful at the beginning of a project, taking the time to gather, validate, and rank user-centered requirements is beneficial for projects at any stage. Early in the development process, user-centered requirements provide a road map and decision-making framework on which to base the initial design. Even if a project is well into the design stage, applying user-centered requirements techniques to gather user feedback provides designers with a preview of how an application will be accepted. Having established the important tasks and expectations, users can evaluate prototype designs within a work context.

Several techniques can be used to develop user-centered requirements, including:

- *Direct observation.* Direct observation is spending time with users who are performing the work tasks the application will affect.

- *Task-based requirements sessions.* These sessions can be conducted formally or informally and are similar to focus groups. Groups of users describe the work they do, assign priorities to their tasks, and develop the measurements that can be used to assess how well an application meets their expectations.
- *Gathering baseline measures of existing applications and processes.* Baseline measures are gathered by usability laboratory testing techniques.

Use of Electronic Meeting Rooms. Gathering user-centered requirements does not necessarily mean that a project will take more time. Tools such as Group Systems V allow significant reductions in the amount of time required to collect and analyze end-user feedback. Group Systems V is an electronic meeting room system that allows trained facilitators to gather user input in the users' own words and to structure meetings with greater consistency across sessions. With the electronic meeting room, users can provide ideas, organize them, and rank them in a single session.

The structure of a task-based requirements session depends on the project. Typically, cognitive psychologists play a role in developing a set of meeting agendas that let users express their needs in a way that is useful to the IT group. Cognitive psychologists are trained in methods of capturing users' perceptions, in creating an atmosphere that is conducive to honest user feedback, and in analyzing feedback in a way that preserves the users' point of view.

User-centered requirements must preserve the users' point of view, as opposed to framing users' ideas solely within the context of the IT group's view of the application. User-centered requirements provide a bridge between the end-user view of an application and the IT group. The bridge is in the form of the tasks, their priorities, and the measurable expectations users have of an application.

DESIGN AND DEVELOPMENT PROCESS

Another primary focus area of human factors professionals is testing or usability evaluation. These evaluations must be performed throughout the development cycle and not just before rollout, when the only remedies are adjustments to training and support or even delaying the rollout.

Usability Evaluation: Techniques for Software Developers

Software development organizations often display a somewhat ambivalent attitude toward usability: They acknowledge that usability is critical to the success of modern software applications, but they conduct few meaningful activities directed specifically at evaluating usability. Software developers may be unaware of usability evaluation techniques, or they may simply hold a mistaken belief that usability evaluation requires laboratory facilities or that usability cannot be measured.

In fact, usability evaluation is within the reach of most software development organizations. Techniques can vary, depending on organizational goals and resources, product capabilities, and the skills available. Applying these techniques can result in high-quality applications, which are more likely to generate client satisfaction.

Usability evaluation approaches fall into three major categories:

- *Heuristic reviews.* An application's usability can be analyzed by a usability or human factors expert.
- *Walkthroughs.* Potential users or project team members review all or part of an application and make judgments after seeing a demonstration of the system.
- *Laboratory tests.* An application is evaluated by potential users who make judgments based on hands-on experience with the application for a defined set of scenarios or functional area.

The choice of the most appropriate technique depends on such factors as:

- Project objectives.
- Desired level of precision.
- Development phase.
- Fidelity of application prototype to the end product.
- Time and schedule constraints.
- Cost.
- Available skills.
- Available facilities.

Heuristic Reviews

Heuristic reviews are the analysis of an application's usability. Heuristic reviews can be used at any stage of the development process—from initial design (when only paper descriptions of the application are available) to the prototype stage to the final product. They can range from informal to highly structured.

The informal review is judgment-based. An expert looks at an application and provides quick feedback about its overall usability. Factors such as consistency, navigation techniques, and screen design are considered. Execution of function focuses on the most common tasks.

At the other end of the spectrum, in a highly structured expert review, the reviewer bases the evaluation on a set of heuristics derived from user-interface design guidelines. These heuristics can be tailored to the particular requirements of an application. Sample heuristics include the following:

- Simple and natural dialogs.
- Use of user's language.
- Minimal memory load.
- Consistency.
- Feedback mechanisms.
- Clearly identified exit paths.
- Expert path.

- Effective error message.
- Prevention of catastrophic errors.
- Ability of the application to be specially tailored.

A highly structured and complete heuristic review presents an opportunity to analyze all functions, whereas walkthroughs or usability tests typically focus on major tasks or functions. The heuristic review may uncover problems spread out over the entire application instead of those focused on primary tasks, which might not be covered in usability laboratory testing. Heuristic reviews can also be an effective technique for comparing several similar applications.

Another important use of the heuristic review is identifying major usability problems before taking the application to users for their evaluation, thereby allowing them to focus on problems related to their expertise. For example, an expert review of an application might reveal a problem such as inconsistent exit techniques. Ideally, this problem would be corrected before involving users in extensive usability laboratory testing. Otherwise, users may become distracted by the difficulty of completing a function and might not provide feedback on how well the function corresponded to their business process.

Walkthroughs

Usability walkthroughs provide an opportunity for a set of participants to step through a series of screens for a task or a number of tasks. The set of participants can be one or more users of the application, programmers, or other members of the development team. The screens can be represented on paper, in a prototype application, or the application itself.

User Walkthroughs. In its simplest form, a user walkthrough occurs when a programmer sits down with a single user and step by step goes through a task or set of tasks. More often, the technique is used as an efficient means of gathering input from a group of users.

The evaluator is able to ask direct questions of users about what action they would take in response to the information on the screen, why they would take that action, and what they expect to happen onscreen next. The evaluator also asks if the users notice any problems, or if they have any questions about the information onscreen, being specific to gather comments from users about terminology, layout, navigation, and icons. Users should also be asked about the changes they would recommend.

The user walkthrough technique can yield qualitative data on the overall conceptual metaphor, screen layout, task flow, navigation issues, terminology, icons, and error-prone tasks. It does not provide precise quantitative data on task completion rates or error recovery rates. Because users are walked through a correct path, this technique does not let the evaluator see the kinds of errors that users might stumble into with a hands-on evaluation.

It is important that evaluators using the walkthrough methodology understand the rationale behind any user-suggested changes. Users are not designers, so initial analysis may reveal problems with the user-supplied solution. Further analysis of the underlying problem may supply a workable solution.

It is usually best to represent the system to walkthrough participants as faithfully as possible. It is best to use a prototype rather than paper renderings, or the product rather than the prototype. However, circumstances can dictate exceptions to this guideline. For example, in one study that involved both experienced and novice users of a complex product, novice users preferred using paper representations of screens during the initial walkthrough. When the walkthrough was done online, the experienced users requested many "side trips," which were confusing for novices. Using the paper screens allowed novices to focus on the functions under discussion. Another approach might have been to walk through the tasks online and have the facilitator defer any side-trip discussions until the end of the program.

Although assembling a group of users for a walkthrough offers the evaluation team a vast amount of data, much of that data is often lost because it is difficult to capture during an interactive session. The use of an electronic decision support center can assist the walkthrough session. Evaluators can capture verbatim comments from participants and structure activities for categorization and prioritization. The approach has been judged to be both efficient and cost-effective.

Walkthroughs by Project Team Members. The cognitive walkthrough and formal usability inspection are two specific implementations of the walkthrough methodology that assume participation by professionals including programmers and human factors specialists.

Based on a formal theory of exploratory learning, the cognitive walkthrough includes three components:

- *Problem solving.* Users select among alternatives according to the perceived similarity between their expectation of the consequences of an action and their current goal.
- *Learning.* After an action is executed, the user evaluates the system response and makes a decision about whether or not progress has been made. If so, the user stores the step in memory as a rule. If not, the user attempts to undo the action.
- *Execution.* If a rule is available for the given context, the user executes it; otherwise, the user engages in problem solving.

A sample worksheet to be completed by a participant in a cognitive walkthrough might include the following:

- Description of the user's task goals.
- Description of the series of actions a user might take, along with judgments about whether the actions are obviously available and appropriate.
- Description of how the user will learn what the action is. (For example, are all other available actions less appropriate? Why or why not?)
- Description of how the user will use the action and how the system will respond.
- Judgments about whether the user will understand the results of the action.

The formal usability inspection is a methodology to assist nonexperts in detecting usability problems. It assumes that there are four stages in the user's approach to a task and assigns analytic tasks to inspectors for each stage, as follows:

- *Perceiving.* Judgment about whether or not the user sees the need for information and can judge goal achievement.
- *Planning.* Description of previous knowledge on which the user would base actions.
- *Selecting.* Description of potential problems during selection process.
- *Acting.* Description of problems the user might incur while performing actions.

Both cognitive walkthroughs and formal usability inspections provide the kind of disciplined approach that can help a team assume the user's viewpoint for an objective evaluation of product usability. These techniques can be used alone or as a preamble to other usability activities in which the users themselves perform the evaluation. Like user walkthroughs, these techniques can be performed as standard group activities or conducted with the support of a trained facilitator and group decision-making software. In either case, success depends on the conscientious preparation of all participants.

Laboratory Usability Testing

Usability laboratory testing is an evaluation method that allows the test conductor or team to observe typical users performing defined scenarios in a simulated work environment. The test format allows for controlled scenarios, repetition of tasks across users/evaluators or applications, and precise measurement of factors such as time spent on the task, task completion rates, and error recovery times.

Usability laboratory testing is usually conducted in a two-room test suite. One of the rooms simulates the office or other work environment of the typical user; the other is a control and observation room for the test conductor or test team. The two rooms are divided by two-way glass, which appears as a mirror to the user/evaluator but allows the test team to look into the work room.

In the work room, one to three video cameras focus on the user/evaluator who is performing tasks that represent major activities for users of the system. In the observation room, a team member uses an electronic logging program that provides a time-stamped record of user/evaluator actions. The log, synchronized with the videotape, is an index to events on the tape and is an important tool in data analysis. If other observers participate in the control room, they may narrate the tape or control video mixing and switching equipment.

Laboratory usability can be a compelling educational tool for members of the development team because it allows unobtrusive observation of real users at work on typical tasks. By observing the users, programmers may modify their own expectations about what actions are simple or apparent to end users.

The laboratory testing method also achieves measurable usability objectives (e.g., an objective to be met might be that an employee can schedule a meeting using the electronic calendar in less than five minutes on the first attempt). If scenarios remain consistent, comparative measures can be made on successive releases of a product or across products when comparisons are necessary.

Strengths and Tradeoffs of the Techniques. Although usability laboratory testing is likely to uncover more usability problems and is better able to measure achievement of usability objectives, the other techniques offer different strengths—including the capability of a relatively quick evaluation or the ability to gather input from a group of users in a single session. The walkthrough/inspection methods may be superior for making numerous, low-level design tradeoffs.

Usability laboratory testing and expert review require the participation of trained human factors personnel. Walkthroughs can be pursued with limited or, in some cases, no human factors participation, although at least coaching-level participation from human factors personnel may be recommended.

The measurable results of usability laboratory tests and the videotapes of users at work are extremely useful tools for convincing software organizations that the usability problem data is valid. However, other methods can also be effective when the development team is appropriately involved.

Any of these methods will aid a project in producing a usable application. Exercising a combination of the methods at various times in the development cycle and choosing techniques, refinements, and tools that are best suited to the project and user organization, can greatly enhance the development organization's ability to implement a high-quality application for a satisfied user community.

ASSESSING DEVELOPER'S READINESS TO PERFORM USABILITY ENGINEERING

Whether a development organization is interested in causing a cultural shift in their focus on end users or simply wants to know how it compares to other organizations, an assessment may be helpful.

Early in the life cycle of any software development project, decisions are made about the methods, tools, and techniques to be employed. These decisions are crucial to the project and they are made with great care. For example, no organization would leave the choice of a development language to chance. At the very least, programmer skills and experience, along with an understanding of application, should influence this decision. Choosing an underlying technology, library system, change control system, tracking/reporting mechanism, life cycle methodology, or project standards and guidelines are all areas where explicit decisions based on knowledge of the organization's needs, skills, and capabilities have a profound effect on success.

To make equally informed decisions about the application of usability engineering methods—including about the level of specific usability skills, the efficiency with which these skills are applied, and the effectiveness of usability-related development processes—the following Usability Leadership Assessment is recommended. This technique has been applied successfully at more than 53 organizations worldwide.

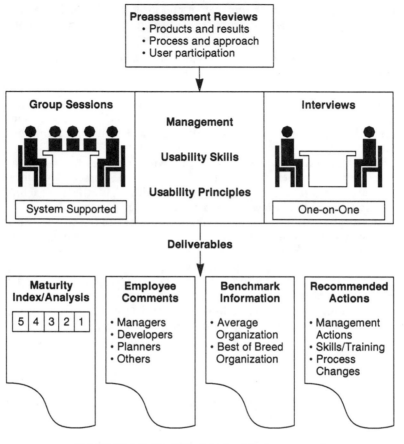

Exhibit IV-2-2. Usability Leadership Assessment

Usability Leadership Assessment Overview

The assessment (outlined in Exhibit IV-2-2) is conducted by a team of skilled assessors who have a thorough knowledge of both software development and usability engineering. The assessment can be done for an entire organization, a cross-organization group, or a specific project group.

The assessors evaluate effectiveness in nine categories that are indicative of an organization's attention to usability issues. These categories are:

- Usability awareness.
- Usability activities.
- Usability improvement actions.

(These three categories are collectively referred to as management issues.)

		Usability Management Maturity					Action Plan Potential
		5	4	3	2	1	
Management Understanding and Awareness	Awareness						
	Activities						
	Improvement Actions						
Usability Skills and Resources	Character, Vitality, and Impact						
	Resources						
Usability Principles Applied in Development	Early and Continual User Focus						
	Integrated Design						
	Early and Continual User Tests						
	Iterative Design						
Overall Assessment							H - high M - moderate L - low
Participant Ratings							

Exhibit IV-2-3. Usability Management Maturity Grid (UMMG)

- Character, vitality, and impact.
- Resource application.

(These two categories collectively are referred to as usability skills.)

- Early and continual user focus.
- Integrated design.
- Early and continual user testing.
- Iterative design.

(These last four categories are referred to as usability principles.)

On the basis of their evaluation, the assessors develop an individual and an overall rating for each of the nine categories and a set of recommendations tailored to the assessed organization. The ratings are summarized on a grid (see Exhibit IV-2-3) and the recommendations, which are linked to these ratings, are prepared to assist in action plan development. Typically, a follow-up assessment is scheduled for six to nine months to reassess usability attention and evaluate progress and the effectiveness of action plans, as well as provide additional recommendations.

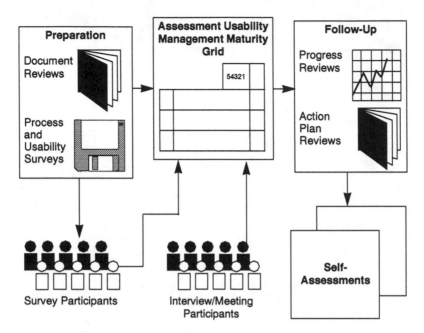

Exhibit IV-2-4. Assessment Process in Four Phases

The Assessment Process

The assessment is conducted in four phases (see Exhibit IV-2-4):

- *Preparation.* Before the actual assessment begins, a two-day visit by members of the assessment team occurs. During this visit, the team develops detailed plans and logistics and initiates a process analysis survey and a data-gathering survey when appropriate.
- *Assessment.* The assessment is typically conducted during an intense one-week period (see Exhibit IV-2-5), culminating in a meeting with the sponsoring executive. During this week, the team completes all major assessment activities, summarizes the data, and prepares and presents a final report.
- *Assessment follow-up.* This activity is similar to the assessment week because similar techniques are used to gather, summarize, and report findings and recommendations. However, this third phase focuses on action-plan effectiveness and potential and the progress made since the assessment.
- *Self-assessment.* Some organizations choose to institute an ongoing, self-assessment program to ensure continued attention to usability and ongoing evolution of the action plans.

	Monday	Tuesday	Wednesday	Thursday	Friday
AM		Organization A and B • One-on-one interviews • Usability presentation • Survey kickoff meeting	Organizations A, B, and C • Survey results collected and summarized • DSC worksession Organization C • Management interviews • DSC worksession	Assessment Team • Data analysis and evaluation • Report preparation	• Executive Presentation
PM	Organization A • Survey kickoff with remote locations Organizations A, B, and C • Group interview with managers	Organizations A and B • One-on-one interviews • DSC worksession Organization C • One-on-one interviews • Survey results analysis	Organization A • Management interviews • Interview data analysis • Final survey collection and summarization	Assessment Team • Data analysis and evaluation • Report presentation	

EVE	Assessment Team
	• Review meeting
	• Identify focus team
		Assessment Team
		• Review meeting
			Assessment Team
			• Review meeting
			  -Interview highlights
			  -Survey summary
			  -Initial findings
				Assessment Team
				• Report preparation	

Contacts by Organization	Total
Design/Architecture	4
Development	25
Quality Assurance	7
End User Representatives	5
Delivery and Support	4
Total	45

Contacts by Type	Manager	Non-manager	Total
Interview	7	10	17
Survey	1	23	24
Other	3	1	4
Total	11	34	45

Exhibit IV-2-5. Final Report to the Sponsoring Executive on Organization's Readiness to Perform Usability Engineering

The Assessment Activities

Depending on the organization being assessed, the process will include some or all of the following activities:

- *Documentation analysis.* This review of past or current project experiences and existing process documentation often occurs before the assessment week and is based on materials supplied by the organization being assessed.

- *One-on-one interviews.* Individuals meet with an assessor for about one hour to discuss experiences and perspectives on how usability engineering practices are applied. Participants represent all relevant disciplines and are a cross-organizational mix that includes both managers and non-managers.

- *Group interviews.* Small, functionally aligned teams meet with one or more assessors to discuss process and practice. During this interview, participants are free to discuss or debate the effectiveness of the existing approach.

- *Process-analysis surveys.* The process-analysis survey is used with organizations having a defined development process. It is completed by three to five team leaders who have considerable knowledge of the organization's development process and procedures.

- *Data-gathering surveys.* Size or geography may dictate the need for a survey where users participate in an introductory presentation (in person or over the telephone) and then complete a survey on existing process and practice.

- *System-facilitated group meetings.* Using laptop personal computers, a LAN, and decision support system software, groups of participants provide their comments and evaluations anonymously. During this session, the assessor chooses from several different tools to facilitate group interaction or brainstorming.

The Assessment Deliverables

The assessment's value is linked to the action taken by the sponsoring executive as a result of the assessment findings. Embodiment of the findings and recommendations are in the following deliverables:

- *Assessment team evaluation.* The assessment team's findings and conclusions form the assessment report and presentation. Included with the narrative is an assessment-team ranking displayed on grid (such as the matrix shown in Exhibit IV-2-3) that is used for benchmark comparisons.

- *Participant evaluations.* Quite often participant evaluations provide the most compelling incentive for management to accept the need for a usability improvement program. The insights gained through the many interviews, direct quotations, and examples of specific activities are all included in the final report.

- *Benchmark data.* The summarized evaluations of all past assessments form a substantial comparison base for organizations to use as a benchmark. This information is included in the report presentation, and the assessed organization's ratings are compared to past assessments. Key characteristics of the top organizations are also presented at this time.

• *Recommendations.* Specific actions that the organization should take to improve its overall attention to usability engineering are presented.

CONCLUSION

There are proven, timely, and cost-effective principles, methodologies, and techniques in the field of usability engineering that can substantially assist development organizations in creating usable, productive systems that meet the needs of end users. Two areas that can produce significant benefits are the requirements process and usability evaluation.

Although the use of trained personnel is most beneficial, many inexpensive techniques can be learned and practiced by every development organization under the occasional guidance of a usability professional. A user-centered cultural change in a development organization can be accelerated by a usability leadership assessment. The primary result of this focus is the creation of application solutions that will truly assist a business in achieving its goals.

IV-3

Building and Using a Data Warehouse

Robert E. Typanski

INTERVIEWS WITH BUSINESS PERSONNEL, which are used to determine their major concerns about the effectiveness of an existing information environment, yield surprisingly consistent results. Management, business professionals, and administrative personnel express dissatisfaction with data availability and development of computer systems to support fundamental company operations, such as manufacturing, order entry, inventory control, and payroll. Their primary needs are:

- Total solutions that met their needs, with a lower price tag.
- Flexible access to company data for decision support purposes.
- Greater accuracy in the applications systems they use to run their daily operations.
- Easier identification of data contained within the company.
- Friendly, easy-to-use software tools, along with responsive training and support.
- Delivery of the applications system within the time constraints of the business need.

To address these needs, information systems organizations must change the way they do things. The entire process of identifying requirements, developing and maintaining applications systems, and making data available to company employees requires a major overhaul.

WHAT NEEDS TO BE DONE

First, a strategic planning process involving all of IS management and many company management and professional personnel must be established. This effort is needed to determine overall goals and objectives and to obtain buy-in.

Only then can specific strategies be developed to address the applications systems and data availability concerns.

Key strategies relating to the major needs are:

- Eliminate bottlenecks that make applications systems development and support unresponsive.
- Maximize applications systems development, productivity, and accuracy by using high-level tools that emphasize design-level development.
- Involve company personnel who will need to work in conjunction with the new environment.
- Provide company personnel with a unified view of the company's data, regardless of its physical location.
- Create a data directory for easy access to company data definitions and descriptions.
- Provide easy access to data through supported, direct-use software tools.

These strategies are the guiding principles. Once they are firmly established it is time to put a structure to these goals.

THE STRUCTURE TO SUPPORT THE STRATEGIES

To implement these strategies, an architecture must be built to provide overall guidance for the next stages—that of changing the IS development and support functions. This architecture is illustrated in Exhibit IV-3-1.

The key characteristics of this architecture is the creation of two distinctive information systems types: information retrieval/decision support systems and applications systems.

Information Retrieval/Decision Support. IR/DS systems are the basis for the informational processing activities of the company. They perform business analysis and reporting functions using data extracted from the applications systems data bases and other sources. This data is then placed in a new data base focused and oriented to the specific needs of the users. This data base is called a data warehouse.

Applications Systems. These types of information systems are the basis for supporting the operational activities of the company. They support fundamental company operations, such as manufacturing, order entry, inventory control, and payroll. These systems enter data into the applications systems data bases, usually through the use of a highly optimized transaction- and batch-processing environment.

THE COOKBOOK: HOW TO DO IT

The processes of the new environment—planning, development, execution, data management, and data enhancement—are created to complement the over-

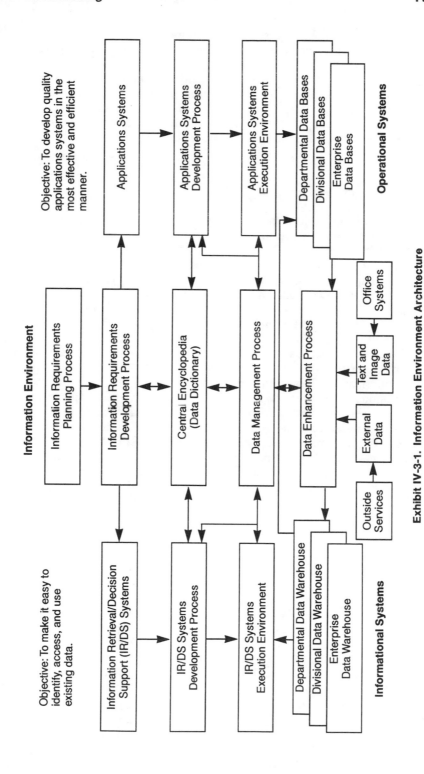

Exhibit IV-3-1. Information Environment Architecture

all architecture. These processes are the recipes that allow IS to produce the products, programs, and data that are required by a company to support its business goals.

Documentation of these processes in a printed or online form must be created as part of the overall effort, to ensure that the information environment achieves its objectives in a way that is structured yet flexible. The documentation of the processes contains the process activities, procedures, and standards as well as guidelines that maximize the benefits of the environment.

DOING IT THE RIGHT WAY

To eliminate the bottlenecks that make IS development and support unresponsive, it is best to select a methodology that ensures responsive development and support activities.

The first criterion used to select a systems development methodology is that it not only complement the information environment architecture but that it also address the needs of the company business personnel. To address the needs of the company, the methodology has to emphasize requirements gathering. This criterion is consistent with the emphasis on quality by companies today, because the definition of quality is "meeting the requirements."

The second criterion used to select a methodology requires that it emphasize data, the major resource of a company's information environment.

The Information Engineering methodology combined with object-oriented (OO) techniques appears to be a good fit as the base methodology for an overall information environment that contains a data warehouse. It places a strong emphasis on gathering data requirements before a solution is attempted.

Inherent to Information Engineering is its emphasis on the development of an information system from a data model. This improves the ability of the IS organization to quickly modify its information systems, adapting them to changing business needs and facilitating the creation of data base designs for the data warehouse.

THE KEY TO SUCCESS

Throughout the process of changing the entire information environment, non-IS business personnel must be involved in steering committees, one-on-one meetings, and group presentations to make sure that the new environment meets their expectations. Feedback from this involvement is used to fine-tune the new environment and verify assumptions regarding the requirements.

A major component of this involvement is the use of pilot projects. Pilot projects determine and ensure the feasibility of the new environment. Company personnel need to be interviewed by IS personnel to apprise them of the new environment and to request their participation in pilot projects used as a final verification.

Finally, success also depends on providing the data administration (DA) and data base administration (DBA) functions with proper staffing and organizational placement.

Organizing and Staffing Data Management Functions

DBA is responsible for the physical placement of data in the data warehouse. It is primarily a technical infrastructure consideration and can be placed in the more technically oriented groups within IS. In some companies the DBA function is outsourced to contract personnel because knowledge of the company business environment is not crucial to its effectiveness.

Data administration is the logical structuring of the data is concert with the business needs of the company and is critical to the usability of the data warehouse by company business personnel. Because of the emphasis on data in a new information environment, this function is created to safeguard that data—the fundamental asset of the information environment—is consistently defined and modeled within the context of company business activities. Just as the financial controllers of the company provide assurance regarding the consistent handling of cash assets, so it is that data administration provides assurance regarding the handling of the data assets.

Although data is usually maintained through the collective actions of many seemingly isolated applications systems efforts, it is the responsibility of the data administration function to define data and model it in a manner that makes it useful to all interested parties.

The data administration function can be optionally placed in several organizations. The most common is to combine DA with DBA in a data management organization within IS. This type of organization helps ensure that the data designs created by the DA function are implemented by DBA if at all possible.

In a company with a strong commitment to data integration and close matching of the data structure to the business needs of the company, both functions need to be staffed with an equal number of employees—this staffing level for data administration is greater than what is usually found in a company, however. It is necessary, though, because of the emphasis on data modeling activities that uncover the business rules and relationships that properly match the data structure to the business. In addition, the data directory usually becomes the responsibility of data administrators, who must establish extensive interactions with company business personnel to accurately define the business definitions for maximum ease of data identification for use by company business analysts.

Other organizational placements for data administrators are in the applications systems development function and, more idealistically, in the company business areas outside of IS. If this last approach is taken, then the DA function truly becomes a company function responsible for the optimal use of the data assets analogous to the financial controller function's responsibility for the optimal use of the cash assets of the company.

In any case, a data administrator must be identified for each company business area. Each of these individuals works with the other data administrators to ensure that all information systems are synchronized to minimize data redundancy. A primary result of the data administrator's activities is the birth of the data warehouse.

THE DATA WAREHOUSE

The data warehouse environment is the final step in the development of a new information environment. It serves as the foundation for strategic and tactical decision making of the company.

A data warehouse is usually defined as a selection of historical data, extracted from the operational data bases, that is organized to facilitate analysis for strategic and tactical decision-making activities. It is time-stamped and contains both detailed and summarized views of the data.

Past efforts to create this environment at most companies have resulted in a fragmentation and duplication of the data. Key components such as data administration and a user-friendly data directory are usually missing. Company personnel can not readily find the data, and costly computer resources are needed to store the duplicated data.

To facilitate use of the data warehouse, the IS organization also creates a data directory, which provides definitions and descriptions of all company data and offers an overview of all data available. This directory is a key factor in making the data warehouse a success.

ROLES AND RESPONSIBILITIES OF COORDINATED TEAMS

Because of the complexity of developing a complete data warehouse environment and the usual preoccupation with technologies and tools rather than data, a separation of the responsibility for selecting components of a data warehouse is sometimes required. Three coordinated teams, each with a special focus, can be created to address the requirements:

- A data team.
- A technical infrastructure team.
- A direct-use tools team.

The teams should comprise both IS and business professionals. The team approach helps to emphasize the focus of attention. A coordinator or steering committee to consolidate the recommendations is also a valuable component. Final recommendations should be tested in a laboratory-style environment (to verify the right technologies were selected) and a pilot data warehouse system developed (to demonstrate the overall capabilities to all company business areas).

The Data Team

This team is the most important of the three teams because it is responsible for ensuring that the data is structured in close coordination with the business needs of the company. Without the proper emphasis on this facet of the data warehouse environment, the contents of the data warehouse would not meet the requirements of the business users. Key objectives of this team are the overall business-oriented data structure, the data enhancement process, and the data warehouse directory.

Data Structure. Business-oriented, logical data structures establish a unified view of company data to users of the data warehouse. Data structure deals with issues of:

- Enterprise versus departmental scope.
- Levels of detail and summarization.
- An overall approach to data modeling.

Pragmatism and Data Integration. The standard approach to data warehouse logical data structures is an enterprisewide, subject-oriented view that helps integrate the functionally oriented, fragmented application data structures. This ideal approach is practical in companies with a strong central IS direction coming from a common company authority body, such as the chief executive or an executive committee that has control over divisional and staff department initiatives.

In today's environment, however, this is seldom the case. Divisions and staff departments are given much more autonomy and are held responsible for their bottom line rather than having to adhere to an overall company policy. Consequently, a more pragmatic approach to data integration is needed.

Generally, although business functions within a company are realizing that they want easier access to their data for information retrieval/decision support purposes, they are not interested in delaying it for the sake of developing an enterprisewide data model that they would then fit into. They want their data to be made available quickly and are not necessarily concerned with standardizing and integrating with other company organizations that have similar needs.

It is the responsibility of the data administration function, therefore, to achieve an enterprisewide view of the data in the data warehouse. This is achieved through close coordination among the various data administrators in standardizing on naming conventions and on business definitions, and in identifying the common sources of data that are requested. The use of a sophisticated data modeling tool that facilitates the sharing of data models and extending them when necessary is also crucial to this approach.

This approach places a large burden or responsibility on the DA function, but the results are a highly integrated view of company data with minimal redundancy. This integration is then accomplished as a by-product of the creation of data warehouses rather than as a preliminary prerequisite activity that delays the warehouse's creation. To achieve data integration in this way requires a DA staff that is:

- Technically proficient in data modeling.
- Skilled in interpersonal and facilitation skills.
- Knowledgeable in the various facets of the company business environment.

These personal attributes, combined with the use of sophisticated data modeling tools, lead to the creation of an integrated data warehouse logical business structure almost in spite of the individual business area demands.

Data Enhancement Process—Gateway to the Data Warehouse. As depicted in the information environment architecture in Exhibit IV-3-1, the data enhancement process is the gateway to the data warehouse. All data must go through this process to ensure that the data warehouse is not only integrated from a logical business perspective but is also operationally maintainable.

The data enhancement process is designed to be used by data warehouse programmers to create the programs that extract data from the operational data bases, outside sources, and office systems. All of these sources must be accessed and data extracted and mapped to the logical business designs of the data warehouse. Programs are created to access the operational data and created timestamps, perform edits, calculate common derived values, and eliminate operationally oriented indicators and flags.

The development of these programs can be accomplished by several methods. Programs can be hand-coded in a procedural language or generated from a computer-aided software engineering (CASE) tool. The most common method used today is to purchase a specially designed data extract tool for creation of a data warehouse. These tools also help to develop the data directory of technical and business information that is used to maintain and access the data warehouse environment.

These programs are then placed in a production environment for scheduling in combination with the business needs that are addressed. Usually these programs are run during the night-time processing schedule.

The Data Warehouse Directory. More than anything else, the success of a data warehouse depends on the content and ease of use of the data warehouse directory. Its purpose is to provide easy identification of the data in the data warehouse to the company's business analysts and managers.

Although the directory contains technical information regarding data lengths, characteristics, and access names, the most important information about data is the business-oriented information that facilitates the identification of data items as the correct ones for the business problem at hand. Key information about this facet of the data warehouse data are:

- Company functional and business area names responsible for the data.
- Subject area names and description.
- Relational data base management systems (RDBMS) table names.
- Data owner name with access authority.
- Access security requirements.
- Frequency of update cycle.
- Update method (e.g., append/replace).
- Physical data base vendor.

Ideally, the directory should provide this information in a manner that is usable by direct-use tools. Recent data warehouse directory products are moving in this direction by including query tools as an integral part of the directory system.

Technical Infrastructure Team

This team is responsible for a more dynamic facet of the information environment—the technical environment that will support the physical data warehouse. The primary requirement placed on this team is that a single image of the data must be provided to users of the data warehouse regardless of its physical distribution. Major components that must be addressed are: physical data base system, hardware/software platform, and middleware.

Physical Data Base System—A Distributed Approach. The prerequisite decision that must be made before a data base management system is selected is whether the data warehouse environment is to be in one physical location or distributed across many hardware environments.

Although one large data warehouse has several advantages, the trend is to the client/server model that allows the implementation of small portions of the data warehouse as business areas of the company develop their requirements and funding approvals. The client/server approach to data warehousing, therefore, dictates a distributed physical data base architecture.

To minimize the complexity of providing a single image of data to users in a distributed architecture, a single data base management system should be used on all distributed hardware platforms. A few such data base management systems are available today that support different hardware vendors and operating systems.

Hardware Platforms. A common operating system should be used, maximizing the flexibility of selecting a hardware vendor to take advantage of periodic vendor price-performance advantages. UNIX is now fulfilling this role; even though all UNIX versions are not identical, they are closer to one another in functionality than proprietary operating systems and require minimal work to migrate from one flavor to another. The actual hardware vendors under consideration should therefore be studied to determine if they provide a UNIX operating system and support the selected data base management system.

Middleware. In today's distributed, client/server environment, middleware has become a major decision. Middleware in the context of this discussion is all the hardware, software, and communications facilities that make it possible to access data across multiple server hardware platforms from multiple client workstations.

Although a myriad of routers, network operating systems, and bridges must be used, two key components need to be selected that have a long-lasting implication in the area of personnel required to develop expertise in their use.

The Communications Protocol—TCP/IP. A single communications protocol is highly desirable to simplify the conductivity issues. Because UNIX is probably the operating system, Transmission Control Protocol/Internet Protocol (TCP/IP) is practically a default choice. TCP/IP's past security deficiencies are being corrected and it is becoming widely available and universally accepted for at least the type of processing incurred in a data warehouse environment.

Data Base-Oriented Middleware. The data base integration middleware decision is very complex. The decision here is influenced by the type of client/server computing used.

In a data warehouse environment, the remote data access type is most common. In this type of client/server computing, the program logic and presentation facets are resident on a client workstation and only data base calls are sent to the server. This type of processing is handled very well by the data base management vendors if their products are used on all hardware platforms.

Middleware supplied by the data base vendors is called data base–oriented middleware. It is usually the responsibility of the data base administration function to maintain the tables within this middleware that are required to provide a single image of the data to users of the data warehouse environment.

Direct-Use Tools Team

Many new tools are available that can access a data warehouse through standard structured query language (SQL) calls. However, because of the very dynamic nature of this facet of the data warehouse environment, it is the most difficult to standardize. In addition, once a particular tool is evaluated and recommended, it usually does not meet the personal preferences of the company's business analysts.

Direct-use tools usually fall into three categories: query and reporting, decision support, and executive information system (EIS). The trend is for tools to combine features of two of the three types. It is now common to have tools classified as either:

- Query and reporting/decision support.
- Decision support/EIS.

In addition, commonly used office software such as spreadsheets are now including query and reporting SQL capabilities, and object-oriented 4GL (fourth-generation language) tools are being used for development of EIS.

IS departments should be careful not to spend too much effort in this area. It is of limited long-term value and the business areas of the company usually want to pick their own favorites.

MARKETING AND DEPLOYMENT OF THE DATA WAREHOUSE

In market-driven environments, new products and services must be packaged and marketed to the customers. So it is with a new product such as the data

Operational Data

Departmental and External Data

Presentations

Servers
Security
Connectivity
Applications

Data Warehouse
(Uses an Open Three-Tier Architecture)

Personalized Business Applications

Shared Business Applications

Warehouse Directory
• Customers
• Products
• Profits

What Is a Data Warehouse? A *data warehouse* is a custom-designed, organized collection of extracted, edited, and summarized periodic, current, and historical data. The data can be extracted from the company's master data bases, can come from external sources, or may contain the someone's own departmental data. The data warehouse contains the data people need to make business decisions.

Why Should the Company's Information Users Have a Data Warehouse?

• Easy data access, from any type of workstation (within company or dial-up using a laptop), is available through several different supported tools or a large selection of individuals' favorite tools.

• Access to purchased and departmental data—by all authorized people.

• Personalized and shared applications can be prepared using a person's tool of choice.

• Presentation of results is customized using the person's tool of choice.

• Periodic data is available, so people do not have to wade through several months of reports to find all the data they need.

• Organized directory means no one has to scroll through irrelevant data.

• Summarized data is available, so users do not have to look at all the details to find exactly the data they need for decision making.

• Customized data means that the data in the warehouse is defined by the users themselves.

• Extracted data is available, so the data a user needs is already there and available.

• The data is current, updated regularly, according to a user's business needs. No one has to wait for the end of the week, month, or quarter.

• New data requests can be addressed by managers or their staffs by using the data warehouse directory.

Exhibit IV-3-2. Sample Marketing Sheet from a Data Warehouse Brochure for Users

warehouse. Marketing is a new world for IS organizations, and it is usually not one of its strengths.

Although the data warehouse benefits the company as a whole, the IS department cannot assume that all company personnel readily appreciate it. The responsibility for convincing the company personnel that they will be better off and more effective in their jobs because of data warehousing lies with the IS staff. Toward this end, IS must develop a marketing plan for its data warehouse product.

The Marketing Brochure. The first step to take is to develop a marketing brochure that provides a more interesting depiction of the data warehouse environment. The use of color and graphic arts–quality artwork, along with down-to-earth wording that whets the audience's appetite, goes a long way toward generating interest. A sample page from a real data warehousing marketing brochure is shown in Exhibit IV-3-2.

This kind of brochure can be mailed to internal company personnel and then followed up with a personal contact by a representative of the IS organization. The brochure can also be distributed during special demonstrations of a pilot data warehouse called a roadshow.

Taking the Data Warehouse on the Road. This roadshow is more than a demonstration of a data warehouse. It is carefully orchestrated to provide potential internal customers with testimonials from satisfied users, introductions to key support personnel who are there to serve them, and an opportunity to test-drive a data warehouse in their own environment. Technology should be discussed only if it is requested and at a level that does not embarrass any of the participants.

CONCLUSION

By taking a structured, planned approach involving personnel from all facets of the business, companies can transform the business of providing IS services from a fragmented, technology-driven approach to a unified, data-driven environment including a data warehouse. This new environment emphasizes the identification of requirements and data as well as the use of appropriate technologies and processes to fulfill those requirements.

Even greater benefits will be realized by the executives, managers, business analysts, and administrative personnel using the data warehouse through:

- Easy identification of data through use of the data warehouse directory.
- Unified data to support strategic and tactical decision making.
- Flexibility in hardware vendor selection.
- Comprehensive training on direct-use software tools.
- Effective management of the data assets of the company.

The data warehouse environment is here. The more it is used, the better it becomes.

IV-4

The Decentralization of Systems Development

Lynn C. Kubeck

PROGRAMMING HAS TRADITIONALLY been an endeavor completed behind closed doors by IS professionals. End users would submit requests for programming and wait, sometimes several years, for their requests to become programmed systems. Often, by the time the systems were developed, they were no longer needed. In the interim, alternative methods were created by the end users out of necessity. End users started relying more on IS to provide the tools that would allow them to develop their own systems. This shift from centralized programming to decentralized end user-developed and maintained systems has been occurring over the past few years and is anticipated to grow.

The end-user development trend has not evolved in isolation. The issues facing many corporations today—how to be more effective with limited resources, how to improve quality, and how to empower employees—are all facets of this trend. When IS management provides end users with the tools they need, the training and support they need to use them and accurate data from which to query and make their decisions, the IS department takes on the role of change agent within the organization. Some of the factors contributing to this trend are: the IS staff's level of business knowledge, the rapidly changing business environment, and the advance of end-user query and development tools.

BUSINESS KNOWLEDGE

For any information system to meet the needs of business, information technology must be linked to the business needs. Traditionally, IS would send systems analysts to study the business function to understand what the users needed. This was (and is) a time-consuming task and does not provide the level of understanding that the individuals working within the various business functions have of their own needs.

IS must spend a great deal of time to understand the needs of a specific business function or unit, and the level of understanding may never surpass that of the individuals within that unit (e.g., the accountants, financial analysts, and production engineers). As various professions become more specialized, the ability of an IS systems analyst to understand, in detail, the functions of every unit of the business becomes a more unrealistic goal. Systems development must be a joint effort between individuals with the appropriate business expertise and those with the requisite information technology expertise.

Computer Literacy Versus Information Systems Literacy. End users need to be not only computer literate, but information systems literate. Computer literacy consists of knowing how to use information technology. It involves understanding how to use software packages, communications, and various hardware components and peripherals.

Information systems literacy takes a broader perspective and includes information technology knowledge, as well as how to apply the technology to a business situation or problem. IS managers, through training and consulting, must promote this type of literacy rather than just computer literacy. If end users can couple information technology with business needs, the benefits provided can far outweigh any system that is developed solely by either IS or the end users.

IS can understand the business needs of the end users only to a certain point, and the end users can understand the capabilities of technology only at a certain level. It makes sense to use a team approach, with each member of the team bringing their expertise to the project. Many organizations, through adoption of total quality management (TQM) or a variation of these principles, are moving toward the team approach to problem solving and process improvement or business process redesign (BPR). Information technology is an enabling tool that can facilitate these changes within an organization.

CHANGING BUSINESS ENVIRONMENT AND CUSTOMER-DEFINED QUALITY

Another factor contributing to the trend of end user–developed systems is the accelerated change in the environment. Competition is swifter, government regulations change more frequently, and the need to alter existing information systems to meet these changing business needs is paramount. The growing IS programming backlog, coupled with the accelerated environmental changes, has forced business units to determine alternative methods to obtain the information they need to make business decisions.

Out of necessity end users must develop methods to stay abreast of the changing demands placed on them. In many instances, they have developed and maintained isolated systems that capture, use, and maintain decentralized data. This has created pockets of information within the company that are not connected and do not provide the information senior executives need to strategically position the business. The need for ad hoc reporting and the ability to modify systems quickly to adapt to the changing environment is a concern of all end users.

The organizational context of many companies is being reassessed. The change is toward empowering employees, measuring quality by the customer, and continually changing and adapting to customize services and improve processes. The adoption of TQM, quality improvement, and all the variations require an organization that can quickly adapt to changing market demands, governmental regulations, and competition. This emphasis on quality and service is affecting internal service providers.

In most businesses, IS departments, or areas that provide service to other departments, have been or will be asked whether the service can be provided by an outside firm at less cost and with higher-quality service. Outsourcing has been a recent trend companies have used to improve quality, reduce costs, and adapt to change rapidly. How can IS stay ahead of direct competition (i.e., the outsourcing companies providing end-user technical support, training, or programming)? One way is to become an integral part of the business—changing the traditional IS service model from one of delivering a specific deliverable (e.g., a training class or a customer-written program) to a team approach, with IS management and the end-user departments jointly developing and supporting their information technology needs. A new paradigm for IS managers should be helping end users develop their own systems and empowering them through training and support to be able to adjust to changing demands from the business environment themselves.

The emphasis should be on empowering end users to be self-sufficient. At first glance, the response from many IS departments is that this tactic may eliminate the department. This goal is a moving target, however, and the more end users know and can do for themselves, the more increased support in the form of consulting, training, and help desk services will be needed. It creates a win-win situation for all concerned.

ADVANCEMENT OF END-USER TOOLS

In the past, end users who wanted to control their own data or have access to corporate data would need to develop a program in a third-generation language (e.g., COBOL or C). This required a great deal of training in programming standards, syntax, and documentation. COBOL, which was purported to be self-documenting, has proved otherwise.

Internal programming documentation is critical for maintenance purposes. With the advent of data base query tools and report generators, end users can access the information they need without using traditional programming languages. Object-oriented technologies (e.g., Object Vision), Windows-based data base management systems (e.g., Microsoft Corp.'s ACCESS), or query tools that are available for many data bases (e.g., ORACLE) allow users to create applications and effectively manipulate and use data without writing a line of code.

Windows technology with object linking and embedding (OLE), macro programming or scripting in Windows spreadsheets, and word processing tools provide a great deal of flexibility for end users to view, manipulate, query, and report on not only local or departmental data but centralized data using structured query language (SQL) links. The windows technologies allow end users to develop sophisticated applications by connecting standard software packages through OLE and DDE analogous to building a structure out of Lego blocks. (OLE and DDE are Microsoft Windows terms discussed in Windows reference and technical manuals; they relate to methods for connecting the programs and are beyond the scope of this chapter.) The power of these tools opens up several possibilities for end users to control their own systems.

This also opens up several opportunities for IS as well. End users can use these tools to access their information. However, installing, configuring, and learning these tools requires some assistance. Also, the commitment to provide users with the necessary knowledge to be information systems literate rather than solely computer literate must be made. In other words, they need to know how and when to apply a specific technology to a business problem or need. This requires ongoing consulting, training, and help desk support to bring their knowledge to an appropriate level for the decisions they need to make.

CHANGING ROLE OF IS AND END USERS

Factors contributing to the end user–developed system trends have been identified, but it should be noted that not all systems can be decentralized. Guidelines to determine which systems should be centrally developed using traditional programming languages or fourth-generation data base languages and tools must be defined. One way to determine whether a system should be centrally developed is by the number of departments that are involved in the system and the need for senior management to have access to the data captured in the system. Systems that capture the raw data for an organization are called transaction processing systems. These systems provide the data from which all other reporting and querying systems obtain their initial data.

In an educational environment, a system that captures student data is a transaction processing system and should be centrally supported. For example, the registrar office could not develop its own system for registering students and the bursary develop another system for accepting payment. Advisers who wish to see or query the data captured, however, should be able to view that data in a way that is convenient for them. This type of adviser's view of the student data is an example where windows-based query tools, IS support through consulting and training, and a team effort between the users and IS technical staff would provide an end-user system customized to their needs. If the advisers learned how to create their own view or interface to the data, they would be empowered to adjust their view as their needs changed without affecting the central transaction processing system.

This trend also causes some changes in the roles of the end user. Aside from developing their own systems, users must take ownership of their systems and data. In the student information example, the advisers may wish to maintain notes on individual students they have advised. These notes would not be maintained in the central student information data base and this function would not be supported by centralized IS. These notes could be maintained on their computer locally, or decentralized, and through IS support linked only for their view with the centralized student data base information.

The systems developed through this team approach are not the same as those traditionally programmed by IS. They do not contain lines of programming code and intricate file structures. The systems the end users develop are created using macros in word processing and spreadsheet applications. They are developed by accessing centrally maintained data and manipulating and arranging that data into meaningful information. In a windows environment, links created between standard applications can provide a custom application for end users without writing any code. The same is true for windows query and report generation tools. The users must understand more about the data and look toward IS management to provide them with the training and knowledge about these new tools. The role of IS will shift from a concentration on programming development to consulting services, data base integrity issues, end-user training, and support issues. Each of these issues is discussed in the following sections.

Consulting Services

A new role that IS must assume is one of internal consultant. End users need assistance to take control of their own information and their own systems to access that information. Information access allows employees to be empowered. Without the necessary assistance and technological advice IS can provide, however, the benefits gained from empowerment are lost. To assist the organization in determining which systems should be centrally developed and which should be developed and maintained by individuals or departments, the following sections provide a list of issues that must be addressed for data and information systems developed and maintained by end users.

Software Standards. To provide adequate support, software standards must be addressed. Can help desk, training, and consulting support for 10 word processors, 7 spreadsheets, and 5 data base query tools be adequately provided? One useful approach for setting standards is the establishment of an advisory committee that can evaluate and establish software standards for the organization. Hardware standards would also be helpful, particularly when dealing with distributed processing and networking issues.

Centralized Development Versus End-User Development. To ensure that mission-critical systems are centrally developed, maintained, and supported, guidelines must be established. One possible guideline mentioned earlier in this chapter was applications that are transaction processing systems (gathering the data that other systems query from) should be centrally developed. This also helps address the issue of data integrity.

End User–Developed Systems. IS should ensure that projects that meet the organization's mission are advanced and those that do not meet that mission are not allocated resources. It is wise to analyze any given problem within a broad organizational context. One possible way to ensure that unnecessary systems do not usurp valuable resources is to initiate a problem-solving or brainstorming session between IS management and the end users before the development of a system begins.

Many times problems can be resolved by simply eliminating steps in a procedure or changing a policy. Emphasis must be placed on identifying a problem before automated solutions are applied. Caution must be taken when end users are developing their own systems. Supplying them with tools and the ability to use them without helping them analyze the system they want to develop is a mistake. The emphasis should be on helping them become information systems literate.

Maximizing Resources. Even if a system is developed by a department, there should be some dialogue in the form of a consultation between central IS management and that department. Information related to the type of system, the data being captured or needed, and the purpose of the system should be maintained in a central repository or encyclopedia. It is important to maintain an accurate account of the data bases and systems within an organization. Without this central listing, the proliferation of redundant data and systems is imminent.

IS can support not only the trend of end user-developed systems, but also the trend toward effective maximization of existing resources by providing business process redesign consulting. Before any system—either a central traditionally developed system or an end-user system developed through query tools and ad hoc reporting—is developed, the business process that it supports should be analyzed.

One helpful method is to ask if a real need for the task exists or if it is an assumed requirement because that is the way it has been done in the past. Not every problem requires a technical solution. The ability to eliminate steps that do not add value to any business process provides an opportunity to link IS more closely with the business functions of the organization, and fosters the team approach necessary in today's business environment.

Data Base Development

For end users to be able to use Windows- or DOS-based query tools and report generators, central data must be created and maintained as well as a repository indicating the types of departmental data that are captured. All of the data belongs to the organization and must be identified to eliminate duplicate information being maintained by various departments. IBM Corp. has developed an information warehouse strategy emphasizing this concept of providing transparent access to corporate data using various tools. As previously mentioned, systems that capture data central to the organization should still be developed, maintained, and supported centrally. The emphasis is shifting from IS focusing on the applications programs to focusing on the data. The integrity of central data is a primary concern and must be ensured.

If IS can provide accurate, timely data that end users can access through various query tools, much of the need for traditional programming can be replaced by an infrastructure to support end user-developed systems. This infrastructure would need to include a records management function applied to electronic data within corporate data bases as part of its support. In other words, the data being captured and maintained must have various characteristics associated with it. The length of time it should be retained must be determined, and when the data is no longer current the decision must be made as to whether it should be archived or destroyed. These are issues that must be addressed for not only an organization's paper documents, but electronic data as well. The Association of Record Managers and Administrators (ARMA) can provide guidelines for data retention, both paper and electronic.

Training

IS must also provide more technical support in the form of end-user training and help desk or information center services. If the focus shifts from end users simply using applications that were built for them to end users using query tools and report generators to access and maintain local as well as corporate data, different skills are needed. End users must be aware of data base concepts and data integrity issues and have a basic understanding of the tools they are using.

In addition, the entire process of developing a data base and a system must be taught. Because the end users are not using traditional programming languages, traditional training does not suffice. Training users to code in C does not mean they can necessarily produce a system that meets their needs—the syntax may be correct and it may be a masterpiece in programming code, but if it does not solve the business problem it is a failure.

Emphasizing Problem-Solving Skills. The important issue that must be incorporated into a training program is problem solving, or systems analysis. It is important that the broader view of problem solving be incorporated into the training; this view incorporates the computer system as a subcomponent. A system consists not only of hardware, software, and data but of the procedures and personnel involved in the entire process.

A helpful model is one that incorporates the training into the consulting services. Although IS management is providing consulting services on business process redesign to determine whether a system should even be developed, just-in-time training could be used to provide end users with the problem-solving skills they need to address similar issues in the future. The goal is to develop self-sufficient end users who would be able to determine when IS must be involved and how to solve many of their problems on their own. This approach is in line with the earlier discussion of computer literacy versus information systems literacy. Training should provide the end users with the ability to know not only which keys to press in a given software application or query tool, but also when to apply information technology or at least know what questions to ask. Awareness comes in incremental steps. The goal should be a joint effort or team approach so that the next call or request the end users makes to IS management is more sophisticated.

After an end user is certain that the system they need is necessary (i.e., they have identified the problem and not a symptom), they need appropriate training in a development tool. IS managers should provide not only classes but departmental training on the various end-user tools that the organization supports. Various video and audio training is available to augment the training classes. In this way, a department can obtain the training it needs when it needs it.

Documentation. Documentation is an integral part of any system and IS managers should provide training to end users to assist them in developing their own documentation. If end users develop their system with traditional programming (which may be necessary until sufficient support is available for the newer tools), they must learn how to document the program. Documentation with a traditional program includes remark lines within the program that are not part of the program but simply supply information about the purpose of the program.

External documentation regarding the file structures, indexes, and programs must also be compiled. Most programming tools that end users employ should provide this documentation. If this is not provided by the vendor, IS managers may want to provide a third-party tool that can determine the data bases, programs, and hierarchy charts used as well as cross references of fields and indexes. These tools are available through software houses. Training would have to be developed for these tools.

Technical Support

An important function that IS management must provide to end users is one of evaluating and recommending current technologies. End users need tools to develop their own systems. If they are required to develop systems using traditional programming languages and methodologies, the organization has not benefited by maximizing resources; it has just shifted programming responsibilities from IS to the various departments.

IS managers must stay abreast of current advances in technology. The issue of providing standards for end-user development tools must also be considered. If no standard or recommended alternative that management fully supports exists, it may be difficult to leverage training through classes.

The questions that end users ask IS managers are also different. They are more technical—related to how they can obtain certain information they need from the existing data bases available to them. End users want to know how to set up a personal or departmental data base. They need to know how to use client/server technologies and how to access several different data bases transparently. The expertise required by IS staff may increase. As end users become empowered to take responsibility and accountability for their systems, their technical questions become more complex and sophisticated.

CONCLUSION

As end users take on more responsibility for their systems, and technology advances to provide simpler user interfaces, the need for IS management to not only support end users but to become integral members of their teams will increase. To continually improve processes in the organization, this team approach is mandatory. Just as the focus of traditional programming is shifting from applications to data, so too is the focus of IS management shifting. IS managers must provide consulting services to end users to assist them in developing their own systems.

IS managers must also provide training when it is needed and go beyond the basics of skills training (i.e., information systems literacy versus computer literacy). The demands facing managers because of the trend toward decentralized systems development are numerous. By changing the focus from applications programs to data, many benefits can be obtained. If the traditional systems development is shifted to end users without changing the paradigm of traditional development, only existing responsibilities have moved and the situation is not improved. If a balance is maintained between centralized and decentralized systems and managers focus on the data rather than the applications programs, substantial improvements in productivity can be made. Organizations can adapt to changes quickly, to learn and improve their processes. It is an exciting time, filled with opportunities for those IS departments that are ready for the challenge.

IV-5

Client/Server Architecture and Implementation

Nathan J. Muller

FOR MUCH OF THE 1990S, the client/server architecture has dominated corporate efforts to downsize, restructure, and otherwise reengineer for survival in an increasingly global economy. Born of the frustration with the slow pace of centralized, mainframe-based applications development, client/server computing brings computing power and decision making down to the user, so that businesses can respond faster to customer needs, competitive pressures, and market dynamics.

THE CLIENT/SERVER MODEL

In the client/server model, an application program is broken out into two parts on the network. The client portion of the program, or front end, is run by individual users at their desktops and performs such tasks as querying a data base, producing a printed report, or entering a new record. These functions are carried out through structured query language (SQL), which operates in conjunction with existing applications. The front-end part of the program executes on the user's workstation.

The server portion of the program, or back end, resides on a computer configured to support multiple clients, offering them shared access to a variety of application programs as well as to printers, file storage, data base management, communications, and other resources. The server must not only handle simultaneous requests from multiple clients but perform such administrative tasks as transaction management, security, logging, data base creation and updating, concurrency management, and maintaining the data dictionary. The data dictionary standardizes terminology so data base records can be maintained across a broad base of users.

DISTRIBUTED NETWORKS

The migration from central to distributed computing has produced two types of networks involving servers: the traditional hierarchical architecture employed by mainframe vendors and the distributed architecture employed by LANs. The hierarchical approach uses layers of servers that are subordinate to a central server. In this case, PCs and workstations are connected to servers that are connected to a remote server or servers. This server contains extensive files of addresses of individuals, data bases and programs, as well as a corporate SQL data base or a file of common read-only information (e.g., a data dictionary). A terminal on a LAN making a request for data not on the LAN has its request routed to the central server. The server adds any pertinent information from its own data base and sends the combined message to the end user as a unit. To the end user it appears to be a single, integrated request. Also, the local LAN server passes data base updates to the central server, where the most recent files reside.

This type of server network maintains the hierarchical relationship of mainframe communications architectures (e.g., IBM Corp.'s SNA), thereby simplifying software development. An added benefit is that more programming expertise is available here than in the distributed environment. The disadvantage is in the vulnerability of the hierarchical network to congestion or failure of the central server, unless standby links and redundant server subsystems are employed at considerable expense.

In contrast, the distributed server architecture maintains the peer-to-peer relationship employed in LANs. Each microcomputer on the LAN can connect to multiple specialized servers as needed, regardless of where the servers are located. Local servers enable such services as data bases, user authentication, facsimile, and electronic mail, among others. There are also servers responsible for managing connections to servers outside the LAN. The microcomputer merges data from the server with its own local data and presents the data as a composite whole to the user (see Exhibit IV-5-1).

Distributed client/server architectures can be difficult to apply because the software to manage them still needs considerable development. After all, microcomputers must know where to find each necessary service, know the access codes, and have the software to access each service. Such software has been developed for traditional mainframe-based architectures. But centralized architectures are beginning to run out of steam and distributed server networks, when combined with LAN internetworking devices like routers and bridges, are coming into the mainstream of LAN internets. Keeping the network and computing process transparent to the end user is rapidly becoming a requirement for future networks.

The network also consists of the transmission medium and communications protocol used between clients and servers. The transmission medium used is no different from that found in any other computing environment. Among the commonly used media for LANs is coaxial cable (thick and thin), twisted-pair wiring (shielded and unshielded), and optical fiber (single- and multimode). Emerging media include infrared and radio signals for wireless transmission.

A medium-access protocol is used to convey information over the transmission facility. Ethernet and token ring are the two most popular medium-access pro-

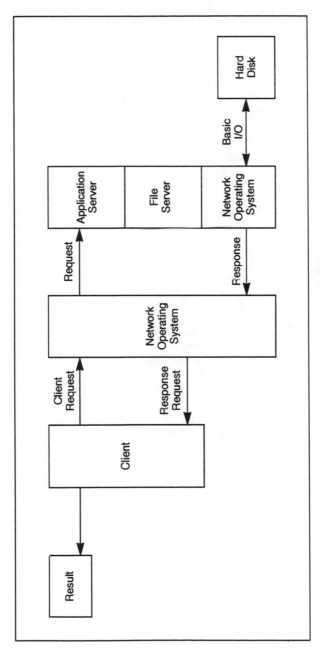

Exhibit IV-5-1. Data Retrieval in a Matrix Server Environment

tocols used over LANs. When linking client/server computing environments over long distances, other communications protocols used over private facilities (e.g., point-to-point T1 links) come into play, which provide a transmission rate of as much as 1.544M bps. Frame relay and other fast-packet technologies are rapidly emerging communications protocols for carrying LAN traffic over the wide area network (WAN), while the Internet protocol (IP) remains the most common transport protocol used over the WAN. Each of these must be evaluated for their price-performance and reliability factors.

OBSTACLES TO IMPLEMENTATION

Although many organizations have embraced the concepts behind client/server, many are still wary about entrusting their mission-critical applications to the new architecture. Despite all the vendor hoopla, today's client/server environment is still burdened by the lack of diagnostic and applications development tools, which are readily available for the mainframe environment. In particular, such troubleshooting tools as debuggers and other diagnostic programs, although more powerful than they were just two years ago, are still less robust than those found in the mainframe world.

Another important concern is the inability to centrally control client/server networks linked to one another and to larger host machines. For the most part, the tools used locally on LANs to diagnose, correct, and troubleshoot problems cannot be used remotely or across different platforms. The lack of such tools as autobackup and autorecovery in current client/server configurations is a big concern for many organizations, as is the lack of integrated data dictionaries and development tools.

Because PC-based networks can be unstable, the hardware platform should be thoroughly tested before it is entrusted with mission-critical applications. Isolating a problem on the LAN can be a very time-consuming task because of the difficulty in tracking it down, whereas on a mainframe, the cause of a problem is often immediately apparent.

Although management tools are emerging for the client/server environment, they are still few and far between. Therefore, it is important to have knowledgeable staff who understand the nuts-and-bolts of the operating system, the interrelationships of the applications, and the networking environment. The reason for this is disaster recovery. Many companies do not fully appreciate or understand how they are going to manage distributed data on multiple servers that may be scattered all over the country. When disaster strikes, it is important to have a recovery plan already in place so that vital data does not get lost. An effective disaster recovery plan will not only be thoroughly scripted to correspond with various disaster scenarios, but be tested periodically, and refined if necessary.

INTEGRATION ISSUES

Even where development and diagnostic tools do exist, users often find they must wrestle with complex integration issues, as well as learn a whole new language. Until there is more collaboration among mainframe and microcomputer vendors to make integration easier, users may want to consider a systems integrator to facilitate the transition to client/server and use them for knowledge transfer.

Integrators can play a key role in managing the complex relationships and problems—both technical and administrative—that arise in a multivendor environment. They not only help integrate products from many different vendors, but smooth out incompatibilities between communications protocols. Integrators also lend valuable assistance in negotiating service and support contracts.

When choosing a systems integrator, it is important that it share the organization's vision. One area in which shared vision is particularly important is that of technological innovation. If the organization wants to implement leading-edge solutions, the integrator must be willing and capable to support that vision with investments in emerging technologies and cutting-edge concepts.

Most integrators are concerned with the problem at hand so they can get on with solving the next problem. They generally do not help clients articulate a vision that will provide essential guideposts into the future. An integrator that can provide business process reengineering services for select applications, for example, is a better choice than a vendor that provides basic service only.

Until corporate management sees the client/server platform providing equivalent stability, performance, consistency, and reliability to the mainframe, they will not be comfortable moving mission-critical applications. Pilot programs can help organizations evaluate the client/server platform in terms of performance, reliability, and disaster recovery. The time to initiate a pilot program and try new techniques is when new programs must be developed. Small, manageable applications allow time for staff to get through the learning curve, which can greatly facilitate the transition to client/server.

NETWORK SUPPORT

Those who have taken the plunge into client/server applications have noted that their organizations are becoming increasingly dependent on enterprise networks. Managers must now invest more heavily in internal and external network support.

Behind the growth in expenditures for network support is the confusing array of communications challenges that are confronting managers, from evaluating new internetworking technologies needed to create enterprise networks to maintaining, managing, and leveraging far-flung corporate data bases. The hope of saving money by using client/server applications may be dashed once managers realize that the cost required to make each program work remains the same. Although it does make end users more productive, by itself, the client/server approach does not really save money. For faster applications development and cost savings, the client/server approach may have to be coupled with computer assisted software engineering (CASE) or object-oriented programming (OOP).

THE CASE FOR OBJECTS

The basic premise of OOP is that business functions and applications can be broken up into classes of objects that can be reused. This greatly reduces applications development time, simplifies maintenance, and increases reliability.

Objects provide functional ability by tightly coupling the traditionally separate domains of programming code and data. As separate domains, it is difficult to maintain systems over time. Eventually the point is reached when the entire system must be scrapped and a new one put into place at great expense and disruption to business processes. In the object-oriented approach, data structures are more closely coupled with the code, which is allowed to modify that structure. This permits more frequent enhancements of applications, while resulting in less disruption to end users' work habits. With each object viewed as a separate functional entity, reliability is improved because there is less chance that a change will produce new bugs in previously stable sections of code.

The object-oriented approach also improves the productivity of programmers in that the various objects are reusable. Each instance of an object draws on the same piece of error-free code, resulting in less applications development time. Once the object method of programming is learned, developers can bring applications and enhancements to users more quickly, thereby realizing the full potential of client/server networks. This approach also makes it easier to maintain program integrity with changes in personnel.

CASE tools, fourth-generation languages (4GLs) and various code generators have been used over the years to help improve applications development, but they have yet to offer the breakthrough improvements demanded by an increasingly competitive business environment. In fact, although these tools have been around for years, the applications development backlog has not diminished appreciably.

This is because many CASE tools are too confining, forcing programmers to build applications in one structured way. This can result in redesign efforts often falling behind schedule and over budget. In addition, CASE tools are not typically compatible with each other. This would require a CASE framework with widely disclosed integration interfaces. Complicating matters is the growing number of government and industry standards organizations that are offering proposals that supersede and overlap one another.

Object-oriented technologies, however, are starting to deliver on the breakthrough promise. In some instances, OOP technology has brought an order of magnitude improvement in productivity and systems development time over that of CASE tools—it is not unheard of for some IS staffs to compress five or six months of applications development time to only five or six weeks. Few technologies available to the applications development community hold as much promise as object orientation.

Transitioning to Objects

Several concrete steps can be taken to ensure the successful transition to object technology in the applications development environment. The success of any large-scale project hinges on the support and financial commitment of senior management, who must be made aware of the benefits as well as the return on investment. Fortunately, this not hard to do with object-oriented technology.

To demonstrate the potential advantages of object-oriented technology, IS managers should seize the opportunity to apply it to a new project. Projects that lend themselves to object-oriented technology include any applications that are being downsized from the mainframe to the client/server environment, because the applications will have to be rewritten anyway.

It is a wise idea to prepare for object-oriented technology now by determining the availability of training and consulting services and reference materials. If senior management wants to know about object-oriented technology and its potential advantages, IS managers will elicit more trust and confidence by demonstrating immediate knowledge and understanding of the topic, rather than begging off until they can become more informed.

Like the move from mainframes to client/server, the skills mix necessary in object-oriented technology differs from that in conventional methods of applications development, if only because the shift in activities is toward the front-end of the applications development cycle. This means there is more emphasis on such things as needs assessment and understanding the workflow processes in various workgroups and departments. This affects the design of the applications in terms of the modularity and reusability of various objects.

New incentives may be needed to encourage systems analysts and programmers to learn and adhere to object-oriented analysis and design methods, and to reward those who create and implement reusable code. The object-oriented paradigm signals a fundamental shift in the way networks, applications, data bases, and operating systems are put together as well as how they are used, upgraded, and managed. The ability to create new objects from existing objects, change them to suit specific needs, and otherwise reuse them across different applications promises compelling new efficiencies and economies, especially in the client/server environment.

STAFFING

A potential obstacle that may hinder the smooth migration to client/server is the apparent lack of skills among programmers outside of the traditional mainframe or standalone personal computer environments. Those with PC experience typically have never worked in an IS shop and are not familiar with the control procedures and testing rigor that a formal shop expects in mission-critical applications development. Alternatively, moving from the mainframe to client/server environment requires knowledge of multiple platforms, rapid applications development tools, relational data base design, and, ultimately, the principles of object-oriented programming.

One difference for many mainframe operators switching to client/server environments is the heavier involvement with users. This often means reconciling system wants and needs among end users. Also, IS managers usually have more contact with senior management, who want to know how the new technology can benefit the company and what to anticipate in terms of return on investment.

TRAINING

The shortage of skilled object-oriented programmers is perhaps the biggest obstacle to speedy implementation. Object-oriented methods require that programmers think about things differently, and thinking in terms of objects as opposed to lines of code is certainly different. Even with the three to four months needed to train staff members in the object-oriented approach, once programming staff are up to speed they can develop applications faster and more efficiently.

The client/server environment tends to require more staff experts, because no one person typically understands all of the pieces. As if to underscore this point, there seems to be more trial and error in developing applications for client/server than on older, well-understood legacy mainframe systems.

One of the trends driving training for the rest of this decade is that the whole information technology arena is undergoing a paradigm shift—from mainframe to client/server systems. This means a major retooling of organizations around the country, requiring substantial increases in training dollars.

Only two possibilities exist for obtaining the necessary skill base required for moving from a mainframe orientation to a client/server orientation. One is to fire everyone and start over, which is not viewed by very many companies as even remotely feasible. The other is to work with existing professionals to upgrade their skill base with training. Within the context of most companies, training is the best solution because the people already have the underlying knowledge base. The object of training is to apply an existing knowledge base in a new direction.

Managers should endeavor to become as well versed in client/server issues as they are about mainframe issues. Such people will be a greater asset to companies making the transition to smaller, diverse platforms. The person most valuable is somebody who has the desire to learn the client/server environment and has the flexibility to go back and do work on the mainframe when it is appropriate.

Outside consultants can be brought in to provide the necessary training, but programmers and systems analysts must learn more than just the details of how products and software work. Working closely with senior management doing strategic planning and getting hands-on experience with how applications are being used can be equally important.

EMERGING ROLE OF AGENT SOFTWARE

The increasing number of professionals working away from a main office is driving the need to tie mobile systems into the client/server environment. This is effectively accomplished through the use of agents—special software that acts on behalf of the mobile client to access and retrieve information on corporate servers through wireless networks.

Within a client/agent/server architecture, agents respond to electronic requests from their mobile clients by accessing the corporate data source and, on receipt of information, automatically providing the response to the mobile client. The software agents are even active when the client is disconnected. For example, a mobile user can request notification for a specific inventory target of a popular product, instruct an agent to monitor the inventory data base, and automatically receive notification when that event occurs. Because the agent performs the monitoring and reporting functions, the mobile user does not have to waste time calling into the data base to check on the current inventory level.

Among the vendors offering this kind of software is Oracle Corp. Its Mobile Agents technology lets applications run offline or disconnected from the main servers. This means that a field representative can use E-mail or order-entry applications while on the road and then reconnect and send any pending messages at any time. Conversely, all data waiting to be sent to the representative is held by Mobile Agents until the individual reconnects to the system. Mobile Agents's security system dismembers and reassembles the message content, as well as passwords and unique IDs, to protect corporate data from unauthorized access.

The client/agent/server architecture enhances overall transaction performance by minimizing use of the mobile client's wireless communications link and taking advantage of the high-speed link between the agent and the server. This results in the fastest possible performance to the mobile client, regardless of the complexity of the task.

AN ASSESSMENT OF CLIENT/SERVER TECHNOLOGY

Systems Costs and Management

The promise of client/server technology has not yet been realized. Because client/server systems are distributed, costs become nearly impossible to track, and administration and management difficulties are multiplied. According to the Gartner Group of Stamford CT, the total cost of owning a client/server system is about three to six times greater than the total cost of owning a comparable mainframe system; the software tools to manage and administer a client/server system cost two-and-a-half times more than mainframe tools.

The Standish Group, a Dennis MA-consulting firm, found that only 16% of IS managers say their client/server projects were on time and on budget. Other studies indicate that only about 40% of companies view the client/server architecture as a worthwhile investment.

As a result of these challenges, corporate intranets, the Internet, and network-centric computing have become the new focus of many corporate IS departments. Others view client/server computing as a stepping stone to network-centric computing.

Internet-Based Computing

Internet-based computing provides new opportunities for integration and synthesis between networks and applications inside the company and those outside the company. With proper security precautions, intranets can easily be set up to reach international divisions, customers, suppliers, distributors, and strategic partners. The greater Internet is used for worldwide connectivity between distributed intranets.

The greatest potential of corporate intranets, however, may be in making client/server systems truly open. The average desktop PC has from seven to 15 applications—50 to 100 software components—each of which has different versions. With the single universal Web client, IS managers need not worry about configuring hundreds or thousands of desktops with appropriate drivers. On the flip side, however, the ease with which users can install Web-based technology may result in out-of-control situations reminiscent of the turbulent days of client/server computing.

Web-based technology is certainly less costly, because businesses do not need to standardize desktops and operating systems for Internet technology to work. An existing client/server infrastructure can serve as the backbone to an intranet. In addition, the availability of free or low-cost applications development tools on the Internet means that the creation of distributed systems and applications is much cheaper than in the client/server environment. This allows IS managers to get more done within available budgets while preserving existing investments in legacy and client/server systems and applications. A Web browser provides a single window to all data, regardless of location. There are even gateway products that allow legacy data on mainframes to be viewed from within a Web browser with no modification to existing data bases.

CONCLUSION

Many experts believe that the rise of the global economy has forced businesses to improve their operating efficiency, customer-satisfaction levels, and the product-to-market implementation cycle to a degree rarely seen before. This has contributed to widespread acceptance of client/server technology as the most practical solution for many companies.

The trouble is that most companies are too heavily focused on the technical issues and ill-prepared to deal with equally important management and budgetary issues. Failure to consider such management issues as support requirements, the skills mix of staff, and training needs can sidetrack the best-laid technical plans for making the transition to the client/server environment. Even on the technical side, attention must be paid to the availability of diagnostic and applications development tools, as well as the opportunities presented by such complementary technologies as object-oriented programming, so that the full potential of client/server networks is realized.

An alternative for many companies is to look at how the Internet and corporate intranets can achieve the same objectives, or even complement existing legacy and client/server investments. The openness of the Internet and its continued growth in terms of universal connectivity ensures compatibility among all the components, which not only greatly reduces the cost of implementing corporate intranets but also enables businesses to transcend time and distance, lowering boundaries between markets, cultures, and individuals.

IV-6

Moving to a Client/Server Environment

Eileen Birge

TRANSFORMING CURRENT MAINFRAME development and operations staff to a client/server environment is an intensive process with four major tasks. Foremost among them is establishing the environment necessary for success.

ESTABLISHING AN ENVIRONMENT FOR SUCCESS

Establishing the Technical Architecture and Operating Environment

Client/server technology offers a complex array of choices. A standard client/server architecture may involve more than 100 products from 30 or more vendors. Desktop hardware, server hardware, network operating systems, fourth-generation languages, virus detection, batch schedulers, remote connectivity, software distribution, relational data base management systems, object managers, methodologies, online transaction processors, and server operating systems represent just some of the areas where the organization will make product decisions. Adding to the problem is the pace of change; available choices change daily. Once an organization finds products that appear to meet its needs, it should test them together—this is the integration test of the technical architecture. The first test is usually a simple application, designed to ensure that all products are communicating. The second test should be a more complex application with higher volumes. (The tests are not real applications but tests of all the systems technology.)

The importance of this step should not be underestimated. During the first year of the move to client/server technology, the IS department makes critical decisions—decisions that will affect the systems development and operating environments for years to come. An organization can make decisions after deliberate reflection and testing or as the result of off-the-cuff thinking in its haste to implement its first application. Large mainframe shops probably have legacy systems written before the days of data dictionaries and naming, coding, and documentation standards. The results have been seen. Moving into a new environment provides the opportunity to put standards in place right from the beginning.

An experienced team can typically design a technical architecture within two to four months. Testing the technical architecture requires one month—more if the environment has very complex requirements.

Gaining Management Commitment

Companies have taken several approaches to gaining the commitment of management, including the following:

- Laying off existing resources and hiring personnel with the new skills.
- Using systems integrators to develop the initial applications and transferring knowledge to the current staff.
- Independently retraining the current staff and hiring (or contracting for) a minimum number of new skill resources.

Recruiting fees, reduced productivity while learning the organization's structure, culture and products, management time to recruit and interview, and high turnover from a work force with limited company loyalty are just a few of the costs associated with replacing current employees.

More than half of the companies moving to client/server have looked at the costs and the benefits of transforming their current staff and elected to retrain and retain. What are the benefits of retaining?

- *Employee loyalty.* Systems work often requires extra effort. Employees who recognize the investment the company has made in them will respond appropriately.
- *Retention of knowledge.* This includes retention of knowledge about the company and its values.
- *Knowledge of systems development process, audit trails, security, and controls.* Many of the skills associated with good development are not specifically technology related. It is usually less expensive to retain these skills and add the technology component than to buy pure technology knowledge and build development skills.

One research company has estimated the cost of retraining for a standard 200-person department at $5.5 million. Management must thoroughly understand that transformation is a process and that not all benefits of the new technology will be realized in the first few months. This whole project should be treated as a major capital expenditure with the same types of approvals and reviews that a company would give to a capital project of this magnitude.

Making the Partnering Decision

The next three years will be ones of major change. The IS department will face significant technical, personnel, and political challenges. Early in the process, the organization must determine the extent of involvement of outside resources. At one end of the spectrum is a partnership arrangement where the experienced outside resource has a long-term commitment throughout the process and provides or contracts for nearly all services. At the other extreme, an organization uses outside resources only in a limited manner, with an internal commitment to do it themselves, rather than use outsiders (i.e., make rather than buy) whenever possible. Regardless of the choice, the approach to a partnering decision should be articulated at the beginning. If the organization chooses significant partner involvement, time and effort should be spent in choosing the partner carefully. Key items to look for include the following:

- *Depth of commitment.* Does the prospect practice internally the methods, technologies, and organizational behavior of interest to the organization?
- *Corporate culture.* Is the prospect's culture one with which the organization feels comfortable?
- *Flexibility.* Will the prospect truly study the organization, or is the prospect wedded to one approach and ready to advocate that solution? For example, do they sell one particular vendor's products or only sell their own proprietary approach?
- *Track record.* How has the prospect performed in assisting other companies?
- *Risk acceptance.* Is the prospect willing to tie financial reward to successful outcomes?

Creating an Atmosphere of Change

There is an anecdote on change; the chair of a meeting says: "Change is good. Change is exciting. Let the change begin." Meanwhile all the meeting participants are thinking, Who's getting fired? If this anecdote represents thinking of the organization to be transformed, attitudes must be adjusted before proceeding.

The process of creating the right atmosphere for change is a major project in itself and beyond the scope of this chapter. Key points of creating this atmosphere, however, include the following:

- Allowing room for participation.
- Leaving choices.
- Providing a clear picture.
- Sharing information.
- Taking a small step first.
- Minimizing surprises.
- Allowing for digestion.
- Demonstrating IS management commitment repeatedly.
- Making standards and requirements clear.
- Offering positive reinforcement.
- Looking for and rewarding pioneers.
- Compensating extra time and energy.

- Avoiding creating obvious losers.
- Creating excitement about the future.

ESTABLISHING THE VISION OF THE IS GROUP OF THE FUTURE

The Vision Statement

What will be the role of IS in the future of the organization? How will success be measured? Is the mission and strategy of the IS department in concert with that of the organization as a whole? Regardless of how client/server technology is implemented, it will be a major financial commitment. Before that commitment is made, the role of IS should be reexamined and optimal use of financial resources ensured.

A Functional Organization Chart for the IS Department

The current organization probably reflects many assumptions from mainframe technology roots, with offshoots reflective of the growth of LAN technology. For the transition to client/server to be effective, an organization must do more than merely layer technology onto the existing organization. A sample new functional client/server-legacy organization chart is shown in Exhibit IV-6-1.

The sample organization emphasizes tools and methods. The client/server environment can be an intensively productive environment—some companies have reported throughput improvements of two to four times in the effort to design and implement systems, with similar increases in user satisfaction. Companies can only realize these gains when the developers have an understanding of the tools they are to use, the design guidelines to follow, data base standards, and version control procedures, among other issues. It is also important to note the position of Manager—IS Professional Development in the exhibit. Transitioning the staff, updating the plan, tracking progress, and managing a multimillion-dollar budget requires full-time commitment.

Position Descriptions for the Future Department

IS managers should create a picture of what their staff will be doing in the not-so-distant future. To create the transformation plan, they must know where they are going. They should write descriptions that are as clear as possible, giving serious thought to how performance will be measured for each item on the position description (a sample is included in Exhibit IV-6-2). Managers should make a preliminary estimate of how many of each type of skill set they will need. They should focus on what the person in the position will be doing as opposed to the place in the organizational hierarchy.

In developing position descriptions and estimated staffing quantities, IS managers should consider the following:

- Client/server development teams should be kept as small as possible. Teams of five to seven are optimal, eight to 10 are manageable.

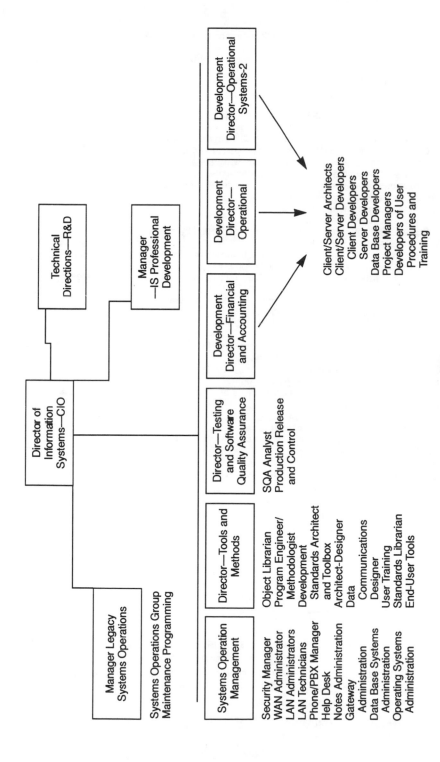

Exhibit IV-6-1. Sample Functional Client/Server Legacy Organization Chart

Company ABC
Position Description

Date: 8/1/96   Position Title: Client/Server Developer

PRIMARY FUNCTION
Develop and maintain ABC business applications.

GENERAL DESCRIPTION OF WORK PERFORMED:
1. Design logical data bases using ABC development platforms.
2. Work with data administration staff to implement and tune data bases.
3. Conduct and record facilitated joint application sessions.
4. Lead prototyping sessions and implement user requests.
5. Interview users and document user requirements.
6. Create detailed program specification packages and test plans.
7. Code and unit test programs.
8. Create and implement systems test plans.
9. Implement conversion plans.
10. Provide quality assurance testing of programs coded by others.
11. Create systems documentation.
12. Design and implement user interfaces in compliance with CUA and ABC standards.
13. Provide user support on ABC business applications.
14. Provide assistance to user procedure/training analysts in the development of user training and documentation manuals.
15. Interview and assist in the recruiting of IS professional staff.
16. Participate in staff and status meetings.

KEY MEASURES OF SUCCESS:
1. Customer satisfaction with usability and quality of delivered software.
2. Programming accuracy and completeness.
3. Timely completion of assigned tasks.
4. Compliance with ABC design and development standards.
5. Customer and project manager assessment of ability to function successfully as a member of a team.

EDUCATION AND EXPERIENCE:
BA or BS in Computer Science or Business Degree with IS minor
2 years development experience
ABC's Client/Server Technology Series
Organizational Development training (as specified by Human Resources)

Exhibit IV-6-2. Sample Position Description: Client/Server Developer

---

- A development team requires a full-time manager, three to four developers, a part-time or full-time data base analyst, a part-time or full-time client/server architect, and (in the latter stages of the project) a full-time user procedure analyst/training specialist. (If the project affects a large number of users and has a significant training component, these metrics do not apply. The project to design and deliver the training material may dwarf the systems project.) One architect and one data base analyst can be assumed for eight developers.
- Client/server developers work with users to identify requirements, reengi-

neer processes, design, code, and test. Development methodologies in this environment are most effective when they stress iterative prototyping and refinement. For optimal efficiency, the roles of analyst and programmer should not be separated.

- A client/server architect works with members of the tools and methods group to set up development environments and to identify components of production environments. If a production environment will involve use of previously untested products or new releases of products, the architect tests the technical architecture before significant development efforts are expended.

- The tools and methods group can be highly leveraged. One tool builder-designer should be assumed for each 50 developers-architects-analysts.

- If software is being developed for internal use only, one tester should be assumed for every two developers.

- One LAN administrator should be assumed for every 200 users and one groupware administrator for every 400 users. Growth in management tools should reduce staffing requirements over time.

This is an area where most organizations look to consultants for assistance—both for identifying the responsibilities and to work with human resources to determine the impact on pay scales and incentive plans.

Skills and Performance Needed

In the preceding step, responsibilities and work performed were defined by each position. In this step, they are translated into the types of skills and levels of proficiency required. For example, a client/server developer position description may include some of the following responsibilities:

- Interviewing users and documenting user requirements.
- Preparing logical data models.
- Preparing unit-test plans and executing plans.
- Managing own time against task deadlines.

Those responsibilities can then be translated into skills. Interviewing users requires the following:

- Interviewing techniques.
- Effective listening.
- Interview planning.

Documenting user requirements requires:

- Effective writing skills.
- Understanding of the selected methodology and associated documentation techniques.

It is tempting to declare that all staff be expert in all skills required for their positions, but that is both unrealistic and unaffordable. It is necessary to identify the minimal skill level needed to perform in the position in a satisfactory manner.

The translation of job responsibilities into skills requirements is difficult. Doing this job well, however, has a tremendous payoff. It affects the evaluation process and has a dramatic affect on explaining the role of training and staff development to senior management. It focuses the work force. Setting the expected skill levels assists staff in planning their own self-study activities.

Hiring and Arranging for Skills from the Outside

At this stage, IS managers should have an initial feel for the skills that are likely candidates for acquisition from new hires or consultants. They should start the process for acquiring these skills now so that the candidates will be on board when they are ready to start implementing the plan. The hiring and acquisition process will continue at the same time as the next seven steps.

CREATING THE PLAN

After establishing the environment needed for success and also establishing the vision of the IS group of the future, companies are ready to tackle the nuts and bolts of the transformation to a client/server environment.

Mapping Current Staff to Future Positions

For this task, managers must look at their current organization and personnel and attempt to fit the individuals and positions into their future roles. During the mapping operation, interests and individuals should be matched rather than it being assumed that programmers will become client/server developers. All staff members should have access to materials developed so far. The position descriptions help staff identify what roles they believe are most suited to their own abilities.

How do managers decide who is best suited to be a client/server developer? A client/server architect? No fixed rules exist, but there are some guidelines. The client/server architect will typically be more technical than his or her developer counterparts. The architect usually has less user involvement and interacts more with the development team. Managers should look for persons and development positions that focus on the technical versus the functional aspects of the work.

Aptitude should be considered. One report concluded that 26% of existing mainframe personnel could not be converted to client/server. In other experience, this figure has been closer to 35%. These persons are clearly the candidates to maintain legacy systems during a transition period.

Skills Analysis

The positions of the future and the skills they require have been identified. The probable candidates for each position have also been identified. In this step, managers identify what skills the candidates have already to avoid wasting training dollars.

A skills assessment document should be prepared. The document asks individuals to rate themselves as 0 (no knowledge), 1 (conceptual), and so forth. Guidelines are given so that the staff understands the indicators for these levels. Guidelines are not meant to be all-inclusive (i.e., the guidelines should help individuals assess themselves, not delineate the total knowledge requirements for that level). Sample guidelines for a developer's knowledge of Microsoft Windows are:

- *Novice.* Can explain the purpose for Windows. Knows which version of Windows is running. Can reset, minimize, and open windows. Can explain the concept of the active window.
- *Needs supervision.* Can explain the functions of the three main Windows programs, can differentiate the various kinds of memory, can explain and use the terms DDE and OLE appropriately. Can list and define the various components of a window. Has developed at least one application program in the Windows environment using a standard 4GL tool (i.e., PowerBuilder or Visual Basic). Can explain the function of the WIN.INI.
- *Works independently.* Can explain the purpose of the Windows API. Has written at least three production programs in the Windows environment. Has written at least one application involving the use of OLE or DDE.
- *Expert.* Has written three or more native Windows programs.

First, each staff member should self-assess without knowledge of the suggested skill levels for the proposed positions. Next, an independent party should assess each individual. Managers should meet with the staff member to discuss any significant variations between the self-assessment and the independent assessment. The final, agreed-upon skill level should be documented.

The manager responsible for implementing the transition plan now has a picture of each individual: current skills, current skill levels, and the target skills and levels needed to perform in the future. Managers must analyze and summarize to develop the profile of the typical staff person slotted for any position for which six or more staff will be assigned. These positions may include: project managers, client/server developers, data base administrators, client/server architects, LAN administrators, and software quality assurance testers. For such positions as WAN administrator or object librarian, plans can be tailored to individual needs. Little benefit will be gained from summarization because there will be only a few candidates for each position.

The Training or Job Assignment Plan

Armed with the profile of typical current skill levels and the target skill levels, a training assignment plan designed to raise skill levels to the targets can be created. The following rules apply:

- Training alone cannot create level 3 (works independently) or level 4 (expert) personnel. Although training often only creates a level 1 (conceptual understanding) rating, effective training can create a level 2 (works under supervision) rating. Training also helps level 3 and 4 performers maintain currency in their skills. For example, training should help a level 4 Oracle V 6.0 data base administrator move to Oracle V 7.0.

- Within two months after training and using a major new tool on a daily basis, productivity should be at 50% of target. By six months, productivity should be at 100% (and skill level should be at 3).
- Training not followed quickly by job-reinforcing experience is wasted.

Given these rules, it is helpful to look at a sample training or assignment plan for an organization's first group of client/server developers.

1. *Form a team of 6–8 people.* Provide a high-level description of the system to be implemented (people learn best when they can relate the knowledge to what they need to know—so as they take classes, they can relate the concepts taught to the system they will work on). Plan on training the team together.

2. *Provide initial technology awareness training and needed soft skills training in the following areas:*
   - Client/server and LAN basics (e.g., terminology and theory).
   - Client and server operating systems.
   - Office suite productivity training (e.g., for word processors, spreadsheets, and graphics tools—tailored to the documentation and probable uses).
   - Data analysis and documentation tool (relational or object oriented).
   - Effective listening and writing.
   - Methodology orientation.
   - Facilitation (developers only).
   - LAN administration (LAN administrators only).

3. *Assign the first job.* The first job assignment within the department should not be a mission-critical system nor have a deadline that is critical to success. One company elected to make its first client/server implementation a companywide budgeting system, to be delivered in September to coincide with the beginning of the annual budget cycle. Not surprisingly, the system failed to make the deadline: everyone in the company knew and IS's judgment was seriously questioned. Preferably, the project should have some kind of high impact when delivered so that the first client/server application helps fuel the excitement about the change to this architecture. The assignment should be sufficiently complex to test most aspects of the technical architecture and reinforce the needed technical skills.

   Plan to have this team conduct initial user-requirements definition—using the data design tool, methodology, and writing techniques. The LAN administrators should set up the development environment and productivity tools.

4. *Provide second-level training after the requirements.* Typically, the developers and architects will need training in user-interface design, the specific development tool, and prototyping. The architect may require additional training on the components of the technical architecture assumed for use in this application. Other members of the team may require training in data base administration and performance tuning.

1. Establishing the environment needed for success, including:
   - Making the partnering decision.
   - Establishing the technical architecture and operating environment.
   - Gaining management commitment to the staff transformation approach.
   - Creating an atmosphere of change.
2. Establishing the vision of the IS group of the future, including:
   - Creating a vision statement for IS.
   - Creating an organization chart for the EUC department in the future.
   - Creating position descriptions for the future department.
   - Identifying the skills and performance levels that will be needed.
3. Creating the plan, including:
   - Making a preliminary cap of current personnel to future positions.
   - Performing a skills analysis
   - Determining which skills must be acquired versus built.
   - Creating a training and job assignment plan.
   - Creating an infrastructure to help manage the plan.
   - Initiating hiring or arranging for skills from the outside.
4. Testing and executing the plan, including:
   - Testing the plan.
   - Executing the plan with the remaining staff.
   - Evaluating and refining the training plan.
   - Incorporating continuing change and development into the culture.

Exhibit IV-6-3. Major Tasks in Moving to a Client/Server Environment

5. *Build prototypes and the technical architecture.* Plan to have the team return to the project and conduct prototyping sessions with the user. Critical functional ability and performance features should be developed to the point where the technical architecture can be tested and modified, if necessary.
6. *Complete the system.* Integrate the completion with methodology training appropriate for the project phase.

An Infrastructure to Manage the Plan

With a 200+ person department, a three-year plan, dozens of vendors, more than 2,500 person-skill combinations, as much as $2 million in hard costs to budget, classrooms to equip and schedule, and as many as 300 official and unofficial training courses to track, management is a challenge. For the typical Fortune 1,000 company, executing this plan requires a full-time commitment and appropriate systems support.

TESTING AND EXECUTING THE PLAN

Testing and Refining the Plan

A plan has already been developed for the first team. Now, the plan must be executed. The manager responsible for all professional development should par-

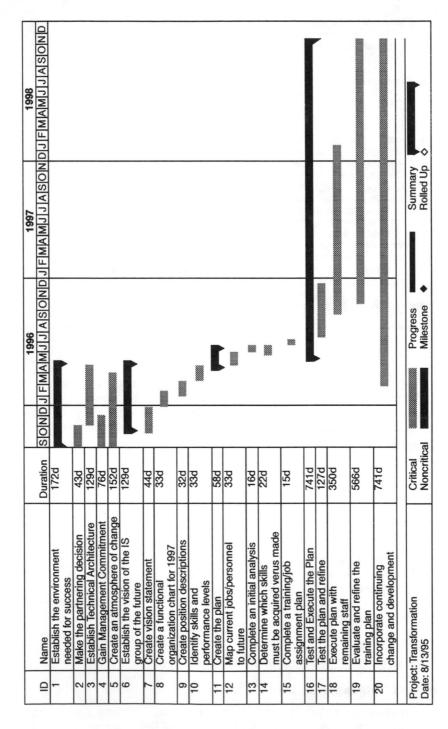

Exhibit IV-6-4. GANTT Chart for Transformation Process

ticipate in as much of the training as possible to observe the participants and the material. Participants should be debriefed after each training session to determine strengths and weaknesses. Participants should also be interviewed at intervals after the training to find out what worked and what did not. The project managers can help determine how job-ready the participants were after returning from training. The suggestions from the first project team should be incorporated into the training plan.

Executing the Plan with the Remaining Staff

The remaining staff should be scheduled into the refined training plan. Again, the mode of training development teams together should be followed once a project has been identified. It should be assumed that a new group can be started through the process every four to eight weeks. The schedule should be modified to reflect current assignments.

Refining the Training Plan

Feedback should be collected from system users regarding satisfaction levels. Additional feedback can be obtained from staff going through the plan and incorporated into the plan on a continuing basis. Staff who went through the training earlier may need refresher or catch-up topics to reflect new thinking or new technology. The plan should be verified at least quarterly with the group responsible for technology to update tools and architecture information.

Incorporating Continuing Change and Development into the Culture

The environment will continue to change. Business and strategies will change faster. It is important for the IS staff to realize that the transformation process will never end. At least semi-annually, each staff member should reassess current skill levels, create new targets for performance, and identify the combinations of job experience, self-development, and formal training needed to achieve those skills. Preferably, achievement of skill development goals should be a component of the bonus or raise process.

CONCLUSION

As can be seen, moving a staff to client/server is just like most systems projects: managers must determine what they want, refine the requirements, create a plan to implement the requirements, and then test and refine the plan. What seems to be a monumental task can be broken down into manageable and measurable steps; see Exhibit IV-6-3 for a review of the steps and Exhibit IV-6-4 for a GANTT chart depicting their timeline.

IV-7

Object-Oriented Programming: Problems and Solutions

Louis Fried

THE SEARCH FOR TOOLS to accomplish higher quality software development has produced data base management systems, query systems, screen development tools, fourth-generation languages, graphic programming aids, and code generators. The ultimate dream, however, is to find ways to avoid programming altogether. The best way to not program is to reuse existing code.

The tools that developers already use are effective because they reuse code in some sense. For example, using data base management systems, programmers need not develop their own access routines as they were forced to do many years ago.

The developers of object-oriented programming (OOP) languages and tools promise to take the reuse of code to new levels, but there are ongoing debates about the benefits and the potential problems associated with object-oriented programming. For each argument there are various responses.

REUSABLE SOFTWARE ASSURES QUALITY

One concern is that objects require continuous maintenance and enhancement to keep up with the changing needs of the business. However, software has always required maintenance.

Another concern is that the analysis task required to identify and define appropriate objects is formidable. Advocates of object-oriented programming respond that the best software development efforts result from spending more time in the definition and specification phases; in addition, developers can reuse objects for long-term savings.

In fact, proponents of object-oriented programming point out that the need to define classes and subclasses of objects, the objects themselves, and the attributes, messages, methods, and interrelationships of objects forces a better model of the system to be developed. Many objects developed in object-oriented programming code will not be reused; however, the real benefit is that OOP code is usually more lucid and well organized than traditional coding methods. The process that forces analysts to define the object hierarchies makes the analysts more familiar with the business in which the application will be used.

When these problems and objections are analyzed, many of them can be discounted: however, some remain. Viewed in isolation, object-oriented programming is simply an attractive way to facilitate structured, self-documenting, highly maintainable, and reusable code. In the context of enterprisewide application building, however, object-oriented programming presents unique challenges whose solutions require additional tools and management methods.

OBJECT-ORIENTED PROGRAMMING ENVIRONMENT

As object-oriented techniques gradually find a place in corporate programming departments, there will be attempts to expand the use of this technology from single applications to broad suites of applications. To accomplish the expansion of use, object-oriented programming will need to be used within a development framework that is composed of computer-aided software engineering (CASE) tools implemented in a distributed, cooperative processing environment.

A likely scenario of the way in which organizations will want to use the object-oriented programming in the future is as follows:

- Objects will be used by decentralized development groups to create applications that are logically related to one another and for which common definitions (i.e., standards) are imposed by various levels of the organization.
- Users will employ objects to develop limited extensions of basic applications or to build local applications, in much the same way spreadsheets and query systems are currently used. Users may access corporate data bases in this environment through objects that encapsulate permitted user views of information.
- Object-oriented programming will become integrated with CASE platforms not only through the inclusion of object-capable languages but through repositories of objects that contain both the objects themselves and the definitions of the objects and their permitted use. Improved CASE tools that can manage and control versions and releases of objects as well as programs will be needed.

This scenario envisions optimum use and benefit from object-oriented programming through extensive reuse of proven code within a framework that allows authorized access to objects.

The current status of object-oriented programming makes this scenario difficult to implement. The effective use of object-oriented programming depends on the ability to solve problems related to two major areas of concern: the management of the object inventory and the preservation of information security in an object-oriented development environment.

MANAGING THE OBJECT INVENTORY

Objects in the inventory must reside in a repository that uses an object-oriented data base management system. Objects are identified by classes and subclasses. (Object class definitions are themselves objects). This identification provides a means of inventory management. For example, retrieving an object within a class called Accounts Payable would help to narrow the domain being searched for the object. A further narrowing can be done by finding a subclass called Vendor's Invoice, and so forth. Polymorphism allows the same object name to be used in different contexts, so the object Unit Price could be used within the context of the Vendor's Invoice subclass and the Purchase Order subclass. (Some relational data base management systems also allow polymorphism.)

Several problems arise as a result of this organizational method. To take advantage of the reusability of objects, the user must be able to find the object with a little effort as possible. Within the classification scheme for a relatively straightforward application, this does not appear to present a substantial problem.

Most companies undertake the development of applications on an incremental basis, however. That is, they do not attempt to develop all applications at once. Furthermore, retroactively analyzing and describing the data and process flows of the entire organization has failed repeatedly. By the time all the analysis is completed, the users have lost patience with the IS department.

It is feasible to limit objects to an application domain. However, limiting objects to use within the narrow domain of a single application may substantially reduce the opportunities for reuse. This means that developers will have to predict, to the extent possible, the potential use of an object to ensure its maximum utility.

Cross-Application Issues

It is possible to establish a class of objects that may be called cross-application objects. Such objects would be the same regardless of the context within which they were designed to be used. For example, the treatment of data related to a specific account in the corporate chart of accounts may always be the same, yet the word *account* appears in many contexts and uses throughout a business. Therefore another approach to this problem is that some objects may be assigned an attribute of cross-application usability.

As more object-oriented applications are created, the typical data dictionary or repository will not be able to serve the needs of users for retrieving objects. Analysts and programmers who are required to move from one application to another to perform their work may find the proliferation of objects to be overwhelming. The IS function would need to develop taxonomies of names and definitions to permit effective retrieval.

Problem of Naming and Defining Objects

Developing and maintaining a taxonomy is in itself a massive effort. For example, a large nuclear engineering company realized that the nuclear power plants it had designed would be decommissioned and dismantled in 50 years. The personnel responsible for dismantling a plant needed to know all about the plant's 50 years of maintenance in order to avoid potential contamination of the environment and injury to themselves.

The company discovered that various names were used for identical parts, materials, and processes (all of which are objects) in the average plant. Furthermore, because the plants were built throughout the world, these objects had names in many different languages. If personnel could not name an object, they could not find the engineering drawings or documents that described the object. Furthermore, if they searched for only the most likely names, they would overlook information that was stored under a name they did not think of.

A taxonomy project was initiated to adopt and use standard terminology for all components of the plant and all information relating to those components. Within two years, a massive volume was assembled, then several problems surfaced. It was impossible to know when the taxonomy would be complete. New terms had to be created to avoid duplication. The taxonomy manual was so large that engineers and other employees refused to use it.

This example illustrates one dimension of the problem of naming and defining objects. In a world of increasingly distributed processing and decentralized use of computing, IS must also consider that:

- Analysts and programmers will not be under centralized control in all instances.
- Other personnel, such as engineers, clerical staff, and knowledge workers, will use objects to create their own programs.

Retrieval Methods to Facilitate Reuse

The ability of users to develop their own programs and applications is one of the greatest benefits that can be obtained from object-oriented programming; it cannot be ignored. Nor can the demands of an increasingly computer-literate clientele be refused. This means that the methods for retrieving objects must be available to all users for a relatively small amount of effort. If not, objects will not be reused.

With users as a recognized component of the management problem, another concern emerges. Objects must not only cross application domains, they must exist at various levels of the corporation. For example, an object may be defined as applicable throughout the corporation in a given context (i.e., a Standard object). Such an object may be called a Corporate object either through being a class of corporate objects or by having a standard attribute as a corporate object. Another object may be applicable only within a specific strategic business unit and may be called, for example, an Engine Manufacturing Company object. Objects can be described in this manner down to the level of the desktop or the computer-controlled machine tool.

Two types of tools may come to partial rescue in resolving this problem. Text search and retrieval systems may provide the ability to allow users to search for objects within various contexts. The result, however, could be the retrieval of many possible objects from a repository, compelling the user to evaluate them before a selection is possible.

An approach is needed that allows the user to obtain a limited number of possible objects to solve a problem and yet does not force the organization to develop a taxonomy or limit the use of terms. Self-indexing of files for nonhierarchical search may prove helpful, but this may mean using the object-oriented DBMS repository in a manner not compatible with its inherent structure.

Regardless of the method used, there is a clear need to establish and maintain conformance to documentation standards for objects so that searches for objects will return meaningful results. One possible solution is to use an expert system in conjunction with a text search and retrieval system. Expert systems can accomplish classification and are capable of supporting natural-language interfaces. Ideally, the user could describe to the system the nature of the object needed and the system could find the most appropriate object. The user could then describe the application at a high level and the system would find and assemble all appropriate objects that fit the system context.

Object Maintenance

When objects are used throughout a large corporation it must be assumed that they will reside in repositories on a variety of machines in many locations. Each of these repositories must be maintained in synchronization with the master depository of approved objects for the corporation and its divisions. Distributed environments imply additional problems that must be solved before object-oriented techniques can work successfully.

For example, if objects are automatically replaced with new versions, there must be a mechanism for scheduling the recompilation or relinking of programs that used the affected objects. If objects are used in an interpretive mode (rather than being compiled into machine code), replacements will automatically affect their use in existing procedures, perhaps to the detriment of the application. Some of the methods currently used to maintain distributed data concurrency and to control the distribution of microcomputer programs throughout a network may be adapted to solve part of this problem. Another approach may adapt the messaging capabilities of objects to send notification of a potential change to any subject within the hierarchy of the object being replaced.

Another problem is that identical objects may need to be developed in different languages to meet the needs of users of different computers. Even if objects are developed in the same language, the options are to use either a restricted subset of the language compatible with all potential environments or a language that allows compiler flags to be placed on code and alternative versions of the code embedded in the object. Neither of these choices is attractive, and the first may require other classes of objects to differentiate between identical objects used on different machines (although polymorphism can help in this respect). As a result, the testing process for new or replacement objects becomes more complex.

Organizations will also need to assign someone the job of deciding which objects should be distributed to which of the distributed repositories. Standard corporate objects may have wide distribution, whereas others may require more circumscribed distribution. Object and object-class management becomes a major administrative job.

OBJECT SECURITY

For users, analysts, and programmers alike to use objects in developing programs or applications, they need access to them. Such indiscriminate access provides a real threat to the security of objects.

Information security has been defined as consisting of three primary properties: availability, confidentiality, and integrity. As applied to the object inventory, these may be defined as:

- Making objects available to those who need to use them, when they need to use them.
- Ensuring the integrity of objects by preventing unauthorized changes.
- Ensuring the confidentiality of objects by preventing unauthorized access.

Current repositories and directories generally assume that all persons authorized to access the directory are authorized to access any item in the directory. This line of thinking does not satisfy the requirements of an object inventory.

Access Control

An object inventory requires an extended set of security controls to make its use safe for the organization. Such controls, required to preserve integrity, need to be implemented at the object attribute level. For example, in a payroll file the individual salary rate (an attribute) may be restricted to certain users. The attribute must therefore have an attached attribute (sometimes called a facet) that specifies which programs are allowed to read the attribute Salary Rate. Alternatively, the salary rate attribute could have a facet that is a function that returns an empty field or no data to nonauthorized callers. In essence, each object defined in the inventory may need to be individually controlled as well as controlled within a set or class of objects.

A solution is to ensure that each object in the inventory can be separately locked to prevent change. When an object is accepted into inventory, the lock is activated. A system that truly intends to protect the integrity of the objects would not permit any change to a locked object. If an object needed to be changed, it would have to be deleted and replaced by an approved, tested replacement. Furthermore, a limited group of authorized inventory managers would be the only personnel able to delete an object. Finally, a safeguard system would automatically file all deleted objects in a locked, backup repository file so that they may be retrieved in the event of incorrect removal.

Locking logic itself is a problem. In current data base management systems, the problem referred to as a deadly embrace—that is, two parties concurrently attempting to update a record by different logical paths—has been solved. When the locking mechanism must deal with atomic objects rather than transactions or records, the solution may be more difficult.

Ownership

In current security practice, the levels of security assigned to information are designated by the application owner. Each application owner has the duty to specify who may access application information and under what conditions. When objects are in common use, new ways of designating ownership become necessary. Who owns an object that is used across many applications? Who owns a corporate object?

When the ownership decision is made, the next issue is how to assign access permission. Some access permissions may be assigned by sets or classes of workers. (In the new alliance model of business operations, it is not only employees who work with a company's systems, but the corporation's suppliers and customers.) Permissions may be granted by levels in the management hierarchy, by sets of people in specific functional areas, by organization unit, and by individual. Permissions need to include (as they do today) the authorization to perform certain functions with an object. Functions for which authorization may need to be defined include real only, delete, add, copy, use, and lock.

Integrity

Integrity may also be addressed by attaching rule-based logic to classes, subclasses, and objects to describe the conditions under which they may be used. The marriage of artificial intelligence techniques and object data base structure may be necessary to prevent misuse of objects.

Availability of objects partially depends on systems availability and network availability, for example. Another concern is that the object is appropriately distributed throughout the organization's processing resources so that it can be conveniently accessed by authorized personnel regardless of the time or location. In large corporations, objects may be distributed in repositories on a variety of machines in various locations, so the potential for erroneous use is multiplied.

Confidentiality

Confidentiality may require that two levels of information access are designated for objects. One level of access may be to permit a user to determine whether a desired object or reasonable facsimile exists in the inventory. This level may permit only authorized personnel to learn of the existence of objects and to obtain a brief description. A second level of access control may be needed to permit users to actually read the object content itself.

Confidentiality can be breached in another way. The aggregation of intelligence through repeated access to selected data bases of information is a threat to current systems. When the atomic level of applications is downsized to objects, a significant change occurs. The aggregation of objects into new relationships may permit combinations of information that would not usually be available to users, thereby enabling unauthorized users to assemble intelligence to which they are not entitled. The property of inheritance—in which an object subclass contains information about the methods and structure of the superclass it is related to—presents special concerns. A classification mechanism may be needed that defines permitted relationships among objects and establishes authorization for object relationships, perhaps as a facet or attribute. Alternatively, it is possible to maintain independence between data and code that permits access controls to be placed on the data at the user view or field levels within a data base.

CONCLUSION

Many of the potential problems associated with object-oriented programming are similar to those of systems development tools of the past. Such problems are usually addressed as a result of market pressure on vendors to supply acceptable solutions.

Under current conditions, the lack of inventory management of objects poses a potential problem to effective widespread reusability. IS managers are therefore advised to examine the inventory management capabilities of proposed object-oriented development systems and limit use of the tools to a scope under which available management methods will work.

Without solving the problems related to object security, it may not be possible to ensure the protection of the company's information assets under conditions of widespread use. Vendors of object-oriented DBMS are attempting to develop systems that reproduce the controls available in hierarchical or relational data base management systems.

Object-oriented techniques are analogous to an automobile that may be safe to drive at 70 miles per hour but unsafe at 110 miles per hour. The potential benefits of object-oriented programming appear to be substantial; however, until object-oriented technology enables users to manage and protect their information assets, it should be used under tightly controlled circumstances.

IV-8

Supporting Legacy Systems

Nancy Blumenstalk Mingus

Legacy systems continue to be used because they continue to add some value to the business and because they are so expensive to replace. Yet staff members responsible for supporting and maintaining 10- to 20-year-old mission-critical systems may feel they are being denied a chance to acquire new skills. IS managers can protect staff morale in their departments by conveying the business reasons for using legacy systems to the staff assigned the job of maintaining and supporting them. This chapter outlines various ways to minimize apathy among support staff, with tips for helping maintenance employees keep the job fresh.

RECOGNIZING LEGACY SYSTEMS

Any of the following conditions may be true of a legacy system:

- It was programmed by someone who left the company 10 years ago.
- It uses data base technology or a programming language rooted in the 1970s.
- It was translated from one language or operating system to another by an automated translating utility.
- It has been modified so many times, the code looks like spaghetti; there are pages of variable names that are never referenced and whole sections never executed.
- It is so old, the source code is missing.

Although physical age and old technology are hallmarks of a legacy system, another equally important identifier is that a company still uses the system in production. If the system is a purchasing, order entry, inventory, human resources, or even an executive information system, odds are the company still trains or supports the users and maintains or enhances these systems even though they are past their prime.

Many companies call their older systems legacy systems. Legacy is a nicer name than dinosaur, though the idea is basically the same. More specifically, however, legacy systems are usually financial or human resources systems left over from the 1970s. Legacy systems are mainframe-based. Whereas most dinosaur systems are mainframe-based, they need not be. A 10-year-old dBase II or Lotus 1-2-3 application can also be a dinosaur, especially if the corporate standards for PC software packages have changed.

WHY LEGACY SYSTEMS LINGER ON

These systems got to be legacy systems in a variety of ways. Here are just a few:

- They were state-of-the-art when implemented. Many of the systems still running today were ahead of their time when they were written. They used the latest technology, and in ways the product developers never intended.
- The systems were passed from person to person, until finally ending up with the one or two employees who know them now. Unfortunately, the documentation never made it past the third or fourth handoff.
- The systems continued to be patched instead of redesigned as corporate conditions changed.

Many systems that became legacy for these reasons continue to be used for similar reasons. The most common rationale is: If it's not broken, why fix it?

Business and Financial Reasons

Even though code is patched, often several layers thick, the system continues to serve its function. A company may be leery about investing time in a redesign of something that still works. What companies often fail to realize, however, is that in many cases, when these systems do break, they might not be fixable. This is especially true of those systems with hard-code date routines that will not be able to handle the year 2000.

Another reason why these systems linger on is that they meet a need no preprogrammed package can meet, even today. Legacy systems are almost universally in-house written and fit the organization so well that no one wants to change procedures so they can upgrade the system.

In addition, other higher-priority jobs continue to push legacy system rewrites to the back burner. This reasoning ties in with the first one: As long as these systems continue to perform their function, a company may determine that it is better to invest in systems that can save money or make money.

Converting legacy systems to new technology is costly in terms of redesigning and recoding the software, as well as purchasing new hardware. Furthermore, current staff may not have the required skill sets for recoding the systems, so new staff, consultants, or extensive training become necessary. Also, the terminals, controllers, and mainframes generally used with existing systems are paid for. Networks and PCs require additional capital.

The company's legacy systems were often the first automated systems, so they usually contain sensitive corporate financial, personnel, and product data. Some companies maintain that rewriting the systems could jeopardize data confidentiality and potentially affect data integrity and for this reason opt to keep their legacy systems.

OBJECTIVELY EVALUATING LEGACY SYSTEMS

Drawbacks of Legacy Maintenance

The drawbacks to maintaining existing systems are fairly obvious. IS professionals confront them daily. Some of the major drawbacks are that:

- *Only a few people know the system, so they end up doing all the training, support, and maintenance.* This deprives the staff of time to learn new systems, which virtually every employee would rather be doing.
- *These systems were installed before data entry validity checks were popular, so bad data is often entered.* It takes extra support time to fix errors, extra maintenance time to fix the system, and extra training time to tell people how to enter data correctly. Furthermore, these systems are often used in departments with high turnover, which adds to the problem of training and support.
- *Training generally must be designed and delivered in-house.* Because the systems are not standard packages, there are no canned or vendor-delivered courses available. Even if they were originally written in a standard, supported product, the product may be so old that it is difficult to find anyone who teaches courses on it anymore.
- *Every time operating systems, file structures, or other support programs change, the existing systems might not work.* This often requires significant support time to get them running again, if they will run at all.

Benefits of Legacy Maintenance

There are some benefits to providing quality training, support, and maintenance on existing systems. Among them:

- *Maintenance can add another five or more years to their life span.* Although this strategy often makes the technology of the legacy system even further removed from state-of-the-art, it also buys time for the company as it makes plans to migrate to newer strategic systems and downsized platforms.
- *Proper maintenance minimizes system errors.* Because legacy systems are often complicated to use, it is easy to create errors and difficult to correct them. Proper training and support minimize the chance of operator error.
- *Unlike off-the-shelf PC or mainframe packages, legacy systems do not change significantly.* The benefit in this case is that there is no need to update the training materials or support procedures every time the vendor introduces a change.

MINIMIZING APATHY TOWARD LEGACY SYSTEMS

Burnout is probably the number-one problem where maintenance staff is concerned. IS managers attentive to the morale of their departments can encourage IS maintenance personnel to regain interest in their work in several ways.

Maintenance staff should be encouraged to dig through the fossil record. For example, one method is to allow the people responsible for legacy systems to investigate the history of the systems, to find out who wrote them and what particular design or programming problems they overcame. This information can give the staff new respect for the systems. By compiling a list of the ages of the various production systems, the people charged with these systems may also be surprised to find that these systems are not as old as some others.

It helps to explain to staff members doing legacy maintenance assignments why the systems are still being used. There is a strong inclination to want to scrap systems simply because they are old. This thinking obscures the real reasons that management and users may be reluctant to bury their existing information systems. It also erodes the sense of purpose maintenance staff members need to remain motivated.

IS managers can take several measures to keep legacy systems and their remaining business value in perspective. At the same time, these practices will help to minimize low morale among employees responsible for legacy systems maintenance.

Do Not Give Responsibility for Legacy Systems as a Punishment. If IS employees have had problems adjusting to new systems, the manager's first urge might be to pull them off that assignment and relegate them to the job of providing maintenance or support for existing systems. This action will always be seen as punishment. In many companies, junior-level people with limited experience or senior-level people nearing retirement are also saddled with older systems. Whether this decision is intentional, this action is generally viewed as a punishment, too. IS managers may be inviting trouble if they give legacy maintenance assignments as a punishment.

Consider Adding a GUI Front End. Adding a graphical user interface (GUI) to an existing system is a common first step in migrating to new platforms. Customers (i.e., users) reap the benefits of a consistent look and feel; at the same time, the GUI front end gives maintenance staff a taste of new technology, trainers new screens to work with, and support people fewer data entry errors to contend with.

Suggest That Trainers Change the Exercises in the Training Course. This simple suggestion takes less time to implement than developing a completely new course, yet it will make the course feel new and give trainers a chance to propose new ideas.

Allow Staff to Find an Undocumented Feature of the System and Publicize It. This is like uncovering a new type of dinosaur bone. It can stimulate more interest in the whole system and give the department some additional visibility.

Encourage People to Train Others as Backup. For maintenance staff, backup usually comes from a colleague, but for trainers and support staff, backup might come from an end user who favors the system, a colleague, or even a consultant. Regardless of the source, having a backup gives the regular maintenance, training, and support staff room to breathe.

Perform a Full Systems Analysis on Existing Systems. The purpose is to find out why the system is still in use. A full systems analysis usually reveals that the documentation for the company's legacy systems is missing, dated, or incomplete. That is why only one or two people know the system. A full systems analysis will provide documentation or bring it up to date.

A full systems analysis that includes a cost/benefit section for replacing the existing system can be used to explain to the staff why management has decided to keep and maintain the existing system. Conversely, a cost/benefit analysis may convince management that the benefits of keeping the system are overstated and that it is time to retire the system, thus paving the way for redeploying current training, support, and maintenance personnel to learn new systems.

Allow Training, Support, and Maintenance Staff to Learn New Technology. The entire staff should be granted time to keep up with new systems. Consider having each person on the maintenance staff pick a technology area they would like to learn about and present the latest developments at weekly staff meetings.

Encourage More Business Knowledge. If IS personnel are working on accounting systems, for example, suggest that they take an accounting course so they can better understand how their system fits in the overall business scheme.

Outsource the Training, Support, and Maintenance of Existing Systems. Although in-house people might view legacy assignments as punishment or a dead-end job, consultants tend to view them simply as work. Outsiders taking over existing maintenance are also likely to lend a fresh perspective to uncovering opportunities to improve efficiency.

CONCLUSION

Even companies committed to downsizing need people to provide training, support, and maintenance for their existing systems that keep the business functioning. But that does not have to mean a step backward for the IS staff involved. By viewing legacy systems objectively, and helping employees do what they can to keep their job fresh, IS managers can help their employees avoid becoming dinosaurs themselves.

IV-9

Open Systems Integration: Issues in Design and Implementation

Leora Frocht

APPLICATIONS DEVELOPMENT in the open systems environment involves numerous and varied integration issues. As previously unusable ideas are translated into entirely new client/server systems, open systems tools make possible the development of applications that could not be developed on mainframes or on pre-windows PCs. Designing and implementing open systems therefore presents new and often daunting challenges.

Development team members generally do not have much experience in the open systems environment. The technology is relatively new, and the tools are not fully mature. There is no foolproof methodology, and many choices regarding software must be made amid a plethora of conflicting opinions. Experienced developers, who are generally familiar with mainframe or pre-windows tools, must change their entire pattern of thought to adapt to open systems and the client/server approach.

This chapter is based on the experiences of one project manager in integrating legacy systems into a new open systems environment. The seven rules of integration developed from that experience address the many components of integration in the open systems environment and should help IS managers anticipate and thus mitigate potential problem areas.

ISSUES IN OPEN SYSTEMS INTEGRATION

Integrated systems development involves issues of performance, communication, and compatibility. Integrating legacy systems into an open environment focused on seven major factors that are generally applicable to systems integration projects:

427

- User specifications.
- Business alignment.
- Communications.
- Data accuracy.
- Resource contention.
- Ergonomics.
- Legacy systems.

IS managers involved in such projects are challenged to manage these components and integrate them into a fluid, sensible, and intuitively usable system.

User Specifications

A reasonable understanding of user requirements makes the integration of the other six components much easier. Accurate interpretation of user specifications depends on the IS manager's ability to bring together project team members with the appropriate training and experience level.

Alignment to Business Objectives

A thorough understanding of the business objective also helps derive the end user's requirements, needs, and goals and ensures system performance. No decisions, whether software- or hardware-related, can be made without a thorough understanding of how the business is conducted.

Many development choices depend on what the business requires. The system must be flexible enough so that users are able to do more than they expected or initially thought of. To the extent permissible by today's technology, the system should be portable so that an application can be moved from one platform to another, from one version of an operating system to its next generation, or from a PC to a workstation. It should be able to support current data needs as well as anticipated future needs.

Performance. Performance, both in accuracy and speed, should reflect the end user's identified business needs. It is hardware- and software-related. Appropriate hardware must be selected. For example, if high-volume communication is an aspect of the system, communications hardware, such as wide area network (WAN) and local area network (LAN) routers and cables, must be selected and implemented to handle higher volumes. Volume is generally identified by users because they are the ones who generate transaction activity. Server machines must be designed to handle large volumes of data or applications software, or at least be scalable and configurable to do so.

Communications

Communications software must also be robust enough to send out and receive large volumes of data. Market data services, which provide information like stock prices, generally address these issues; they transmit hundreds of thousands if not millions of messages a day. Applications software designed to accept high volumes of messages must not only receive the messages but also interpret and process them efficiently. The source code must be able to receive each individual record, determine what to do with it (e.g., based on information in header data), and send it off to the designated recipient. The recipient could be a data base, a flat file, or a window panel. All of these transactions must be completed in a reasonable time frame.

Accuracy

Accuracy is a critical issue in systems development. Appropriate business decisions are made only if data is correct all the time. Because accuracy is sometimes overlooked in the development of an integrated open system, project team members must be vigilant in resisting the temptation to do so.

Resource Contention

Applications developers must consider how data is used because in an integrated system two users could contend for resources. Resource contention frequently occurs when requests for data are made to the data base. Standard procedure dictates that one user's retrieval of data from a data base not restrict other users' access to the data.

Resource contention can also occur in the area of communications, particularly regarding interprocess communication or data broadcasting to a large group. The application must not require that the broadcasting of one message depend on the successful delivery of another message. Also, the successfully integrated system allows sent messages to be buffered or stored and processed separately from the communications portion of the application. That way, messages can continue to be received instead of being backed up because the communication protocol is still processing the previous message.

Ergonomics

Screen real estate is a lesser, but still important, component of integrated systems design. Developers should ensure that the user's screen does not become overcrowded with panels, screens, or other kinds of activity. A large number of panels, or newly generated panels that cover up others, becomes unmanageable. Screen management and screen real estate can make or break the integrated system, even though they do not directly affect the system's actual functioning.

Similar consideration should be given to the interior layout of each window. When the widgets (e.g., menus, buttons, scroll bars, etc.) inside a window are arranged counterintuitively on the screen, the user needs more time to sort out all the information on the window. When this happens, the user ends up spending more time trying to comprehend what he or she is looking at than actually dealing with the business at hand. Common types of information should be grouped together.

Team members must also consider the function of the widgets—the actual conduits to the functions of the integrated system—when choosing from the various types of widgets available in graphics packages. If these conduits either implicitly or explicitly mislead the user, the system will not function properly. Books on windows and window design explain the appropriate use of the different types of widgets.

Appropriate color choice provides the finishing touch to an open system. Color helps separate the different functions implemented in the integrated system. For example, the functions to add, modify, or delete records in a data base can be color-coded in the following manner: buttons for the add function can be colored yellow to indicate that the user can perform the function but should proceed with care; the button for the modify function can be orange to indicate that anything done can be undone; and red can be used for the delete function to indicate that a record removed is gone forever and using this function may be dangerous.

Colors should also be selected with aesthetics in mind. Human beings work more easily and happily when what they are looking at is pleasing to the eye. Color can also delineate separate sections of a window to aid the user in interacting with the system. Finally, colors selected should be easy on the eyes. Screens with strong, bold colors or highly contrasted colors next to each other are difficult to view for long periods of time and can cause eye strain. In more ways than one, the integrated system should not create headaches for the user.

Legacy Systems

Replacing a legacy system with an open system solution further complicates systems integration. For the purposes of this discussion, legacy systems are considered those that were developed in a nonwindows environment. So a legacy system is one that relies on linear menu selection rather than on the ranges of selection available on window-type panels. The platform may be mainframe or PC, but it does not allow for the branching out of applications functions or for simultaneous display of panels.

By definition, a legacy system has been around for a long time. If it did the job reasonably well for the user, the new open system must do it better; if the old system never fulfilled the user's needs, a successful replacement is critical. The worst thing that can happen for users and developers alike is for a new application to have less functionality than its predecessor.

It therefore becomes essential that navigation of open systems windows be a primary consideration in the design of a new system or of a legacy system's replacement. Where once users had no choice but to follow their own linear thinking when running a mainframe system, the open, integrated system gives users many path choices. The system designer should consider logical flow: namely, which panels can parent a subsequent panel (the child process), and which panels would be a final child in what becomes the tree structure of insatiable window-upon-window navigation. At all costs, the development team should avoid creating a confusing mass of windows that users will never be able to navigate.

Another consideration related to window navigation is the actual number of window processes that should be running at any given time. Although integrated systems can support several simultaneously running processes, a benchmark should be established for the number of processes that should be running within the system at any given time. The system should never crash, but it should perform within reasonable expectations. Consideration of the these points can make the difference between whether users accept or reject a legacy system replacement.

Replacing a legacy system also involves consideration of how much, if not all, of the system to replace, even if the original hardware remains. In some cases, a legacy mainframe system is ideal for housing large volumes of data, for example, but it is not really suited to providing that data to the user in a digestible way. The integrated, open system then becomes the suitable front end, and the remaining issue is that of communication between the client/server hardware and the mainframe.

A similar scenario results if the existing system contains a mainframe component whose accuracy, reliability, and performance cannot be replicated and should therefore not be replaced. Work done well on the mainframe can be communicated to the client/server platform and vice versa.

SEVEN RULES OF INTEGRATION

The intricate, laborious, and expensive process of integrating legacy systems into an open system environment yielded seven rules of integration:

- Understanding the technology.
- Understanding the application.
- Creating a solid development team.
- Making users part of the team.
- Gathering detailed specifications.
- Organizing and interpreting specifications.
- Understanding the critical components of integrated systems.

Rule 1: Understanding the Technology

Systems integrators must select from a large range of products and unite their selections into a system that solves the problem at hand for the user. The challenge for the systems integrator is knowing which criteria should be used in selecting the software for each application and ensuring that all the choices work together in a smoothly running system. At a minimum, software choices must be made for the graphical user interface (GUI), the data base, interprocess and intraprocess communication, and the programming languages.

Generally, however, senior management, technology committees, or infrastructure specialists in the individual shops make the software choices and give the development team one or two options from each software category. In this case, team members should assess the qualities of the software tools under consideration: particularly, how the qualities affect the end result and how the tools will interact with each other.

Rule 2: Understanding the Application

The key to integration of open systems for both the experienced analyst as well as the novice developer is to completely understand the goals of the application. All members of the development team have to understand how the user's business works and the user's desired goal for the system. This point cannot be stressed enough. Users often have a specific approach to the way they do business. Whether this approach is based on habit or on business constraints, members of the development team must pay attention to it. New approaches and ideas should be considered, but ultimately, the way the system approaches the business is always up to the user.

Rule 3: Creating a Solid Development Team

The third important component of a successful integration process is a solid development team. The mainframe environment allowed a single developer to successfully create an integrated system because the components were already integrated by the mainframe provider. When a technology shop was set up on an IBM mainframe, IBM Corp. also provided the software. There was certainly systems software, which was centralized, and other data base software products like SQL/DS or DB2. There was one text editor available. The mainframe manufacturer usually supported only one copy of utilities and operating systems, so much of the software choices were already made, installed, and known to work together. The only thing the programmer/analyst was ultimately responsible for was the function of the system for the user. In addition, the mainframe architecture rendered it often more convenient for one person or a very limited number of people to work on a single integrated system.

At the opposite extreme, client/server systems are built on hardware platforms like workstations or PC LANs—platforms that give systems analysts and developers much more freedom and flexibility in tool selection, design, and implementation. Because so much more flexibility and processing alternatives can be built into an integrated system, a team of two or more analysts and developers, as well as component specialists, is required to create an integrated, open system.

The sum total of knowledge of the development team must encompass two disciplines: technology and the user's business. Within the technology discipline, the team must include members with strengths in data base software, GUI software, communications software, and applications software. Analysts and software developers are often experienced in more than one of these areas. Similarly, specialists in the hardware technology in use should be part of the team effort. They can advise on interprocess and interplatform communication and performance, as well as on programming for them.

The technologists' knowledge of the user's business is at least as important as experience with computer technology. Understanding the user's requirements determines how the applications software will function; hardware accommodation will ultimately determine the system's performance.

Rule 4: Making Users Part of the Team

Interpersonal communication and creation of an effective personal rapport between members of the development team and the users are essential to the integration process. The building of working relationships significantly improves the effectiveness of the completed system. When productive working relationships exist between the users and the development team, developers are less intimidated about what they do not know and ask many more questions. Instead of glossing over issues in the fear of appearing naive, developers are more apt to explore them with users.

Conversely, users need to feel comfortable too. Many business people are still very much intimidated by computers. When unfamiliar terminology and technology is suddenly flung at successful people who are used to being in control, they feel uncomfortable, and uncomfortable users do not easily or willingly discuss the system with the development team. Barriers between the users and the developers degrade the quality of the systems integration process. Forging a solid, interpersonal working relationship between users and the development team is therefore integral to the process of systems integration.

Rule 5: Gathering Detailed Specifications

The first technical step in the integration process is gathering specifications for the system. There are several ways to do this. Naturally, the best source is the user. Discussions and interviews with the main user or user liaison and his or her peers are essential, as is careful and organized note taking by members of the development team. Ideally, part or all of the development team should sit with the user for as long as possible, or at least for a minimum of several business days.

Other sources for gathering specifications include other companies, professional associations, and books and articles on the subject. Information from these sources often adds depth and perspective to the goals specified by the user.

Following each session with users, the development team should discuss the information given to it and how it relates to what the team already knows. This exercise helps team members refocus on the goal of the integrated system and to share their own experiences and development ideas. At the end of the systems specification process, the members of the development team should comprehend the user's business on an intuitive level.

Rule 6: Organizing and Interpreting Specifications

Organizing Notes Along Logical Components. Formal documentation of user specifications is important to the project team's effectiveness and efficiency. Team members should bear in mind particular areas when collecting specifications, namely the programming language, the GUI and windowing, data base, and communications. Written notes should be organized according to logical components. Most systems require a main screen from which all other operations are managed. System components can be organized according to what is required on the main screen level and according to what is required for subgroups of the main level. Possible functions to launch from a main screen are data maintenance, reporting, tools such as data item searches, or utilities such as refreshing a screen from a data base.

Painting Up Windows. In the next step in organizing specifications, project team members take the notes arranged by function and paint up a window that corresponds to the main function and as many windows as necessary for the subfunctions. They must resist the temptation to build deep layering into a system. If users have to call up several chained windows just to get to the one they need, the window navigation process becomes unmanageable and most users will likely give up. As flat an implementation as possible is best, but balance should be kept in mind. Generating too many screens on the same level that are required to be displayed at once jeopardizes the issue of optimized screen real estate. Team members should get the input of users on the issue of balance while also drawing on their own intuition and experience. Generally, however, one main window with the subfunctions spawned from a main menu bar or other widgets gives the best chance of functional success.

Reassessing Goals and Implementation Feasibility. At this point in the systems development process, team members should consider whether the goals of the system are too broad. They should also address the question of implementation feasibility.

Oftentimes, when a new integrated system, as opposed to a legacy system replacement, has been proposed, users' requests are not entirely focused, yet alone finalized with any degree of precision or confidence. Users have a broad idea of what they want, but meeting all their business needs may actually require the development of more than one integrated system, or of several smaller subsystems chained together. Painting the windows clearly demonstrates to users how large or small the system will be, based on the specifications.

Dividing a system into manageable components wherever possible is better not only for the systems developers but for the users themselves. In *Object-Oriented Analysis and Design with Applications* (Redwood City CA: Benjamin/Cummings, 1994), Grady Booch discusses the fact that the human brain can only manage a limited number of processes at a time while the computer can manage as many processes as the hardware can accommodate. Because the computer can handle many, many more processes than the human brain, functions should be separated into different components. In this way a system that accommodates the physical limitations of the human brain is implemented rather than one that can only be comprehended by a computer. Separating out specifications into different components also results in flatter, simpler, and targeted individual systems that are easier to develop concurrently, enhance, and maintain.

Defining Additional Components. At this point, developers have a prototype that can be critiqued and revised. It now becomes apparent where the other components of the integrated system come into play. Interaction with the data base can now be defined because the origination point of data is known. The data types and sizes are also known because they have been designed into the windows. Determination of which windows and functions will communicate with each other becomes obvious. When all the windows are visually displayed, it is easy to identify which windows should publish data and which windows should receive it. Any other roles relevant to the systems integration process can be hooked in once the system is visually prototyped.

Understanding the Critical Components of Integrated Systems

There are six major components of an integrated open systems environment:

- Programming language.
- Communications protocols and network.
- Graphical user interfaces.
- Data base.
- Applications management.
- Applications systems.

Programming Language. The programming language is the component of the system that actually does the integration. As the glue between all the other components, it facilitates the interaction between the screens and the data base, between the screens and the communication components, and so on. It must provide functions that help users do their jobs as smoothly and easily as possible. In addition to providing flexibility and performance, it must hook into the other components of the integrated open system.

Many information technology organizations use C or C++ programming languages because these languages meet the previously mentioned qualifications; when properly implemented, they are flexible, precise, and high-performing. Oddly enough, the very flexibility of the language creates a managerial challenge. When using these languages, developers are often tempted to reprogram and hone the programs—a process that can become unnecessarily time-consuming. Project managers must be able to clearly define and limit the extent of the implementation and provide feedback to the developer.

Communications Protocols and Network. Communication, especially real-time communication, is critical to a successfully integrated system. The development team must consider various types of communication: interprocess communication, communication between the open system and a mainframe, and broadcasting or publishing information through the open system. The communication method is selected based on the end-user's goal.

Interprocess communication methods (a point-to-point approach) is used for discreet communications, such as when one user wants to send information only to certain individuals. When a user needs to communicate from an open system to a mainframe, special communications software must be invoked because workstations, PCs, and mainframes all speak different languages themselves (i.e., they have different operating systems). When large amounts of data need to be distributed across a group of users on a network, a broadcast method is used. Many market data vendors provide software that facilitates this type of communication.

In every type of communication, performance by the hardware and response time by the application is key. User specifications determine how the application will handle its processes, which must be designed so as not to overload the hardware and slow performance. Here, bandwidth must be determined. If an organization anticipates a large number of transactions, it should plan for accommodating them. Users can be consulted about how many transactions they anticipate executing in current conditions; they should also try to estimate future business volume. Then benchmarking can be done on existing hardware using tools available for this purpose.

A final but important aspect of the communications protocol is scalability. Whatever product is used to develop communication should be able to accommodate a growing number of users. Of course, no open system can handle an infinite number of users, but a number reasonable to the business should be accommodated. Different protocols have different limitations, so the project team must examine specifications for those products.

Graphical User Interfaces. Design of the windows can be the most important and most complicated part of the integration process. At a minimum, the team should consider how the programming language will make the windows perform, navigation through the various windows, how large the panels are, what color they are, and how information is organized on them.

A GUI painter should be used to create the windows for the proposed system. The window attributes are clear, readily available, and easily modifiable.

Once the windows are designed, team members can often populate certain types of widgets with sample data, which makes the functional aspects of the system clearer without doing a coded implementation. Pull-down lists are good examples of this feature. For example, if a limited and thus predefinable set of standard entries, such as the regions of the United States, is required, the information can easily be entered as a pull-down list and be immediately visible to users. As many of the widgets as possible should be predefined and implemented before taking a finished screen prototype to the users for approval. At this stage, users can tell if the development team has integrated the business requirements into the system.

Data Base Design. Design and implementation of the data base is also determined by the business flow. Although the gathering of the individual pieces of information is relatively straightforward, organization of the data may not always be. The goal in data base design is to provide the best performance possible by implementing the optimum data organization. Whether the data model is relational or hierarchical, the designer/developer must resist the temptation to design a perfectly normalized, not-a-byte-wasted data base, even if it means decreased performance for the user. It is often worthwhile to repeat fields to save the user an additional data retrieval.

Effective data base design must consider hardware organization. Attention must be paid to the size of the data base. It is possible for a component of a data base to require an entire machine by itself; the various pieces of data, which could end up on different machines, need to be able to interact with each other.

Applications Management. The success of integrated open systems also depends on having a centralized location for all applications source and executable code. Because of the larger team size required to develop an integrated system, the project manager must ensure that tight control is kept over the code and that only single copies of any program or function are resident in the system.

The first step in applications management is to install a file librarian or source code control system (SCCS). SCCS creates a central repository for all the source code of the system. It can maintain version control and does not allow more than one copy of source code to be checked out at a time. It also provides a layer of security. Any developer not given permission cannot pull source code out of the repository.

Third-party software libraries are also effective, and many software vendors provide for standard functions like date calculations or matrix manipulations.

Separating the development environment from the prototype environment, and the prototype environment from the actual production environment, is important to a successful system. The development environment is meant strictly for the writing and testing of code; it allows developers to try out different ideas or to rework ideas. Only programs still in flux, or those that are still deemed modifiable, reside in this environment.

In the prototype environment, tested and debugged components are installed, linked together, and run as a whole. In this environment, a system that is not yet ready for the users is built up. Full systems and integration testing is conducted, and software components are replaced only when bugs are found in the testing process. Only executable code resides in this environment. Replacement of system components and recompiling is done in the development environment.

The production environment also contains only executable code and is the area from which the user executes the system. It should be bug-free and should never be affected by anything that goes on in the development or prototype environments.

These different environments make change management vital. Careful scheduling must be observed when modifications are made to existing systems, be they in the development, prototype, or production stages. Developers and systems administrators should be aware that to avoid problems, changes to a system must be scheduled and implemented as scheduled. The procedures outlined in the change management process ensure that the development team, system administrators, and data base administrators communicate with each other and that each knows what the other is doing with respect to the system. Systems modifications or upgrades should always be implemented at a time that is least inconvenient to the user. Unfortunately for the systems personnel, that time is usually at night, after business hours, or on the weekends.

Notification of changes to an existing production system can be as easy as sending electronic mail. However, more sophisticated notification products are available on the open market.

Applications System. The applications system is the enabling component of the integration process. As noted, the application must be process supporting, easily ported and scaled to other needs, and functionally superior to its legacy predecessor. Above all, it must be maintainable.

RECOMMENDED COURSE OF ACTION

Systems integration projects are enormously challenging and rewarding. Users are more demanding, and systems personnel are learning new roles as well as new technology. Regardless of how much experience an IS or project manager has in mainframe and mid-range systems development and integration projects, each integration effort is a learning experience.

Because the project management environment is an ever-changing landscape of technical, personnel, and user issues, managers need an orderly and effective methodology that addresses the numerous components of integration and anticipates problem areas. The seven rules of integration presented in this chapter should better prepare IS and project managers for the rigors and unknowns of systems integration. During what is often a tedious and demanding process, managers should remember that the tangible results of a successful integration are felt almost immediately by the organization.

IV-10

Using Hypertext for Group Decision Support Systems

Vatcharaporn Esichaikul
Yaowaluk Chadbunchachai
Gregory R. Madey

THE ABILITY TO EFFECTIVELY AND EFFICIENTLY GATHER, manage, and retrieve voluminous amounts of information is essential to decision makers in today's information age. Because of time constraints, decision makers face not only the problem of information access but also of information overload. The ability to integrate ideas and knowledge from various sources is particularly important for the effectiveness of group decision support systems (GDSSs).

One promising technology for the group-decision making environment is hypertext. In a hypertext system, information is grouped and linked to allow users easy access to related pieces of information. A hypertext-based GDSS therefore facilitates the collection of group process data and makes the tasks of idea generation, problem structuring, member-to-member communication, and voting/issue resolution more effective.

This chapter describes the design and the development of a hypertext-based GDSS called HyperGroup. The system is designed to support communication in small, face-to-face, decision-making groups. Three decision support tools—issue identification, criteria identification, and alternative identification—are built into the system. HyperGroup also has the capability to capture group decision-making processes.

HYPERTEXT FOR GROUP DECISION MAKING

Hypertext is an alternative approach to organizing and managing data. A hypertext system consists of two basic elements: nodes and links. A node represents a unit of information. Related nodes are then connected by links.

A hypertext system can be viewed as a network of nodes and links with no specific or global structure regarding sizes of nodes or the number of links to or from a specific node. Users interactively take control of a set of dynamic links among units of information that allow them to jump from one piece of information to other related units of information.

Hypertext is a promising technology for group decision making because it provides an environment suitable for supporting the functions of an effective GDSS. In a group decision-making environment, information elicited during the course of meetings must be gathered, organized, recorded, and accessed. Each piece of information is related to rather than isolated from others. Decision makers usually need to browse through the information base to see connections between pieces of information. In addition, each decision maker contributes information to the group piece by piece. The information is therefore grouped into a form equivalent to a node in hypertext systems.

CONCEPTUAL DATA MODEL

The conceptual data model proposed for the hypertext-based GDSS accommodates the various kinds of group decision-making techniques implemented by GDSSs. The model consists of two node types:

- List nodes, each of which contains a list of ideas on a certain topic.
- Comment nodes, each of which contains comments on certain ideas.

Both types of nodes are semistructured to facilitate the organization and management of information in GDSSs. Links in the model are not typed or structured.

Exhibit IV-10-1 presents the conceptual model on which the HyperGroup is based. In the model, a list node may contain a list of alternatives, a list of issues or problem statements, or a list of criteria. A list node containing a list of issues is linked to a corresponding list node containing a list of alternatives or criteria, or a certain comment node.

SYSTEM FUNCTIONS

The functional theory of communication and a conceptual GDSS framework were used to identify a set of system functions. Based on the functional theory, the system must assist decision makers in satisfying the following four requisite functions of effective group decision making:

1. Thorough and accurate understanding of a problem.
2. Identification of a range of realistic and acceptable alternatives.
3. Thorough and accurate assessment of the positive consequences or qualities associated with each alternative.
4. Thorough and accurate assessment of the negative consequences or qualities associated with each alternative.

These requirements helped identify the following set of minimum functions to be supported by the system:

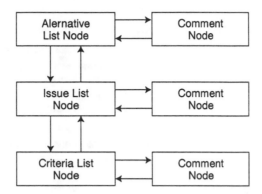

Exhibit IV-10-1. A Conceptual Model of HyperGroup

- Problem identification and formulation.
- Criteria generation.
- Alternative generation.
- Alternative assessment.

Exhibit IV-10-2 presents a summary of HyperGroup's features.

SYSTEM TOOLS

System functions and features as well as the implementation of selected group techniques were used to design and develop the system tools. Based on a review of the literature and findings on the effectiveness of different group decision-making techniques, two group techniques were selected: goal orientation and idea writing. Implementation of the two selected techniques to satisfy the system functions resulted in the development of three system tools displayed as three options on the main menu. The tools are:

1. Issue/problem identification.
2. Criteria identification.
3. Alternative identification.

Issue/Problem Identification. The issue/problem-identification tool assists users in satisfying the first system function: problem identification and formulation. It implements the goal-orientation technique for idea generation and the idea-writing concept for idea reasoning.

Criteria Identification. The second system tool helps users satisfy the second system function, i.e., criteria generation. The tool assists decision makers in

Group Activity Supported	HyperGroup Features
1. Information Generation	• Ability to write ideas into shared database
	• Ability to create, edit, delete, and add to the list at personal input devices
	• Ability to transfer the generated ideas to the public screen for open discussion
	• Enforces a process-related structure
	• Ability to enter ideas anonymously
	• Ability to enter comments
2. Information Retrieval	• Ability to preserve and redisplay critical events
	• Provides a common viewing screen or a public screen at each group member's terminal
	• Ability to save information between meetings
3. Information Share/Use	• Computer terminal for each group member with connection to a central computer
	• Ability to provide a summary and display of ideas
	• Ability to provide group process tools, i.e., goal orientation and idea writing
	• Ability to edit, delete, and add to the database file

Exhibit IV-10-2. Main Features of HyperGroup

generating and reasoning out a set of criteria that will be used to identify whether a certain alternative is realistic and acceptable when they apply the idea-writing technique.

Alternative Identification.   The alternative identification tool supports users in satisfying the third and fourth system functions, i.e., alternative generation and alternative assessment. The idea-writing technique not only helps users generate the alternatives but also assess the positive and negative consequences associated with each alternative.

Exhibit IV-10-3 shows the relationships among the system functions, decision-making techniques implemented, system tools, requisite decision-making functions, and group activities supported.

COMMUNICATION AMONG GROUP MEMBERS

HyperGroup uses a public window, which is similar to a bulletin board facility to the extent that group members send their comments to the public data base.

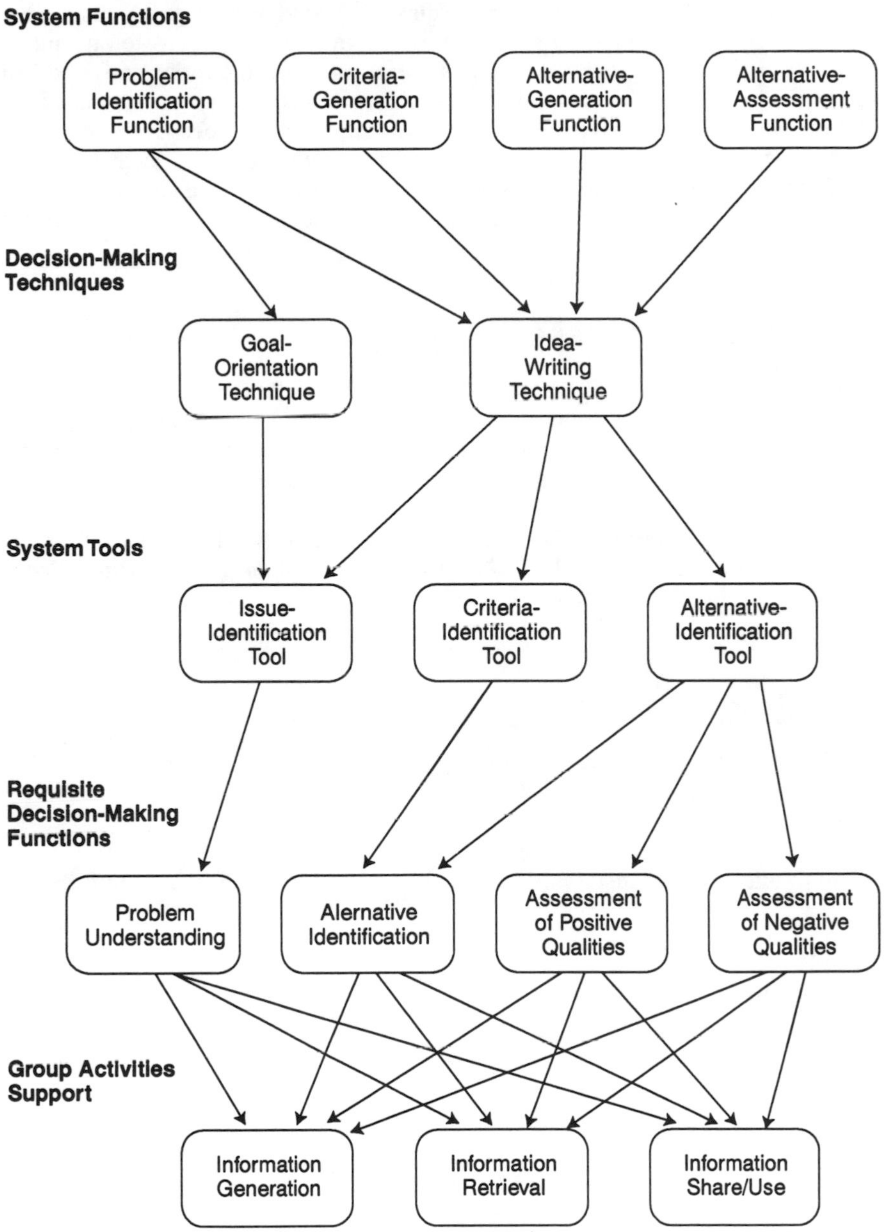

System Functions

Problem-Identification Function	Criteria-Generation Function	Alternative-Generation Function	Alternative-Assessment Function

Decision-Making Techniques

Goal-Orientation Technique

Idea-Writing Technique

System Tools

Issue-Identification Tool

Criteria-Identification Tool

Alternative-Identification Tool

Requisite Decision-Making Functions

Problem Understanding

Alernative Identification

Assessment of Positive Qualities

Assessment of Negative Qualities

Group Activities Support

Information Generation

Information Retrieval

Information Share/Use

Exhibit IV-10-3. Relationship among the Elements of HyperGroup

443

The system provides a medium for recording private ideas and group discussion. The information in the public data base is retrieved at both private and public terminals; as a result, group discussion focuses more on the topic under consideration than on finding or recalling pieces of information in previous discussions.

The availability of a private terminal for each group member allows members to simultaneously contribute their ideas to the group instead of each having to wait for a turn. Because this system feature reduces the time needed to elicit ideas from group members, group communication and discussion are more productive.

DESCRIPTION OF HYPERGROUP

HyperGroup was developed using HyperCard on a Macintosh computer network comprising five computers. One computer acts as a server (i.e., a public machine) and the others as private machines. HyperGroup supports up to four decision makers. HyperTalk, a scripting language, is used to extend the functionality of HyperCard.

The software component of HyperGroup consists of two folders: a public and a private folder (see Exhibit IV-10-4). The private folder, which is installed in each of the four private machines, contains five stacks:

1. MainStack.
2. HomeMenu.
3. HGTools.
4. Public View.
5. epsiTalk.

The public folder is installed in the public machine. It contains the following 10 stacks:

1. HGPublic.
2. HGView.
3. PublicMenu.
4. NeedDB.
5. ConstraintDB.
6. ObstacleDB.
7. IssueDB.
8. CriteriaDB.
9. AlternativeDB.
10. epsiTalk.

The conceptual model is implemented by imposing templates on cards, each of which represents a node. HyperGroup consists of two types of cards: one implements the structure of a list node, and the other implements the structure of a comment node.

444

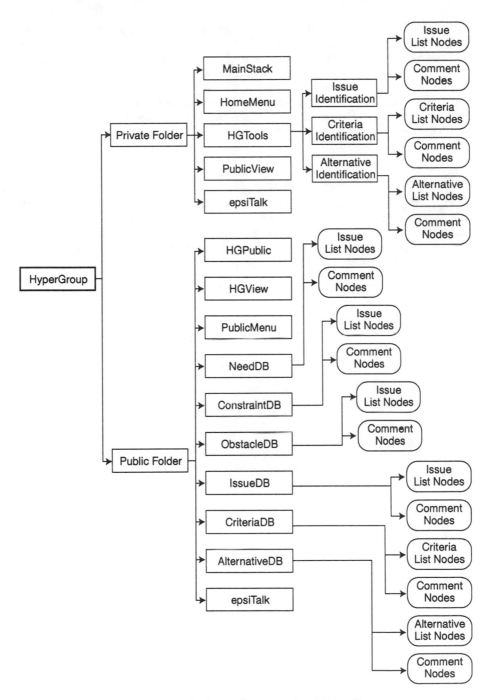

Exhibit IV-10-4. Software Components of HyperGroup

HyperGroup uses scrolling fields to accommodate variable lengths of idea lists and comments. Organization of cards, or of the information in HyperGroup, follows a hierarchical approach from a topic to its detail. Each text item of cards containing a list of ideas in HyperGroup connects to another card, which may contain a comment on that text item or another list of ideas.

The destination of a link in HyperGroup is always a whole card. Links in HyperGroup are not typed or structured. The role of links is to provide navigation through the system's information base.

CONCLUSION

HyperGroup captures group decision-making processes and enhances the effectiveness of group decision making. As a communications and recording medium for group decision making, it provides an example of how hypertext is incorporated into a group decision support system to support the decision-making tasks of small, face-to-face groups.

Section V

Delivering Products and Services

<hr>

THIS SECTION FOCUSES ON HELPING IS managers deliver services to the customer through distributed computing and networks. Managing networks has become a major responsibility for most IS managers, not all of whom have the background to prepare them for this task. Some basic management skills apply, but specific knowledge of network planning and control also helps.

Companies around the world are working to build an information infrastructure that will provide global access and open new markets. Establishing an infrastructure that integrates communications, computers, and entertainment technology requires the cooperation of industry, government, and standards-setting bodies. Chapter V-1, "The Information Superhighway Infrastructure," discusses the role of and impact on corporate network users of this extended capability.

Just as many organizations are reaching a level of comfort with hub and router technology, vendors are espousing the benefits of switched virtual networks. "Virtual Networking Management and Planning," Chapter V-2, examines the issues IS managers must deal with as the technology evolves from physical networks built using hubs, routers, and bridges to virtual nets built using true switches.

As companies expand into new geographic locations, single-site data bases become too slow and limiting. Distributed data base management systems allow data to be spread among different locations, increasing accuracy and timeliness of data. Chapter V-3, "Data Communications and Distributed Data," explains the key element of distributed computing—the communications networks that tie remote data bases together.

Networks today support voice processing, the increased capabilities of desktop operations, and resource sharing throughout the organization. The resulting complexity of the networked environment has caused many organizations to explore offloading network management responsibilities. Virtual private networks (VPNs) are one solution for managers seeking this type of offloading. Chapter V-4 examines the basics of "Virtual Private Networks," benefits to the client, and several aspects of VPN services.

"Enterprise Network Strategies" discusses some of the challenges network managers face when attempting to establish an enterprise network. Chapter V-5 recommends that time and effort be spent thoroughly analyzing the networking options and vendor offerings. Such a time investment, and the production of a blueprint for the enterprise network architecture, can prevent expensive surprises and disruptions in communications.

IS managers must ensure that adequate and effective controls are in place for LAN applications. The focus is to match risks with controls and create a recommended control architecture. Chapter V-6, "Evaluating LAN Management and Control," discusses an approach for the security review of LAN applications. This approach focuses on understanding the environment, evaluating the risks, determining the effective controls, and making recommendations for improved operation.

"Adding Multimedia to the Corporate Network" affects the processing environment, and Chapter V-7 discusses the implications of the increasing use of digitized audio and video on LAN servers. The chapter outlines the transmission requirements IS managers need to know about to deliver multimedia applications without adversely affecting other network users; it also discusses the methods of restructuring an existing network to cost-effectively accommodate transportation of multimedia.

Client/server systems are becoming the rule and not the exception in today's business organizations. "Managing Hardware in a Client/Server Environment" provides some valuable advice for IS managers trying to survive in this new environment. Many of the issues that need to be resolved in managing client/server hardware are similar to mainframe management issues. Chapter V-8 addresses these issues.

Improving standards should be an ongoing objective of the IS manager. Operating standards for LANs offer certain advantages for keeping expenses for procurement and maintenance under control. Chapter V-9, "Operating Standards for LANs," reviews the basics to include in a LAN standards document.

The difficulty of monitoring both local area networks and distributed data services usage has hindered efforts to allocate costs outside the host environment. "Cost Allocation for Enterprisewide Network Chargeout," Chapter V-10, explores strategies for defining network server domains and bundling data packet resource units to enhance cost recovery and achieve optimum efficiencies for each business unit served.

V-1

The Information Superhighway Infrastructure

Keith G. Knightson

ACHIEVING A SINGULAR, seamless information highway is going to be a challenge, and whether users can influence development remains to be seen. Unless all interested parties act in harmony on the technical specifications (i.e., standards), market sharing, and partnering issues, the end user may be the biggest loser.

For provision of a given service (e.g., voice or data), it should not matter whether a user's access is through the telephone company, the cable company, or the satellite company. Similarly, it should not matter whether the remote party with whom a user wants to communicate has the same access method or a different one.

Many countries and organizations have developed initiatives aimed at establishing an electronic highway such as the National Information Infrastructure (NII) in the US and the European Information Infrastructure (EII). To cover global aspects, a Global Information Infrastructure (GII) is being developed. The outcome of these initiatives depends on the changes taking place in the information and communications industries because of converging technologies, deregulation, and business restructuring or reorganization based on economic considerations. This chapter explores some of the possibilities and problems associated with information infrastructures.

The information infrastructure is important because it provides an opportunity to integrate technologies that have traditionally belonged to specific industry domains, such as telecommunications, computers, and entertainment. (Integration details are discussed later in this article.) The information infrastructure also presents an opportunity to greatly improve the sharing and transferring of information. New business opportunities abound related to the delivery of new and innovative services to users.

INFORMATION INFRASTRUCTURE

The term *information infrastructure*, which is used interchangeably with the term *information superhighway* in this chapter, describes a collection of technologies that

449

relate to the storage and transfer of electronic information, including voice, data, and images. It is often illustrated as a technology cloud with user devices attached, including broadband networks, the Internet, and high-definition TV.

However, problems emerge when users attempt to fit technologies together. For example, in the case of videophone service and on-demand video service, it is not clear whether the same display screen technology can be used, or whether a videophone call can be recorded on a locally available VCR. This example illustrates the need for consistency between similar technologies and functions.

Goals and Objectives of Information Infrastructures

The goals of most information infrastructures are to achieve universal access and global interoperability. Without corporate initiatives, the information infrastructure could result in conflicting and localized services, inefficient use of technology, and greater costs for fewer services. Some of the elements necessary to achieve such goals, including standards and open technical specifications that ensure fair competition and safeguard user interests, have yet to be adequately addressed.

BACKGROUND: TECHNOLOGY TRENDS

Two factors are often cited as driving the technology boom: the increase in computer processing power and the increase in the amount of available memory. Advances in these areas make a greater number of electronic services available for lower costs. This trend is expected to continue.

Bandwidth Pricing Issues

Unfortunately, comparable gains of higher bandwidths and decreasing costs are not as evident in the communications arena. Whether this is because of the actual price of technology or because of pricing strategies is debatable. Many applications requiring relatively high bandwidths have yet to be tariffed.

On-demand video is an interesting test case for the pricing issue. To be attractive, this service would have to be priced to compete with the cost of renting a videotape. However, such a relatively low price for high bandwidth would make the price of traditional low-bandwidth phone services seem extremely expensive by comparison. ATM-based broadband ISDN is likely to emerge as the vehicle for high-speed, real-time applications where constant propagation delay is required.

The lack of higher bandwidths at inexpensive prices has inhibited the growth of certain applications that are in demand. The availability of inexpensive high bandwidth could revolutionize real-time, on-demand applications, not only in the video entertainment area but also in the electronic publishing area.

Decoupling Networks and Their Payloads

One factor that is influencing the shape of the superhighway is the move toward digitization of information, particularly audio and video. Digitization represents a total decoupling between networks and their payloads.

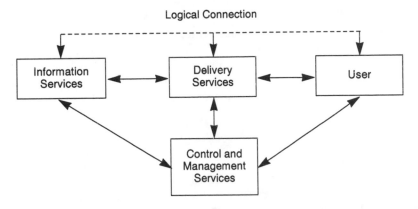

Exhibit V-1-1. Service-Oriented Architecture

Traditionally, networks have been designed for specific payloads, such as voice, video, or data. Digital networks may become general-purpose carriers of bit streams. In theory, any type of digital network can carry any and all types of information in digital format, such as voice, video, or computer data, thus banishing the tradition of video being carried on special-purpose cable TV networks and telephone service being carried only over phone company networks. All forms of information are simply reduced to bit streams.

The Service-Oriented Architecture

The separation of information services from bit-delivery services leads to the concept of a new service-oriented architecture as shown in Exhibit V-1-1. The most striking aspect of this service-oriented architecture is that the control and management entity may be provided by a separate service organization or by a distributed set of cooperating entities from different service organizations. The architecture represents a move away from the current world of vertical integration toward one of horizontal integration.

Deregulation of communications also plays a part in this scenario. Deregulation often forces an unbundling of components and services, which creates a business environment ideally suited to a service-oriented architecture.

KEY ISSUES IN CREATING THE SUPERHIGHWAY

Several common elements exist in any end-to-end service. For example, there is a need for agreed-on access mechanisms, network platforms, addressing schemes, resolution of inter-provider requirements, and definition of universal services. The development of a generic framework would help to ensure that service requirements are developed equitably and to introduce innovative new services.

The User's Role

Users are becoming more technology literate. The use of technology in the home in recent years has increased. Many users already benefit from what can be achieved through the convergence and integration of user-friendly technologies. User perspectives, rather than those of a single industry or company, should be thoroughly considered in the development of infrastructure initiatives.

Government's Role

The private sector takes most of the risks and reaps most of the rewards for development of the information superhighway. However, government should assert some influence over the development of universally beneficial user services. The role of the government mainly involves:

- Encouraging industry to collaborate and develop universally beneficial user services.
- Mediating between competing industry factions.
- Solving problems involving cultural content, cross-border and customs issues, protection of the individual, obscene or illegal material, and intellectual property and copyrights.

Industry's Role

Three dominant technology areas—telecommunications, computers and related communications, and the entertainment industry—are converging. Although there has already been some sharing of technology among industries, a single integrated system has not been created.

For example, many existing or planned implementations of videophone service invariably involve a special-purpose terminal with its own display screen and camera. For a home or office already equipped with screens and loudspeakers for use with multimedia-capable computers, the need for yet another imaging system with speakers is a waste of technology. Apart from the cost of duplication associated with the industry separations, there is the problem of the lack of flexibility. For example, if a VCR is connected to a regular TV, it should also be able to be used to record the videophone calls.

A plug-and-play solution may soon be possible in which the components are all part of an integrated system. In such a case, screens, speakers, recording devices, computers, and printers could be used in combination for a specific application. The components would be networked and addressed for the purposes of directing and exchanging information among them. Similar considerations apply to computing components and security systems. Using the videophone example, if the remote videophone user puts a document in front of the camera, the receiving party should be able to capture the image and print it on the laser printer.

Plug-and-play integration is not simple; yet if the convergence is not addressed, the result will be disastrous for end users, who will be faced with a plethora of similar but incompatible equipment that still fails to satisfy their needs.

The Dream Integration Scenario

Exhibit V-1-2 shows what the ideal configuration might look like when a high degree of convergence has been achieved. Ideally, there would be only one pipe into the customer's premises, over which all services—voice, video, and data—are delivered. User appliances can be used interchangeably. In this scenario, videophone calls could be received on the home theater or personal computer and recorded on the VCR.

The Nightmare Scenario

Exhibit V-1-3 shows what user networks might look like if convergence is not achieved. Customer premises would include many pipes. Some services would only be available on certain pipes and not others. The premises would have duplicate appliances for generating, displaying, and recording information. End-to-end services would be extremely difficult to achieve because all service providers would not choose to use the same local- or long-distance delivery services. In addition, all the local- and long-distance networks would not be fully interconnected.

Purveyors of technology and services may argue that this means they can all sell more of their particular offerings, which is good for business. Users, on the other hand, are more likely to feel cheated, because they are being forced to subscribe to different suppliers for slightly different services.

Corporate Networks

Large corporations create networks that are based on their preferred supplier of technology. They are usually extremely conservative in their technology choices because many of their business operations depend totally on the corporate network.

Two factors are causing this traditional, conservative approach to be questioned:

- *The cost of maintaining private networks.* In many cases, several private networks operate within a single corporation, such as one for voice, one for IBM's SNA network, and one for a private internet using TCP/IP or Novell's IPX. The change taking place is sometimes referred to as consolidation. Consolidation involves network sharing by operating the different systems protocols over the same physical network.
- *The need for global communications.* Corporations cannot afford to remain electronically isolated from their customers. As every business tackles cost cutting by increasing the use of information technology, the need for intercompany communication increases. Companies now need to communicate electronically with the banking industry, their suppliers, their customers, and the government to carry out their business. The GII is going to increase in importance for corporations, particularly in terms of availability and reliability.

THE INTERNET AND B-ISDN

Many users consider the Internet the only true information highway. In many ways, this is true—the Internet is the only highway, at least in the sense that it is the

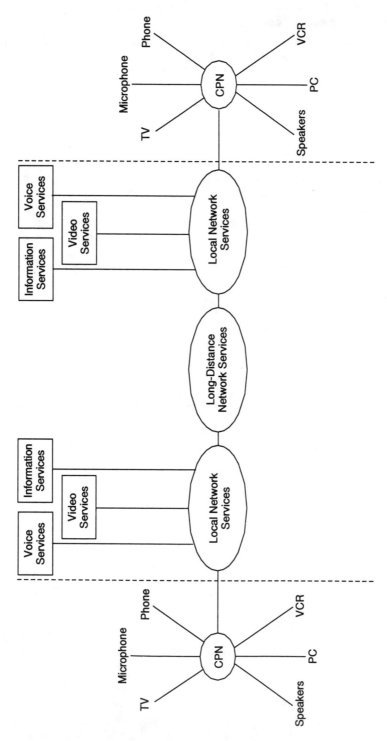

Exhibit V-1-2. The Dream Integration Scenario

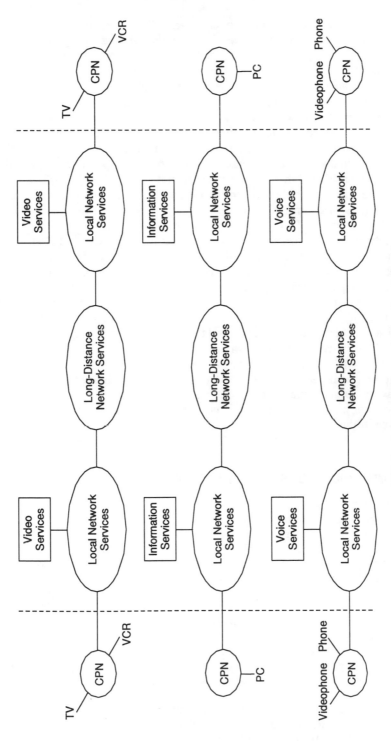

Exhibit V-1-3. The Nightmare Integration Scenario

only worldwide, seamless, and consistent end-to-end digital networking facility available. In addition, it has become a place where certain standardized applications can be used. It has a globally unique, centrally administered address space. The Internet provides national and international switched data services on a scale that would usually be associated with the major telecommunications carriers.

Not surprisingly, not everyone agrees that the Internet is the only highway. Technically, the Internet is a connectionless packet network overlaid on a variety of network technologies, such as leased lines, frame relay, asynchronous transfer mode (ATM), and LANs . However, it is difficult to imagine that at some point in the future, all voice and video traffic would be carried over such a network rather than directly over a broadband integrated services digital network (B-ISDN).

Thus, there may be a battle between the Internet and the traditional telecommunications carriers for control of the primary switching of data. The carriers may try to establish broadband ISDN as the primary method of switching data end to end, using telephone company-oriented number/addressing plans such as E.164.

The Internet community is interested in the use of broadband ISDN, primarily as a replacement for leased lines between Internet switching nodes (i.e., routers) where the real switching occurs. The deployment of broadband ISDN within the Internet may result in the migration of routers to the edges of the Internet, eliminating the need for intermediate routers. In any event, the interaction between the traditional router-based Internet style of operation and the emerging broadband ISDN switched services will be closely watched by corporate users.

The anarchic nature of the Internet will also be put to the test by commercial users who will want better service guarantees and accountability for maintenance and recovery. Despite these known deficiencies, the Internet remains the predominant information highway and it is difficult to imagine that it will lose its dominance in the near future.

TELECOMMUNICATIONS AND CABLE TV

Deregulation in many countries now permits cable TV companies to offer services traditionally offered by the telephone companies. One of the scenarios under consideration in many countries is shown in Exhibit V-1-4.

The cable companies are just beginning to form plans on how new two-way services should be offered. Access to the telephone company network would also provide access to other services, such as the Internet.

A major issue is the kind of interface to be provided on the cable network for associated telephone apparatus. It is not clear whether a traditional phone could simply be plugged into the cable system. Other issues, such as numbering and access to 800 service, need to be resolved. Whether traditional modem, telephony, or ISDN interfaces could be used or whether new cable-specific interfaces would be developed is also under consideration. Both solutions could coexist through provision of appropriate conversion units.

Cable systems usually consist of a head end with a one-way subtending tree and branch structure. Whether the head end would provide local switching within the residential area has not been determined. Other topologies, such as rings, may be more appropriate for new services.

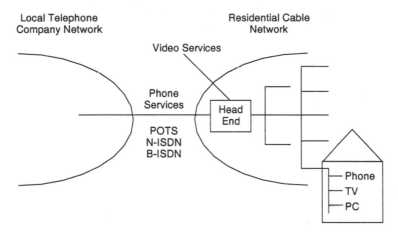

Exhibit V-1-4. Telephone Company and Cable TV Network Interconnection

Conversely, deregulation also permits the telephone companies to offer services previously offered by the cable companies. In such a case, a video server would be accessed by the telephone company network, probably using broadband ISDN and ATM technology, as shown in Exhibit V-1-5.

COMPUTER-INTEGRATED TELEPHONY

Computing and telecommunications are coming together in several ways. Computers can now be attached to telecommunications lines to become sophisticated answering machines, autodialers, and fax machines.

The availability of calling and called-line identification permits data bases to be associated with telephone calls. For example, the calling line identification can be used to automatically extract the appropriate customer record from a data base so that when the call is answered the appropriate customer information becomes available on a screen.

Computer-integrated telephony allows a variety of telephone service features to be controlled by the customer's computers. Intelligent network architectures that facilitate the separation of management and control are ideally suited to external computer control.

Public switched data networks have not been very efficient because of the costs of building separate networks and because the scale and demand for data proved nothing like that for voice services. A single digital network such as narrowband ISDN (N-ISDN) or broadband ISDN changes the picture significantly when coupled with the new demand for digital services.

COMPUTING AND ENTERTAINMENT

Most personal computers on the market have audiovisual capabilities. Movies and audio clips can be combined with text for a variety of multimedia applications. Video or images can be edited as easily as text.

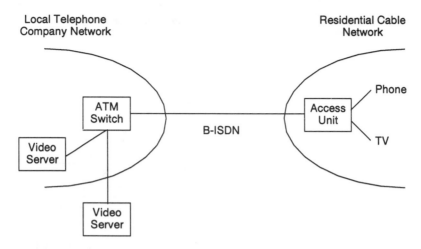

Exhibit V-1-5. Telephone Company-Provided Video Services

With the advent of high-definition TV and digital encoding of TV signals, it is easy to imagine a system in which the traditional TV screen and the PC monitor would be interchangeable. Computers are already being used to produce movies and as a playback medium, even providing the possibility of real-time interaction with the users.

Integrating all the appliances into a single architecture is the difficult part. Home theater systems provide simple forms of switching between components—for example, video to TV or VCR, or audio from TV to remote speakers. Soon, no doubt, the personal computer will be part of this system.

NATIONAL AND INTERNATIONAL INITIATIVES

Many countries have prepared recommendations for their respective national information infrastructures, including the US, Canada, Europe, Japan, Korea, and Australia, among others. The major differences in each country's initiatives seem to revolve around to what extent government will fund and regulate the information infrastructure.

The US

The Information Infrastructure Task Force (IITF) launched the National Information Infrastructure (NII) initiative in early 1993. The IITF is composed of an advisory council and committees on security, information policy, telecommunications policy, applications, and technology. Government funding is being made available for the development of NII applications.

The IITF's goal is that the information infrastructure become a seamless web of communications networks, computers, data bases, and consumer electronics. The NII initiative is also closely associated with the passage of a new communications act, which outlines principles for the involvement of the government in the communications industry. According to the communications act, the government should:

- Promote private sector investment.
- Extend the universal service concept to ensure that information resources are available at affordable prices.
- Promote technological innovation and new applications.
- Promote seamless, interactive, user-driven operation.
- Ensure information security and reliability.
- Improve management of the radio frequency spectrum.
- Protect intellectual property rights.
- Coordinate with other levels of government and with other nations.
- Provide access to government information and improve government procurement.

International Initiatives

The G7 countries (Britain, Canada, France, Germany, Italy, Japan, and the US) are considering developing an information infrastructure that would offer, among others, the following services:

- Global inventory.
- Global interoperability for broadband networks.
- Cross-cultural education and training.
- Electronic museums and galleries.
- Environment and natural resources management.
- Global emergency management.
- Global health care applications.
- Government services online.
- Maritime information systems.

STANDARDS AND STANDARDS ORGANIZATIONS

It is difficult to imagine how objectives such as universal access, universal service, and global interoperability can be achieved without an agreed-on set of standards. However, some sectors of industry prefer that fewer standards are established because this gives them the opportunity to capture a share of the market with proprietary solutions. Regardless, several national and international standards development organizations (SDOs) throughout the world are initiating activities related to the information infrastructure.

ISO and ITU

Both the International Standards Organization (ISO) and the International Telecommunications Union (ITU—formerly the CCITT) are embarking on in-

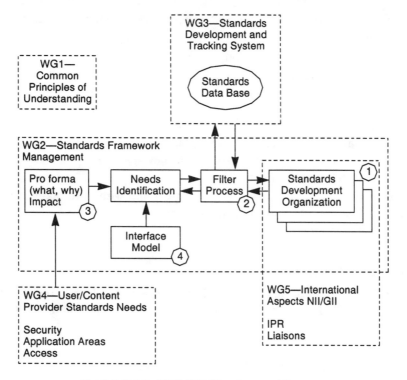

Exhibit V-1-6. ANSI IISP Structure and Process

formation infrastructure standards initiatives. The ISO and ITU have planned a joint workshop to address standards issues.

American National Standards Institute Information Infrastructure Standards Panel (ANSI IISP)

The ANSI IISP goals are to identify the requirements for standardization of critical interfaces (i.e., connection points) and other attributes and compare them to national and international standards already in place. Where standards gaps exist, standards development organizations will be asked to develop new standards or update existing standards as required.

ANSI IISP is developing a data base to make standards information publicly available. The process of identification and the structure of the IISP is illustrated in Exhibit V-1-6. In its deliberations, the ANSI IISP has been reluctant to identify specific networking architectures or interconnection arrangements and appears to be confining its efforts to a cataloguing process.

Telecommunications Standards Advisory Council of Canada (TSACC)

The TSACC is an umbrella organization for all the standards organizations in Canada. It is a forum where all parties can meet to discuss strategic issues. The objectives of TSACC, in respect to the Canadian Information Infrastructure and the GII, are similar to those of the ANSI IISP. However, TSACC considers the identification of specific networking architectures and associated specific access and interconnection points essential to achieving the goals of universal access, universal service, and interoperability.

European Telecommunications Standards Institute (ETSI)

The Sixth Review Committee (SRC6) of ETSI published a report on the European Information Infrastructure (EII) that emphasizes the standardization of the EII. Many of the recommendations in the report concern the development of reference models for defining the particular services and identifying important standards-based interface points. Broadband ISDN is recommended as the core technology for the EII.

The Digital Audio Visual Council (DAVIC)

DAVIC was established in Switzerland to promote emerging digital audiovisual applications and services for broadcast and interactive use. DAVIC, which has a very pro-consumer slant, believes that these services will only be affordable through sufficient standardization. The council has formed technical committees in the following five areas:

- Set-top units.
- Video servers.
- Networks.
- Systems and applications.
- General technology.

DAVIC may be the only forum in which home convergence issues can be solved.

ALTERNATIVE INITIATIVES

Following are two interesting US-based information infrastructure initiatives.

EIA/TIA

The Electronic Industries Association (EIA) and its affiliate Telecommunications Industry Association (TIA) have released version 2 of their white paper titled "Global Information Infrastructure: Principles and Promise." The basic principles conclude that:

- The private sector must play the lead role in development.
- Enlightened regulation is essential.
- The role of global standards is critical.

- Universal service and access must support competitive, market-driven solutions.
- Security and privacy are essential.
- Intellectual property rights must support new technologies.

The Computer Systems Policy Project (CSPP)

The CSPP is not a standards organization but an affiliation of the chief executive officers of several American computer companies. The CSPP has published a document titled "Perspectives on the National Information Infrastructure: Ensuring Interoperability." The CSPP document identifies the following four key points-of-presence as candidates for standardization:

- The interface between an information appliance and a network service provider.
- The applications programming interface between an information appliance and emerging NII applications.
- The protocols that one NII application, service, or system uses to communicate with another application, service, or system.
- The interfaces among and between network service providers.

CONCLUSION

The technical challenges of creating a Global Information Infrastructure are not insurmountable. The main difficulties arise from industries competing for the same business rather than sharing an expanding business, and from the lack of agreement on necessary open standards to achieve universal access and global interoperability that would expand the total business.

Interoperability requires agreed-on network architectures and the associated standards that could, in some cases, stifle innovation. A balance must also be struck between government regulation and private sector control over GII development. However, if each camp can cooperate, it is possible that in the future the communications, information, and entertainment industries could merge technology to provide plug-and-play components integrated into a single, coherent system that offers exciting new services that exist now in only the wildest imaginations.

V-2

Virtual Networking Management and Planning

Trenton Waterhouse

FROM THE USER'S PERSPECTIVE, a virtual network is a data communications system that provides access control and network configuration changes using software control. It functions like a traditional network but is built using switches.

The switched virtual network offers all the performance of the bridge with the value of the router. The constraints of physical networking are removed by the logical intelligence that structures and enforces policies of operation to ensure stability and security. Regardless of access technology or geographic location, any-to-any communications is the goal.

The switch could be considered a third-generation internetworking device. First-generation devices, or bridges, offered a high degree of performance throughput but relatively little value, because the bridge's limited decision intelligence resulted in broadcast storms that produced network instability. Routers, the second generation of internetworking devices, increased network reliability and offered great value with firewalling capabilities, but the trade off was in performance. When routers are used in combination with each other, bandwidth suffers, which is detrimental for delay-sensitive applications such as multimedia.

THE BUSINESS CASE FOR VIRTUAL NETWORKING

Both the business manager and the technical manager should find interest in this new virtual networking scheme. The business manager is usually interested in cost-of-ownership issues. Numerous studies from organizations such as the Gartner Group and Forrester Research have found that only 20% of networking costs are associated with capital equipment acquisition. The other 80% of annual budgets are dedicated to items such as wide area networking charges, personnel, training, maintenance and vendor support, as well as the traditional equipment moves, adds, and changes.

It is important for IS managers to remember that capital expenditure happens in year one, even though the equipment may be operating for another four years. Wide area network (WAN) charges can account for up to 40% of an organization's networking budget. For every dollar that the technical staff spends on new equipment, another four dollars is spent on the operation of that equipment. Therefore, focus should be on the cost-of-ownership issues, not necessarily the cost of the network devices.

Network Reliability. Business managers are also looking for increased reliability as the network plays a major role in the core operations of the organization. Networks have become a business tool to gain competitive advantage—they are mission critical and, much like a utility, must provide a highly reliable and available means of communications. Every office today includes an electrical outlet, a phone jack, and a network connection. Electrical and phone service are generally regarded as stable utilities that can be relied on daily. Networks, however, do not always provide comparable levels of service.

Network Accountability. Managers also can benefit from the increased accountability that virtual networks are able to offer. Organizational networking budgets can range from hundreds of thousands of dollars to hundreds of millions per year. Accounting for the use of the network that consumes those funds is a critical issue. There is no better example than WAN access charges. Remote site connectivity can consume a great deal of the budget, and the questions of who, what, when, and where in regards to network use are impossible to determine. Most users consider the network to be free, and the tools to manage and account for its use are increasingly a requirement, not an option.

THE TECHNOLOGY CASE FOR VIRTUAL NETWORKING

The IS manager's needs for higher capacity, greater performance, and increased efficiency can be met through the deployment of switched virtual networks. Each user is offered dedicated bandwidth to the desktop with uplinks of increasing bandwidth to servers or other enterprise networks. Rather than contending for bandwidth in shared access environments, all users are provided with their own private link. This degree of privacy allows for increased security because data is sent only to intended recipients, rather than seen by all.

The most attractive feature to the technical manager, however, may be the benefits gained through increased ease of operation and administration of virtual networks. A long-standing objective has been to deliver network services to users without continually having to reconfigure the devices that make up that network. Furthermore, many of the costs associated with moves, adds, and changes of users can be alleviated as the constraints of physical networking are removed. Regardless of user location, they can remain part of the same virtual network. Through the use of graphical tools, users are added and deleted from work groups. In the same manner, policies of operation and security filters can be applied. In a sense, the virtual network accomplishes the goal of managing the individual users and individual conversations, rather than the devices that make up the network.

VIRTUAL NETWORKING DEFINED

The ideal virtual network does not restrict access to a particular topology or protocol. A virtual network that can only support Ethernet users with transmission control protocol/Internet protocol (TCP/IP) applications is limited. The ultimate virtual network allows any-to-any connectivity between Ethernet, Token Ring, fiber distributed data interface (FDDI), asynchronous transfer mode (ATM), Internet protocol (IP), Internetwork Packet Exchange (IPX), Apple-Talk, or Systems Network Architecture (SNA) networks. A single virtual network infrastructure under a single management architecture is the goal.

Network management software becomes a key enabling requirement for the construction of switched virtual networks. The greatest challenge network designers face is the separation of the physical network connectivity from the logical connection services it can provide. Many of the design issues associated with networks can be attributed to the physical parameters of protocols and the routers used as the interconnection device. A challenge for any manager is to remain compatible with existing layer 3 protocols and routers and still preserve the investment in existing local area network (LAN) equipment to the greatest extent possible.

Using Telephony as a Model

The principles of operation for switched virtual networks are concretely founded in the success of the global communications systems. Without doubt, the phone system is the world's largest and most reliable network. Built using advanced digital switches controlled by software, extensive accounting and management tools ensure the success of this highly effective means of communication. The connection-oriented switch is the key. End-to-end connections across multiple switches and various transmission types ranging from copper to fiber optics to microwave to satellites allow millions of calls per day to be successfully completed, regardless of the type of phone or where the user is calling from. The telephony model is used throughout this chapter to help illustrate the workings of a virtual network.

SWITCHING DEFINED

One of the more confusing terms in the networking industry today is the word *switch*. For the purpose of this chapter, switching can be broken down into three fundamental areas:

- Configuration switching.
- Packet switching.
- Cell switching.

The earliest form of switching enabled the network manager to assign an individual port or an entire group of ports to a particular backplane segment within an intelligent hub device. This port configuration switching allowed the logical grouping of users onto a particular segment without the need to physically travel to the wiring closet to move cables or connectors. In a sense, this offers an electronic patch panel function. Although the benefit is a reduction of moves, adds, and change costs, this advantage can only be realized within the confines of a single hub. The application of this type of switching is limited because it cannot extend beyond one intelligent concentrator. Although beneficial in the work group, the enterprise needs cannot be met.

Phone system operators in the 1940s manually patched user connections through to destinations and recorded call time and duration. Using configuration switching is similar to patching phone lines together. Just as the phone network grew at a pace that required the switching to be performed automatically without operator intervention, so too have data networks outgrown the limitations of configuration switching.

Packet switching isolates each port to deliver dedicated bandwidth to each user in the network. Fundamentally, a packet switch is any device that accepts an incoming packet on one port and then makes a decision whether to filter or forward the packet out another interface. There are two types of packet switch transports: connectionless and connection-oriented.

Connectionless Packet Switching

Connectionless devices are probably more familiar to IS professionals when described as bridges or routers. A bridge is a layer 2 (of the Open Systems Interconnection [OSI] reference model) switch that bases its decisions on the media access control (MAC) address of attached workstations. What many vendors describe as a switch is actually a wire-speed MAC layer bridge. Three methods of decision making in these types of devices are cut-through, modified cut-through, and store-and-forward.

The Cut-Through Switch. This switch reads a packet to the destination address before it starts forwarding to the outbound interface. The benefit is an extremely low latency or delay in the forwarding of packets. The penalty is the propagation of errors, because the frame is being forwarded before it can be verified as valid, and the inability to support interfaces of different speeds that prevents high-bandwidth uplinks of FDDI or ATM on these type of devices.

The Modified Cut-Through Switch. This switch reads the first 64 bytes of a frame and then starts forwarding to the outbound interface, which greatly reduces the chances of propagating errored frames throughout the network. However, this method still requires all ports to be of the same type and speed.

Store-and-Forward Switch. The most flexible switch design uses a store-and-forward methodology that reads the entire frame before any filtering or forwarding decisions are made, thus ensuring that only packets that are error free are forwarded on the network. This method also allows packets to be buffered when transferring data between networks of different types, such as Ethernet to FDDI or ATM.

Bridges and Routers. A router is a layer 3 switch that bases its decisions on the network protocol address of attached workstations. Bridges and routers are considered connectionless because they forward and forget, requiring a decision to be made on every single inbound packet. The performance implications are that even though two communicating nodes on opposite sides of a bridge or router may be the only devices on their respective networks, the bridge or router must continuously make filter or forward decisions on every packet sent between the two nodes. If the phone network were built using bridges or routers, users would have to hang up and redial their destination after every word, which is not a very practical proposition.

A connectionless transport is not capable of defining which path its payload will take, cannot guarantee delivery, and is generally slower than a connection-oriented system. When a node sends a packet through a bridged or routed network, it is analogous to dropping a letter into a mailbox. It is not apparent how the letter got to its destination. The arrival of a letter cannot be guaranteed (protocol prioritization techniques are comparable to sending a letter by express mail). If a letter is lost (or a packet dropped), determining where it was lost is often difficult. The only way the sender knows that the letter was received is if the recipient sends another letter back to the sender (i.e., frame acknowledgment).

In a sense, today's shared-access networks are like the party lines of the early telephone network. But just as the phone network evolved from party lines to dedicated lines as usage and deployment grew, so too must the data networks offer this same level of service guarantee and broad adoption.

Connection-Oriented Switches

The connection-oriented switch that the phone systems use offers immediate acknowledgment of communications when the person picks up at the other end. The exact path the call took as well as its time and duration can be logged. The destination needs to be dialed only once and information is exchanged until both parties hang up.

The idea of connection-oriented communications is not new. This type of switching provides a high degree of reliability and reduces operational costs. Multiple classes of service can be defined to support voice, video, and data transfer. Excellent bandwidth management through congestion control techniques are possible and security and access control are greatly improved. Connection-oriented switching, along with easy-to-implement policy-based management and accounting facilities, have enabled the phone system to become universally accessible.

Frame relay technology is centered around connection-oriented communications, as is the most promising future networking technology—ATM. ATM is the most desirable networking technology because it offers dedicated, scalable bandwidth solutions for voice, video, and data.

ATM Switching. ATM switching is connection-oriented. Communications in an ATM network can be broken down into three phases: call setup (analogous to dialing a phone), data transfer (talking on the phone), and call teardown (hanging up the phone). The use of fixed-length 53-byte cells for data transfer delivers fixed latency transfer times for constant bit rate applications such as voice and video. ATM addressing schemes are similar to a telephone number. In fact, the original designers of ATM technology had their roots in the telephony arena, so many analogies to the operation of the phone system can be made when referring to an ATM network.

Although the benefits of ATM networking are attractive, there are currently nearly 100 million networked personal computers that do not have ATM interfaces. Few organizations can afford to replace all of their existing desktop and server interfaces, not to mention network analyzers and troubleshooting equipment.

Through the preservation of existing interface technology, by merely changing the internetworking devices from being connectionless to connection-oriented, many of the benefits of ATM may be realized without requiring the investment in all new ATM equipment. If LANs were designed to operate using the same principles as ATM, rather than making ATM compatible with LANs, users would benefit without significant capital investments in new equipment. By adding switch technology to the middle of the network, network administrators can be spared the trouble of upgrading numerous user devices, and users can be spared the inconvenience of rewiring and disruptions at their work site during an upgrade.

FEATURES OF SWITCHING SOFTWARE

The software that runs on switches is just as important as the switches themselves. A salesperson from AT&T, Fujitsu, or Northern Telecom does not focus the potential customer on the hardware aspects of the telephone switches. On the contrary, the salesperson conveys the benefits of the call management software, accounting, and automatic call distributor (ACD) functions. Switched virtual networks should also be evaluated for their ability to deliver value because of the software features.

The Virtual Network Server

Network management software has traditionally been thought of as software that passively reports the status and operation of devices in the network. In the switched virtual network, the network management software takes on a new role as an active participant in operations as well as configuration and reporting. A new middleware component known as the virtual network server (VNS) enforces the policies of operation defined by the network administrator through management software applications. The switches provide the data transport for the users of the network.

Directory Service. One of the software features in the VNS is the directory service. The directory service allows the identification of a device by logical name, MAC address, network protocol address, and ATM address, along with the switch and port that the user is connected to within the virtual network domain. The directory listing could be populated manually or dynamically as addresses are discovered. To fully realize the benefits of switched virtual networking, automatic configuration is absolutely essential. The directory service allows end nodes to be located and identified.

Security Service. The VNS security service will be used during call setup phases to determine whether users or groups of users were allowed to connect to each other. On a user-by-user and conversation-by-conversation basis, the network manager would have control. This communications policy management is analogous to call management on a telephone private branch exchange (PBX) where 900 numbers, long-distance, or international calls can be blocked. Users could be grouped together to form policy groups in which rules could be applied to individual users, groups, or even nested groups. Policies could be defined as open or secure, inclusive or exclusive.

A sample default policy can ensure that all communications are specifically defined to the VNS in order to be authorized. Policy groups can be manipulated either through drag-and-drop graphical user interfaces or programatically through simple network management protocol (SNMP) commands.

Finally, and most important, the directory service can work in conjunction with the security service to ensure that policies follow the users as they move throughout the network. This feature alone could save time spent maintaining a router access list, as occurs when a user changes location in the traditional network. However, it is important to realize that switched virtual networks ease administrative chores, they do not eliminate them.

Connection Management Service. The VNS connection management service is used to define the path communications would take through the switch fabric. A site may be linked by a relatively high-speed ATM link and a parallel but relatively low-speed Ethernet link. Network connections with a defined high quality of service (QOS) could traverse the ATM link and lower QOS connections could traverse the Ethernet. This connection management service allows for the transparent rerouting of calls in the event of a network fault. Connection management could also provide ongoing network monitoring in which individual user conversations could be tapped or traced for easy troubleshooting.

Bandwidth Service. The VNS bandwidth service is used during the call setup when a connection request is made. Video teleconferencing users may require a committed information rate (CIR) of 10M bps whereas the terminal emulation users may only require 1M bps. This is where ATM end stations and ATM switches negotiate the amount of bandwidth dedicated to a particular virtual circuit using user-to-network interface (UNI) signaling. Ethernet, Token Ring, and FDDI nodes do not recognize UNI signaling, but the switches they attach to could proxy signal for the end station, thus allowing a single bandwidth manager for the entire network, not just the ATM portion.

Broadcast Service. The VNS broadcast service uses as its base the concept of the broadcast unknown server (BUS) that is part of the ATM Forum's LAN emulation draft standard. This is how broadcasts are flooded through the network to remain compatible with the operation of many of today's protocols and network operating systems. A degree of intelligence can be assigned to the VNS that would allow for broadcasts or multicasts based on protocol type or even policy group.

Virtual Routing Service. The VNS virtual routing service is one of the most critical components of a virtual network. Just as traditional networks required traditional routers for interconnection, virtual LANs will require virtual routers for internetworking between virtual LANs. In other words, routing is required, but routers may not be. Some protocols such as TCP/IP actually require a router for users on two different subnetworks to speak with each other. In addition, most networks today are logically divided based on network layer protocol addresses with routers acting as the building block between segments.

The difference in operation between a virtual router and a traditional router goes back to the connection-oriented versus connectionless distinction. Routing allows for address resolution between the layer 3 protocol address and the layer 2 MAC address just as it happens through the address resolution protocol (ARP) process in TCP/IP networks. The VNS virtual routing service performs the address resolution function, but once the end station addresses are resolved, establishes a virtual connection between the two users. Two users separated by a traditional router would always have the router intervening on every single packet because the router would have resolved the protocol addresses to its own MAC address rather than the actual end station's MAC address. This VNS routing service allows the network to route once for connection setup and switch all successive packets.

Accounting Service. The VNS accounting service is beneficial because it allows the creation of the network bill. Similar to the way a telephone bill is broken down, the accounting service details connection duration with date and time stamp along with bandwidth consumption details. This is most directly applicable in the WAN. For many managers, WAN usage is never really accounted for on an individual user basis, yet it can consume up to 40% of the operations budget.

As usage-based WAN service options such as integrated services digital network (ISDN) gain popularity, accounting becomes that much more critical. Interexchange carriers (IXCs), competitive access providers, and the regional Bell operating companies (RBOCs) continue to deliver higher-bandwidth links with usage-based tariffs. In the future, they could install a 155M-bps synchronous optical network (SONET) OC3 link and only charge for the actual bandwidth used. Unless managers have tools to control access to and account for usage of WAN links, WAN costs will continue to rise. This service lets IS managers know who is using the WAN.

VIRTUAL NETWORKS VERSUS VIRTUAL LANs

Throughout this discussion, words have been carefully chosen to describe the operation of switched virtual networks. Many of the current vendor offerings on the market have as their goal the construction of a switched virtual LAN. These virtual LANs are interconnected using a traditional router device. However, the router has been viewed as the performance bottleneck. Routers should be deployed when segmentation or separation is the need; switches should be used to deliver more bandwidth. The virtual LAN (VLAN) concept is merely an interim step along the way to realizing the fully virtual network.

The ATM Forum's draft LAN emulation standard allows ATM devices to internetwork with traditional LAN networks such as Ethernet and Token Ring. However, it seems ironic that it essentially tries to make ATM networks operate like a traditional shared-access LAN segment. Although it is required for near-term deployment of ATM solutions into existing LAN architectures, its position as an end-all solution is questionable. A more logical approach uses ATM as the model that LANs must emulate.

CONCLUSION

Each vendor's approach to virtual networking features will vary slightly in implementation. Most vendors have agreed, however, that the router is moving to the periphery of the network and the core will be based on switching technologies with virtual network capabilities. The three critical success factors that a virtual network vendor must display to effectively deliver on all the promise of virtual networks are connectivity, internetworking, and network management.

Connectivity expertise through a demonstrated leadership in the intelligent hub industry ensures the user a broad product line with numerous options in regards to topology and media types. The product should fit the network, rather than the network design being dictated by the capability of the product. This indicates a vendor's willingness to embrace standards-based connectivity solutions as well as SNMP management and remote monitoring (RMON) analyzer capabilities.

Internetworking expertise ensures that the vendor is fully equipped to deal with layer 2 as well as layer 3 switching issues through an understanding of protocols and their operation. This is not something that can be learned overnight. The integration of these technologies is still unattainable.

Network management software is crucial—virtual networks do not exist or operate without it. The virtual network services provide all the value to the switch fabric. Users should look for a vendor that has delivered distributed management capabilities. Just as the telephone network relies on a distributed software intelligence for its operations, so too must the switched virtual network provide the same degree of redundancy and fault tolerance. Users should also consider whether the vendor embraces all of the popular network management platforms (e.g., SunNet Manager, HP OpenView, Cabletron SPECTRUM, and IBM NetView for AIX) or only one. Finally, users should make sure the vendor has experience managing multiple types of devices from vendors other than itself. It would be naive to think that all of the components that make up a network are of one type from one vendor.

V-3

Data Communications and Distributed Data

Dave Brueggen
Sooun Lee

AS COMPANIES MOVE INTO NEW MARKETS and geographic locations, it has become increasingly difficult and costly to access data in a traditional centralized data base environment. Users need systems that transcend the usual barriers by making all information sources available to all users in real time.

Distributed data represents a way of overcoming these barriers. By allowing access to data bases in physically separate areas, corporations and their users are thus provided with a flexibility not attainable in centralized environments. Key to distributed data are the data communications facilities and networks that underlie them. They provide the vital link between users and data in different locations.

DISTRIBUTED DATA

Large centralized computer systems have traditionally been the mainstay of large corporations. Centralized data bases developed as a result of large-scale computing systems. Under this configuration, all processing, storage, and retrieval of data is done at a central site.

The growth of distributed data has paralleled the growth of relational data base technology. Because data is stored at separate sites, the need to move groups of data is critical. Relational technology is ideally suited for this task because queries can be expressed in a nonprocedural mathematical form that can be transformed into other equivalent queries. Also, the case of use of relational designs is better suited to the growing number of end users manipulating data.

Because distributed data base management systems (DBMS) technology is still new, determining an exact definition of its composition and functions has been difficult. A distributed data base has been described as a collection of data distributed over different computers of a computer network. Each site of the network has autonomous capabilities and can perform local applications. Each site also participates in the execution of at least one global application, which requires accessing data at several sites using a communication subsystem.

Key to this definition is the ideal that data be located in physically separate sites, close to the users who are actually manipulating the data. Users have access to data at each site through a communications network linking the sites. Managing distributed data is a distributed DBMS, which monitors such functions as query optimization, concurrency control, and transaction handling.

Implementing Distributed Data

Three approaches can be used in implementing distributed data, and the adoption of each approach presents a unique data communications requirement. The first approach is fragmentation, which deals with how a table is broken up and divided among different locations. Horizontal fragmentation breaks a table into rows, storing all fields (i.e., columns) within the table at a separate location but only a subset of its rows. Vertical fragmentation stores a subset of a table's column among different sites. Mixed fragmentation combines both vertical and horizontal fragmentation. With fragmentation, only site-specific data is stored at each location. If data is needed from other locations, the distributed DBMS retrieves it through a communications link.

The second approach is replication, which stores copies of the total file around the network. Replication achieves maximum availability because all data is stored in every location. It is also used to provide backup copies of data in case a particular network node fails.

The third approach—allocation—is a combination of replication and fragmentation. With allocation, records are stored at the nodes exhibiting the highest use of that data, thus maximizing local processing. The degree to which these three approaches are combined to form a distributed data base structure is a critical factor in determining the data communications requirement for any distributed DBMS.

Another key feature of a distributed DBMS is the notion of transparency. Performance transparency means that a query can be made from any node in the distributed DBMS and runs with performance comparable to that of a centralized DBMS. Performance transparency depends on a distributed query optimizer—which finds a heuristically optimized plan to execute the command—and the network equipment over which data is transferred. This issue is also closely related to the determination of the data communications requirements of distributed DBMSs; three approaches to this issue are presented further in this chapter.

DATA COMMUNICATIONS REQUIREMENTS

A model of the decision-making steps for determining distributed data base communications requirements is shown in Exhibit V-3-1. The first step in this process is to choose the applications that the data communications networks will support. Determining a single set of communications requirements for all distributed DBMSs is difficult, because architectures and features depend on whether they are aimed at high-volume, online transaction environments, or decision-support and revenue-generating applications. Each of these environments requires different network characteristics, depending on the number of distributed queries or updates made and the level of fragmentation and replication. Because of the complexity and difficulty of establishing the relationships between these factors and data communications requirements, detailed discussion of this matter is not provided in this chapter. Once the type of application has been identified, performance considerations can be evaluated.

PERFORMANCE CONSIDERATIONS

To provide an adequate level of performance, efficient response time is critical. Response time consists of processing time, transmission time, and delay. In a distributed environment, transmission time has a large impact on response time. The first step in evaluating network performance is to set response time objectives. If data is fully replicated or is fragmented at many sites requiring frequent accesses between sites, a low response time is difficult to achieve. For systems that have low fragmentation and do not replicate data fully, response time is not as critical. Once response time objectives have been determined, factors that directly affect performance (e.g., propagation delay, network capacity, media, and communications equipment) should be examined.

Propagation Delay. Propagation delay is perhaps the most significant obstacle to distributed DBMSs. Low delay is especially critical in systems that replicate data at different locations, or in situations in which a query is made that brings together fragments from several data bases. Circuit propagation times of 20 milliseconds or more will cause throughput delays for data. With full replicated systems, any update made at a local level must be propagated to the central site to maintain the integrity of the data base. If updates are made frequently, the network can become congested, further increasing response time.

Network Capacity. If a network does not have the capacity relative to the amount of data it carries, the network will become congested, and response time will increase. When evaluating network capacity, the number of current locations over which data will be distributed should be considered. In addition, future locations should be accounted for as well as the number of users at each location. Users should be classified as to the type of data they can access. Those that have rights to update or query information at remote nodes will have a direct impact on the amount of traffic generated on the network. By identifying the type and frequency of remote queries made at each node in the network, the total network capacity can be determined.

General Media. The media used to transfer data affects both capacity and performance. A chart showing different media available for different network configuration appears in Exhibit V-3-2. The choices here include twisted pair,

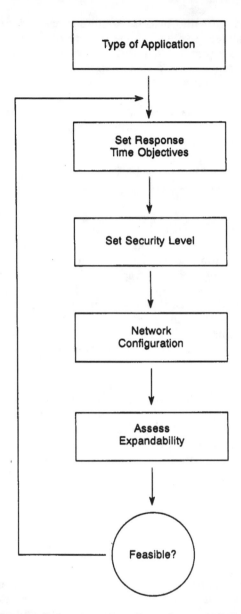

Exhibit V-3-1. Steps for Determining Data Base Communications Requirements

	Configuration	**Medium**
Long Distance	Wide Area Networks Mainframe Minicomputer LANs	T1, T2, T3, T4 Lines Satellite Microwave Leased Lines Switched Lines
Short Distance	Interconnected Local Area Networks	Fiber Optics Coaxial Cable Twisted Pair

Exhibit V-3-2. Network Media Configuration

coaxial cable and fiber optics, as well as radiated signal from microwaves and broadcast signals from satellites. For high-end performance, fiber optics should be employed. Fiber has the highest data rate (1,000M bps and higher) and the lowest error rate; it also allows the greatest bandwidth. It is the most costly of the three, however. Coaxial cable has a high data rate (500M bps and higher) but is bulky and easy to tap. Twisted pair has the lowest data rate but is relatively cheap and not as bulky as coaxial cable.

Microwave Transmission. Microwave transmission allows data to be transferred to locations at which running cable would normally be difficult (e.g., in crowded metropolitan areas). Microwave transmission rates can reach 45M bps. Microwave transmission also has several disadvantages: microwave relay stations have a line of sight limitation of 30 miles and are susceptible to interference from other radio waves and weather. Satellites, while allowing significant flexibility in terms of location, introduce significant propagation delay because signals have to be relayed 22,300 miles above the earth.

Switched Connections, Leased Lines, and High Capacity Lines. Other media choices for long-distance communications include switched connections, leased lines, and high capacity T1, T2, T3, and T4 lines. Of these, only conditioned leased lines and high-capacity lines provide the performance and capacity needed for highly replicated or fragmented distributed DBMS applications. T1, T2, T3, and T4 lines have respective transfer rates of 1.5, 6.3, 46 and 281M bps. Because these lines have a large bandwidth, they can be broken down into several lower-speed channels, increasing network capacity.

Modems and Multiplexers. Other communications equipment (e.g., modems and multiplexers) can affect network performance. Multiplexers allow two or more simultaneous transmissions over a single communications circuit, thereby saving communications costs. Multiplexers can introduce delay, however, as they switch a circuit from a channel interface to an inter-nodal link or bypass a circuit from one internodal link to another. A network with as few as five nodes may cause circuit delay if data must traverse many nodes to complete the transmission path. There are devices available (e.g., echo cancelers) to compensate for such delays. High-speed digital modems can also be used to guarantee maximum throughput on the line.

Compatibility. Another issue affecting network performance is compatibility among different vendors' networks. If network protocols are different, gateways must be used to translate messages across networks. Gateways handle any conversions necessary to go from one set of protocols to another, including message format conversion, address translation, and protocol conversion. Protocol conversion increases processing overhead, further reducing throughput and performance.

Availability. In a distributed environment, high availability requires that all necessary components be operational when they are needed. Availability is measured in mean time between failures and mean time between repair. The level of availability is guided by the level of replication and the number of nonlocal queries made. If data is manipulated on a local level and updates are done in batches, availability may not be a critical requirement. Data availability becomes an issue when data is not replicated between sites, and there are frequent queries or updates of nonlocal data. Maximum availability requires that data be fully replicated at all sites. In such a case, if a network node failed, processing could continue until the failed node is recovered.

SECURITY CONSIDERATIONS

In a distributed data base environment, data becomes more vulnerable because it exists in more than one location and is transferred by a communications network. Data traveling over communications circuits is especially vulnerable to tampering. Areas in which data may be tampered with include the media itself and the hardware involved in transmitting the data.

At a local level, twisted pair and coaxial cable are very easy to tap. Fiber, on the other hand, is very difficult to tap. When communicating over longer distances, microwave and satellite communications are very insecure. Because they employ broadcast media, anyone with the proper receiver can intercept the signals. Encryption can be used to ensure that if the data is intercepted, it will not be in a readable form. Although such precautions virtually guarantee security, there is a performance trade-off in the overhead required to first encrypt, then decrypt the data. In highly fragmented or fully replicated data bases, the additional overhead may affect system performance.

The IS manager must also take actions to protect the data communications equipment used to transfer the data. This includes actions to ensure the physical security of communications equipment (e.g., modems, encryption devices, earth stations for satellites or microwave, and any other key equipment).

NETWORK DESIGN CONSIDERATIONS

Distance is the major determinant of the network configuration chosen. Several configurations are available. If the distance between sites is short, a local area network (LAN) configuration can be used. The high throughput and low cost of LANs make them an attractive alternative for a communications network for distributed data bases. Speeds of as high as 100M bps are attainable on LANs using a fiber-optic backbone.

A single LAN often will not reflect the geographic distribution of an organization. One way of extending the distance over which a LAN may be used is to employ repeaters, which strengthen and relay messages across a network. The disadvantage of configurations that use repeaters is that as new nodes are added, network performance decreases.

Another way of extending the physical distance of a network is to use a bridge to connect two or more LANs. Unlike repeaters, bridges filter traffic between networks, keeping local traffic local while forwarding only traffic destined for other networks. In this configuration, data would be distributed between two or more LANs connected by a bridge.

If several LANs are involved, brouters may be used. Brouters—a type of smart bridge—not only filter data but perform routing functions as well. By deciding which route data will take through the network, brouters minimize traffic and increase throughput.

Employing any of the previously described devices allows network designers to extend the distance of a network. In essence, it enables them to create a wide area network (WAN) using several LANs. But this strategy could not be employed over extremely long distances such as between cities. In these situations, WANs using leased lines would be chosen. For maximum throughput, a network employing a high-speed service (e.g., T1, T2, T3, or T4) could be used. Because cost increases with performance, careful consideration should be paid to the level of service desired.

In a WAN configuration, leased T1 lines would act as a backbone for LANs at different physical locations. Using a LAN at each physically distinct site would provide cost and performance advantages, because most data would be manipulated at a local level. Accessing data on high-speed LANs allows maximum performance. Using a T1 line also allows voice to be run along with data, further increasing savings. Although performance does not match that of a LAN configuration, an acceptable level of throughput can be achieved.

NETWORK EXPANSION CONSIDERATIONS

As organizations grow and change, data requirements at new and existing sites also change. The communications networks and hardware that support distributed data bases should be capable of changing along with the data they support. To assess possible changes in data communications requirements, IS managers should have a thorough understanding of the company's business and strategic goals. Equipment bought today should provide a basis for the expanding technology of tomorrow.

Equipment based on open systems interconnection (OSI) provides this flexibility. OSI represents a group of international architectural standards for data communications. It is meant to be a step toward truly open systems. A common OSI architecture enables networks built by different vendors and based on separate technologies to be connected across wide geographic areas. This will become more important as LANs play a larger role in processing at a local level. The effect of OSI on network expansion will be a phasing out of equipment based on proprietary technology.

CONCLUSION

This chapter has focused on data communications issues of distributed data bases. There are still many communications as well as other problems that must be addressed before true distributed data bases can become reality. Key to this will be further technical advancement of high-speed communications on local and wide area networks.

In the near term, distributed DBMSs will be used only for decision support applications that are not highly fragmented or are fully replicated. Key to this will be to keep data access on a local basis, therefore taking advantage of savings in communications and computer resource costs. As communications technology continues to improve, distributed DBMSs will support more online, baseline applications. If distributed DBMSs are part of a company's long-term planning horizon, IS managers should begin to analyze their current communications networks to avoid costly changes in the future.

V-4

Virtual Private Networks

Nathan J. Muller

CARRIER-PROVIDED NETWORKS that function like private networks are referred to as virtual private networks or VPNs. By relying more on VPNs, corporations minimize the operating costs and staffing requirements associated with private networks. In addition, they gain the advantages of dealing with a single carrier instead of with the multiple carriers and vendors required for a typical private network. This relieves organizations of the costs associated with staffing, maintenance, and inventory without sacrificing control, service quality, and configuration flexibility.

AT&T introduced the first VPN service in 1985. Its Software Defined Network (SDN) was offered as an inexpensive alternative to private lines. Since then, VPNs have added more functionality and expanded globally. Today, the Big Three carriers—AT&T (SDN), MCI (Vnet), and Sprint (VPN Service)—each offer virtual private networks. In the case of AT&T, various services—including high-speed data and cellular calls—may be combined under one service umbrella, expanding opportunities for cost savings within a single discount plan.

THE VPN CONCEPT

VPNs let users create their own private networks by drawing on the intelligence embedded in the carrier's network. This intelligence is actually derived from software programs residing in various switch points throughout the network. Services and features are defined in software, giving users greater flexibility in configuring their networks than is possible with hardware-based services. In fact, an entire network can be reconfigured by changing a few parameters in a network data base.

The intelligence inherent in virtual private networks lets network managers control many operating parameters and features within their communications environments. For example, the flexible-routing feature allows the network manager to reroute calls to alternate locations when a node experiences an outage or peak-hour traffic congestion. This feature is also used to extend customer service business hours across multiple time zones. The location-screening feature lets network managers define a list of numbers that cannot be called from a given VPN location. This helps contain call costs by disallowing certain types of outbound calls.

Originating call screening is a feature the gives network managers the means to create caller groups and screening groups. Caller groups identify individual users who have similar call restrictions, and screening groups identify particular telephone numbers that are allowed or blocked for each caller. Time intervals are also used as a call-screening mechanism, allowing or blocking calls according to time-of-day and day-of-week parameters.

With a feature called NNX sharing, VPN customers reuse NNXs (i.e., exchange numbers) at different network locations to set up their seven-digit on-net numbering plans. This provides dialing consistency across multiple corporate locations. Another feature, partitioned data base management, lets corporations add subsidiaries to the VPN network while providing for flexible, autonomous management when required by the subsidiaries to address local needs. The VPN can even transparently interface with the company's private network or with the private network of a strategic partner. In this case, the VPN caller is not aware that the dialed number is a VPN or private network location, because the numbering plan is uniform across both networks.

VPNs provide several other useful features, including automatic number identification (ANI) data, which is matched to information in a data base containing the computer and telecommunications assets assigned to each employee, for example. When a call comes through to the corporate help desk, the ANI data is sent to a host, where it is matched with the employee's file. The help desk operator then has all relevant data available immediately to assist the caller in resolving the problem.

MAKING THE BUSINESS CASE FOR VPNs

An increasing number of companies are finding virtual private networks to be a practical alternative method for obtaining private network functionality without the overhead associated with acquiring and managing dedicated private lines. There are several other advantages to opting for a virtual private network, including:

- The ability to assign access codes and corresponding class-of-service restrictions to users; these codes are used for internal billing, to limit the potential for misuse of the telecommunications system, and to facilitate overall communications management.
- The ability to consolidate billing, resulting in only one bill for the entire network.
- The ability to tie small remote locations to the corporate network economically, instead of using expensive dial-up facilities.
- The ability to meet a variety of needs (e.g., switched voice and data, travel cards, toll-free service, and international and cellular calls) using a single carrier.

- The availability of a variety of access methods, including switched and dedicated access, 700 and 800 dial access, and remote calling card access.
- The availability of digit translation capabilities that permit corporations to build global networks using a single carrier. Digit translation services perform seven-to-10-digit, 10-to-seven-digit, and seven-to-seven-digit translations and convert domestic telephone numbers to International Direct Distance Dialed (IDDD) numbers through 10-to-IDDD and seven-to-IDDD translation.
- The ability to have the carrier monitor network performance and reroute around failures and points of congestion.
- The ability to have the carrier control network maintenance and management, reducing the need for high-priced in-house technical personnel, diagnostic tools, and spares inventory.
- The ability to configure the network flexibly, through on-site management terminals that enable users to meet bandwidth application needs and control costs.
- The ability to access enhanced transmission facilities, with speeds ranging from 56K bps to 384K and 1.536M bps, and plan for emerging broadband services.
- The ability to combine network-services pricing typically based on distance and usage with pricing for other services to qualify for further volume discounts.
- The ability to customize dialing plans to streamline corporate operations. A dealership network, for example, assigns a unique four-digit code for the parts department. Then, to call any dealership across the country to find a part, a user simply dials the telephone-number prefix of that location.

The intelligence embedded in the virtual network at the carriers' serving offices also gives users more flexibility in selecting equipment. PBXs from various vendors connect to a VPN service provider's point of presence (POP) through various local access arrangements. The private network exists as a separate entity on the VPN service provider's backbone network, with the service provider assuming responsibility for translating digits from a customer-specific numbering plan to the service provider's own numbering plan, and vice versa. All routing and any failures are transparent to the customer and, consequently, to each individual user on the network.

Billing Options

One of the most attractive aspects of virtual private network services is customized billing. Typically, users select from among the following billing options:

1. The main account accrues all discounts under the program. In some cases, even the use of wireless voice and data messaging services qualifies for the volume discount.
2. Discounts are assigned to each location according to its prorated share of traffic.

3. A portion of the discounts is assigned to each location based on its prorated share of traffic, with a specified percentage assigned to the headquarters location.

4. Usage and access rates are billed to each location, or subsidiaries are billed separately from main accounts.

5. Billing information and customized reports are accessed at customer premises terminals or provided by the carrier on diskette, microfiche, magnetic tape, tape cassette, or CD-ROM as well as in paper form.

6. A name substitution feature allows authorization codes, billing groups, telephone numbers, master account numbers, dialed numbers, originating numbers, and credit card numbers to be substituted with the names of individuals, resulting in a virtually numberless bill for internal distribution. This prevents sensitive information from falling into inappropriate hands.

AT&T, MCI, and Sprint all offer rebilling capabilities that use a percentage or flat-rate formula to mark up or discount internal telephone bills. Billing information is even summarized in graphical reports, such as bar and pie charts.

Carrier-provided software is available that allows users to work with call detail and billing information to generate reports in a variety of formats. Some software even illustrates calling patterns with maps.

Electronic invoicing also is available. AT&T, for example, provides this capability by linking its SDN Billing Advantage with EDIView, its electronic data interchange (EDI) offering.

NETWORK MANAGEMENT

Each of the major VPN service providers offers various management and reporting capabilities through a network management data base that enables users to perform numerous tasks without carrier involvement.

The network management data base contains information about the network configuration, usage, equipment inventory, and call restrictions. On gaining access to the data base, the telecommunications manager sets up, changes, and deletes authorization codes and approves the use of capabilities such as international dialing by caller, workgroup, or department. The manager also redirects calls from one VPN site to another to allow, for example, calls to an East Coast sales office to be answered by the West Coast sales office after the East Coast office closes for the day. Once the manager is satisfied with the changes, they are uploaded to the carrier's network data base and take effect within minutes.

Telecommunications managers access call detail and network usage summaries, which are used to identify network traffic trends and assess network performance. In addition to being able to download traffic statistics about dedicated VPN trunk groups, users receive five-, 10-, and 15-minute trunk group usage statistics an hour after they occur; these statistics are then used to monitor network performance and carry out traffic engineering tasks. Usage is broken down and summarized in a variety of ways—such as by location, type of service, and time of day. This information is used to spot exceptional traffic patterns that may indicate either abuse or the need for service reconfiguration.

Through a network management station, the carrier provides network alarms and traffic status alerts for VPN locations using dedicated access facilities. These alarms indicate potential service outages (e.g., conditions that impair traffic and could lead to service disruption). Alert messages are routed to customers in accordance with preprogrammed priority levels, ensuring that critical faults are reviewed first. The system furnishes the customer with data on the specific type of alarm, direction, location, and priority level, along with details about the cause of the alarm (e.g., signal loss, upstream failed signal, or frame slippage). The availability of such detail permits customers to isolate faults immediately.

In addition, telecommunications managers can request access-line status information and schedule transmission tests with the carrier. The network management data base describes common network problems in detail and offers specific advice on how to resolve them. The manager submits service orders and trouble reports to the carrier electronically through the management station. Telecommuncations managers can also test network designs and add new corporate locations to the VPN.

ACCESS ARRANGEMENTS

A variety of access arrangements available from the VPN service providers are targeted for specific levels of traffic, including a single-voice frequency channel, 24-voice channels through a DS1 link, and 44-voice channels through a T1 link equipped with bit-compression multiplexers, in addition to a capability that splits a DS1 link into its component DS0s at the VPN serving office for connection to off-net services. The same DS1 link is used for a variety of applications, from 800 service to videoconferencing, thereby reducing access costs. Depending on the carrier, there may be optional cellular and messaging links to the VPN as well. Even phone card users can dial into the VPN, with specific calling privileges defined for each card. All of a company's usage can be tied into a single invoicing structure, regardless of access method.

The architecture of the VPN makes use of software-defined intelligence residing in strategic points of the network. AT&T's SDN, for example, consists of a network action point (ACP) connected to the PBX through dedicated or switched lines. The ACPs connect with the carrier's network control point (NCP), where the customer's seven-digit on-net number is converted to the appropriate code for routing through the virtual network.

Instead of charging for multiple local access lines to support different usage-based services, the carriers allow users to consolidate multiple services over a single T1 access line. A user who needs only 384K bps for a data application, for example, fills the unused portion of the access pipe with 18 channels of voice traffic to justify the cost of the access line. At the carrier's cross-connect system, the dedicated 384K-bps channel and 18 switched channels are split out from the incoming DS1 signal. The 384K-bps DS0 bundle is then routed to its destination, whereas the voice channels are handed off to the carrier's Class 4 switch, which distributes the voice channels to the appropriate service.

DATA NETWORKING OVER THE VPN

Although obtaining economical voice traffic has traditionally been the primary motivation behind the move to VPN service, a variety of low-speed and high-speed VPN data services are available as well.

Low-Speed Data Services

AT&T has been especially aggressive in offering its SDN customers the means to access a wide array of AT&T EasyLink messaging services. The offering, AT&T SDN EasyLink Solutions, enables customers to use their SDN networks to connect directly to electronic messaging features from AT&T EasyLink Services, including electronic mail, shared folders, text-to-fax (MailFAX), electronic data interchange, Telex, and a variety of information services.

SDN EasyLink Solutions includes the following services:

- *AT&T SDN Electronic Mail.* This worldwide public messaging service offers secure transport and feature-rich functionality that supports access from a variety of computer platforms. The service provides worldwide electronic mail delivery to many systems, including X.400 gateways and the Internet.
- *AT&T SDN Shared Folder.* This service is an electronic bulletin board service within SDN Electronic Mail that automatically downloads new information to each user's mailbox whenever the user queries the system for new mail.
- *AT&T SDN MailFAX.* This service lets users send electronic mail documents from a computer to a receiver's fax machine. Service features include automatic retry, fax broadcast, and customized logos and signatures.
- *AT&T SDN Enhanced FAX.* The enhanced fax service provides secure store-and-forward delivery of facsimile transmissions worldwide.
- *AT&T SDN Electronic Data Interchange.* This EDI service provides message transport, storage, and tracking for the electronic exchange of business documents such as purchase orders or invoices in standard data formats.
- *AT&T SDN Telex.* This service connects users to the worldwide Telex network and the extensive community of Telex subscribers in numerous industries, including international commerce, banking, and shipping.
- *AT&T SDN Information Services.* These information services give users fast and cost-effective access to a broad spectrum of online news services, interactive research data bases, bulletin boards, and research-on-demand services.

High-Speed Data Services

VPNs also are capable of supporting such bandwidth-intensive applications as local area network (LAN) interconnection, image transfers, and videoconferencing. These services are offered under AT&T's Software Defined Data Network (SDDN), MCI's Virtual Private Data Service (VPDS), and Sprint's VPN Premiere.

AT&T's SDDN, for example, offers high-speed data networking in conjunction with SDN's advanced call-handling capabilities. SDDN shares the network capabilities of ACCUNET switched digital services (SDS) for reliable transport of data at rates of 56K bps and higher. (Low-speed data is transported over SDN using dial-up modems or PBX data connections.) SDN supports low-speed dial-up modem connections and higher-speed connections through a PBX, T1 multiplexer, or D4 channel bank. AT&T's SDDN offering supports 56K- and 64K-bps service, 64K-bps clear channel, and 384K- and 1.536M-bps connections utilizing the ISDN PRI (primary rate interface). These high transmission speeds are achieved by stacking contiguous 64K-bps clear channels. Users take full advantage of virtual networking by combining and routing their voice and data traffic in a single T1 access line to the SDN/SDDN network.

Users access SDDN with DDS (dataphone digital service) lines for data transmission at rates of up to 56K bps using dial-up modems or DSUs (digital service units) with an optional auto-dial or re-dial capability; alternatively, access is obtained through AT&T ACCUNET T1.5 lines. Customer premises equipment (e.g., intelligent multiplexers and PBXs) interprets ISDN PRI messages for call setup, detection of facility failures, and reinitiation of call setup in response to abnormal call disconnects. Real-time restoration is achieved within seconds of a service disruption so that critical data applications remain operational; SDDN also supports SDN network management capabilities such as call screening, flexible routing, periodic traffic reports, and customer-initiated testing. SDDN is well-suited for applications that:

- Have high-speed or high-volume data transmission requirements.
- Have a time window for completion (e.g., applications performed during the night, morning, or other specified time periods).
- Benefit from bandwidth-on-demand and usage-based pricing (e.g., applications active for a limited duration, used infrequently, or required for unscheduled events).
- Have restoration requirements (e.g., critical applications that must remain operational in the event of a network failure). Such applications are currently protected through a dial backup capability or spare bandwidth and alternate routing in a T1 multiplexer network.
- Have multiple endpoint destinations (e.g., applications requiring serial or nonsimultaneous communications between an originating point and several endpoints).
- Benefit from networking flexibility (e.g., applications with traffic patterns that demonstrate daily or seasonal variations or that change drastically as

the network grows). SDDN eliminates the time and expense required to install additional private lines.

Specific applications that benefit from SDDN include remote job entry (RJE) and network job entry (NJE), CAD/CAE (computer-aided design/computer-aided engineering) and medical imaging, distributed/shared computing, LAN interconnection, high-speed mainframe communications, PC-to-host and PC-to-PC transfer, peak traffic overflow and private line backup, videoconferencing, and Group IV facsimile.

Performance Objectives

VPN performance standards for data are comparable to private line services through the use of high-quality digital transport and an automatic restoration capability. With AT&T's SDDN, for example, performance is measured in terms of network interface availability, network reliability, post dialing delay, call blocking, restoration, and service availability.

Network Interface Availability. For SDDN connections, an availability number indicates the percentage of time that all SDDN components are usable for customer applications. The target SDDN network interface-to-network interface (NI-to-NI) availability is 99.9% and includes ACCUNET T1.5 or DDS access links. Without the service restoration feature, the availability figure drops to 99.75%.

Network Reliability. Network reliability for SDDN is a measure of line transmission performance given in terms of error-free seconds (EFS) and severe errored seconds (SES). An EFS is a second with no bit errors, and an SES is a second with more than one error per 1,000 bits. Reliability objectives for EFS and SES are, respectively, 99.9% and 30 SES per day between AT&T serving offices (SO-to-SO), and 99.75% and 38 SES per day between network interfaces (NI-to-NI).

Post Dialing Delay (PDD). Post dialing delay refers to the amount of time from call initiation to call setup and is measured at the originating network interface. The SDDN objective is a four-second average post dialing delay, with 95% of all calls receiving network response within six seconds.

Call Blocking. Call blocking is the probability that an unsuccessful call attempt is due to network congestion. The target blocking is 1% during peak busy hours and 0.5% with planned improvements, with fewer than one call per 200 attempts blocked during peak traffic hours.

Restoration. The SDDN restoration objective for an NI-to-NI connection is less than 20 seconds for an estimated 99% of all restoration attempts. The stated restoration performance may not be guaranteed, however, in the event of a catastrophic or widespread network failure. The stated objective includes time spent in detection, redialing, and call setup stages in the SDDN network and customer premises equipment (CPE). A maximum of six seconds is allocated to CPE for failure detection and redialing an SDDN connection in response to a disconnect message.

Service Availability. Service availability objectives are improved with the use of diverse access arrangements. The Split Access Flexible Egress Routing (SAFER) capability, available with AT&T 4ESS software, allows origination and distribution of calls between two toll switches, which minimizes the vulnerability to access link and nodal failures.

LOCAL VPN SERVICE

A new development in the VPN market is the emergence of local services. Bell Atlantic, for example, is offering a local VPN service in the mid-Atlantic region under the name of Bell Atlantic Allonce Virtual Private Network Solutions (VPNS). VPNS service allows companies to manage their local and intraLATA calls and save money on interLATA calls using Bell Atlantic's public network as if it were their own private network.

With VPNS, customers can do such things as access their voice network remotely, make business calls from the road or home at business rates, originate calls from remote locations and bill them to the office, and block calls to certain telephone numbers or regions. Uniform pricing and billing plans are also arranged for all of the customer's locations to reduce the administrative costs involved with reviewing billing statements, even if each location uses a different carrier.

Bell Atlantic's service lets large business customers configure components of the public network like a customized private network without the expense of dedicated lines or equipment. Until now, services of this kind could not be used for local calls because they were offered through long-distance companies. Bell Atlantic's service, however, is used for local calls and also works with the customer's existing long-distance services. The VPNS service also is compatible with Centrex services, PBX systems, or other customer premises equipment.

Once Bell Atlantic has achieved regulatory approval to deliver interLATA long-distance services, the company plans to expand the VPNS service to include in-region and out-of-region long-distance service.

As an integral part of its All at once approach for large businesses, Bell Atlantic consults with companies to assess their communications needs and analyze their local calling patterns to design a solution that optimizes use of the public network. Bell Atlantic estimates that 25% to 75% of local phone traffic could be on a local VPN, where it could be subject to the lower rates that make VPNs attractive to users.

CONCLUSION

VPNs permit the creation of networks that combine the advantages of both private facilities and public services, drawing on the intelligence embedded in the carrier's network. With services and features defined in software and implemented through out-of-band signaling methods, users have greater flexibility in configuring their networks from on-premises terminals and management systems than is possible with services implemented with manual patch panels and hardwired equipment. These capabilities make VPNs attractive for data as well as for voice—for regional, national, and international corporate locations—and portend success for VPNs long into the future.

V-5

Enterprise Network Strategies

Keith G. Knightson

MANY PRODUCTS ARE GLOWINGLY DESCRIBED with the phrases "enterprise networks" and "enterprise networking." These are buzz-phrases on every salesperson's lips, together of course, with "open" and "open systems." Creating an enterprise network, however, requires more than just a knowledge of the buzzwords. This chapter explains the basic subtleties of an enterprise network and the challenges of establishing one.

WHAT IS AN ENTERPRISE NETWORK?

In the same sense that the word enterprise conveys the totality of an organization's operations, the phrase *enterprise network* means combining all the networking and information technology and applications within a given enterprise into a single, seamless, consolidated, integrated network.

The degree of integration and consolidation may vary: Total integration and consolidation may not be always achievable, as this chapter will show.

The first example is an organization that has an SNA network from IBM Corp. and a DECnet from Digital Equipment Corp. In all probability, these two networks have their own communications components. There might be one set of leased lines serving the SNA network and another completely independent set of leased lines serving the DECnet.

It would be useful if all the IBM users could intercommunicate with all DEC users, but a first and evolutionary step might be to have both the SNA network and DECnet share the same leased lines. Now, only one physical network has to be managed instead of two separate ones, and more efficient and cost-effective sharing of the physical communications plant can be achieved.

A second step might be to interconnect the mail systems of the two networks to achieve at least the appearance of a single enterprisewide electronic mail system.

A third step might be to unify the data and information and its representation as used within the organization. This would enable basic forms of data to be operated on by many applications.

The challenges of building an enterprise network fall into two distinct categories: getting the data (i.e., information) from A to B; and enabling B to understand the data when it receives it from A. (Getting the data from A to B is addressed in the section titled "The Networking Challenge," and enabling B to understand the data is addressed in the section titled "Beyond the Networking Challenge.")

THE NETWORKING CHALLENGE

The networking part of the problem has three major components:

- Choosing from and integrating the various network technologies.
- Selecting from the many vendor solutions.
- Moving information from a local to a global environment.

Network Technologies

The first basic problem with networks is that there are so many of them. In this context, networks are taken to mean the raw network technologies—leased lines (i.e., T1 and T3), X.25, ISDN, frame relay, asynchronous transfer mode (ATM), and the various LAN access methods.

If all the users in an enterprise are connected to the same network technology, there is no problem. Unfortunately, this is never likely to be the case, and communication between users on dissimilar networks (e.g., two different LANs) is where the problem occurs.

Each network technology has its own characteristics and inherent protocols. From an enterprise viewpoint, this is very bad news. For example, users connected to an X.25 network cannot easily be connected to those already connected to a LAN. How, for example, would the X.25 user indicate the destination's media access control (MAC) address, and vice-versa? X.25 networks understand only X.25 addresses, and LANs understand only MAC addresses. The differences between network technologies and native protocols almost invariably prevent their direct interconnection. Differences in addressing schemes present another difficulty. Addressing considerations alone will typically dictate the use of a network interconnection device (NID) at the point at which two network technologies come together.

Exhibit V-5-1 illustrates several network technologies, represented by N_1, N_2, N_3, N_4. Each of these technologies has its own native protocol (i.e., P_1, P_2, P_3, P_4).

A way must be found to integrate all these disparate technologies into a single supernetwork, with globally uniform and globally understood characteristics and a single addressing scheme.

This is achieved by operating an integrating, unifying protocol (shown in Exhibit V-5-2 as P_x), sometimes known as an internet protocol, over the top of

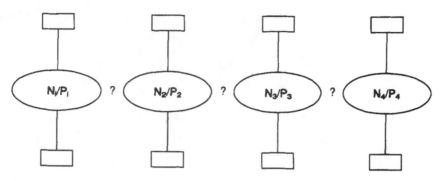

Notes:
- For every (sub)network N_i, there is a network-specific protocol P_i.
- All P_i's are different and thus not directly interconnectable.
- A way to make all the different networks look the same must be found, if the enterprise network is to appear as a single seamless network.

Exhibit V-5-1. The Interoperability Problem

---

all the possible basic communications networks. The Internet Protocol (IP) of TCP/IP is one such protocol. The Connectionless Network Layer Protocol (CNLP) specified in the OSI International Standard (IS) 8473 is another. Proprietary systems have their own internet protocols (e.g., Novell uses its Internetwork Packet Exchange (IPX) and Banyan uses Vines).

From the architectural standpoint, the technical term for such an internet protocol is *subnetwork independent convergence protocol* (SNICP). The protocols used on real-world communications networks (e.g., leased lines, X.25, frame relay, LANs) are known as subnetwork access control protocols (SNACP). Readers interested in obtaining a more detailed technical understanding about the network layer architecture and principles of internetworking should consult IS 8648, Internal Organization of the Network Layer. The basic internetworking architecture is shown in Exhibit V-5-3.

Unification does not mean simplification. Two protocols operating over a given subnetwork still require two address schemes. Routing tables are then needed in the network interconnection device to map the global enterprise address to the address to be used by the network interconnection device for the next link in the composite path. Exhibit V-5-4 is a simplification of how the two addresses are used. In practice, the "next" address may be more complex, depending on the internetworking protocols under consideration. A network interconnection device of this type is called a router.

Vendor Solutions

The second basic problem is that each system vendor has a vendor-specific idea of how to build the supernetwork—the type of supernetwork protocol, the global

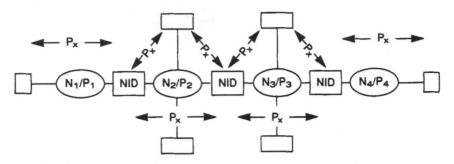

Notes:

NID   Network interconnection device

• A neutral protocol P_x is run over the top of every subnetwork.
• Because P_x is now the basis of the enterprise network, the individual networks are called subnetworks.
• P_x operates at the top of the network layer and is usually called an internet protocol.
• Several Px's exist (i.e., IP, CNLP, IPX).

Exhibit V-5-2. The Interoperability Solution

addressing scheme, and the internal routing protocols to be used. At worst, this leads to a multiprotocol network, which amounts to several separate internets operating in parallel over the same physical communications plant.

An alternative to the multiprotocol network is to choose a single protocol for the entire enterprise supernetwork. This inevitably requires finding techniques to accommodate the systems that do not inherently operate this chosen protocol. Such techniques include encapsulation (sometimes called tunneling) at the edges of the single-protocol network, or other techniques (e.g., transport service interfaces and application gateways).

However, even with a single protocol, tunneling permits only the coexistence of incompatible systems; there can be little or no interaction between each of the tunneled applications. The major advantage of tunneling is that the core of the network is unified, optimizing network management and networking skills. The disadvantage is the effort required to set up to the tunneling configurations at the edges.

The best solution is for all vendors to use the same internet protocol. This solution is, however, some way off, and in the meantime some degree of multiprotocol networking must be tolerated. Many vendors do offer versions of their own internetworking protocols over tunneled TCP/IP. Thus, an appropriate choice for the enterprise network might be a dual-protocol network comprising the native routing of both IP (of TCP/IP) and OSI's CNLP.

Going Global

Many LAN-based systems include internal protocols that advertise the existence of various LAN-based servers. Such a protocol is sometimes known as a

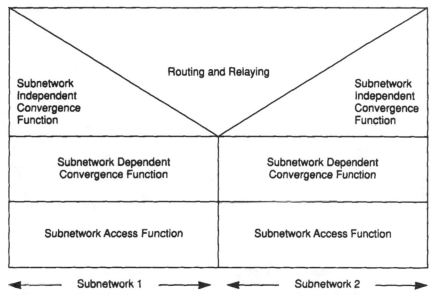

Exhibit V-5-3. Network Layer Architecture

service advertising protocol (SAP). Such protocol exchanges, frequently broadcast over the LAN, ensure that the availability and addresses of various servers are known throughout the LAN user community. This is useful when the geographic area is confined to a work group or a floor of a building; for example, the knowledge of a set of available printers is useful only in the area that has ready access to one of them. Thus local messages must be constrained to local environments by putting adequate filtering at the point of access to the wide area portion of the enterprise network. There is no point in telling a user on a LAN in New York that there is a printer available on a LAN in Seattle.

Another global problem relates to the extra transit delay involved in transport over a WAN, particularly for nonroutable protocols. Many protocol stacks used in local environments do not contain a network layer protocol; in other words they have no routing layer. Such protocols cannot be routed directly in a router-based enterprise network. Where it is necessary for such an application to be networked outside a particular local environment, the local protocol stack must be encapsulated within an internetworking protocol. Then it can be launched onto the wide area part of the enterprise network.

Many of the local or nonroutable protocols are designed for very rapid acknowledgment. The transfer of these types of protocols across a wide area may cause problems; applications may prematurely time-out, or suffer poor throughput because of lack of a windowing mechanism adequate for the wide area transit delay. To accommodate such applications, it is necessary to "spoof" the acknowl-

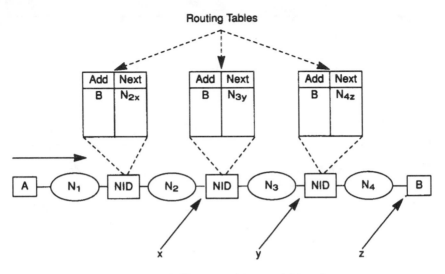

Exhibit V-5-4. Simplified View of Addressing

edgments. This means that acknowledgments must be generated by the local encapsulation device. This requires the extra complication of adding a reliable transport protocol on top of the internetworking protocol across the wide area portion of the enterprise network. Once a local acknowledgment has been given, the originator will discard the original so it is no longer available for retransmission. Having given the local acknowledgment, the spoofing device must ensure reliable delivery to the remote end by employing a transport protocol of some sort (e.g., TCP or OSI Transport Class 4). The scheme is shown in Exhibit V-5-5. This avoids the end-to-end round trip delay T_r for every packet of data, by providing an acknowledgment at time T_1.

Going global also poses some challenges in the area of network layer addressing, particularly with regard to achieving enterprisewide uniqueness and structuring addresses for scalability and ease of routing.

Addressing. Usually, addresses of stations within a local work group are allocated locally. This can present problems when subsequently the local work groups must be integrated into a single enterprisewide address scheme. If several work group addresses, or parts of an address (e.g., an area or server name), are the same, clearly some changes will have to be made. From an operational perspective, changing addresses is not a trivial matter. It is best to avoid address allocation clashes right from the outset by having an enterprisewide address registration authority set up within the organization.

Some addressing schemes do have some hierarchy associated within them. This can be used to avoid address encoding clashes by ensuring that local ad-

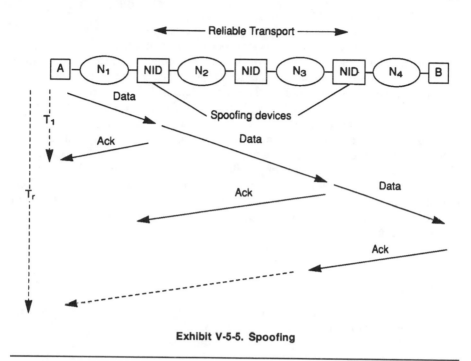

Exhibit V-5-5. Spoofing

dresses are only the low-order part of the total address. Even in this case, however, an enterprise perspective is necessary to avoid clashes in the high-order part
of the address.

Some vendors achieve uniqueness by allocating unique addresses when the
equipment is shipped. However, this usually results in a flat, random address
space that makes routing considerably more complex because there is no structure in the address to help "scale" the enterprise network from the routing
perspective.

If the enterprise network is to be interconnected to the Internet, IP addresses
must be obtained from a central authority (i.e., the Internet Society). This is not
so simple either, because the Internet is running out of addresses and the form of
an Internet address dictates to some extent the degree of scaling that is achievable. Furthermore, Internet addresses are not hierarchical in a meaningful way,
because they are more or less randomly allocated.

The most widely recognized and hierarchically administered address available
today is the OSI address space available for OSI Network Service Access Point
(NSAP) addresses. This permits an address of as many as 40 digits, which allows
good scaling potential and simplified routing.

The reason that a hierarchical (i.e., scaled) address scheme is so important has
to do with the way that routers operate and the size of the associated routing
tables. If addresses were allocated completely randomly, but uniquely from a
large address space, every router would need a table with every address in it. Not

only would the table be extremely large, but the time needed to find an entry could also be a problem. Routing is thus better arranged on the basis of hierarchical distinctions that are implicit in the address scheme. So, to service a local work group or other limited geographical area, a local router must know only whether the destination address is internal or external. If it is internal, the router knows how to get the message to the destination; if it is external, the router can pass it on to the next-level router. This leads to the concept of areas, groups of areas, domains, and countries being components of a hierarchical address.

When legacy systems must be accommodated with conflicting address schemes and re-allocation of addresses is impossible, tunneling may have to be employed merely to avoid interaction between the conflicting addresses. Because conflicting networks are divided into separate virtual private networks, the protocol under consideration cannot be routed natively even if the backbone routers are capable of doing so.

ROUTING

To reduce the amount of time devoted to setting up routing tables manually, and to allow dynamic rerouting and a degree of self-healing, routing protocols are often employed to distribute routing information throughout the enterprise network. These protocols are in addition to the internetworking protocol itself but are related to it. For every internetwork protocol routed in a multiprotocol network, there may be a specific routing protocol (or set of protocols). This also means in general that there will also be a separate routing table for each internetworking protocol. The situation in which several routing protocols are used simultaneously, but independently, is sometimes known as a ships-in-the-night situation because sets of routing information pass each other and are seemingly oblivious to each other even though there is only one physical network.

Some router manufacturers operate a single proprietary routing protocol between their own routers and convert to individual protocols at the edges of the network. There have been some attempts to define a single standard routing protocol based on the International Standards Organization's (ISO) intermediate system to intermediate system (IS-IS) standard.

In an enterprise network, end systems (e.g., terminals. workstations, mainframes) usually announce their presence and their own addresses to the nearest local router. The local routers record all the local addresses within their area and inform all neighboring higher-level routers of their own area address. In this way, a router at the next and higher level in the hierarchy only needs to know about areas. Recursive application of these principles to a hierarchical configuration can lead to efficient routing by minimizing the amount of routing information to be transmitted and by keeping the size of routing tables small.

As the process of promulgating routing information proceeds across the network, every router in the enterprise network will obtain a table of reachability that it can then use for choosing optimum routes. Route optimality may be based on a number of independent metrics (e.g., transmit delay, throughput, monetary cost). Invariably, a shortest path first (SPF) algorithm is used to determine the optimal route for any particular metric chosen as the basis for routing. Both the Internet and OSI routing protocols use an SPF algorithm.

Routers are the key interconnection devices in the enterprise network; subsequently, the router market has been one of the key growth areas during this decade. Some router vendors have grown from small $10 million companies to $1 billion companies.

In most cases, routers are purpose-built communications processor platforms with hardware architectures specifically designed for high-speed switching. Several possible pitfalls await the unwary purchaser of routers. Such a purchase involves four important considerations:

- The capacity and architecture, in terms of the number of ports accommodated and throughput achievable.
- Internetwork protocols supported and their associated routing protocols.
- Support of technologies for the connected subnetworks.
- Interoperability between different vendors.

Capacity and Architecture

The number of ports required determines to a large extent the size of the router required, which in turn affects the architecture and throughput of the router. Physical size of circuit boards dictates how many ports can be placed on a single board. The greater the number of ports, the greater the number of boards required and the more critical the architecture.

Routing between ports on the same board is usually faster than routing between ports on different boards, assuming that there are on-board routing functions. Boards are usually interconnected by means of some kind of backplane. Backplane speeds can vary greatly between vendors. Routing functions and tables may be distributed across all boards or may be centralized. The bottom line is that the architecture affects the performance, and performance figures are sometimes slanted toward some particular facet of the architecture. Thus, some routers may be optimal for certain configurations and not so good for others.

When making comparisons, the IS manager must carefully analyze vendor throughput and transit delay figures. Although worst cases are helpful for the user and network designer, some vendors specify either the best cases or averages. Other metrics involved in measurement may also be different (e.g., packet size assumed, particular internetwork protocol, particular subnetwork).

Other architectural considerations include extensibility and reliability. For example, is hot-swapping of boards possible? If the router must be powered down and reconfigured to change or add new boards, the disruption to a live network can have severe ripple effects elsewhere in the network. Can additional routing horsepower be added easily, as loads increase, by simply inserting an additional routing processor?

Standalone or Hub-Based Routers. The question of using standalone or hub-based routers may also be relevant. This is a difficult problem because of the traditional split between the hub and router manufacturers. Hub vendors tend not to be routing specialists, and router vendors tend not to be experts at hub design. Alliances between some vendors have been made, but the difference in form factors (of circuit boards) can result in some baroque architectures and poor performance. Except in the simple, low-end cases, purpose-built standalone routers usually perform between and are more easily integrated with the rest of the network.

Some standalone routers can directly handle the multiplexed input streams from T1 and T3 links, making voice and data integration possible. This is unlikely to be the case for a hub that has been designed mainly for operation in a LAN.

Many of the router manufacturers make several sizes of router, which could be referred to as small, medium, and large. All of one vendor's routers may, regardless of size, offer the same functions, but the circuit boards may not be interchangeable between the different models. This can make a big difference when it comes to stocking an inventory of spare parts. There may also be differences in network management capabilities.

Internetwork Protocols Supported

Most router vendors claim that they support a large number of internetworking protocols. In some cases, however, there may be restrictions on the number of protocols that can be supported simultaneously. There may also be restrictions on the use of multiple protocols over certain network technologies, or hidden subnetwork requirements. An example of the latter might be the need for a separate X.25 permanent virtual circuit (PVC) for every individual protocol, as opposed to operating all the protocols over a single PVC.

Some vendors may also use a proprietary routing protocol scheme for internal routing, only making the standard protocols available at the periphery of the network. This will make it difficult to mix different vendors' router products on the same backbone or within a single routing domain.

Network Technologies Supported

Most manufacturers provide interfaces to a large number of network technologies (e.g., X.25, ISDN, frame relay, T1, T3, Ethernet, Token Ring). The method of support may also vary. For example, in the case of leased circuits, it may or may not be possible to directly connect the carrier's line to the router. Some routers may accept the carrier's framing mechanism directly, others may require an external converter to provide a simple serial interface (e.g., V.35) before connection can be achieved. As stated previously, however, the interaction between these interfaces and the multiple internetwork protocols may not be clearly reported by the vendor.

Interoperability

In general, there is little interoperability between routers from different vendors. The reason often cited is lack of standards for operating multiple protocols over a given subnetwork technology. There is a need in these cases to specify how the various protocols are encapsulated in the subnetwork protocol and how each protocol is to be separately distinguished.

The only case in which interoperability may be more or less guaranteed is over a LAN, using OSI's Subnetwork Access Protocol (SNAP) identification scheme, so LANs are frequently used to interconnect islands of routers. An example is shown in Exhibit V-5-6. Two islands of routers from different manufacturers, A and B, are interconnected over a LAN. This scheme works well when the respective backbone routers can be physically located near an existing LAN cluster, as depicted by the dotted ellipse.

Efforts are under way to improve standardization. For example, the Internet community has defined the Point-to-Point Protocol, designated PPP, which defines encapsulation and discrimination methods for multiprotocol operation over leased circuits. Unfortunately, this standard is not fully complete with respect to many of the internetwork protocols already in existence. In theory, at some future date it will be possible to use PPP between a backbone comprising routers from one vendor and access routers from another vendor. An arrangement such as that shown in Exhibit V-5-7 then could be configured. Such an arrangement may be practical now if the protocols routed are restricted to those that have been standardized for use with PPP.

Where no standard wide area interoperability solution can be used, an alternative for large numbers of access routers is a variation of the LAN interconnect technique. Use of this technique may increase the number of routers required (as shown in Exhibit V-5-8) but may be necessary where existing local islands that contain a set of routers of type R_B must be accommodated.

A standard for using frame relay is also under development in the Internet community, and a generic standard for encapsulation and discrimination is under development by the ISO. The latter is intended for use over most subnetwork technologies.

Filtering

Filtering was mentioned in the context of preventing local traffic uselessly flooding the enterprise network. The degree of filtering that can be applied may vary with the manufacturer. Various parameters can be used as the basis for filtering, for example, source address, destination address, protocol type, and security codes. The disadvantage of using filtering is the labor involved in setting up the filter tables in all the routers.

Network Management

It is extremely unlikely that a common set of management features will apply to all vendors' routers. Thus, if several manufacturers' routers are deployed in a given enterprise network, several management systems probably will be required. In the best case, these systems can be run on the same hardware platform. In the worst case, different hardware platforms may be required.

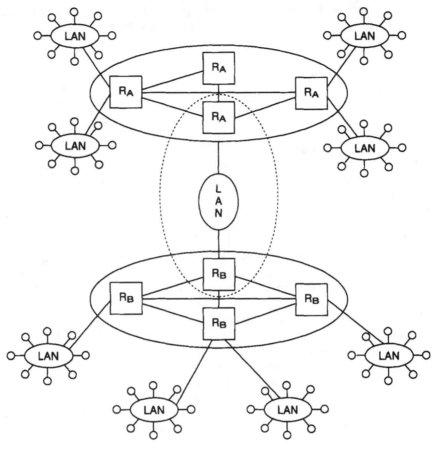

Notes:
R_A Router from manufacturer A
R_B Router from manufacturer B

Exhibit V-5-6. LAN Interconnect Approach

BEYOND THE NETWORKING CHALLENGE—THE APPLICATIONS

All the considerations discussed so far apply to the internetworking protocols. Mulitprotocol networks serve only to share bandwidth, they do not allow applications to interoperate. Where that is necessary, with completely different stacks of protocols, an application gateways must be used, as shown in Exhibit V-5-9. This exhibit shows an OSI-based mail (X.400) application interoperating with a TCP/IP-based mail application, over an application gateway.

Such gateways may be sited either centrally or locally. The use of local gateways makes it possible to deploy an application backbone with a single standard

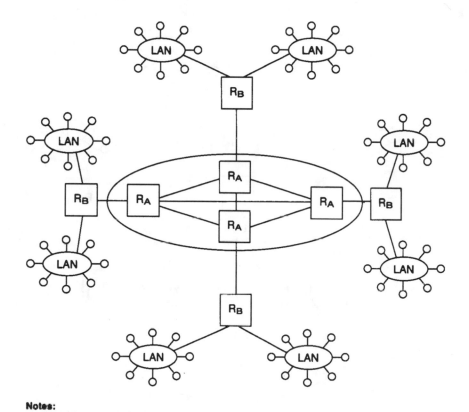

Notes:
R_A Router from manufacturer A
R_B Router from manufacturer B

Exhibit V-5-7. Access Routers Approach

---

application operating over the wide area portion of the enterprise network (e.g., an X.400 mail backbone). This reduces the number of gateways needed for conversion between all the different applications. Only one conversion is necessary for each application (i.e., to the one used on the backbone). A considerable number of different local systems could interoperate through the "standard" backbone application.

The encapsulation technique already mentioned in the context of IP tunneling allows all applications that can be so configured to operate across the enterprise network. A tunneled SNA application is show in Exhibit V-5-10.

Another solution that may help in the future is the availability of transport service interfaces for end systems (e.g., workstations, terminals, servers). A transport service interface allows a given application to be operated over any underlying communications protocol stack. In other words, applications and commu-

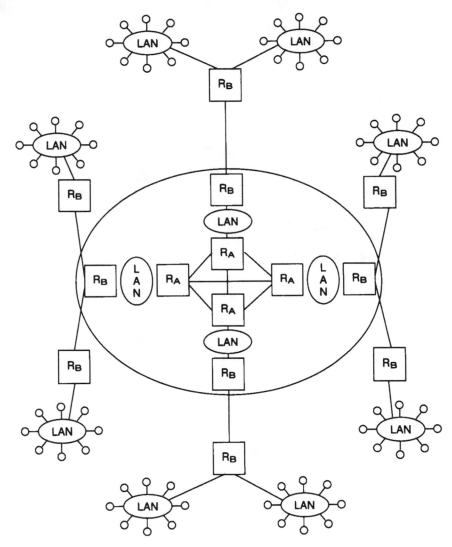

Notes:
R_A Router from manufacturer A
R_B Router from manufacturer B

Exhibit V-5-8. Access with LAN Interconnect

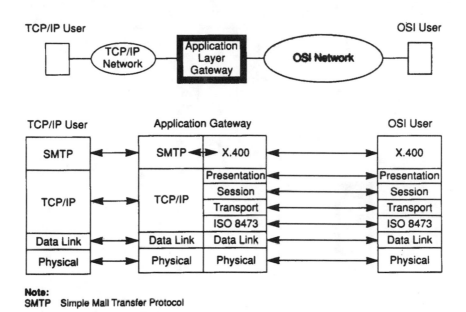

Note:
SMTP   Simple Mail Transfer Protocol

Exhibit V-5-9. Mail Application Gateway

nications stacks can be mixed and matched as necessary. The so-called open operating systems (e.g., POSIX and X/Open) adopt this approach.

The transport layer is a fundamental dividing line in the system architecture. Network-related functions are separate from application-related functions, so that applications work with many communications protocols. Exhibit V-5-11 shows an end system containing both an open OSI/TCP/IP stack (shaded) and a proprietary stack (unshaded). Within an end system, protocol stacks can generally be separated into the communications-specific lower-layer parts and the application-specific upper-layer parts. The two stacks communicate through a transport layer interface (TLI).

CONCLUSION

Setting up an enterprise network rarely involves starting with a blank piece of paper. In practice, legacy systems or other requirements result in the existence of a variety of heterogeneous systems. Several techniques can be applied to, at least, make the heterogeneous systems networkable over a single physical network. Varying degrees of interoperability between them may also be possible.

This chapter has addressed only the bare essentials of enterprise networking. The subject is highly technical. Setting up a large enterprise network is not for the

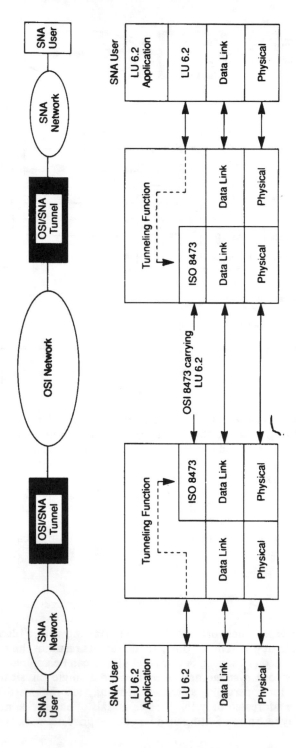

Exhibit V-5-10. Tunneled SNA Application

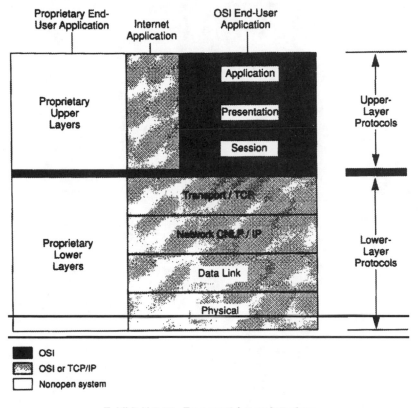

Exhibit V-5-11. Transport Layer Interface

---

fainthearted. The pitfalls are many and are difficult to anticipate because of vendors' exaggerations and sheer technical complexity.

For the moment, a single "pure" multivendor network is out of reach. A successful convergence of the ISO's CNLP and the next generation IP into a single common internetworking protocol would be a tremendous advance. A project within the Internet Society, known as TCP and UPD with Bigger Addresses (TUBA), is currently focusing on this issue.

However, it is essential to create an enterprise network architecture. An overall plan for the network will minimize confusion and put in place a timed migration strategy toward a completely integrated network. Central control has lately fallen into disrepute, but without some control over how networking is to be achieved, all the real benefits of an enterprise network will not be realized. Finally, it is probably fair to say that enterprise networking is still something of a black art and is bound to present all IS managers with some surprises and disappointments.

V-6

Evaluating LAN Management and Control

Paul Cullen

LOCAL AREA NETWORKS (LANs) enable users to efficiently share information while allowing them to maintain individual control of processing. PC LAN applications provide good price/performance but must be reviewed so that they get the maximum return. This distribution of processing power is part of a growing end-user computing trend that brings the management of internal controls out of centralized information processing and shifts the responsibility to end users. These users are not generally proficient in implementing controls over processing. In addition, PC LAN vendors tend to focus on product capabilities rather than controls. Thus, controls have typically taken a back seat in the development of LAN applications, resulting in critical applications being implemented without adequate controls in the areas of security, contingency planning, and change control. Users are faced with the challenge of assessing risk and implementing controls in a new arena. Evaluating PC LANs begins with an understanding of the standards and practices established by management and with the risk inherent in the applications running on the LANs.

THE PROCESSING ENVIRONMENT

PC LAN processing controls should be based on standards developed by users and information systems management. These standards define the practices and controls adopted by management to ensure that processing is performed with a prudent level of risk. Standards should be developed in information security, systems development, program change control, problem management, and disaster recovery planning. Early DOS-based PC LANs had limited control capabilities, and frequently, separate standards needed to be developed. These standards focused on manual controls, limited data security controls, and the assessment of risk to determine whether high-risk applications should even run in a PC LAN environment. Recent releases of LAN software, such as IBM Corp.'s OS/2 LAN Manager and Novell, Inc.'s Netware, have made improve-

509

Risks	Related Controls
Inappropriate Disclosure of Information	Physical Security, PC LAN Logical Security
Loss of Data or Software	Contingency Plan
Processing Errors	Edits Checks, Test Plans

Exhibit V-6-1. Risk Analysis Table

ments in logical security and systems recovery, making their features similar to those found in mainframe environments. This reduces and sometimes eliminates the need for separate PC LAN standards.

Evaluating Risk

PC LAN controls should be implemented on the basis of the risk inherent in the application. Risk should be assessed in information confidentiality, loss of data and software, and processing integrity. Users typically confuse applications with the software used to run the application. A typical error is the misconception that Lotus spreadsheet software is an application that could be easily replaced. This does not consider the consequences of errors in the spreadsheets or the effort required to reconstruct a spreadsheet if it is lost. In addition, users have frequently assumed that mainframe and minicomputer applications were of high risk and applications running on PC LANs held little risk. This paradigm is changing as LAN applications have been developed for financial reporting, processing payroll, and handling banking transactions. IS managers must verify that PC LAN application risks have been accurately assessed and that controls are in place to address the risks.

For example, Exhibit V-6-1 presents a simplified risk analysis table showing risks and related controls. The following sections describe steps that can be taken to ensure that risks are properly assessed and controls are in place to address those risks.

Evaluating Program Steps

The evaluation program steps in this chapter begin with obtaining an overview of the LAN, including how it is used and managed. This first step is to identify the LAN administrator and the designated backup. This person is responsible for resolving LAN problems, adding new users to the LAN, making changes to LAN software, and backing up programs and data to ensure that the applications could be recovered in the event of a disaster.

The next step reviews what the LAN is used for. This requires documenting the applications as well as the risks and exposures associated with the applications. The risks inherent in the LAN applications determine the necessary controls and resulting review procedures required to verify the adequacy and effectiveness of those controls.

Many types of equipment are available to monitor LAN traffic for tuning operations and diagnosing problems. However, these products can also view and capture sensitive data being transmitted across the network. Although these are necessary tools for LAN administration, their use should be properly controlled by authorized individuals.

Gateways are used to connect networks running on different architectures. They are commonly used to connect LANs to mainframes. PCs on the LAN can then act like smart terminals, allowing users to have mainframe sessions. Some gateway products can capture data flowing from the PC to the mainframe, and there is the potential for sensitive information, such as passwords and other confidential data, to be captured. Gateways should be physically secure and restricted to authorized individuals.

The IS manager also should determine whether public domain or shareware software is being used. This software may not be properly licensed, and the company may be infringing on a copyright by using it. In addition, public domain software is arguably more likely to contain computer viruses that could infect and destroy data in individual PCs or an entire LAN.

Finally, a review should be performed to determine the inventory and fixed assets process for controlling hardware and software. PC LAN equipment consists of expensive and typically marketable fixed assets and should be controlled to prevent loss or theft. The inventory review should include tracing equipment and software to inventory records and license agreements to verify their accuracy and to verify that license agreements and copyrights are not being violated.

In summary, the objective is to determine whether a process has been established for planning, organizing, directing, and controlling the activities related to the LAN. The risks that should be evaluated include:

- Disclosure of confidential information because of inappropriate data security controls.
- Loss of hardware, data, or software because of viruses or inappropriate fixed-asset controls.

The review steps include:

1. Identifying the LAN administrator and designated backup.
2. Reviewing what the LAN is used for (e.g., application systems, data, files, documents, spreadsheets).
3. Identifying what data is stored on the server and workstations.
4. Documenting the LAN hardware and software. Can it provide the level of access control necessary for the applications and data used on the LAN? Are additional products used to improve LAN security and operations?
5. Determining whether LAN monitoring equipment is used for problem resolution. If so, is this equipment secure enough to prevent the inappropriate capture or disclosure of sensitive information?
6. Determining whether public domain or shareware software is used on the LAN. If so, is there virus detection and prevention?

7. Determining whether an inventory of hardware is kept and whether there is a process for physically controlling the assets.

8. Determining whether there is a written policy and enforcement process regarding copyright infringement.

DATA AND APPLICATIONS REVIEW

The purpose of this review module is to determine whether the LAN applications and data are adequately secured. To begin with, the layout of the LAN should be documented, and physical security over the LAN should be evaluated to determine logical controls required to compensate for weaknesses in physical security.

Dial-access capabilities should also be controlled. Many companies believe that LAN operating system security alone is not adequate to control dial access, which could allow access to anyone with a PC, modem, and software. PC products allow a user to dial into a LAN workstation from any location and control the workstation, which removes the ability to rely on physical security. Additional features, such as callback or encryption modems, should be considered to improve dial-access security.

The reviewer must also understand the access controls available within the LAN operating system. This typically begins with a review of the administrator's manual for the PC operating system.

Security can be cumbersome if a log-on is required for the personal computer, the network operating system, the application, and the mainframe interface. Vendors are beginning to work together to interface log-ons to provide integrated PC, LAN, and mainframe security solutions. This trend will likely become more refined as processing is distributed in the end-user computing environment.

The review should next identify the particular platform on which processing is performed and where data is stored. Application software and data can be stored on the PC workstation, LAN server, or mainframe. In addition, the application software itself can provide internal security. This situation creates the potential for up to four information security functions—workstation, LAN operating systems, mainframe, and application security. The strengths and weaknesses of each of these platforms must be assessed to ensure that the applications programs and data are adequately secure, based on corporate standards and risk assessment.

The final steps in this section refer to the administration processes for adding new users, reviewing access capabilities, approving new access, and removing access when it is no longer needed. These steps are the same for PC LAN-based systems as they are for applications processing on other platforms. The reporting features may differ, however, as was discussed previously.

In summary, the objective is to determine whether the systems and data on the LAN are adequately secure, considering the nature of the data and LAN system capabilities. The risks to be evaluated include the loss or inappropriate disclosure of sensitive information because of inadequate or ineffective controls.

The review steps include:

1. Documenting the physical layout of the LAN:
   — What physical units have access to the LAN? Is access to all nodes on the LAN physically controlled?
   — Is dial-up access to workstations or servers allowed? If so, are products such as Microcom, Inc.'s Carbon Copy used?
   — Are the logical access controls implemented on the LAN system documented?
   — How do workstations access data? Who has access to the servers' logical drives?
2. Documenting the LAN security administration process:
   — Is there a process for authorizing new users, authorizing and updating user access capabilities, and deleting access when it is no longer needed?
   — Is there a user listing and a periodic review process established to ensure that access capabilities remain commensurate with job accountabilities?

SYSTEM ACCESS TESTING

The purpose of this review module is to verify that access capabilities are authorized and appropriate based on the user's job function. The first step is to review log-on password protection.

Passwords can be scripted in .bat files, logging on the workstation when it is started. Then, everyone who has access to the workstation can access the LAN. If passwords are not encrypted, as is true with early LAN systems or possibly application software, users who can access the password file can access the entire system. The IS manager must also determine whether the LAN software allows peer-to-peer network capabilities. If so, logical access controls over each PC through the network must be assessed. If the network allows access only to a dedicated server, then it is only necessary to identify the servers and review network logical access controls over the servers. There are software packages, such as Brightwork Development, Inc.'s Netremote, that enable administrators to access and control a user's workstation, typically for problem resolution. The use of these packages and controls preventing inappropriate access must be evaluated.

The final steps in this section are to review access capabilities and verify that they are appropriate, based on job function. This includes reviewing user IDs and the groups and resources to which they have access, tracing them to authorized individuals, and verifying their appropriateness. Administrator capabilities and supervisory authority should be limited, as they permit access to all system resources.

In summary, access testing includes the following steps:

1. Determining whether passwords are printed during log-on.
2. Determining whether the password file is encrypted.
3. Determining whether there is violation and access capability reporting.
4. Determining whether file access passwords can be properly controlled if file

7/5/93 3:56:05 PM Severity = 0.

1.1.60 Bindery open requested by the SERVER

7/5/93 4:17:19 PM Severity = 0.

1.1.62 Bindery close requested by the SERVER

7/5/93 4:17:20 PM Severity = 4.

1.1.72 SERVER TTS shut down because backout volums SYS was dismounted

7/5/93 4:38:39 PM Severity = 0.

1.1.60 Bindery open requested by the SERVER

Exhibit V-6-2. Sample File Server Error Log

---

security is based on passwords. What is the process for distributing and periodically changing file access passwords?

5. Determining whether workstation drives can be accessed from other workstations.

6. Determining whether access control can be based on user or group IDs.

7. Reviewing access capabilities on the PC and LAN server:
   — Tracing IDs to an organization chart to verify that the ID belongs to a valid employee.
   — Tracing access capabilities to employee functions to verify that the level of access is appropriate.

User IDs with administrator authority should also be listed if such a capability exists in order to verify that the administrative capability is limited and appropriate for the individual's job function.

PROBLEM MANAGEMENT

The purpose of this review module is to determine whether LAN system problems are being recorded, analyzed, and resolved.

The DOS operating system does not have an automated error-tracking file system similar to mainframe systems, such as IBM's System Management Facility (SMF). Thus, there are no automated problem logs for the early DOS-based LANs. Other networks, such as Novell's Netware, OS/2 LAN Manager and LAN Server, and certain implementations of UNIX, do have logging capabilities for operating system and hardware errors. Exhibit V-6-2 presents a sample file server error log that shows operating system errors and incorrect log-ins. Application software and user errors, however, would not be logged. Therefore, LAN administrators should keep a manual or automated log of all problems with users, applications, operating systems, and hardware.

This log should contain the date the problem occurred, the problem description, length of downtime, priority of resolution, actions taken to resolve the problem, and date the problem was resolved. The problem log should be analyzed on a periodic basis to find weaknesses in testing, hardware reliability, and user training requirements. Another concern is whether problems are resolved within the appropriate time frames. For critical time frame applications, companies may need to contract LAN equipment support within specified time frames and have backup equipment to ensure that the network is performing according to business needs. In addition, a complete problem log can show problem trends with hardware, operating system software, application software, and communications software.

The review steps include:

1. Documenting the process used for tracking and reporting problems:
   — How are problems identified, tracked, rated for severity, and resolved?
   — Is there a process for problem history and trend analysis?
   — Is there a process for reviewing the criteria for assigning resolution deadlines for problems?
2. Obtaining and reviewing the problem listing. LAN users should be questioned regarding LAN problems to see that they are recorded.
3. Reviewing open problems to verify that they are being responded to in a timely manner.
4. Reviewing closed problems and verifying with users that they are in fact properly resolved.

MANAGING SYSTEMS CHANGES

The purpose of this review module is to determine whether systems changes are authorized, tested, documented, communicated, and controlled. A process should also be in place to ensure that production source and executable code are synchronized and secured in order to prevent inappropriate program alteration and to be able to reconstruct system files. LAN operating systems typically do not have the change management facilities that are found in mainframe systems. Therefore, LAN administrators manually control production code on restricted subdirectories on the LAN servers. After a change has been tested and approved, an independent party should move the code to a restricted library on the server or distribute the software to workstations. This process would ensure the integrity of the software version that is running in production and that unauthorized changes are not introduced into the system.

LAN software is also frequently developed by software vendors outside the company. Because users are not involved in program development, they do not feel the need to test the software. User acceptance testing, however, should be performed before each LAN software release is propagated to ensure that the system is performing according to user expectations and to prevent the corruption of user files.

The IS manager should review the process for unit, system, and user acceptance testing. There should be evidence showing that users have signed off approving the change, that the systems documentation was updated, and that the risk factors were updated, if appropriate.

The risks to be evaluated include:

- Fraud and embezzlement resulting from unauthorized change.
- Incorrect or incomplete information caused by improperly installed changes.
- Business interruptions caused by improper change coordination.

The review steps include:

1. Reviewing the process used for authorizing, communicating, and documenting changes to the LAN systems and applications.
2. Reviewing who makes the changes and how programs and files are managed during the change process.
3. Verifying that there is a separation of duties.
4. Reviewing the change testing process.
5. Reviewing test plans to ensure that they are complete and that any inadequate test results have been resolved.
6. Reviewing the change fallback process, which would enable a system to use the previous version of the software to facilitate the recovery of the system in the event of an error.
7. Reviewing directories on the workstations and servers to verify that production programs and data are adequately secure to prevent inappropriate alteration.
8. Verifying that production source and executable software are synchronized to provide the ability to reconstruct the processing environment in the event of problems.
9. Reviewing the process for updating systems documentation for each change.
10. Reviewing documentation to verify that it is current.

DISASTER RECOVERY AND CONTINGENCY PREPAREDNESS

Disaster recovery planning is basically no different for LAN systems than for any other system. The business unit must identify the risk of the applications and the time frame required for recovery. Based on this assessment, a disaster recovery plan should be developed identifying who will be responsible for recovering the system, where recovery will be performed, and what equipment is required for recovery, including communications equipment. Backups of workstation and server programs and data are frequently kept in the form of backup tapes. A common approach is for users to back up workstation data to the server and then back up the server to tape. The backup tapes should be kept in a secure off-site location, and the frequency of the backups should be assessed based on how often the data is updated and the amount of effort required to reconstruct the files. The security of the off-site location is critical for LAN systems because logical security provided by the LAN operating system does not protect data on backup tapes.

The disaster recovery plan should be tested to ensure that it is sufficient to recover all critical functions. This test should include recovery from the backup tapes to verify that the backup process is effective.

The review steps include:

1. Reviewing the backup and recovery process:
   - Has the area documented the time frame required for recovery and the critical software, data, and hardware to be used in the event of a disaster?
   - What is the process for backing up LAN workstations and servers?
   - What is the frequency of backups?
   - Has recovery been tested?
   - What hardware and software are used for the backup process?
   - Is redundancy built into the system to avoid downtime for systems that need high availability (e.g., are there duplicate disk drives and servers to be used as a fallback)?
2. Reviewing procedures for off-site storage of essential software, documentation, and data.
3. Verifying that the backup media is properly stored in an off-site location.
4. Verifying that the media is effectively secured.

SYSTEMS DEVELOPMENT AND PROJECT MANAGEMENT

This review module is also basically the same for LAN systems as it is for other systems, and is similar to the change management module previously discussed. This module may need to be tailored for certain situations. The LAN environment may not have a dedicated development staff as is found in mainframe environments. A development methodology, project management, and user involvement and training, however, should be required components of LAN systems development, as they are with the development of any other systems.

Strategic planning should be performed to ensure that the LAN systems and applications are compatible with the corporation's wide area networks and technology direction. The systems strategic plan should be based on the business units' plans. In addition, the operating system and hardware should provide for migration to future platforms.

In the development of LAN systems, a cost/benefit analysis should be performed to ensure that the payback from the system justifies its cost; competitive bidding should be required to ensure that the prices paid for the hardware and software are reasonable. It should be verified that purchases are properly approved by the appropriate level of management. Finally, if software is developed by vendors or contract programmers, copyrights and software ownership must be clearly defined.

The review steps include:

1. Reviewing the planning process for LAN systems:

- Has management devised a strategic plan for the development of applications and the selection of the optimum processing platform based on user requirements?
- Was a cost/benefit analysis performed?
- Is there evidence of user management approval?

2. Reviewing the project management process used for systems development, including project budgets, schedules, timekeeping, and issue tracking.

3. Verifying that purchases were cost-effective through the use of competitive bids, vendor analysis, and thorough requests for information.

4. Documenting the process for user training and involvement in LAN systems development.

5. Reviewing and evaluating the systems development documentation, including evidence of approvals and cost/benefit analysis.

6. Verifying that copyrights have been registered for new software.

PROCESSING INTEGRITY REVIEW

Processing integrity for LAN-based systems has some unique control concerns in addition to the basic input, processing, and output controls that apply to all application systems. Users typically install applications on a PC LAN platform because it is a low-cost alternative to mainframe development. The ability for the LAN application to grow on this platform is limited, however, particularly for DOS-based systems. Equipment should be benchmark tested before it is propagated to ensure that it has enough capacity.

Another concern for LAN-based systems is the control over concurrent updates; that is, the ability to prevent two people from updating the same file or record at the same time. This can be controlled by the LAN operating system and by the application software. LAN servers typically fail in one of two ways. First, they can be too restrictive by preventing two or more people from accessing the same disk drive or subdirectory at the same time. This can obviously be inefficient when the desire is to share data. The second way LANs can fail is by letting two or more individuals into the same file or even the same record but not preventing the loss of one person's update or the destruction of a file.

Another concern regarding LAN applications is that they are typically used to process data that is downloaded from a mainframe or minicomputer environment. Processes need to be developed to ensure that the distributed data bases are current and that users are all working from the same data.

The risks to be evaluated include:

1. Reviewing system capability and user procedures for editing input data, accurate processing, and reconciliation of output. The input, processing, and output controls are not unique to the LAN platform and are not covered in this article.

2. Interviewing the LAN administrator regarding the consideration given to the controls in place to prevent concurrent updates of data.

3. Reviewing software manuals to verify that controls are in place.

4. Reviewing procedures relating to the cataloging and updating of application programs.
5. Reviewing program documentation for accuracy and currency. Are system operating instructions documented?
6. Verifying that the equipment has been adequately tested, including benchmark testing to ensure that the systems have adequate capacity for the number of users and transaction volume.
7. Verifying that processes are in place to ensure that distributed data bases are current.

CONCLUSION

High-risk applications are being developed on LAN platforms. Management must assess the risk of these applications and implement appropriate controls. IS managers must verify that this assessment is performed and that adequate and effective controls are established based on the application risk. Although the steps required for assessing LAN controls are basically the same as those for assessing controls on most platforms, they should be tailored to match the unique characteristics of the LAN. The steps included in this chapter provide a starting point for the development of a security evaluation program required to assess controls on this important platform.

V-7

Adding Multimedia to the Corporate Network

Gilbert Held

THE USE OF IMAGES has recently moved off the individual PC workstation, where the images were most likely incorporated into word processing or desktop publishing applications, and onto network servers and mainframes. Thus images are now available for retrieval by virtually any employee with a PC connected to a local area network or to the corporate network.

Although most people associate Web servers with the Internet, the use of such servers has rapidly expanded to internal corporate networks. The term *intranet* refers to the application of TCP/IP (Transmission Control Protocol/Internet Protocol) protocols to include the World Wide Web Hypertext Transfer Protocol (HTTP) on internal corporate networks. Because almost all Web servers include a mixture of images and text, and some Web servers add audio and video clips for user access, the Web server represents a rapidly evolving source of multimedia activity on the corporate network.

In addition, the standardization of multimedia data storage has increased the ability of organizations to purchase or develop applications that merge audio, video, and data. There is therefore a growing trend of organizations adding images, as well as multimedia data, to applications. This chapter discusses methods of restructuring an existing network to cost-effectively accommodate the transportation of images and multimedia.

MULTIMEDIA

The term *multimedia* is a catchall phrase that refers to the use of two or more methods for conveying information. Thus, multimedia can include voice or sound (both collectively referred to as audio), still images, moving images, and fax images, as well as text documents. This means that multimedia can be considered an extension of image storage. To understand how multimedia data storage requirements differ from conventional data storage requirements, this chapter first focuses on the storage requirements of images.

Image Acquisition

Until recently, the scanner provided the primary method of acquiring images in a digital format that could be stored on and retrieved from computers. In the mid-1990s, a digital camera was introduced by several manufacturers. This type of camera uses a charged coupled device (CCD) to convert viewed images directly into pixels capable of being stored on a PC card inserted into the camera.

Although the cost of digital cameras was relatively high when they first reached the market, by late 1996 the retail cost for several medium-resolution models capable of acquiring images for use on Web pages was less than $1,000. This substantial decline in price, as well as the cameras' capability of capturing images with sufficient detail for display on computer monitors requiring only 72 lines per inch resolution, has considerably expanded use of digital cameras and obviated the need for scanners.

Image Storage Requirements

Images are converted into a series of pixels or picture elements by a digital camera or scanner. Software used to control the scanner will place the resulting pixels into a particular order based on the file format selected from the scanning software menu. When a digital camera is used, the captured image will be stored using a predefined image format supported by the camera. Some file storage formats require the use of compression before the image can be stored. Compression typically reduces data storage requirements by 50% or more.

Text and image data storage requirements differ greatly. A full page of text, such as a one-page letter, might contain approximately 300 words, with an average of five characters per word. Thus, a typical one-page text document would require 1,500 characters of data storage. Adding formatting characters used by a word processor, a one-page text document might require up to 2,000 characters, or 16,000 bits of data storage.

When an image is scanned, the data storage requirements of the resulting file depend on four factors. Those factors include the size of the image, the scan resolution used by a digital camera or by a scanner during the scanning process, the type of image being scanned—color or black and white—and whether the selected file format results in the compression of pixels before their storage.

To illustrate the data storage requirements of different types of images, let us examine the scanning of a 3×5 photograph. That photograph contains a total of 15 square inches that must be scanned.

A low-resolution black-and-white scan normally occurs using 150 lines per inch and 150 pixels per line per inch, where a pixel with a zero value represents white and a pixel whose value is one represents black. Thus, the total number of bits required to store a 3×5 black-and-white photograph using a low-resolution scan without first compressing the data would be 337,500 bits, which would result in a requirement to store 42,188 bytes of data. Thus, a 3×5 black-and-white photograph would require 42,188/2000, or approximately 21 times the amount of storage required by a one-page document.

Most scanners now consider 300 lines per inch with 300 pixels per line per inch, to represent a high-resolution scan. However, some newly introduced scan-

Type of Document/Image	Data Storage (Bytes)
Text of document containing 300 words	2,000
3"×5" B&W photograph scanned at 150 pixels/inch	42,188
3"×5" B&W photograph scanned at 300 pixels/inch	84,375
3"×5" B&W photograph scanned at 450 pixels/inch	126,563
3"×5" B&W photograph scanned at 600 pixels/inch	168,750

Exhibit V-7-1. Document Versus Image Storage Requirements

ning products now consider 300 lines per inch with 300 pixels per line to represent a medium- or high-resolution scan. Regardless of the pixel density considered to represent a medium- or high-resolution scan, the computation of the resulting data storage requirement is performed in the same manner. That is, storing the photograph would entail multiplying the number of square inches of the document—in this example, 15—by the pixel density squared. Exhibit V-7-1 compares the data storage requirements of a one-page text document to the data storage required to store a 3 x 5 black-and-white photograph at different scan resolutions.

To store an image in color, data storage requirements would increase significantly. For example, if a scanner supports color, each pixel would require one byte to represent each possible color of the pixel. Thus, a color image would require eight times the data storage of a black-and-white image when a scanner supports up to 256 colors per pixel. This means that the 3×5 photograph scanned at 300 pixels per inch would require 675,000 bytes of storage. Similarly, a 3×5 color photograph scanned at a resolution of 600 pixels per inch would require 1.35M bytes of storage, or 675 times the amount of storage required for a 300-word one-page document.

When considering the effect of color, it is important to note that the default scan or camera resolution often is for what is referred to as *true color*, in which 24 bits are used to represent the color that can be assigned to each pixel. This represents approximately 16.7 million color combinations. Because use of true color triples the storage requirements of an image stored using 256 colors per pixel, a 3×5 true-color photograph would require more than 200 times the amount of storage required by a 300-word one-page document. In addition, when the image is downloaded, the file transfer time is also tripled. Although true color may be valuable for displaying images in a glossy magazine, it is usually beyond the viewability of most users whose monitors are limited to displaying 16 or 256 distinct colors.

Without considering the effect of data compression, the transmission of images on a network can require from 20 to more than 2,000 times the amount of time required to transmit a one-page standard text document. Obviously, the transmission of images by themselves or as a part of a multimedia application can adversely affect the capability of a network to support other users in an efficient manner unless proper planning precedes the support of the multimedia data transfer.

Audio Storage Requirements

The most popular method of voice digitization is known as pulse code modulation (PCM), in which analog speech is digitized at a rate of 64K bps. Thus, one minute of speech would require 480,000 bytes of data storage. At this digitization rate, data storage of digitized speech can easily expand to require a significant portion of a hard disk for just 10 to 20 minutes of audio.

Multimedia applications developers do not store audio using PCM. Instead, they store audio using a standardized audio compression technique that results in a lower level of audio fidelity but significantly lowers the data storage requirement of digitized audio.

Today, several competing multimedia voice digitization standards permit speech to be digitized at 8K bps. Although this is a significant improvement over PCM used by telephone companies to digitize voice, it still requires a substantial amount of disk space to store a meaningful amount of voice. For example, one hour of voice would require 2.8M bytes of data storage. Thus, data storage of audio is similar to video, in that a meaningful data base of images and sound must either be placed on a CD-ROM or on the hard disk of a network server or mainframe computer that usually has a larger data storage capacity than individual personal computers.

Image Utilization

In spite of the vast increase in the amount of data that must be transported to support image applications, the use of imaging is rapidly increasing. The old adage "a picture is worth a thousand words" is especially true when considering many computer applications. Today, several network-compliant data base programs support the attachment of image files to data base records. Using a Canon digital camera or similar product, real estate agents can photograph the exterior and interior of homes and transfer the digitized images to the network server upon their return to the office. When a potential client comes into the office, an agent can enter the client's home criteria, such as the number of bedrooms, baths, price range, school district, and similar information, and have both textual information as well as photographs of the suitable homes meeting the client's criteria displayed on the screen. This capability significantly reduces the time required to develop a list of homes that the client may wish to view firsthand.

Audio Utilization

The primary use of audio is to supplement images and text with sound. Unlike a conventional PC, which can display any image supported by the resolution of the computer's monitor, the use of audio requires specialized equipment. First, the computer must have a sound board or specialized adapter card that supports the method used to digitize audio. Second, each computer must have one or more speakers connected to the sound board or speech adapter card to broadcast the resulting reconverted analog signal.

STORING IMAGES ON A LAN SERVER OR MAINFRAME

There are three methods for storing images on a LAN server or mainframe. First, images can be transferred to a computer's hard disk using either a personal computer and scanner, or, by connecting the PC to a digital camera. Another method for storing images involves forwarding each image after it is scanned or transferred from a digital camera or similar device. A third method for placing images on a file server or mainframe is based on premastering a CD-ROM or another type of optical disk.

Transferring a large number of images at one time can adversely affect network users. To minimize the effect, images can be transferred from a computer to a server or mainframe after normal work hours. As an alternative, a removable hard disk can be used, permitting movement of the disk to a similarly equipped network server to transfer images without affecting network users.

Forwarding Images After Scanning

This method of storing images on the LAN has the greatest potential for negative impact on network users. Some document scanners are capable of scanning several pages per minute. If a large number of images were scanned, transferring the digitized images through a local or wide area network connection could saturate a network during the scanning and transferring process. This happens because, as other network users are transferring a few thousand bytes, transferring images containing 20 to 2,000 times more data would consume most of the network bandwidth for relatively long periods of time. Thus, the effect of image transfer can be compared to the addition of 20 to 2,000 network users transferring one-page documents, with the actual addition based on the resolution of the images, and whether they are in black and white or color.

Premastering

Premastering a CD-ROM or other type of optical disk permits images to become accessible by other network users without adversely affecting network operations. However, cost and ease of image modification or replacement must be weighed against the advantage of this method.

From a cost perspective, equipment required to master images on a CD-ROM will cost between $750 and $1,500. In comparison, the use of conventional magnetic storage on the server or mainframe can avoid that equipment cost as well as the cost of a CD-ROM drive connected to a file server.

Concerning ease of image modification or replacement, CD-ROM data cannot be modified once a disk is mastered. This means that a new CD-ROM disk must be mastered each time previously stored images or data must be modified or replaced.

If real-time or daily updates are not required, this method of image placement on a network server or mainframe should be considered. The time required to master a CD-ROM disk has been reduced to a few hours, when mastering occurs on a 486- or Pentium-based personal computer. Also, write-once CD-ROM disks now cost under $10. Thus, a weekly update of an image data base could be performed for a one-time cost of between $750 and $1,500, and $10 per week for each write-once CD-ROM disk used. Because this method would have no negative impact on network operations, its cost would be minor in comparison to the cost of modifying a network.

The next section focuses on methods used to provide access to images stored in a central repository. Use of both LAN and WAN transmission facilities are examined, with several strategies for minimizing the effect of image transfers on network users.

ACCESSING IMAGES

Once a decision is made to add images to a data base or other application, the potential effect of the retrieval of images by network users against the current organizational network infrastructure must be examined. Doing so provides the information necessary to determine if an existing network should be modified to support the transfer of images, as well as data.

To illustrate the factors that must be considered, this example assumes that images are to be added to a 50-node LAN server. The network server is attached to a 10M-bps Ethernet network as illustrated in Exhibit V-7-2a, with images placed on a CD-ROM jukebox connected to the server.

Based on an analysis of the expected use of the LAN, 15 stations were identified that are expected to primarily use the images stored on the CD-ROM jukebox. The other 35 network users are expected to casually use the CD-ROM jukebox, primarily using the LAN to access other applications on the server, such as workgroup software and other application programs, including a conventional text-based data base and electronic mail.

One method of minimizing the contention for network resources between network users is obtained by segmenting the network. Exhibit V-7-2b illustrates the use of a local bridge to link separate networks. In this illustration the 35 network users expected to have a minimal requirement for image transfers are located on one network, and the remaining 15 network users who have a significant requirement to transfer images are placed on a second network. The use of the bridge permits users of each network to access applications stored on the file server on the other network. However, this new network structure segments network stations by their expected usage, minimizing the adverse effect of heavy image transfer by 15 users on what was a total network of 50 users.

INTERNETWORKING

The segmentation of an Ethernet LAN into two networks linked together by a local bridge created an internetwork. Although the network structure was created to

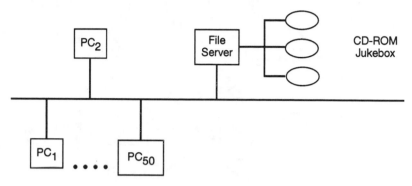

a. Initial Network

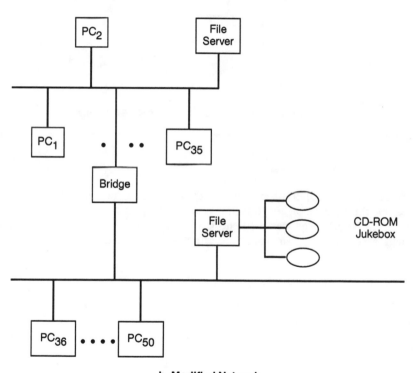

b. Modified Network

Exhibit V-7-2. Modifying an Ethernet Network

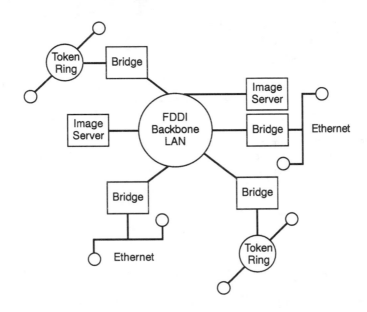

Exhibit V-7-3. Using a High-Speed FDDI Backbone

minimize the effect of transporting images on a larger network, this method of increasing the volume of image traffic through bridges that directly interconnect separate LANs can produce a bottleneck and inhibit the flow of other traffic, such as client/server queues, E-mail, and other network applications.

Placing Images on Image Servers

When constructing a local internetwork consisting of several linked LANs within a building, one method to minimize the effect of image traffic on other network applications is to place image applications on image servers located on a separate high-speed network.

Exhibit V-7-3 illustrates the use of an FDDI backbone ring consisting of two image servers whose access is obtainable from workstations located on several Ethernet and token ring networks through local bridges linking those networks to the FDDI ring. By using the FDDI ring for image applications, the 100M-bps operating rate of FDDI provides a delivery mechanism that enables workstation users on multiple lower operating rate LANs to simultaneously access image applications without experiencing network delays.

For example, one network user on each LAN illustrated in Exhibit V-7-3 accesses the same image application on an image server connected to the FDDI backbone LAN. If each token ring network operates at 16M bps and each

Ethernet operates at 10M bps, the composite transfer rate from the FDDI network to each of the lower operating rate LANs bridged to that network is 52M bps. Since the FDDI network operates at 100M bps, it can simultaneously present images to network users on each of the four LANs without any internetwork bottlenecks occurring.

Another advantage associated with using an FDDI backbone restricted to supporting image servers and bridges is economics. This configuration minimizes the requirement for using more expensive FDDI adapter cards to one card per image server and one card per bridge. In comparison, upgrading an existing network to FDDI would require replacing each workstation's existing network adapter card with a more expensive FDDI adapter card.

To illustrate the potential cost savings, assume each Ethernet and token ring network has 100 workstations, resulting in a total of 400 adapter cards, including two image servers that would require replacement if each existing LAN was replaced by a common FDDI network. Since FDDI adapter cards cost approximately $800, this replacement would result in the expenditure of $320,000. In comparison, the acquisition of four bridges and six FDDI adapter cards would cost less than $20,000.

Using Switching Hubs

Another option for minimizing the effect of image transfers on network users is offered by switching hubs. Unlike a shared media network in which only one network user can communicate at any one time, use of a switching hub permits multiple communications through the hub as long as the source and destinations of each communications session differ. Thus an n-port switching hub could theoretically support n/2 sessions. This means that an eight-port Fast Ethernet hub with each port operating at 100M bps could support a maximum throughput of 400M bps, because four simultaneous communications sessions could occur through that switching port.

Exhibit V-7-4 illustrates the use of a six-port switching hub to provide a communications capability for users on three Ethernet LANs to an image server and a conventional network server. Note that the image server has two connections to the hub. This lets two network users on different LAN segments simultaneously access images on the image server while another network user accesses the conventional server.

The cost of four- to eight-port switching hubs ranges from $500 to $1,000 per port, making switching hubs an economical alternative to using a backbone LAN to enhance access to LAN-based images.

TRANSFERRING IMAGES THROUGH WIDE AREA NETWORKS

In the next example, a group of PC users requires the use of a WAN to access images on a data base at a remote location. Images are placed on a CD-ROM jukebox connected to a server on LAN A, which in turn is connected to LAN B through a pair of remote bridges operating at 64K bps. This network configuration is illustrated in Exhibit V-7-5.

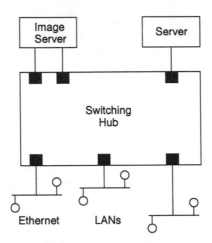

Exhibit V-7-4. Use of a Switching Hub

If users on network A access several applications on network B and vice versa, in addition to accessing the images stored on the CD-ROM jukebox on network A, what happens when a user on network B attempts to access text data on network A during an image transfer? If another network B user requested an image transfer, the user requesting a text transfer is now contending for network resources with the user performing the image transfer. This means that alternate frames of data flow over the 64K-bps transmission facility—first a frame containing a portion of an image, then a frame containing a portion of the text transfer. This alternate frame transmission continues until one transfer is completed, prior to all network resources becoming devoted to the remaining transfer.

Thus, not only is the 64K-bps transmission rate a significant bottleneck to the transfer of images, but WAN users must contend for access to that resource. A 640K-byte image would require 80 seconds to transfer between remotely located LANs on a digital circuit operating at 64K bps and devoted to a single remote user. If that remote user had to share the use of the WAN link with another user performing another image transfer, each transfer would require 160 seconds. Thus, transferring images through a WAN connection can result in a relatively long waiting time. Although the WAN connection could be upgraded to a T1 or a fractional T1 circuit, the monthly incremental cost of a 500-mile 64K-bps digital circuit is approximately $600. In comparison, the monthly cost of a 500-mile 1.544M-bps digital circuit would exceed $4,200.

Localizing Images

One alternative to problems associated with the transfer of images through a WAN can be obtained by localizing images to each LAN to remove or substantially reduce the necessity to transfer images through a WAN. To do so with

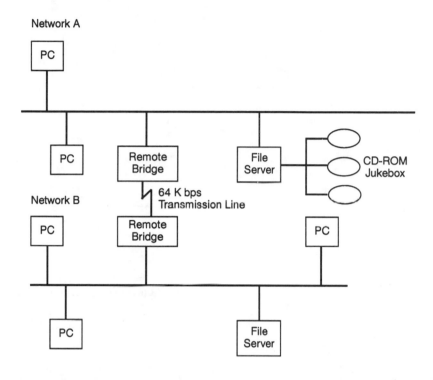

Exhibit V-7-5. Image Transfers Using a WAN Link

respect to the network configuration illustrated in Exhibit V-7-5 would require the installation of either a single CD-ROM drive or a CD-ROM jukebox onto network B's file server. This would enable network users on each LAN to obtain the majority of the images they require through a LAN transmission facility that normally operates at 10 to 100 times the operating rate of most WAN transmission facilities. The placement of additional image storage facilities on each LAN can substantially reduce potential WAN bottlenecks by reducing the need to transfer images via the WAN.

Bandwidth-on-Demand Inverse Multiplexers

A second method of reducing WAN bottlenecks caused by the transfer of images is obtained by the use of bandwidth-on-demand inverse multiplexers. Several vendors market bandwidth-on-demand inverse multiplexers that can monitor the utilization of a leased line and initiate a switched network call when a predefined lease line utilization threshold is reached.

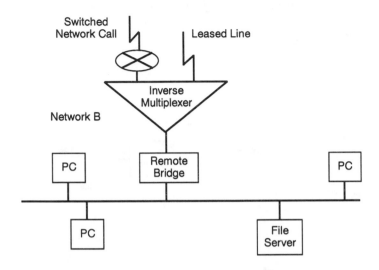

Exhibit V-7-6. Using a Bandwidth-on-Demand Inverse Multiplexer

Exhibit V-7-6 illustrates the use of a bandwidth-on-demand inverse multiplexer at the network B location shown in Exhibit V-7-5. Under normal operating conditions, a 64K-bps leased line connects network A to network B. When the transfer of images begins to saturate the use of the leased line, one inverse multiplexer will automatically initiate a call over the switched network to the other multiplexer. That call can be a switched digital call at 56/64K bps or a call over the public switched telephone network, in which the data transfer operating rate depends on the type of modems used with each inverse multiplexer. Because a switched digital or analog call costs between 10 and 25 cents per minute, the use of inverse multiplexers can represent an economical alternative to the use of additional or higher-speed leased lines when image transfers only occur periodically during the workday.

CONCLUSION

Since multimedia includes either digitized images, digitized speech, or both, the methods and techniques described in this chapter for handling images are applicable for multimedia. Thus, the segmentation of a local area network, use of a high-speed backbone network for providing access to image servers, use of switching hubs, or addition of multimedia storage facilities on individual LANs to reduce WAN traffic are all applicable to the transfer of multimedia information.

Placing images and multimedia on the corporate network can be considered equivalent to the addition of a very large number of network users. When planning to add access to image and multimedia data bases, IS managers should use the same planning process required to support conventional access to file servers and mainframe data bases. When data transfer requirements begin to adversely affect network performance, managers should consider transferring multimedia data to storage repositories and accessing it through the methods suggested in this chapter. The goal at all times is to avoid burdening network users while remaining able to support an organization's image and multimedia data base access requirements in an efficient and cost-effective manner.

V-8

Managing Hardware in a Client/Server Environment

John P. Murray

MANY IS PROFESSIONALS believe that client/server processing is less expensive than processing data on a mainframe. Client/server computing can ultimately be less costly, but moving to a client/server environment requires a considerable up-front investment. IS managers should carefully analyze the investment involved in moving to client/server processing. Estimates of implementation costs, lacking an objective analysis of all factors involved, are almost always lower than actual costs. Preparing senior management for the actual cost of client/server processing makes good business sense.

For IS managers, a key issue in moving to client/server processing is the use of hardware. The important questions concerning the management of hardware resources are not only about the client/server environment but about the mainframe environment during the transition to client/server computing. The answers to these questions carry consequences that will affect an organization long term.

IS managers should be aware of the negative consequence of rushing into client/server processing and bring these risks to the attention of senior management. The potential to incur excessive project expense is only one risk; others are:

- Failure of the client/server project.
- A disruption of customer service.
- Selecting the wrong technology and therefore having to begin the project again.
- A negative effect on morale throughout the organizations.

IS managers should not try to stall the introduction of client/server technology, but present a realistic appraisal of the expense and risk associated with moving to it. By rushing into client/server processing, an organization can increase expense rather than reduce it. The watchword for moving to client/server processing is planning. IS managers must actively participate in planning the move to a client/server platform. This chapter discusses hardware issues that IS managers must consider as part of a well-thought-out plan for moving to client/server processing.

INITIAL HARDWARE COSTS

Hardware is the largest single expense in the move to client/server technology. This hardware includes:

- File servers.
- Workstations.
- File backup equipment.
- Printers.

A client/server system is often linked to a mainframe, and this connection requires such equipment as bridges, routers, and gateways.

Mainframe Costs

The expense associated with the current data center mainframe is not the baseline cost for the client/server environment. For example, additional software must be added to the mainframe operating system to handle the transfer of data between the mainframe and the client/server hardware. Additional hardware and software may also be required to accommodate the transfer of mainframe data.

Also to be carefully considered is the adequacy of the installed mainframe's processing capacity. Because of additional software loaded onto it, the mainframe may not be able to accommodate data transfers between it and the client/server environment, and the mainframe may have to be upgraded. Such an upgrade must be calculated into the cost of the transition.

During the transition and afterward, at least the current mainframe expense will have to be carried as well as the cost of client/server hardware. So, the total IT-related expense increases rather than decreases. There are many sound business reasons for moving to a client/server environment, but an immediate savings in data center hardware costs is not one of them.

Client/Server Costs

Although the cost of client/server hardware is less than the price of a mainframe with comparable processing power, the total cost of a transition to client/server processing is still substantial. The total expense may shock senior managers who are responsible for approving expenditures. Data center managers have a responsibility to senior management to provide information as accurate as possible about the cost of moving to a client/server environment. The only way those expenses can be identified is through hard work and careful investigation.

ANTICIPATING GROWTH

IS managers should make certain that hardware proposed can handle rapid growth. Excess capacity does not have to be installed at the onset of a project, but provisions for future growth must be made. The budget for a client/server project must include funds for the purchase of increased processing capacity when required. Recommendable is an approach that increases processing capacity incrementally.

There are several reasons for taking this phased approach. One is avoiding additional expense. Another is a lack of experience in determining the level of processing capacity required of a client/server system.

Future Client/Server Needs

In many instances, methods for producing estimates are, at best, educated guesses. Client/server technology is still new. As a result, many tools and methods used for client/server capacity planning and response time are still immature, though vendors are constantly improving methods for forecasting client/server capacity loading. IS managers should include growth figures into any client/server capacity planning model. The best approach is to double or triple the original estimate. IS managers should not bet their careers on client/server capacity forecasting tools currently available.

Work load processing capacities should not be underestimated because of rapid changes in microcomputer technology. New microcomputer software often requires more processing power. Although the additional expense of a new or advanced version of a software product must be considered, the decision to acquire the new software may not rest with the IS manager. However, IS managers should anticipate the growth of the client/server system. If department managers cannot purchase more powerful software because the cost of the additional hardware required to run that software was not adequately anticipated, IS managers will find themselves open to criticism. Also, if additional processing capacity has not been adequately anticipated, data center service levels will be lowered or senior management will have to be asked for additional money. Neither situation bodes well for any IS manager.

Managers must also understand different client/server configurations. Costs differ if a mainframe does the majority of processing (i.e., a host-based system) or if most of the processing is done by the workstation or server (i.e., a two- or three-tiered system).

IS managers have been able to successfully forecast mainframe work load capacities. The combined use of forecasting tools, experience, and intuition has enabled organizations to address mainframe work load growth. However, it has taken some IS managers a long time to learn how to avoid running out of processing power and going to senior management with unplanned demands for more funds. Because of rapid growth of processing demands in many client/server environments, the topic of adequate hardware expense forecasting and control is often a highly charged, emotional issue. IS managers should explore the use of techniques for forecasting hardware capacity requirements in client/server environments.

STANDARDS FOR CLIENT/SERVER HARDWARE

IS managers should establish a set of hardware standards to be used in forming decisions about hardware purchases and use. These standards should be based on the criteria examined in the following sections.

State of Technology. Moving to and maintaining a state-of-the art environment is a practical business decision for those companies that want to use client/server processing. To develop that environment, state-of-the-art microcomputer hardware and software must be made available within the organization.

Redundancy. Redundancy can affect expense, service levels, and customer service because it involves systems availability and network response time. Because client/server hardware costs less than traditional mainframe hardware, managers should push for hardware redundancy.

Redundancy can be provided at the server level by duplicating the number of servers and disk mirroring, which double the cost of the server hardware. However, that cost must be weighed against the value of the organization's ability to continue computing in the case of a server failure. Another way to provide redundancy is to purchase spare servers and extra disk capacity. One or more extra servers that meet the data center's standards are purchased and stored for future use. If a server fails, the data center staff can quickly install a spare server to bring the LAN back online as quickly as possible. This process is not as effective as the use of fully redundant servers, and it causes a higher level of disruption. However, if expense is a constraint to full redundancy, a spare server is a reasonable alternative.

Inventory Control. In many client/server sites, staff members have no more than a vague notion about the number and types of PCs installed. Although some information about an organization's microcomputers can be obtained by consulting the fixed asset records in its accounting department, the IS manager should develop and maintain sound PC inventory records. At a minimum, a microcomputer inventory record should contain the following information:

- Make, model, and size of each microprocessor.
- Date of purchase.
- Cost and depreciation schedule.
- The location of the unit.
- The department and the name of the person who uses the microcomputer.
- Serial number.

Inventory records must be developed and accurately maintained in the data center for two reasons. First, microcomputers can easily be moved or stolen, and inventory records can ensure that all microcomputers can be accounted for and have a better chance of recovery if stolen. Second, inventory records enable the data center to place in other departments microcomputers that one department has replaced. This reuse of microcomputer hardware is not only a sound business practice but improves the reputation of IS.

Controlling Purchases of Microcomputers. A manager should identify not more than two microcomputer hardware vendors with which the organization will do business. The simple issue of compatibility can become complex if an organization mixes vendors. Also, an IS manager should select vendors that have a good chance of remaining in business. To control the hardware selection process, a manager must start with an agreement with senior management that all microcomputer hardware purchases must be approved by the IS manager.

Hardware Portability and Scalability. The ability to shift various pieces of processing hardware (e.g., workstation and servers) from one location to another provides great flexibility in having processing power wherever it may be needed. The ability to scale up hardware in increments as more server processing power is needed provides increased flexibility and controls expense. In addition, because of rapid advances in processing power, usually at less overall cost, moving the existing server to some other location in the company and replacing it with a larger unit can prove to be a good business decision.

The development and implementation of microcomputer standards mark the start of controlling PC hardware. Before the standards are published, IS managers should gain the support of senior managers in anticipation of resistance to the enforcement of the standards. Senior management's authority is needed to overcome any such resistance. Some organizations are currently attempting to deal with the failure in enforcing microcomputer standards and have realized the value of such standards.

Client/Server Infrastructure Expenses

Another aspect of client/server hardware costs are infrastructure expenses involved in moving to client/server processing. These expenses are tied to physically preparing a facility where the hardware will be located. Some infrastructure issues to be considered include:

- Is the area appropriately wired to accommodate the placement of additional workstations and servers? If not, what actions will have to be taken to prepare for the client/server hardware?
- What standards should be set for the wiring of the location? The wiring that is installed should be able to meet the anticipated growth of user demands as well as accommodate the demands of new technology. Because the largest expense in wiring is not the cost of the wire but the labor associated with its installation, an organization should carefully plan for this activity. Wiring also involves network response and reliability issues. For example, the

higher the capacity of the wiring, the more likely the network is to perform better.

- Can the installed telephone switch handle both current and probable future demands for increased communications capability? Part of the planning process for the client/server environment should be to carry out telephone traffic studies. On the basis of those studies, increases in traffic demand during the next several years should be predicted.
- Although server hardware usually does not require a great deal of space, the hardware should be located in a secured area, where access can be restricted. A provision must be made for adequate wiring closet space for each LAN installed. Again, these closets, which may occupy the same area as the server hardware, must be designed to limit access.

CONCLUSION

Although experienced IS managers readily understand the hardware issues involved in the management of a mainframe environment, moving to a client/server system presents new challenges and opportunities. Much knowledge about appropriately managing client/server systems is gained through trial and error. Necessary management tools will become available over time just as they have for mainframes. In the meantime, IS managers are advised to take a cautious approach in the development of a client/server environment.

Controlling client/server hardware expense and developing, implementing, and enforcing appropriate hardware standards are mandatory. Some of the same battles that had to be fought in building strong mainframe functions will be fought again. Establishing an effective client/server environment is not easy, but IS managers who do the job right will find that doing so has been worth the struggle.

V-9

Operating Standards for LANs

Leo Wrobel

THE FOLLOWING SCENARIO is common in many organizations: There are 200 local area networks (LANs) located across the country, in everything from small sales offices with a handful of people to regional distribution centers. The company does not know if these outlying locations handle mission-critical data. The company does not know with certainty who is running these LANs, because staffing ranges from office managers and clerical employees right up to seasoned IS professionals. A site that once had 10 salespeople now has 9 salespeople and a LAN administrator. The company does not know how these sites are buying equipment, yet it is reasonably sure that they are paying too much, because they are not buying in bulk or enjoying any economies of scale in equipment purchases.

Locations are beginning to lean on IS for help desk support because there is no way they can keep up with the rapid proliferation of hardware platforms, software, and special equipment being installed in the field. The telecommunications department is worried about connecting all of these locations together.

Although some attempts at standardization of these locations may be made, LAN managers in the field invariably consider standards to be an attempt by the IS department to regain control of the LAN administrators' environment. Because LAN managers seldom have had any input into what these standards would be, they were soundly rejected.

Today, there are literally thousands of companies fighting this same battle. This chapter offers some solutions to these problems. First, however, it is important to understand why standards are required and how IS can implement standards without stifling productivity or adversely affecting the organization.

WHY LANS REQUIRE STANDARDS

In an ideal environment, the LAN administrator can select exactly the type of equipment best tailored to do the job. LAN managers are historically close to the core business. For example, if the company is involved in trading stock, the LAN operations department can go out and buy equipment tailored exactly to trading stock. If the organization is engaged in engineering, the LAN administrator can buy equipment exactly tailored to engineering.

From the standpoint of operational characteristics, LANs are far more desirable than mainframes because they are closer to the business, they empower people, and they make people enormously productive by being close to the core business. This is not the whole story, however. It is equally as important to support LANs once they are in place. This is where the trade-offs come in.

Lessons from Mainframe Experience

Because mainframes have been around so long, there is a high degree of support available. When users in the mainframe environment call the help desk with a hardware or a software problem, the help desk knows what they are talking about. Help desk staff are well trained in the hardware and the software packages and can quickly solve the users' problems.

As another example, in an IBM 3070 terminal environment, 100 terminals or more could be supported by a single technician. When those terminals became PCs, the ratio perhaps dropped to 50 PCs per technician. When those PCs became high-end workstations, the ratio dropped even further. The value of a mainframe level of technical support cannot be underestimated.

Mainframe professionals had 20 years to write effective operating and security standards. These standards cover a number of preventive safeguards that should be adopted in the operational environment to ensure smooth operation. These range from:

- How often to change passwords.
- How often to make backups.
- What equipment should be locked up.
- Who is responsible for change control.
- Defining the standards for interconnecting between environments.

In the mainframe world it was also easy to make very large bulk purchases. Because the mainframe has been around for so long, many advanced network management systems exist that provide a high degree of support and fault isolation.

Balancing Productivity and Support Requirements for LANs

Because LAN platforms are relatively new, in comparison to mainframes, there has not been as much time to develop operating and security standards. This is especially irritating to auditors when mission-critical applications move from the traditional mainframe environment onto LANs and the protective safeguards around them do not follow. Something as simple as transporting a tape backup copy of a file between LAN departments can be extremely complicated without standards. What if everyone buys a different type of tape backup unit? Without standards on what type of equipment to use, bulk purchases of equipment become difficult or impossible.

Even though major improvements have been made in network management systems over the past five years, the management systems associated with LANs often lag behind those associated with mainframe computers. Again, this causes the company to pay penalties in the area of maintenance and ease of use.

One answer, of course, is to force users into rigid standards. Although this pays a handsome dividend in the area of support, it stifles the users' productivity. They need equipment well suited to their core business purpose.

An alternative is to let users install whatever they want. This may increase productivity greatly, though it is doubtful that a company could ever hire and support enough people to maintain this type of configuration. Worse, mission-critical applications could be damaged or lost altogether if users are not expected to take reasonable and prudent safeguards for their protection.

It is the responsibility of both users and technical staff to find the middle ground between the regimented mainframe environment and the seat-of-the-pants LAN environment. Through careful planning, it is possible to configure a set of standards that offers the advantage of greater productivity that is afforded by LANs, but also the advantages learned through 20 years of mainframe operations in the areas of support, bulk purchases, and network management.

The remainder of this chapter concentrates on exactly what constitutes reasonable operating and security procedures for both LANs and telecommunications.

STANDARDS COMMITTEES

One method for establishing LAN standards is through the formation of a communications and LAN operating and security standards committee. An ideal size for a standards committee would be 10 to 12 people, with representatives from sales, marketing, engineering, support, technical services (including LANs), IS and telecommunications, and other departments. It is important to broaden this committee to include not only technical staff, but also people engaged in the core business, since enhancement of productivity will be a key concern.

The actual standards document that this committee produces must deal with issues for both the operation and protection of a company's automated platforms (Exhibit V-9-1 provides a working table of contents from which to begin to write a document). Subjects include:

- Basic physical standards, including access to equipment rooms, where PBX equipment is kept, what type of fire protection should be employed, standards for new construction, standards for housekeeping, and standards for electrical power.
- Software security, change control, which people are authorized to make changes, and how these changes are documented.
- The security of information, such as identifying who is allowed to dial into a system, determining how to dispose of confidential materials, determining which telephone conversations should be considered private, and the company's policy on telecommunications privacy.

At a minimum, an operating and security standards document should incorporate the following areas:

I. Objective: Defining Mission Critical
    1. For Non-Mission-Critical Support Systems
    2. For Mission-Critical Support Systems

II. Physical Security
    1. For Non-Mission-Critical
    2. For Mission-Critical

III. Operational Support Issues
    1. Standards for All LAN and Telecommunications Installations
    2. Documentation Standards for Software and Application
    3. Server and PBX Class-of-Service Indicator Backups

IV. Access Control
    1. Procedures for Passwords

V. Change Control Policy and Procedures

VI. Virus Protection Procedures

VII. Disaster Recovery Procedures
    1. For Non-Mission-Critical Equipment
    2. For Mission-Critical Equipment

Exhibit V-9-1. Sample Operating and Security Standards Document Table of Contents

---

- Weighing options with regard to technical support of equipment.
- Resolving issues regarding interconnection standards for the telecommunications network.
- Disaster backup and recovery for both LANs and telecommunications, including defining what users must do to ensure protection of mission-critical company applications.

Defining "Mission Critical"

Before all of this, however, the committee is expected to define and understand what a mission-critical application is. Standards are designed to cover both operational and security issues, so the business processes themselves must be defined to avoid imposing a heavy burden of security on users who are not engaged in mission-critical applications, or not imposing a high enough level of security on users who are.

Standards for equipment that is not mission critical are relatively easy. In practice, this means securing the area in which the equipment resides from unauthorized access by outside persons when there is danger of tampering or theft. It also includes avoiding needless exposures to factors that could damage the equipment, such as water and combustibles, and controlling food items around the equipment, such as soft drinks and coffee.

Mission-critical equipment, however, has a value to the company that far exceeds the value of the equipment itself, because of the type of functions it supports. Determination of what constitutes a mission-critical system should be made at a senior management level.

LAN and telecommunications equipment that supports an in-bound call center for companies such as the Home Shopping Club would definitely be mission-critical equipment, because disruption of the equipment, for whatever cause, would cause a financial hit to the company that far exceeds the value of the equipment. Therefore, mission-critical equipment should be defined as equipment that, if lost, would result in significant loss to the organization, measured in terms of lost sales, lost market share, lost customer confidence, or lost employee productivity.

Monetary cost is not the only measurement concerning mission-critical. If an organization supports a poison-control line, for example, and loss of equipment means a parent cannot get through when a child is in danger, it has other implications. Because financial cost is a meaningful criteria to probably 90% of the companies, it is the measurement used for purposes of this discussion.

There is not necessarily a correlation between physical size and mission criticality. It is easy to look at a LAN of 100 people and say that it is more mission-critical than another LAN that has only 4 people. However, the LAN with 100 people on it may provide purely an administrative function. The LAN with four people on it may have an important financial function.

WRITING THE OPERATING AND SECURITY STANDARDS DOCUMENT

The following approach recommends that two distinct sets of standards be created for mission-critical versus non-mission-critical equipment.

Network Software Security and Change Control Management

One item that should be considered in this section is, Who is authorized to make major changes to LAN or telecommunications equipment?

There is a good reason to consider this question. If everyone is making major changes to a system, a company is inviting disaster, because there is little communication concerning who changed what and whether these changes are compatible with changes made by another person. Standards should therefore include a list of persons authorized to make major changes to a mission-critical technical system. It should also have procedures for changing passwords on a regular basis, both for the maintenance and operation functions of LANs and telecommunications. Procedures should be defined that mandate a back-up before major changes in order to have something to fall back on in case something goes wrong.

Procedures should be established to include DISA (direct inward system access). Unauthorized use of DISA lines is a major cause of telecommunications fraud or theft of long-distance services. Automated attendants, for example, should also be secured and telephone credit cards properly managed. As a minimum, establish a procedure that cancels remote access and telephone credit to employees who leave the company.

Physical and Environmental Security

There should be a set of basic, physical standards for all installations, regardless of their mission-critical status. These might include use of a UPS (uninterruptible power supply) on any LAN server. A UPS not only guards against loss of productivity when the lights flicker, but also cleans up the power somewhat and protects the equipment itself.

There should be standards for physically protecting the equipment, because LAN equipment is frequently stolen and because there is a black market for PBX cards as well. There should be general housekeeping standards as far as prohibitions against eating and drinking in equipment areas and properly disposing of confidential materials through shredding or other means. No-smoking policies should be included. Standards for control of combustibles or flammables in the vicinity of equipment should also be written.

Physical standards for mission-critical applications are more intensive. These might include sign-in logs for visitors requiring access to equipment rooms. They may require additional physical protection, such as sprinkler systems or fire extinguishers. They may require general improvements to the building, such as building fire-resistant walls. They should also include protection against water, since this a frequent cause of disruption, either from drains, building plumbing, sprinklers, roof leaks, or other sources.

Technical Support

The standards committee ideally should provide a forum for users to display new technologies and subject them to a technical evaluation. For example, LAN managers or end users may find a new, innovative use of technology that promises to greatly enhance productivity in their department. They can present this new technology to the standards committee for both productivity and technical evaluations. The technologist on the committee can then advise users of the feasibility of this technology; whether it will create an undue maintenance burden, for example, or whether it is difficult to support.

If it is found that this equipment does indeed increase productivity and that it does not create an undue maintenance burden, it could be accepted by the committee and added to a list of supported services and vendors that is underwritten by the committee. Other issues include what level of support users are required to provide for themselves, what the support level of the help desk should be, and more global issues, such as interconnection standards for a corporate backbone network and policies on virus protection.

CONCLUSION

The LAN operating and securities standards document is designed to be an organization's system of government regarding the conduct and operation of technical platforms supporting the business. A properly written standards document includes input from departments throughout the organization, both to enhance productivity and to keep expenses for procurement, maintenance, and support under control. Standards also ensure that appropriate preventive safeguards are undertaken, especially for mission-critical equipment, to avoid undue loss of productivity, profitability, or equity to the company in the event something goes wrong. In other words, they are designed to prevent disruptions.

Use of a LAN operating and security standards committee is advised to ensure that critical issues are decided by a group of people with wide exposure within the company and to increase ownership of the final document throughout the organization. If properly defined, the standards document will accommodate the advantages of the mainframe environment and needs of LAN administrators by finding the middle ground between these operating environments. By writing and adopting effective standards, an organization can enjoy the productivity afforded by modern LAN environments while at the same time enjoying a high level of support afforded through more traditional environments.

Exhibit V-9-1 lists examples of typical standards for these types of installations. Readers are recommended to use them as a baseline in developing standards.

V-10

Cost Allocation for Enterprisewide Network Chargeout

Keith A. Jones

MOST EFFICIENTLY MANAGED data centers provide services involving controlled access to information system resources. The costs of providing these services to a business enterprise are recovered by a cost allocation or chargeback billing system that distributes these costs to business units. Once suitable cost allocation and chargeback systems are established for the host processing environments, operational expenses can be allocated across data communications network infrastructures in a similar fashion. Such allocation both reinforces and benefits from economies of scale that can be coordinated and controlled from the central-site host data center.

The pricing methodology for the transfer of costs to the business enterprise customer are based on various alternatives. These alternatives range from direct cost recovery based on measured billing by resource to more indirect recovery methods that distribute system cost according to negotiated agreements. The method selected is largely determined by the functional benefit of services to the customer's business enterprise.

This range of pricing methods places most data centers somewhere between a cost center on the one hand and a profit center on the other. It is important to recognize that regardless of the pricing policy used, the ultimate value of any data center operation and its service is based on how well it improves the efficiency and cost-effectiveness of the customer business units it serves.

NETWORK COST ALLOCATION STRATEGIES

In the not-too-distant past, most IS managers were able to regard their internal business enterprise clients as an essentially captive customer base. When central host mainframe computer center operations was the only data processing alternative, the reasonable assumption was that most of the system hardware costs could be recovered using straight-line methods. Such financing methods involved a level of risk that was low enough to consider payback over a relatively long term.

When dedicated links to the central host mainframe were the only option for internal business enterprise customers, the most important decision was what business function to automate, not how to automate a function. Allocation of operational costs was based on transfer pricing methods that encouraged conservation of resources within predictable system capacity levels and no foreseeable drop in information access demand.

With the availability of desktop workstations that have the machine-speed equivalent of large host mainframes and with new telecommunications technology that supports Internet and World Wide Web dial-up access to any open architecture system in the world, the situation has clearly changed. No longer can managers assume that all customer enterprise business units that used the central data center in the past will continue to do so. As costs for local area networks (LANs) and desktop processing configurations continue to drop and the difference in capability between host mainframes and local processors diminishes, customers can no longer ignore LANs and distributed data server solutions.

Mainframe vendors have begun to provide increasingly more powerful network management tools that far exceed the original promises of common standards for application program interfaces. These vendors have come to realize that open architectures and multiple, nonaligned distributed networks are not the threat that they had feared and can instead provide expanded opportunities for penetration into new host mainframe support markets. This is especially true as TCP/IP and UNIX platforms have become increasingly popular.

The ready availability of these new network management tools and common interface standards, when effectively integrated with existing cost allocation systems within host data center operations, can readily enable enterprisewide network cost allocation as effective as any host-based chargeback system. Integration of data center chargeback systems into the network management domain is not, however, automatic or even easy.

Any data center manager who has undergone the pain of installing a new cost allocation system can attest to the difficulty involved. The implementation of an enterprisewide distributed network resource management system and its related cost allocation system can be even more difficult. Even with advanced network management tools and management and staff with years of proven system cost allocation experience, the effort can be a formidable undertaking.

The average business user of a LAN has little interest in performing even simple data center operations tasks. In fact, most technical tasks performed by host data center staff traditionally are transparent to the average end user. In addition, many LAN administrators are unprepared to perform many operations management tasks required to support a full-service, functional data center.

As customer resource demands shift away from centralized host mainframe processors, management can find itself in a rather difficult position if full system use was anticipated and therefore necessary to recover long-term paybacks. Unless the anticipated losses in recoverable costs due to a bypass of the host data center operations can be levied against business enterprise customers in the form of disconnection fees, the only options available to management are either to reduce system capacity or to raise system chargeback rates to spread losses from unrecovered costs to the remaining data center customers.

Liquidating host hardware may not be feasible, depending on the current value of system components, and can be doubly costly if the system capacity is reduced below any future demand. The problem with raising cost allocation rates to internal business enterprise customers is they may be not only unwilling but possibly unable to pay. Under such circumstances, any increase in chargeback rates only encourages more customers to attempt to bypass central host information services.

Fortunately, there is one more acceptable alternative: expand rather than cut back host system functional capabilities. If operational, technical, and financial services expertise is extended within data centers throughout host-linked network domains, greater efficiency can be achieved at lower cost.

It may not be easy to convince enterprisewide LAN users of the benefits of central management. Data centers must be willing to change many existing operational procedures and expand business enterprise customer support services. This includes bundling client domains into data packet networks, decentralizing technical support and other data center resources to address tactical needs, convincing prospective clients of the benefits of packet networks, tracking network resource use without degrading performance, and locking in service links despite open architectures.

NETWORK ALLOCATION DOMAIN

Managers must clearly understand the boundaries that separate each of the various interconnected system environments before integrating under a single chargeback system. The first boundary, between the host and the network, may at first seem obvious, but it is critical that any host equipment or resource that is moved or distributed away from the central-site data center is understood to be a part of the host domain, regardless of its physical location.

With this boundary, the terminology for the manner in which networks relate to the host and other networks is important. The first set of terms relates to local area communications between networks. This includes peer-to-peer bridging of the LANs, which involves direct cable connections at the same site, and resource sharing, which may involve pooling external link controls for multiple LANs on a single file server.

Another set of terms refers to wide area networks (WANs), including dial-up network communications between remote-linked sites over digital telephone lines. Another wide area alternative is UNIX with TCP/IP and its File Transfer Protocol (FTP). Satellite communication is yet another WAN alternative; however, it has numerous technical complications and an extremely high cost. LAN-to-host communications are the most important to the network-linked host data center manager; for the IBM environment, this usually involves some variation of the System Network Architecture (SNA) remote or coaxial gateway. Data centers often already have extensive experience with SNA connectivity.

Still another boundary involves broadband networks, which become increasingly important as video and graphics replace text-based distributed systems. IS managers should familiarize themselves with Frame Relay, SMDS, and asynchronous transfer mode (ATM).

The subordinate network boundaries include such network distinctions as server-versus-peer and serial-versus-shared resources. The concepts of serial or dedicated, one-at-a-time, and shared or concurrent multitask use are similar, and therefore easily related, for host and network environment processing. It is more difficult to relate servers and peers to the host, however, because these are more network-specific terms.

The server can be understood as similar to a smaller or subordinate host, because it is the main control processing unit for a network. A peer, or network node, is any individual processing unit, workstation, or device that is connected to a network and shares equally (or, depending on the nature of the network access method, competes for) the use of information resources.

The nature of the peer network domain is determined by at least two important architectural characteristics: access method and topology. The term *access method,* in the host environment, describes the manner of index used to traverse data base architectures. In the network domain, the access method involves the manner in which data packets traverse network architectures.

The term *topology* refers to geometric point-to-point arrangements that connect lines together in space, which is also the fundamental basis for the mathematical description of a network. Every network involves a series of peers or station nodes that are interconnected and are in some cases also connected to other networks.

Three primary types of network topologies account for most local configurations. These are the bus (or linear, serial-dedicated line), ring, and star topologies. Many names are used for network topologies, depending on the culture associated with each network vendor access method. Nevertheless, all specialized topologies should be classified by the IS manager into one of these three basic topologies, because each has a typical configuration-associated set of support costs.

These three basic topologies can be easily accommodated because they correspond to dedicated host terminal interface configurations. The bus topology is similar to daisy-chained multiple-terminal links by way of a single dedicated port, which is common in time-sharing or remote job entry when the number of incoming dial-up lines is limited. The ring topology is similar to a linear bus linkage, except that there are no dead-end termination line drops, and every terminal joins hands on both sides to have two alternative paths back to the LAN server or gateway port to SNA or TCP/IP host. The star topology involves a direct link to the server as well as every other terminal on the same network. Although this overcomes most of the local network path contention and can also reduce resource contention, the star network topology is by far the most complex and can require extensive network management software as well as network technical support.

Two basic pass methods are generally used by all network access methods to control the transmission and flow of data packets across networks. The first is known as token passing, communications protocol handshaking that passes synchronous token keys that may contain passwords or control parameters. The second method is data contention passing, which is basically a first-come, first-served arrangement that involves asynchronous transmission and often requires multiple or redundant transmission paths.

In general, the least complicated cable and interconnection terminals also have the slowest transmission rate. Although the midrange of network cost has far greater transmission speeds, it also has a greater number of transmission errors, and the subsequent retransmission costs can vary widely, depending on the nature of the cable and network hardware used. The most reliable and fastest methods of network data transfer typically have the greatest capital outlay and support costs. One pleasant surprise is that greater complexity and capital outlay costs do not have a related increase in support costs.

For all intents and purposes, Ethernet has become the dominant network access method. As the middle ground solution of choice, there are now a wide variety of Ethernet implementations.

The technology of network management changes rapidly, especially when compared to the host environment technology with which IS managers are more familiar. Network technology is extremely competitive, and there is still substantial market pressure from some network vendors that do not fully support open architectures.

Apart from the often intimidating complexity of network technology, only three basic considerations must be determined to define enterprisewide network domains properly and decide how network costs can best be recovered. These include whether each resource will be dedicated or shared, how each dedicated resource will be capitalized, and whether to charge shared use according to resource unit measures or equitable distribution methods.

FUNDAMENTAL COST ALLOCATION SUPPORT METHODS

Although most cost allocation methods required for enterprisewide network chargeback are similar to the basic component cost distributions involved in central host chargeout, several essential differences must be understood. These network processes affect the efficient integration of the data network with central host system processing. If they are not properly managed, impact to the host could range from a mild degradation in response times to complete system outages.

Risk Management

Extended network architectures open unlimited potential pathways to the host information and processes. If the planning for enterprisewide network integration does not anticipate every possible impact on each existing host system component with adequate standards, control procedures, and system security measures to limit exposure to the risks from network access, unanticipated host system costs could be incurred.

Every possible risk posed by an enterprisewide network must be identified and quantified in terms of the statistical probability that the host will be affected. The probability of each risk must be routinely reevaluated by a statistical sampling of the same network monitor data that is used to allocate resource use costs. This data can be used to estimate the current range of potential host system costs that could be expected to result from each potential risk. Risk management is a fundamental method of controlling the costs of networks as well as their impact on interconnected systems; it typically requires rigorous contingency planning.

Depending on the complexity of the network, it may be appropriate to involve an underwriter or security specialist in a routine review of enterprise network cost allocation methodology to guarantee reduction or transfer of network risk and help minimize unforeseen costs. Depending on the size of the network, involvement of risk management professionals with expertise in telecommunications may be required for network financing.

Security and Change Control

Security and access audit logs are a very simple and useful source of cost allocation data. They consist of RACF or ACF2 for IBM host mainframe data centers, the network administrative logs for the attached networks. This information tells who logged on (or attempted to log on), when, and for how long they were actively attached to discrete data center and information network resources. This information, when used with proper versioning standards or change control, identifies those resources that are seldom if ever used and allows IS managers to place higher premiums on detailed data storage without some form of generational summarization and backup. One product that can help do this based on the "footprint" of successive timestamps (almost like a network resource "fingerprint") is the ENDEVOR product offered by Legent. IS managers should develop and enforce consistent naming conventions for data and programs so security and access routines can automate clearance to access information network resources.

These audit logs are provided by most vendors with enough detail to determine what data file names were opened or updated and, in some cases, the name of the programs that accessed or processed the data contained in those files. If sufficient standards are in place, it is possible to charge only for the share of data storage and management resources that are accessible to each enterprise business unit. IS managers could even apply level-of-access billing to data resources that are restricted or more likely to be in possible contention for access, which can be a double-edged sword because it may encourage data redundancy or duplication of effort. Like with system logs, security access audit logs capture and maintenance can be simplified using such products as Merrill's MXG, Legent's MICS, or BGS's Best/1.

It is possible to recover costs by chargeback billing of connect time if the total network connect time can be forecasted and priced in cost resource pools or more simply charged as a flat rate per month based on the average cost of one workstation of a particular discrete category as its share of the total workstation capacity and enterprisewide cost pools. It is also possible to charge for information access based on the number of data packets transmitted across a server domain, if this number can similarly be forecasted based on historical trends.

This simple cost allocation strategy is possibly also the fairest, as it will concentrate charges where the most traffic and demands on the network occur.

Traffic Control

Although most IS managers have had exposure to risk management and change control in preparing their host disaster recovery plan, other functions may be less familiar. One of these is traffic control. This is somewhat similar to methods used by operations staff to optimize message streams or input queues, but several processes are usually managed automatically by the host operating system. They must be accounted for on a more fundamental basis when network use cost is allocated.

One such process involves data collision by networks in competition for the same resources at the same time. Such an event is typically prevented by host operating system controls, and if it is not, recovery is automatic. Increasingly, the same is true of network operating systems.

Estimating traffic at each enterprisewide network intersection and the likelihood that such collisions will occur for each type of data packet work load is one of the most important methods of controlling network costs and the continent impact on host work load. Some measures used to reduce the likelihood of data collision are: average packet arrival rates, time spent waiting for service, data packets per arrival, service time durations, and ratio of time busy to time available. These statistics are also used in the allocation of cache buffers and other packet flow controls.

Although IBM enterprise strategy favors access methods that provide handshaking between communications control devices, an internal customer may prefer interconnection by way of less expensive network access methods with only a limited ability to prevent network collisions. Special fees should be considered to discourage this.

Token-passing network access methods are usually preferred by the IS manager but may not be favored by business enterprise customers because of the higher cost of data switching and coaxial cable involved. Unless a business customer procured the existing network with LAN-to-host data communications as a prime consideration, a LAN that was not designed for the host may re-send continuous streams of redundant data packets or access requests to be sure at least one gets through.

Any inefficiency in data packet control from even one of the interconnected networks can degrade all other network access to the host and potentially even the host itself. Even access method topologies which are much more sophisticated than basic peer-to-peer, central file-server, or dedicated client-server LANs can present problems to interconnected enterprisewide networks and the host.

Many such methods listen to line activity until there is an opening in the traffic before beginning to transmit. This relatively sophisticated process nevertheless leads to an increased number of collisions as traffic volume increases. These can be prevented if the methods are localized within restricted domains by limited enterprise network interconnection except across controlled traffic gateways.

LAN Management

It is a practical reality that IS managers who must accommodate external links to client networks must also be prepared to handle a wide variety of LAN access methods. It is advisable, however, that customers be encouraged to accept the long-range benefits of the most sophisticated network access methods, which will be more compatible with SAA and the enterprise network architecture advocated by IBM and Novell or other credible vendors.

Although many IS managers may be locked into one of the other network architectures or access methods, NetView NetWare and SNMF via IBM communications controls and IBM establishment controllers may ultimately be the best choices in terms of efficient management and measurement of network resources as well as long-term cost benefits. The most effective extension of a host chargeback system to network cost allocation should be based on technology similar to host data networks (e.g., SNA or TCP/IP).

Even Novell NetWare, which is widely regarded as the most sophisticated LAN manager available to date, has only a small number of resource measures that can be used to allocate costs of local network services. These measures include disk blocks read, data packets written, storage blocks used, connect time, and number of network service requests. Although these measures constitute far fewer statistics than are available on host system use logs, they are sufficient to support enterprisewide network cost allocation based on resource use.

It is much more common to allocate the costs of LAN software or data bases according to connect time or other use data in network access logs. This may be accomplished by use of simple network file directory management tools bundled with the network log-on administrator software. A related application of LAN file access logs is in building audit trail files for automated record keeping to bill professional services. This technique may also be used to allocate the proportional share of indirect labor costs for network technical support.

Network Monitors

The statistics available from network monitors residing on the network communications controllers provide far more extensive resource measurements than the network management accounting software available for the LAN. Depending on the vendor, the network controller may have as many resource measurements available as the host system, including a large number of highly specialized statistics that can be used to optimize network performance in a way that provides maximum control of the critical parameters that can have the greatest impact on the host system resources.

The IBM NetView product line is designed to centralize management and analysis of most critical network and system statistics to improve network throughput and online response times while maintaining the highest possible level of efficiency for the host system processing. Netview is a critical part of the IBM enterprise architecture and is an important strategic tool available to IS managers who need to integrate complex networks quickly. It has diagnostic, risk management, performance tuning, and access security measures that are useful for enterprise network cost allocation.

Network communications control devices usually have a variety of optional monitors that can be installed to allocate network costs on the basis of a pricing method other than measured resource use. Examples include special pricing to recover priority scheduling of dedicated lines or devices, dynamic reconfiguration or automated remote diagnostic test services, discounts for transmission at low-demand times, premiums for relinquishing access to priority packet transmissions, adjustments for excessive service time awaiting data transmission, and penalties for excessive network data packet collisions.

These network monitors are usually bundled with a wide variety of tools for specifying statistical sampling algorithms and system accounting codes needed to support complex network pricing methods. The monitors also provide extensive service-level accounting and quality control measurements used for cost containment. In addition, they yield management segmentation of demand into classes of packet work loads with common resource needs.

These packet work loads can be statistically grouped into server domains using cluster analysis techniques that help determine the most cost-effective configuration for the local servers and the topology of the interconnected networks. The techniques can also help identify which networks are the best candidates for concurrent resource sharing. This method of optimizing client-server domains is critical to recovering all resource costs.

Two types of specialized monitors—software monitors and hardware monitors—can be used by management to integrate enterprisewide network-resource-use data with the system measurements input to the host chargeback system. Software monitors estimate network use on the basis of statistical methods (e.g., randomized sampling); hardware monitors provide actual resource unit counts. Although software and hardware monitors used to have the potential to degrade system response times, this is no longer the case for either type of monitor—assuming that they are properly installed and administered.

Hardware monitors are usually installed in a file server workstation, the gateway communications controller, or both. They are most often on a specialized logic board; it is increasingly common, however, to provide network resource data acquisition functions on network control boards and on some microprocessor clock chips. Hardware monitors consist of impedance probe circuits, counters, and accumulators as well as a high-level interface for measurement access and control. Probe circuits monitor the number of electronic state changes that occur between memory locations, registers, or data I/O connections and are accumulated against the number of functional process cycles, real-time clock cycles, and interrupt states.

Software monitors provide estimates of the number of system software access requests for tasks or functions that are related to each type of resource managed by the network control programs. This type of network monitor is most similar to the System Management Facility (SMF) logs used for system job accounting in MVS host data center environments. The software monitor can estimate the hardware resource units by polling the supervisor states at time-based intervals or can use statistical software sampling to estimate resource use at appropriate reliability levels.

The selection of system resource usage monitors usually depends on which method is more accessible using the network control utilities that are either bundled with or can be selected for use with the local network configuration. If the data center has some in-house statistical expertise, the software monitor may be preferred. Otherwise, a hardware monitor may be preferable, because this unit involves simply plugging in boards and connectors and reading counts from registers.

Regardless of the type of LAN monitor, the measures that are most important to cost allocation as well as the performance tuning of the enterprisewide network are basically the same. These include arrival time to start of service, wait times, arrival to end of service, queue times, time busy, time that each resource was dedicated or shared, and the critical rate of effective processing for each resource type as well as the rate of effective throughput for each I/O data packet type.

As with host-based chargeback system methods, the choice of how to charge shared resource units (and what resources will be charged) can help direct patterns of enterprise network customer demand and can increase awareness of resources that might otherwise be hidden. This may be especially important for newly integrated network customers or dedicated host customers who are just introducing LANs into their workplace.

CONCLUSION

The implementation of an enterprisewide network cost allocation system is a formidable undertaking, yet the opportunities for broadening the customer base and improving cost-efficiency by increasing economies of scale make network chargeout a valuable strategy for the IS manager and staff. To design and develop an effective network cost allocation system that is fully compatible with the objectives of the host chargeback system, the following course of action is recommended. IS managers should:

1. Define the boundaries between each of the various interconnected system environments.
2. Determine the direct and indirect costs of host system linkages.
3. Identify and quantify possible risks to the host posed by a network.
4. Define the potential need for communications links and software.
5. Define servers, workstation clusters, and terminal support needs.
6. Establish security and access audit logs, as a source of cost-allocation data.
7. Estimate traffic flow and establish traffic control methods.
8. Create and enforce consistent naming conventions, to summarize organizational enterprise unit charges for consolidated billing.
9. Verify charges through automated system logs and network monitors.

V-11

Benefits and Challenges of Intranet Implementation

Nathan J. Muller

COMPANIES ARE ESTABLISHING INTRANETS—internal corporate networks built on World Wide Web protocols—at a fairly rapid pace. According to Forrester Research, 16% of the Fortune 1,000 have intranets and another 50% are either in the consideration or planning stages.

Companies that build intranets to improve internal communication, streamline processes such as purchasing, and simplify transaction processing are finding multiple sources of business value. Intranets help reduce communication costs, increase sales and customer response, and significantly improve work quality and productivity.

The appeal of intranets comes from enabling, enhancing, and extending effective communications within and among organizational entities and communities of interest. Intranets have allowed companies to reach new potential customers, enter untapped markets, and expand elements within their businesses. Mutual benefits are also derived from intranetworking between synergistic businesses. Coupled with such innovations as the Java development language and new zero-administration net-centric computers, intranets can become the base around which businesses reinvent themselves.

BENEFITS OF INTRANETS

Cost-Effective Communications

The foremost benefit that a company derives from an intranet is more cost-effective communications. Attaining cost-effective communications entails making information directly accessible to people who need it without overwhelming the people who do not need it. Intranets provide direct access to information, so people can easily find what they need without involving anyone else, either for permission or direction on how to navigate through the information. At the same time, companies can protect their information from people not entitled to access it.

Efficient Information Management

From the perspective of information management, an intranet extends the reach of distribution and simplifies logistics. For example, it is often cumbersome to maintain the distribution list for a typical quarterly status report sent to a large mailing list. The existence of an intranet simplifies the process of incorporating changes made before the next report is due, because updates are easily developed and posted to give everyone access to the new information.

Easier Information Publishing

Publishing information on an intranet is quite simple, especially because intranets use the same protocols as the greater Internet, including the HyperText Transfer Protocol (HTTP) used by the World Wide Web (WWW). There are a number of Web publishing tools, including Microsoft's FrontPage, that quickly turn individual documents into the HyperText Markup Language (HTML) format. Even documents maintained in a Lotus Notes data base are easily published on the Web with a complementary product called Domino, which renders Notes data in HTML format on-the-fly and serves HTML documents from the file system. With a feature called Notes Access Control, Domino also keeps the information out of the hands of people who are not authorized to see it.

Improved Searching and Retrieval

Not only has Web publishing become much easier, but users are finding it easier to search for the right document just by using key words. It is no longer necessary for people to ask someone for copies of a document or request that their names be put on a distribution list. Users also have more control over what they see. If the big picture is all a person wants, then only that level of information is delivered. When detailed information is desired, such search mechanisms as boolean parameters, context-sensitivity, and fuzzy logic may be employed. Of course, the user also has the option of accessing greater levels of detail by following the hypertext links embedded in documents.

Enhanced Real-Time Collaboration

Enhanced, timely information exchange within the organization is another benefit of intranet adoption. As people from various organizations, functions, and geographic locations increasingly work together, the need for real-time collaboration becomes paramount. Teams need to share information, review and edit documents, incorporate feedback, as well as reuse and consolidate prior work efforts. Intranets that enable collaboration to occur without paper or copies of files can save hours and even days in a project schedule.

Electronic collaboration eliminates many hurdles such as distance between co-workers, multiple versions and paper copies of information, and the need to integrate different work efforts. The intranet becomes a unifying communications infrastructure that greatly simplifies systems management tasks and makes it easy to switch between internal and external communications.

Companies are also finding that by extending their intranets beyond their immediate boundaries, they are able to communicate more directly and efficiently with the communities with which they do business. Establishing electronic connections to suppliers and partners can result in key savings in time and money in communicating inventory levels, tracking orders, announcing new products, and providing ongoing support. Intranets are being used in a range of application areas to leverage access to existing information and extend a company's reach to employees, partners, suppliers, and customers.

As companies progress in such methods of interaction, more sophisticated intranetworking can result in increased responsiveness and shortened order-fulfillment time to customers. Suppliers track inventory levels directly, reduce delays in order fulfillment, and save costs in inventory maintenance. More companies are developing this form of information exchange, where the electronic capability actually drives the process.

INTRANET CHALLENGES

It is important to note that intranets may introduce new chores in managing information. For example, ensuring that all departments have the same updated versions of information requires synchronization across separate departmental servers, including directories and security mechanisms. Providing varying levels of information access to different audiences—engineering, manufacturing, marketing, human resources, suppliers, and customers—is also an issue.

An additional task results from the fact that the hypertext links that facilitate information search and retrieval must be maintained to ensure integrity as information changes or is added to the data base. Fortunately, there are tools that help data base administrators identify broken links so appropriate corrective steps can be taken.

Domino/Notes, for example, supports major data management functions including link management and replication. As more people access data from more locations, the need for adequate security increases. Domino/Notes has a broad range of security facilities that include access control, authentication, and encryption. The product incorporates industry standards for security, such as the secure socket layer (SSL) for encrypting data during transmission.

SECURITY ISSUES

Along with the benefits of enhanced access that corporate intranets provide lie some risks. Increasing the number of people who have access to important data or systems without supplementary protective measures adds vulnerability to an

IT infrastructure. Integrating security mechanisms into an intranet minimizes exposure to misuse of corporate data and to overall systems integrity.

A secure intranet solution implies a seamless and consistent security function integrated among desktop clients, application servers, and distributed networks. It should include policies and procedures and the ability to monitor and enforce them, as well as robust software security tools that work well together and do not leave any gaps in protection.

The following basic functions are necessary for broad security coverage:

- *Access control.* Access control software allows varying degrees of access and different granularities of access to applications and data.
- *Secure transmission.* Mechanisms like encryption impede outside parties from eavesdropping or changing data sent over a network.
- *Authentication.* This software validates that the information that appears to have been originated and sent by a particular individual was actually sent by that person.
- *Repudiation.* Repudiation software prevents people who have bought merchandise or services over the network from claiming they never ordered what they received.
- *Disaster recovery.* Disaster recovery entails both software and procedures that assist recovery from loss of data in an organization's systems.
- *Virus protection.* Antivirus software protects systems from disruptive to destructive viruses by detecting, verifying, and removing them.

Intranets that extend beyond organizational or company boundaries may require integration among various security systems. In addition, because intranets give more people the opportunity to access information, they can cause companies to increase their dependence on computer technology. This increased dependence requires that appropriate backup and emergency recovery measures be in place and that alternative links be available should the network experience an outage.

COSTS

Costs are another important intranet implementation consideration. Beyond the list prices for hardware and software components lie less obvious costs of administration, maintenance, and additional development.

Intranets are most effective if they include the following attributes:

- *Reach.* Reach is the ability to easily connect with one or more employees, a group of contractors, suppliers and other vendors, and even millions of customers.
- *Flexibility.* Flexibility provides the freedom to merge on either a permanent or ad hoc basis with anyone, anywhere, anytime—regardless of hardware or software differences.
- *Scalability.* Scalability means the ability to handle up to the most demand-

ing enterprise-level computing and transaction rates across disparate systems as needs justify.

- *Transparency.* Transparency is the ability to interact with external or remote systems seamlessly and without regard for data location or underlying hardware or software.
- *Security.* Security lets entire organizations come together for maximum synergy without jeopardizing proprietary data or compromising systems integrity.

CASE STUDY

Unisys is one of the many corporations to embrace the intranet concept. The company has already established a public presence on the Internet (www.unisys.com) and has since put into operation two intranets: the Idea Factory and the Marketplace Insider. Both are run from the server that supports the divisional network of the Unisys Computer Systems Group in Blue Bell PA.

Although each intranet was set up for a different purpose, both can be accessed by all employees from any Unisys location over the same communications links used to access the corporate network, including T1 and T3, ISDN (integrated services digital network), and ordinary dial-up connections using such products as NetBlazer and pcAnyWhere. Protection from unauthorized access is provided by Checkpoint Software's FireWall-1 software.

The Idea Factory

The Idea Factory—implemented in December 1995—is a virtual gathering place for members of the Computer Systems Group's (CSG) technology-oriented community. The primary goals of the Idea Factory are to provide nearly effortless access to a vast technology resources library, connect thousands of employees with complementary skills dispersed around the globe, and promote the development of leading-edge technology solutions for Unisys customers.

One of the ways these goals are achieved is through chat rooms. Essentially, a chat room is a real-time bulletin board system. Instead of posting messages and replies on a bulletin board, a chat room allows participants to converse with each other in text mode. Organized by topic, the chat rooms provide a convenient way for employees to establish new contacts within the organization and tap into each other's expertise. Some of the chat rooms have 400 to 500 participants from all over the world who tune in and out at various times during the ongoing conversations.

For example, a customer's configuration problem can be discussed in a chat room, allowing participants to brainstorm a possible solution. If the proposed solution proves effective, it is added to a reference data base. Should another customer experience a similar problem in the future, a data base search will reveal the solution instantly and result in a prompt resolution.

The creators of the Idea Factory focused on a specific community—CSG's own technology-oriented employees, who are mostly engineers and geographically as well as organizationally dispersed. Initially, the focus was on such subjects as object-oriented architectures, multimedia, the Internet, and high-speed broadband networks. Access was provided to internally developed content and external links to other sources of information so users could have one-stop access to a vast library of technical resources.

The Marketplace Insider

The concept embodied in the Idea Factory was expanded in mid-1996 when the Computer Systems Group introduced the Marketplace Insider, which is intended to have a much broader appeal within CSG. It features a newsstand where employees can stay up to date on company- and industry-related developments. A resource center provides employees access to all the libraries within the division, and a human resources section provides a directory listing of CSG employees. An education area provides links to downloadable education tools. There is also an online organization chart and a swap-and-sell bulletin board.

The Marketplace Insider offers Lotus InterNotes links for the sales and marketing people so they can have intranet access to the competitive analysis data base, marketing toolkits, and other resources. It also provides links to the Idea Factory.

Intranet Skills Sets

The skill sets required for developing an intranet are varied and quite specialized. Technical people need knowledge of systems and network architectures, understanding of the Internet Protocol (IP), and experience in developing applications. There is also a need for creative people, particularly graphic artists and HTML coders who excel at making the content visually compelling through the integration of images and text.

The cumulative efforts of about 200 people went into the initial development and implementation of the Idea Factory and the Marketplace Insider. Many of these people, however, were only peripherally involved. For example, the same network managers and technical staff that keep the division's network up and running by default keep the intranets up and running, because they all run off the same server.

The daily maintenance of the two intranets requires only the part-time efforts of three people from the marketing group and three people from the technical group. Although this might not seem like a resource-intensive effort, it is important to stress that the caliber of skills individuals bring to the task is much more critical than the number of people actually involved.

Although it takes people with specialized skills to develop an intranet, it takes a different set of skills to sustain one. What CSG has tried to do is recruit multifunctional people—those who can apply what they normally do on the job to the medium of the intranet. The secret to productivity is to have the best people.

Among the improvements planned for the Marketplace Insider are links to some of CSG's legacy applications. In the future, Unisys customers will be able to access the intranets for such things as global procurement, order tracking, problem reporting, and software ordering. In building these applications, the Computer Systems Group is positioning itself to offer consulting and support services to customers who wish to implement intranets within their organizations.

IMPLICATIONS OF JAVA AND NETWORK-CENTRIC COMPUTING

Lower Cost of Network Ownership

For the past 10 years, companies have been struggling with the ever-growing complexity and cost of computing. First came mainframes, then minicomputers, and then PCs, each bringing with it a higher cost per person. Now the fourth big wave in information technology is about to begin with the introduction of net-centric computers that leverage corporate intranets and the new Java development language to enable mission-critical work to be done at a fraction of the cost.

Sun Microsystems has been the leader in the development of the net-centric computing model, as well as Java. The company's worldwide Java Computing initiative includes the JavaStation network computer, which leverages the power and flexibility of Sun's Java technology.

Aimed at slashing the high cost of networked environments, Java Computing enables companies to ease the burden of network and desktop administration, speed applications development and deployment, and improve network security. Because Java technology allows developers to write applications that will run on any device, Java Computing shifts applications and storage from the desktop to the network and the server. The advantages of this platform-independent approach is that it could save many large companies anywhere from 50% to 80% on the total cost of network ownership.

Fully configured, JavaStation systems will start at less than $1,000 with entry systems priced at around $750. These initial hardware and software costs pale in comparison to what companies will save in the long run on total cost of ownership. Industry estimates put the average annual cost of administering a single PC in a network at about $11,900, or approximately $35,700 over three years. Hardware and software only account for about 21% of that cost, according to industry estimates. By comparison, Sun estimates a JavaStation will require about $2,500 per year to administer, or about $7,500 for three years. For Fortune 1,000 companies with thousands of desktop PCs in their enterprises, the move to a Java platform could slash total cost of ownership by tens-of-millions to hundreds-of-millions of dollars.

Platform-Independent Applications Development

Java technology unlocks the potential of the network by allowing companies to write applications once that will run anywhere, regardless of operating system or hardware. Java applications reside on the server, where they can be easily managed, deployed, and updated by network administrators. Although the zero-administration desktop computer, or JavaStation device, needs no hard drive, floppy, or CD-ROM to make this happen, applications are still executed locally on a powerful RISC processor.

Java technology can be gradually deployed without having to discard current client/server investments. JavaStation devices will run alongside PCs, Macs, workstations, or even dumb terminals. In fact, much of the growth in JavaStation network computers will take place in companies looking to replace their aging 3270 terminals. This is a significant market with an installed base of approximately 35 million nonprogrammable terminals worldwide.

Companies with so-called fixed-function applications will be among the first to reap the benefits of JavaStation network computers. These include companies that offer airline or hotel reservation desks, kiosks, health care systems, and stock brokerage services. These types of companies will benefit from the simplicity, cost-effective systems management, and more efficient applications deployment that Java affords.

Many large companies are actively deploying Java technology. More than 60% of Fortune 1,000 companies surveyed by Forrester Research are already using Java technology for some applications development, and 42% expect Java to play a strategic role in their company within a year. Although JavaStation and other network computers were only rolled out in 1996, the Gartner Group market research firm estimates that 20% of client/server applications in the late 1990s will run on so-called thin clients like the JavaStation.

Other vendors are pursuing the net-centric computing model. Oracle's Network Computing Architecture is a common set of technologies that will allow all PCs, network computers, and other client devices to work with all Web servers, data base servers, and application servers over any network. As the manifestation of Oracle's vision for network computing, the Network Computing Architecture will help companies protect their technology investments by allowing mainframes, client/server, Internet and intranets, and distributed object software to work together.

Like Java, the Network Computing Architecture transcends the Internet/object standards battle, so users and developers can make software programs work together without getting locked into dead-end solutions. The Network Computing Architecture will simplify the problem through the use of open Internet standards and unique bridging software that helps proprietary application programs work together.

In contrast to the PC-centric computing model, which focuses on independent users and computation, the Network Computing Architecture recognizes the increasing importance of Web servers, data base servers, and application servers working together over corporate intranets to enhance communication and deliver a wide array of information on demand to internal and external corporate constituents.

CONCLUSION

Corporate intranets are becoming as significant to the telecommunications industry as the PC has become to the computer industry. They fundamentally change the way people in large organizations communicate with each other. In the process, intranets can improve employee productivity and customer response. They are also being used to connect companies with their business partners, allowing for collaboration in such vital areas as research and development, manufacturing, distribution, sales, and service.

A variety of tools are used for these purposes, including interactive text, audio and video conferencing, file sharing, and whiteboarding. In fact, anything that can be done on the public Internet can also be done on a private intranet—easily, economically, and more securely.

Section VI
Ensuring Quality and Control

THE QUALITY OF SERVICE that IS delivers for the enterprise is gaining attention as competition for resources becomes more intense. IS managers are under scrutiny more than ever before, and maintaining quality is not easy when the direct responsibility for building systems is dispersed and not always under the IS manager's auspices.

In these circumstances, IS managers find that they must be innovative in addressing the stewardship of information technology resources. This section offers ideas for ensuring the quality and control of IS services.

Chapter VI-1, "Quality and Change Management," begins this process by showing how IS planning can provide the basis for an integrated approach to change management and quality improvement for the enterprise.

Continuous improvement is increasingly important in most organizations. Chapter VI-2 demonstrates how to build an improvement plan and shows that installing changes and obtaining benefits is worth the time and effort required. "Improving Productivity and Effectiveness" also shows how the IS manager can develop a strong case that improvements in customer service can more than outweigh any investment required for them.

Outsourcing is an evolving trend in IS. However, the decision to outsource is usually not made by the IS manager, who is often not even involved in the deliberations that lead to the consideration of the outsourcing option. Although it may not be the IS manager's job to make the outsourcing decision, the IS manager is responsible for ensuring that the proper safeguards provide the organization with long-term protection of its assets. Chapter VI-3 reviews why the outsourcing decision is often controversial and presents steps to ensure "Control of Information Systems Outsourcing."

Flexible user interfaces can help people accomplish subordinate tasks more efficiently, but gains in individual productivity need to be balanced against the need to meet overall system goals. Chapter VI-4, "More Productive User Interfaces," shows that providing more flexibility in system interfaces can be productive as well as allow more room to accommodate individual needs.

Chapter VI-5 intends to help the IS manager improve systems development by bringing the voice of the customer to each stage of the systems development process. Although quality has emerged as a formal management function, the way that this function will be applied to the IS organization is still evolving. "Quality Improvement in Systems Development" introduces a customer interaction model that combines elements of process improvement and quality function deployment.

Another area where quality improvement techniques can be applied is the data center. The model office concept described in Chapter VI-6 mirrors the data center's production and runs all applications before they are actually sent to the data center. "Improving the Quality of Production Systems" describes how this added step ensures that applications are thoroughly tested, that they run effectively, and that systems rework caused by errors will not result in serious disruptions to its users.

"Secure Workstations" underscores the need for a security program that protects the unique characteristics of a workstation environment. As workstations become more prolific, more users are developing or acquiring software that performs critical business functions. Chapter VI-7 describes how to establish a program for workstation security, recommends controls, and suggests methods for creating end-user security awareness.

Software quality auditing is a process to determine whether the various elements within a quality improvement system meet their stated objectives. Because the ISO 9000 series of quality review standards can be applied to any quality system, a TQM or CMM program can be integrated into the ISO 9000 certification process. Chapter VI-8, "Software Quality and Auditing," discusses the significance of software quality improvement processes and their relationship to the ISO 9000 auditing process.

Knowing where the IT department stands and how IT adds value to a business becomes doubly important when IS services are outsourced. Benchmarking—a technique companies use to compare themselves against other companies and identify best practices—can uncover ways to show how information technology contributes to business value. Chapter VI-9, "Benchmarking the IT Function," discusses six key components for baselining the IS organization. The chapter's worksheets and graphs can be used by IS managers to create a baseline snapshot of the organization and to measure the performance of outsourced services.

The information gained from a measurement program can provide IS managers with insight on how to apply resources more effectively. Whereas metrics can be gathered easily by automated tools, these metrics need interpretation to enable meaningful decision making. In Chapter VI-10, "Improving Quality with Software Metrics," the IS manager will learn that software metrics are significantly enhanced when they are combined with cost and frequency of change information.

VI-1

Quality and Change Management

Yannis A. Pollalis

BUSINESS PROCESS REENGINEERING (BPR) has become one of this decade's most cited management issues in the managerial, academic, and trade press. It has also been listed as a top priority by most surveys of corporate executives, business planners, and management consultants. The BPR concept was introduced during the late 1980s primarily by a few influential consultants and academics. BPR uses information technology (IT) to radically change (or redesign) the business processes within organizations to dramatically increase their efficiency and effectiveness. Although some of the concepts and methods of previous management practices are similar to those of BPR (e.g., total quality management and activity value analysis), BPR is still perceived by some advocates as a different way of management thinking. Thus many of the mistakes committed with BPR's predecessor concepts and methods have been repeated.

Furthermore, evidence indicates that a great percentage of BPR efforts have failed. Research on these failures produced a list of critical failure factors that include lack of management commitment and leadership, resistance to change, unclear specifications, inadequate resources, technocentricism, a lack of user/customer involvement, and failure to address the human aspect of planned change.

Although BPR reflects a relatively new way of thinking about process change, similar efforts have already taken place in the areas of information systems planning (ISP) and total quality management (TQM). Thus, integrating BPR, ISP, and TQM into a holistic model capitalizes on the lessons learned from ISP and TQM efforts and avoids repetition of past mistakes.

ISP AS PLATFORM FOR INTEGRATING BPR AND TQM

TQM's main goal is to improve the processes within an organization and the organization's ability to meet the needs of the customer by emphasizing continuous quality improvement and responsiveness to customer demands. Overall,

571

TQM activities involve improving business processes and implementing incremental change by:

- Focusing on satisfying customer needs.
- Analyzing business processes continuously to increase efficiency and customer service.
- Emphasizing teamwork and employee empowerment across and within the firm to ensure the previous two activities.

ISP activities include:

- Identifying information resources that support or redefine the goals of the firm and the IS organization.
- Identifying opportunities to use IT and improve the firm's competitive advantage.
- Implementing process change through IT.
- Meeting the systems requirements of internal and external users.

More specifically, ISP aims to reduce the uncertainty associated with the internal and external business environments. Uncertainty in the internal environment is generated by process changes in an organization. Its successful resolution depends on the ability of IS management to understand the interrelationships among the various organizational functions and processes and minimize redundancy and inefficiencies. This type of uncertainty requires that ISP consider process quality improvements along with user satisfaction goals.

Uncertainty in the external environment results from IT developments and competitive market pressures. ISP's role in this arena is to identify opportunities and threats in the environment and successfully integrate them with the IS organization's goals.

Thus, ISP can be defined as a proactive process that emphasizes IT-based process change to improve an organization's ability to:

- Respond successfully to external threats and opportunities.
- Strategically apply its own capabilities and competencies through information resources.

Based on this definition, ISP focuses on three areas common to both BPR and TQM:

- *Technological improvement*—which reflects the IT focus of BPR's process redesign and innovation efforts.
- *Process improvement*—which emphasizes both the redesign of existing organizational processes and the employee empowerment concepts used in TQM's cross-functional and coordination activities.
- *Strategic improvement*—which concentrates on BPR's and TQM's alignment with corporate objectives.

Thus, as illustrated in Exhibit VI-1-1, IS planning can act as a platform that integrates an organization's BPR and TQM process change efforts.

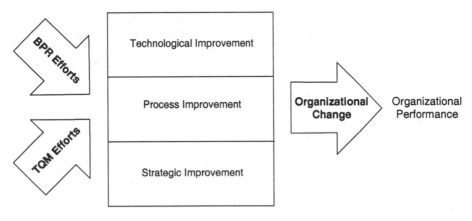

Exhibit VI-1-1. IS Planning as a Platform for Integrating BPR and TQM Efforts

COMMON DIMENSIONS OF ISP, BPR, AND TQM

IS planning has four major components that overlap with the objectives of BPR and TQM.

Alignment of Corporate and IS Goals

In IS planning, information resources are used to support business goals (usually efficiency, effectiveness, and competitive uniqueness), as well as to lead corporate strategic efforts to capitalize on external opportunities and internal competencies derived from IT. For example, Wal-Mart's integrated distribution network and United Services Automobile Association's (USAA) state-of-the-art document-handling system represent two such IT-based distinctive competencies. These competencies were aligned with corporate strategies and brought competitive advantage to the two companies.

Customer Focus

The ultimate goal of ISP should not be to use IT to implement organizational change for the sake of technology's potential capabilities. Rather, ISP should address various concerns of quality and customer needs, and improve and sustain linkages between the organization and its internal and external customers. In general, IT systems that help the customer to order, choose among alternatives, and purchase products and services contribute to both BPR's radical change and TQM's continuous improvement efforts. Thus, IS planning becomes the hub for a value-added network that includes linkages to both external customers (e.g., suppliers, buyers, and competitors) and internal customers (e.g., functional departments and divisions).

For example, the legendary systems of American Airlines' SABRE, American Hospital Supply's ASAP, and McKesson's ECONOMOST have helped to build strong ties with external customers (i.e., travel agents, hospitals, and drugstores, respectively) owing to their user friendliness, convenience, and value-added services offered with the total package.

Similarly, ISP can facilitate relations and linkages among an organization's internal customers (e.g., accounting, purchasing, production, and marketing) by improving the quality and efficiency of IS services. Examples of such cases include Charles Schwab's integrated account environment (called cashiering), which allows faster and more reliable retrieval of customer/investor account information by the various Schwab brokers; and Citicorp's work-group computing environment, which integrates the business divisions of leasing, retail banking, institutional banking, capital markets, and real estate loans to promote reliable information and overall organizational effectiveness.

IT-Based Process Change

By changing, updating, or replacing existing information systems and processes within a firm, ISP facilitates restructuring of a firm's business processes. Prerequisites of such planned IT-based change include management support, strong IT leadership, involvement of IS executives in corporate planning, and systems thinking. The efforts of Pacific Bell, Xerox, and Texas Instruments in this area are discussed in the section on successful integrations.

Organizational Learning

By forcing its participants to understand a firm's various processes, their critical success factors, and the way IT can improve them, the ISP function becomes a facilitator for learning about organizational processes. Various techniques, such as scenario-based planning and simulations of internal or external crises, promote such learning and prevent unexpected disasters. Among the classic examples of such efforts toward organizational learning through strategic planning is Shell's crisis management simulations, in which what-if exercises resulted in major redesign and process changes that helped the company anticipate and prevent market- and technology-based disasters.

Exhibit VI-1-2 depicts the integration of an organization's BPR and TQM process change efforts with IS planning.

INTEGRATED PROCESS CHANGE MANAGEMENT

Recent research and case studies confirm the similarities between ISP, BPR, and TQM. Organizations that engage in uncoordinated and sometimes concurrent efforts for ISP, BPR, and TQM engender the following problems and concerns:

- Different organizational members advocate and participate in often similar

Major Components of IS Planning

Strategic Alignment: Dynamic relationship between corporate and IS goals; focus on capitalizing on external opportunities and internal capabilities and competencies derived from IT.

Customer Focus: Introduction or adoption of IT to inoroaoo ouotomer satisfaction and create value-added services.

Process Change: Changing, updating, or replacing existing IT-based processes to improve organizational effectiveness and efficiency (i.e., change management and systems planning).

Organizational Learning: Understanding the firm's critical success factors, the relationships among its cross-functional processes, and its capacity to prevent crises and disasters (i.e., scenario-based planning).

Interrelated BPR Efforts

Reengineering efforts begin with the corporate objectives and aim to realign operational capabilities with corporate strategic goals.

Reengineering is driven by customer demands and is taking advantage of the market opportunities derived from customer needs.

BPR asks whether organizational processes can be redesigned to increase their effectiveness and looks at the interrelationships among the organizational processes affected by IT.

Reengineering examines the possibility of changing organizational members' business mental models by challenging management's existing assumptions and learning both from past failures and successes and from new IT developments.

Interrelated TQM Efforts

TQM's efforts include alignment with the IS organization's goals to improve IS operations (e.g., introduce new software tools and promote acceptance by organizational users).

TQM is focusing on both internal customers (e.g., business divisions) and external customers (e.g., buyers, suppliers, competitors, support institutions).

TQM emphasizes coordination among IS professionals and the rest of the organizational functions and continuous improvement of the processes across the organization's value chain.

TQM focuses on incremental changes similar to prototyping systems development methodology and learning by completing small stops in implementing change, using work groups, and sharing information across functional areas.

Exhibit VI-1-2. Integrating BPR and TQM with IS Planning

ISP, BPR, and TQM change initiatives, which results in redundancy, inefficiency, and inconsistency in organizational projects and goals.

- Some organizational members participate in more than one of the three initiatives, resulting in confusion and inability to define clear and consistent goals across the organization. In addition, because participants in ISP and TQM activities often fear that the reengineers will eliminate or ignore their efforts in the eagerness to start with a clean slate, they are reluctant to commit needed resources to ISP and TQM activities.

- Because of the confusion and lack of trust among the participants in the preceding two scenarios, very few ISP, BPR, and TQM projects can be successfully implemented.

● There are no clear and compatible measures or criteria of success for ISP, BPR, and TQM projects, resulting in inadequate evaluation of efforts to implement organizational change. Furthermore, although BPR advocates might view organizational change as strategic, TQM and ISP advocates might regard it as simply operational, thus making it almost impossible to set priorities for projects and coordinate change efforts across a firm.

Success Stories

Companies such as Pacific Bell, Xerox, and Texas Instruments (TI) are among the few firms that, under the concept of process management, have integrated traditional TQM procedures with IS process-modeling and BPR techniques. In these organizations, in contrast to what Michael Hammer and James Champy preached in *Reengineering the Corporation* (New York: Harper Business, 1993), BPR and TQM are viewed as two sides of the same coin and IS planning is integrated with their efforts toward process change management. For example, Pacific Bell and TI created central-process-management teams responsible for providing tools and methods to concurrent BPR efforts and for ensuring that IS, TQM, and BPR teams are coordinated and learn from each other's successes and failures.

More specifically, at Pacific Bell, process management efforts include IS projects responsible for aligning systems development strategies and tools with the current needs of BPR projects. At Texas Instruments, various process-capture tools allow continuous improvement methods to be integrated with BPR and IS development processes; similarly, Xerox has created the concept of process owners who decide what kind of change needs to be performed in a broad business process (e.g., tweaking versus a major overhaul) and how IS, TQM, and BPR groups can work together to provide the necessary tools and methodologies.

In contrast to these integrated environments, some companies continue to view BPR as a radically different type of activity for IT planning and TQM. Such companies take a more top-down approach that allows BPR consultants to identify specific projects and procedures and ignore any organizational learning accrued before their involvement. The problem with uncoordinated approaches, however, can be traced back to TQM efforts that attempted to deliver competitive advantage without considering key external and organizational factors (i.e., technology developments, market conditions, and corporate strategy) and focusing instead on internal improvements and process changes.

BPR is not an entirely new activity, different from ISP and TQM. As Exhibit VI-1-3 illustrates, all three are complementary elements of efforts toward process change management. Although ISP, BPR, and TQM should be coordinated within an organization, certain activities are unique to each of them. These activities are shown as the nonoverlapping areas in Exhibit VI-1-3 and include the following:

● TQM usually involves bottom-up, incremental design changes focusing on specific processes.

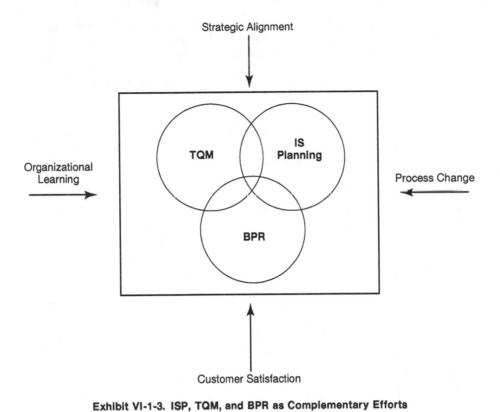

**Exhibit VI-1-3. ISP, TQM, and BPR as Complementary Efforts
of Process Change Management**

- BPR usually involves a top-down orientation, focusing on innovation and radical change.
- ISP can be both top-down (i.e., strategic) and bottom-up (tactical) to identify corporate strategies as well as IS implementation problems. Successful ISP practice involves both top-down and bottom-up orientations to anticipate short-term organizational changes and long-term technology developments and market forces that could affect corporate goals.

STEPS FOR INTEGRATING ISP, BPR, AND TQM

The following steps form the basis for process change management and will help IS professionals integrate ISP, BPR, and TQM.

Step 1: Determining priorities and organizational goals for each initiative. For example, does process change mean the same for ISP, BPR, and TQM groups? Organizational priorities can be established through the techniques of critical success factors (CSFs) and benchmarking, or be based on the organization's distinctive competencies (IT-based or otherwise). ISP, BPR, and TQM task forces should collaborate with top management in this stage.

Bringing participants from each initiative together to discuss their projects and identify similarities and differences among them. The shared dimensions of the three initiatives as delineated in Exhibit VI-1-1 should be used in this step.

Establishing collaborative meetings or groups to discuss organizational learning issues over the course of ISP, BPR, and TQM efforts. Such groups discuss what the organization has learned from each initiative so far to make use of successful processes and avoid repetition of past mistakes.

Avoiding redundant activities. Scheduling regular informative meetings at which teams disclose their findings helps ensure that activities are complementary rather than redundant.

Setting synchronous and clear goals. This step is accomplished by having teams agree on measurement criteria or participate in one another's evaluation procedures.

ROLE OF IS MANAGEMENT

The importance of IT resources in any BPR and TQM effort gives IS management a central and sensitive role in the integration process. By conveying the following key concepts to the BPR and TQM groups, IS managers clarify the necessity of integration as well as the IS function's supportive role in the effort toward organizational change.

Shared Resources and Expertise. The whole organization benefits when teams share resources and expertise while working toward process innovation (BPR), continuous process improvement (TQM), and IT-based strategic advantage (ISP). Stressing the common goals among the three initiatives (e.g., alignment, customer satisfaction, process improvement, and organizational learning) helps clarify this concept. IS managers can also play the role of the outside IT consultant for both BPR and TQM initiatives.

Alignment of Technology with Corporate and Customer Needs. Teamwork and inclusion of IS managers in BPR and TQM decision processes and projects increases the likelihood that the systems delivered by IS will be aligned with corporate goals, as well as with customer (internal and external) specifications. Furthermore, senior executives are more likely to accept and support recommendations for organizational change when participants from all three initiatives are involved in and accountable for final process change results.

Commitment and Accountability. Integrating ISP, BPR, and TQM efforts is not about giving more control to the IS organization and management. It is about commitment and accountability—both at the individual and group levels—to build a platform for shared expertise and organizational goals.

Strategic Advantage. The ISP-BPR-TQM think tank promotes competency-based strategies and creates processes and systems that optimize the organization's unique capabilities and resources.

Prototyping for Success. Prototyping procedures in which pilot systems and processes are tested before full-scale implementation and full commitment of resources increase the success rate for the perspective change and enhance the learning capacity of the ISP, BPR, and TQM groups.

Common Measurement Criteria. Consistent measurement criteria across all three initiatives eases the transition from independence and redundancy to integration and sharing. Common criteria among BPR, ISP, and TQM include process quality, product quality, system quality, customer satisfaction, cost reduction, faster delivery, and value-added service to customers.

Measurement should not be based on a bottom-line approach that continuously monitors costs and benefits. Such an approach eventually results in micro savings and demoralization of the ISP-BPR-TQM alliance. Measurement should be flexible, because there is no guaranteed way to include or predict all benefits and costs from a system or process, and allow for innovation and guided risk taking. This approach gives participants a certain degree of responsibility for decision making and brings better results in the long run. In summary, measurement should be able to see both the forest and the trees in regard to the short- and long-range goals of a BPR, ISP, and TQM alliance.

CONCLUSION

Organizations that quickly jump from one management trend to another without first learning from past experience have high failure rates in their efforts toward process change management. In contrast, organizations that have a universal management philosophy embedded in all activities related to improvement and change compete successfully. In an age when global competition and continuous technology developments are the norm, the ISP-BPR-TQM holistic model capitalizes on and integrates the learning that occurs in IS planning, business process reengineering, and total quality management.

VI-2

Improving Productivity and Effectiveness

John P. Murray

THE INTRODUCTION OF MORE easily understood technology gives managers with limited information technology experience more autonomy and confidence in managing that technology. That in turn has encouraged non-IT staffs, often urged on by vendors with an interest in installing their own brands of hardware and software, to assume increased responsibility for the management of their own processing environments. Given these circumstances, IS managers must show that the function is well managed and will continue that way, or they will become vulnerable to some form of takeover.

Beyond the issue of relinquishing management is the issue of the general trend within organizations to reduce expenses whenever possible. Because the operation of the data center often represents a considerable line item in the IT budget, it can easily become a focal point for expense reduction. The well-prepared data center operations manager anticipates that possibility and begins to reduce expenses before being forced to do so.

Reducing expense through reductions in data center customer service, however, is not the answer. In fact, managers must reduce expenses while improving customer service levels. To meet this challenge, improvements in data center productivity and effectiveness must be made.

BEGINNING THE PROCESS

The issues of improved productivity, efficiency, and expense reduction almost always collide with cultural issues. First, IS staff members likely believe that current work is as productive as possible and if there were a need for improvements, they would already have made them. Second, the IS manager may believe there is a need for more money to make improvements, not less. Whatever the cultural issues, however, they must not hinder taking action to move toward increased productivity and effectiveness.

To begin, managers must make the time and effort to ensure that staff members understand the benefits from a plan to make productivity, effectiveness, and expense management improvements. How the topic is handled depends on the manager's perception of the operation's status. For example, if the perception is there are many opportunities to improve because of the poor quality of the current practices, an aggressive approach would be in order . The manager would clearly state the situation: changes must come about if the operation is to survive. Conversely, if the perception is the department is well run, the approach would be to find opportunities to make improvements that enhance an already strong operation. In this example, the process is more a fine tuning than an overhaul.

The idea is to share with the staff the reasons for moving to improved productivity when possible and to work together to identify areas to reduce expenses. The goal is to gain staff support, which is critical in making the project as successful as possible. If everyone understands the purpose of the project and what its outcome means to each individual, there will be a much stronger likelihood of success. Sometimes, the message being sent is painful, but it is important to send the message. The reason for change should be obvious, no matter how difficult. To succeed, the IS manager needs the support of all the members of the IS staff. The benefit is that the success of the group ensures the success of the individuals.

MAPPING THE EXISTING PROCESS

The starting point for an improvement project must be to gain a clear understanding of current processes. Too often, there is a gap between the staff's perception and the reality of what the department is doing. The concern should be to determine accurately what is occurring, so any proposals produce the most benefit. Also, identifying the current process often reveals areas that are candidates for adjustment to bring about positive results quickly.

One effective method for gaining an understanding of what is occurring is to map the work flow within the data center. The procedure is to build a collection of process maps that track the flow of a particular piece of work from the time it enters the data center until it leaves. Each step in the flow is identified, recorded, and then connected to show how they relate to each other and to other processes. The result is a graphic representation of the entire work process as it moves through the center. One map should be completed for each set of processes.

Doing a thorough job creating and reviewing the maps provides the basis for changes that can both reduce processing time and effort and improve customer service. The value of the mapping process is to produce an objective analysis of the way work is being done.

Mapping need not be an onerous task, as there are microcomputer-based graphical application packages available that both create and modify process maps. These tools are inexpensive, flexible, and easy to use. It is a sound business approach to buy such a tool rather than attempt to draw and revise the process maps by hand, if such a tool is not already available in the department.

When the maps are completed, the staff meets to review them. All staff familiar with the work being reviewed (i.e., IS staff and data center customers) should attend, to contribute to the informed decisions on what is being done and what changes must be considered. Each map should be carefully reviewed within the context of the following questions:

- Is this function necessary? If not, can it be eliminated?
- Is it possible to combine several functions to reduce the work load?
- Are there more effective methods that produce the same, or better, benefits?

As a result of this review, there may be some discussion about the accuracy of the maps because of some staff assumptions about how the work is processed within the installation. Also, the amount of work that is either redundant or no longer necessary will quickly become apparent.

For example, in one data center, in reviewing the maps, the staff found that two production runs for two departments used files and programs that seemed to be identical. After further analysis, it turned out that two applications did almost the same processing for the different departments, the only difference being that the reports, though identical in content, were sorted and printed differently.

In this example, several minor adjustments were made to one of the applications, and the other was eliminated. Even though the initial processing increased the work load a bit, the net results are savings in processing time and increases in productivity and effectiveness.

As the review of the process maps moves forward, one goal should be to identify processes that do not add any value to serving customers. When these non-value-added processes are identified, they should be noted and marked for removal from the work flows. There may be an eagerness to remove the work immediately; however, it is important to make sure there are no links to other processes. Such links may not be obvious, so it will be important to understand what those links mean before any changes are made. This is an example of how it helps to involve people from the business areas of the organization who have an understanding of the applications.

Circumstances that would support not making the changes include having to alter a significant amount of code to effect the change. The effort and time involved might not justify the change. The idea is to get at as much of the unnecessary processes as possible and to understand that there is a cost to making the changes.

Once the links between processes are identified, however, one rule of thumb is that any processes that can be removed, should be. It may be that the removal of a particular process appears to produce a marginal effect on the efficiency of the operation, but it represents a significant cumulative benefit. As a rule, no matter how small the benefit, any candidate for removal should be deleted.

IT installations that have been in operation for some time—10 or more years, for example—have applications that have outlived their usefulness. The development and review of the data center maps hold the promise in some installations for a considerable improvement in productivity and effectiveness. Taking the time for a careful analysis of those old applications is likely to produce several or more candidates for restructure or elimination.

IMPROVING ERROR RATIOS

The next area with potential for improvement in productivity, efficiency, and expense reduction is a review of operational errors. The purpose of such an investigation is not to place blame, although doing a good job here will probably highlight some areas or individuals that would benefit from improvement. Rather, the idea is to locate the causes of errors and to do whatever is necessary to stop those errors from occurring. An analysis of the cause of data center errors will show that there are only a few areas that create conditions for errors. These include:

- Applications defects.
- Operations errors.
- Insufficient operator training and supervision.
- Vendor problems.
- Technical support errors or omissions.
- Inadequate, outdated, or incorrect operations documentation.

Again, depending on the particular operation, one or two categories of problems may be responsible for a large portion of the errors, or errors may be created in varying degree by all of the conditions. Having identified the areas in which errors are occurring, the staff should assess the impact of those errors on performance.

The Impact of Data Center Errors

Several examples of various error conditions and their effect on data center performance illustrate the need to follow up on the errors and their causes. The most dramatic and usually the most expensive example of the negative effect of data center processing errors are those that could be termed *operating disruptions*. These are error conditions that slow, disrupt, or shut down the normal online data processing functions.

Any of the items in the preceding list fall into the category of operating disruptions because they can play a part, either by themselves or in combination with another condition, in disrupting normal processing operations. Sometimes the reasons for such circumstances are obvious and therefore easy to correct. An example is a production application that has insufficient editing capabilities. Because the edits are weak, inaccurate data enters the system and, later in the process, that data causes the application to fail. In other instances, the cause of difficulty may be much less obvious and, as a result, much harder to pinpoint and correct.

To obtain a strong return from addressing and correcting operating disruptions, the staff should identify as many of those instances of operating disruptions as possible. When the operating disruptions are identified, they should be categorized on the basis of their severity and the amount of time and effort required to make the needed corrections. One way to proceed is to develop a grid that includes a description of the error condition, the severity of that condition, and the estimated amount of time and effort required to bring about the desired corrections. In addition, there should be an estimate of the potential benefits to be obtained as the result of the correction. Those benefits should be considered in terms of both hard dollars and intangibles. It may be that improvements in customer satisfaction, as an example, carries as much value to the organization as a dollar saving.

With the data produced in the grid, the error items can be ranked and schedules for the anticipated correction of the error conditions can be developed. In using the grid to both list the items and analyze the effort and timing of the corrections, the IS manager can move the process forward on several levels simultaneously.

DEFENDING THE NEED TO FINANCE CHANGES

When the IS manager begins to focus on correcting operations disruptions, it is not long before the topic of the expense associated with making the corrections arises. Making the corrections often depends on the concentration of resources, because applications development staff may have to correct errors in production applications and vendors must correct errors in purchased software packages.

Applications development managers might resist efforts to direct resources away from projects already in process. When challenged with flaws in its applications, vendors could be reluctant to make any changes, to avoid incurring expenses. This is more likely to be the case when the error condition appears to be isolated to one customer. When moving into these discussions, the manager should expect emotional battles for resources to arise.

An effective method to defuse the issue is to gather the facts beforehand to support the need for action. To do that, the manager should develop a process to assign costs to the organizations responsible for correcting the errors. Again, the use of the grid can be helpful in showing the costs, both tangible and intangible, of not making the corrections. Because many error conditions result in the loss of time, it helps, when practical, to relate the time lost to dollars lost.

Several examples of the effects of error conditions and their associated cost to the organization effectively build a case to gain the resources necessary to correct those conditions. For example, a batch processing system that had been working correctly is modified to add more processing capabilities. When the changed system moves to the production processing environment, the system aborts on occasion. When the problem is brought to the applications manager's attention, the manager installs a fix rather than repairing the problem, to enable the operators to restart the processing sequence when it fails.

The difficulty is that when the system fails, the restart process requires three hours of processing time to correct the error. The results are that data center production processing schedules are disrupted and online customer files are not available at the start of the work data. Placing the fix in the application does not solve the problem; it only provides a way to get around the problem, and the problem remains in the operations disruption category.

In addition, the organizationwide impact is that several departments must simply wait until files become available to start their work. Assuming that the average fully loaded cost of the employees in all departments affected is $15 per hour, that the departments affected have a total of 50 employees, and that the files affected are unavailable for three hours each month, the cost to the organization, in terms of productivity, is $2,150 per month.

MAKING THE NECESSARY CORRECTIONS

Whereas the issue of lost productivity is an important item, it is hardly the only issue. The effect on department morale, coupled with the bad publicity for the IS department, are issues that must be considered. Beyond the internal harm suffered by the organization is the issue of the negative effect on the organization's customers. If these delays result in lower standards of customer service—and in today's world that is almost always the case—the $2,150 in lost wages may represent only a very small sum.

If, however, it is not possible to obtain assistance from the applications department or the vendor, securing additional funding to address these items should be considered. Working with an outside firm to develop a plan to attack the items that create operational disruptions may provide the solution.

Another approach is to request adding programmer positions to the staff to work on the operational disruptions. Again, using the data from the mapping process and the estimates for the expense associated with uncorrected operational errors, the IS manager should not find it difficult to justify one or more programmers to address those problems. It should be made clear that those employees will correct the applications that are causing operations disruptions.

Documentation Difficulties

Errors in applications code are not, however, the only source of operations errors. Another source of errors may be found in the data center's operations documentation, which may be of poor quality or so meager as to be virtually nonexistent. In this case, the applications developers' reasons are usually that, because of impending implementation deadlines, there has not been time to complete the documentation. Under these circumstances, it is unlikely that adequate documentation will ever be provided.

Poor operations documentation can also be found within the data center because changes and updates to operations documents are not maintained or clearly written. True, experienced operators can work with poor documentation because they can process the systems from memory, but new operators must rely on faulty documentation, which causes operational errors. The worst-case scenario occurs when operations documentation has not been properly updated for some time and those operators who understand the processing environment leave the organization.

A third area of documentation difficulty is the delivery of poor quality documentation from outside vendors. To avoid this situation, all installations should include a standard to withhold approval of vendor payments until the documentation has been reviewed and approved by the operations personnel. That condition of purchase should be made clear to the vendor well in advance of any vendor marketing effort. The most effective way to carry out that review and approval is to process several runs of the applications and determine documentation quality.

The most easily resolved of these three documentation problems is faulty documentation from outside vendors. Avoiding difficulty in this area requires cooperation between the IS manager and those purchasing the software. As an example, the development manager purchasing application software should understand and agree that the purchase will not be completed without formal acceptance of the products, including their documentation, from the data center.

Regarding the other two documentation problems, the IS manager may have to consider several options. One effective approach to the problem of less-than-adequate data center documentation is to employ the services of a technical writer. This position can be full or part time, or the technical writer can be engaged on a contract basis. The technical writer's role should go beyond that of maintaining the documentation; the writer should also be willing to conduct periodic reviews to ensure that all data center documentation is maintained in accord with the established standards.

DEVELOPING THE IMPROVEMENT PLAN

The plan to address the topic of operational disruptions should include:

- The identification of difficulties created by the current problems and the probable cause of those problems.
- An estimate of the cost, both in hard dollars and in reduction of customer service levels, associated with the occurrence of the operational disruptions.
- An approach to address the various problem areas and to develop a tentative schedule to remove the areas of difficulty.
- The estimates of the expense required to make the needed corrections.
- The anticipated benefits associated with the success of the plan. When doing this section of the plan, the IS manager must address benefits beyond those of hard dollars (e.g., improvements in customer satisfaction).

A part of the plan should deal with the topic of communicating the results of the improvement effort. The plan, once approved, should be published so anyone interested can find out what is going to be done. As results are achieved, they should be made known so that people throughout the organization can see that progress is being made.

CONCLUSION

Opportunities exist within many IS departments to bring about improvements in productivity, efficiency, and reduction of expense, without the requirement to move to radical changes. Taking the time to think through current practices, to identify areas for improvement, and to install those improvements on the basis of the adoption of an aggressive plan produces tangible results.

Moving to the most effective, efficient operation must become the goal of every IS manager. Finding the time to do this is something the IS manager must accomplish.

VI-3

Control of Information Systems Outsourcing

S. Yvonne Scott

OUTSOURCING IS A REALITY. As an industry force that causes significant change in the established environment, it raises several issues that warrant the IS manager's attention. One of these issues stems from the fact that information—and the systems used to generate it—can differentiate a company from its competitors; therefore, information is a valuable asset worthy of protection. Increasing organizational reliance on outsourcing has thus created a new area of audit responsibility.

A MATURE CONCEPT

Outsourcing IS is not a new trend. The use of service bureaus, contract programmers, disaster recovery sites, data storage vendors, and value-added networks are all examples of outsourcing. There are even examples of entire large IS organizations being outsourced 25 years ago. Orange County CA did this in the mid-1970s.

Functions such as time-sharing, network management, software maintenance, applications processing, limited facilities management, full facility management, and EDI services are now considered potentially outsourced functions. In 1989, Eastman Kodak entered into a 10-year agreement to outsource its entire IS function.

Outsourcing is not a transfer of responsibility. Tasks and duties can be delegated, but responsibility remains with the organization's management. Therefore, outsourcing does not relieve the organization or management of the responsibility to provide IS services for internal operations and, in some cases, for customers.

Outsourcing is not an excuse for substandard customer service, regardless of whether the customers are internal or external to the organization. Customers do not care how or by whom services are provided. Their concern is that they receive the quality services they need, when they are needed.

The most successful outsourcing deals are tailored relationships that are built around specific business needs and strategies. There has been a definite shift from an all-or-nothing approach to a more selective application of outsourcing. In many cases, deals have been structured to more closely resemble partnerships or alliances rather than service agreements. For example, some of these deals include agreements to share in the profits and products that result from the alliance.

Outsourcing does not eliminate the need to audit the outsourced services. It is the auditor's responsibility to safeguard all of the assets of an organization. Because information is clearly an asset, the organization must ensure that information confidentiality, integrity, and availability are preserved.

OUTSOURCING SERVICES

Any agreement to obtain services from an outside vendor rather than to provide them internally meets the definition of outsourcing. The following list includes the types of IS outsourcing service contracts that the IS community is being required to address:

- Time-sharing and applications processing.
- Contract programming.
- Software and hardware maintenance.
- Contingency planning and disaster recovery planning and services.
- Systems development and project management.
- Electronic data interchange services.
- Network management.
- Reengineering services.
- Transitional services.
- Limited facilities management.
- Full facility management.
- Remote LAN management.

It should be noted that the first six services in this list have been outsourced for at least 20 years. The remainder of the list represents expansions of the other services. For example, facilities management is the use of time-sharing on a broader basis, and remote LAN management is hardware maintenance on a distributed basis.

WHY OUTSOURCE?

Outsourcing should be specifically tailored to the business needs of an organization. It appears to be most feasible for those organizations with the following characteristics:

- *Organizations in which IS is not a competitive tool.* If there is little opportunity for an organization to distinguish itself from its competition through systems applications or operations, there is less concern over entrusting the execution of these services to a third party.

- *Organizations in which short-term IS interruptions do not diminish the organization's ability to compete or remain in business.* An outsourcing vendor should be able to recover operations in one to two days. It is probably not reasonable to rely on a third party to recover complex systems within one to two hours. Contracts can be structured to specify that the outsourcer must recover within a one- to two-hour time frame or incur severe penalties. However, if the outsourcer fails to comply with the contract, it is unlikely that the penalty adequately compensates the organization for the long-term effects of losing customers. Therefore, the shorter the tolerable window of exposure, the less viable outsourcing becomes.

- *Organizations in which outsourcing does not eliminate critical internal knowledge.* If outsourcing eliminates internal resources that are key to the future innovations or products of the organization, the risk may be too great to assume.

- *Organizations in which existing IS capabilities are limited or ineffective.* If this is the case and the organization is considering outsourcing, management has probably determined that additional investments must be made in the area of IS. In this situation, it may make more sense to buy the required expertise than to build it.

- *Organizations in which there is a low reward for IS excellence.* In this case, even if the organization developed and operated the most effective and efficient information systems the payback would be minimal. Because every organization must capitalize on its assets to survive, the effort that would be expended could probably be spent more wisely in other areas.

MOTIVATING FACTORS

Companies have various reasons for outsourcing. Just as the outsourcing agreement itself should be tailored to the individual circumstances, the factors that cause an organization to achieve its objectives through outsourcing are unique.

It is important to understand these motivating factors when evaluating whether a particular solution meets an organization's objectives. In addition, as in all cases in which the IS manager has an opportunity to participate in the solution of a business problem (e.g., systems development audits), it is important to understand the overall objectives. To add value to the process, these objectives and their potential shortcomings should be considered when evaluating whether the outsourcing agreement maximizes asset use and maintains the control environment. For this reason, the motivating factors often cited by management, as well as some of the reasons why these objectives may not be readily met, are discussed in the following sections.

Cost Savings. As the global economy grows, management faces increased competition on reduced budgets. The savings are generally believed to be achievable through outsourcing by increasing efficiency (e.g., staff reductions, shared resources). However, several factors may preclude cost savings. Comparable reductions in service levels and product quality may occur, and comparable staff reductions may not be reflected in decreased fees to outsourcers. In addition, vendors may not achieve the economies of scale previously gained through shared hardware because many software vendors have changed licensing agreements to vary with the size of the hardware.

Fixed Cost versus Variable Cost. In some cases, management has been driven to a fixed-cost contract for its predictability. However, service levels may decrease as the cost of providing those services increases. In addition, should business needs dictate a reduction in information systems, the company may be committed to contracted fees.

Flexible IS Costs. Management may have indicated that outsourcing is preferred because it allows management to adjust its IS costs as business circumstances change. However, necessary revisions in service levels and offerings may not be readily available through the vendor at prices comparable to those agreed on for existing services.

Dissatisfaction with Internal Performance. Dissatisfaction is often cited by senior management because it has not seen the increases in revenue and market share nor the increased productivity and cost reductions used to justify projects. Many outsourcing agreements, however, include provisions to transfer employees to the outsourcer. The net result may be that the personnel resources do not change significantly.

Competitive Climate. Speed, flexibility, and efficiency are often considered the keys to competitive advantage. By outsourcing the IS function, personnel resources can be quickly adjusted to respond to business peaks and valleys. However, the personnel assigned to respond to the business needs that determine the organization's competitive position may not be well acquainted with the company's business and its objectives. In addition, short-term cost savings achieved through reactive systems development may lead to long-term deficiencies in the anticipation of the information systems needs of both internal and external customers.

Focus on Core Business. Outsourcing support functions such as IS allows management to focus on its primary business. If IS is integral to the product offering or the competitive advantage of the organization, however, a shift in focus away from this component of the core business may lead to long-term competitive disadvantage.

Capital Availability and Emerging Technologies. Senior management does not want to increase debt or use available capital to improve or maintain the IS function. If IS is proactive and necessary to support the strategic direction of the organization, however, delaying such investments may result in a competitive disadvantage. In addition, precautions must be taken to ensure that the outsourcing vendor continues to provide state-of-the-art technology.

Staff Management and Training. Outsourcing eliminates the need to recruit, retain, and train IS personnel. This becomes the responsibility of the vendor. But regardless of who these individuals report to, IS personnel need to receive training on the latest technologies in order to remain effective. After control over this process is turned over to a vendor, provisions should be made to ensure that training continues. In addition, the cost of this training is not actually eliminated. Because the vendor is in business to turn a profit, the cost of training is included in the price proposal. In addition, this cost is likely to be inflated by the vendor's desired profit margin.

Transition Management. As mergers and acquisitions take place, senior management views outsourcing as a means to facilitate the integration of several different hardware platforms and application programs. In addition, some managers are using outsourcing as a means to facilitate the organization's move to a new processing environment (e.g., client/server). However, knowledge of strategic information systems should not be allowed to shift to an outside vendor if the long-term intention is to retain this expertise within the organization. In such cases, the maintenance of existing systems should be transferred to the outsourcer during the transition period.

Reduction of Risk. Outsourcing can shift some of the business risks associated with capital investment, technological change, and staffing to the vendor. Because of decreased hands-on control, however, security risks may increase.

Accounting Treatment. Outsourcing allows the organization to remove IS assets from the balance sheet and begin to report these resources as a nondepreciable line item (e.g., rent). The organization should ensure that outsourcing is not being used as a means of obtaining a capital infusion that does not appear as balance sheet debt. This can be achieved if the outsource vendor buys the organization's IS assets at book (rather than market) value. The difference is paid back through the contract and, therefore, represents a creative means of borrowing funds.

All of these driving forces can be valid reasons for senior management to enter into an outsourcing arrangement. It should be noted that the cautions discussed in the previous sections are not intended to imply that outsourcing is undesirable. Rather, they are highlighted here to allow the reader to enter into the most advantageous outsourcing agreement possible. As a result, these cautions should be kept in mind when control measures are considered.

CONTROL MEASURES

Although it is desirable to build a business partnership with the outsource vendor, it is incumbent on the organization to ensure that the outsourcer is legally bound to take care of the company's needs. Standard contracts are generally written to protect the originator (i.e., the vendor). Therefore, it is important to critically review these agreements and ensure that they are modified to include provisions that adequately address the following issues.

Retention of Adequate Audit Rights. It is not sufficient to generically specify that the client has the right to audit the vendor. If the specific rights are not detailed in the contract, the scope of a review may be subject to debate. To avoid this confusion and the time delays that it may cause, it is suggested that, at a minimum, the following specific rights be detailed in the contract:

- Who can audit the outsourcer (i.e., client internal auditors, outsourcer internal auditors, independent auditors, user-controlled audit authority).
- What is subject to audit (e.g., vendor invoices, physical security, operating system security, communications costs, and disaster recovery tests).
- When the outsourcer can or cannot be audited.
- Where the audit is to be conducted (e.g., at the outsourcer's facility, remotely by communications).
- How the audit is conducted (i.e., what tools and facilities are available).
- Guaranteed access to the vendor's records, including those that substantiate billing.
- Read-only access to all of the client company's data.
- Assurance that audit software can be executed.
- Access to documentation.
- Long-term retention of vendor records to prevent destruction.

Continuity of Operations and Timely Recovery. The time frames within which specified operations must be recovered, as well as each party's responsibilities to facilitate the recovery, should be specified in the contract. In addition, the contract should specify the recourse that is available to the client, as well as who is responsible for the cost of carrying out any alternative action, should the outsourcer fail to comply with the contract requirements. Special consideration should be given to whether these requirements are reasonable and likely to be carried out successfully.

Cost and Billing Verification. Only those costs applicable to the client's processing should be included in invoices. This issue is particularly important for those entering into outsourcing agreements that are not on a fixed-charge basis. Adequate documentation should be made available to allow the billed client to determine the appropriateness and accuracy of invoices. However, documentation is also important to those clients who enter into a fixed invoice arrangement. In such cases, knowing the actual cost incurred by the outsourcer allows the client to effectively negotiate a fair price when prices are open for renegotiation. It should also be noted that, although long-term fixed costs are beneficial in those cases in which costs and use continue to increase, they are equally detrimental in those situations in which costs and use are declining. Therefore, it is beneficial to include a contract clause that allows rates to be reviewed at specified intervals throughout the life of the contract.

Security Administration. Outsourcing may be used as an agent for change and, therefore, may represent an opportunity to enhance the security environment. In any case, decisions must be made regarding whether the administration (i.e., granting access to data) and the monitoring (i.e., violation reporting and follow-up) should be retained internally or delegated to the outsourcer. In making this decision, it is imperative that the company have confidence that it can maintain control over the determination of who should be granted access and in what capacity (e.g., read, write, delete, execute) to both its data and that of its customers.

Confidentiality, Integrity, and Availability. Care must be taken to ensure that both data and programs are kept confidential, retain their integrity, and are available when needed. These requirements are complicated when the systems are no longer under the physical control of the owning entity. In addition, the concerns that this situation poses are further compounded when applications are stored and executed on systems that are shared with other customers of the outsourcer. Of particular concern is the possibility that proprietary data and programs may be resident on the same physical devices as those of a competitor. Fortunately, technology has provided us with the ability to logically control and separate these environments with virtual machines (e.g., IBM's Processor Resource/System Management). It should also be noted that the importance of confidentiality does not necessarily terminate with the vendor relationship. Therefore, it is important to obtain nondisclosure and noncompete agreements from the vendor as a means of protecting the company after the contract expires. Similarly, adequate data retention and destruction requirements must be specified.

Program Change Control and Testing. The policies and standards surrounding these functions should not be relaxed in the outsourced environment. These controls determine whether confidence can be placed in the integrity of the organization's computer applications.

Vendor Controls. The physical security of the data center should meet the requirements set by the American Society for Industrial Security. In addition, there should be close compatibility between the vendor and the customer with regard to control standards.

Network Controls. Because the network is only as secure as its weakest link, care must be taken to ensure that the network is adequately secured. It should be noted that dial-up capabilities and network monitors can be used to circumvent established controls. Therefore, even if the company's operating data is not proprietary, measures should be taken to ensure that unauthorized users cannot gain access to the system. This should minimize the risks associated with unauthorized data, program modifications, and unauthorized use of company resources (e.g., computer time, phone lines).

Personnel. Measures should be taken to ensure that personnel standards are not relaxed after the function is turned over to a vendor. As was noted earlier, in many cases the same individuals who were employed by the company are hired by the vendor to service that contract. Provided that these individuals are competent, this should not pose any concern. If, however, a reason cited for outsourcing is to improve the quality of personnel, this situation may not be acceptable. In addition, care should be taken to ensure that the client company is notified of any significant personnel changes, security awareness training is continued, and the client company is not held responsible should the vendor make promises (e.g., benefits, salary levels, job security) to the transitional employees that it does not subsequently keep.

Vendor Stability. To protect itself from the possibility that the vendor may withdraw from the business or the contract, it is imperative that the company maintain ownership of its programs and data. Otherwise, the client may experience an unexpected interruption in its ability to service its customers or the loss of proprietary information.

Strategic Planning. Because planning is integral to the success of any organization, this function should be performed by company employees. Although it may be necessary to include vendor representatives in these discussions, it is important to ensure that the company retains control over the use of IS in achieving it objectives. Because many of these contracts are long-term and business climates often change, this requires that some flexibility be built into the agreement to allow for the expansion or contraction of IS resources.

In addition to these specific areas, the following areas should also be addressed in the contract language:

- Definition and assignment of responsibilities.
- Performance requirements and the means by which compliance is measured.
- Recourse for nonperformance.
- Contract termination provisions and vendor support during any related migration to another vendor or in-house party.
- Warranties and limitations of liability.
- Vendor reporting requirements.

PROTECTIVE MEASURES DURING TRANSITION

After it has been determined that the contractual agreement is in order, a third-party review should be performed to verify vendor representations. After the contract has been signed and as functions are being moved from internal departments to the vendor, an organization can enhance the process by performing the following:

- Meeting frequently with the vendor and employees.
- Involving users in the implementation.
- Developing transition teams and providing them with well-defined responsibilities, objectives, and target dates.
- Increasing security awareness programs for both management and employees.
- Considering a phased implementation that includes employee bonuses for phase completion.
- Providing outplacement services and severance pay to displaced employees.

CONTINUING PROTECTIVE MEASURES

As the outsourcing relationship continues, the client should continue to take proactive measures to protect its interests. These measures may include continued security administration involvement, budget reviews, ongoing reviews and testing of environment changes, periodic audits and security reviews, and letters of agreement and supplements to the contract. Each of these client rights should be specified in the contract. In addition, a continuing review and control effort typically includes the following types of audit objectives:

- Establishing the validity of billings (IBM's Systems Management Facility type-30 records can be used).
- Evaluating system effectiveness and performance. (IBM's Resource Management Facility indicates the percentage of time the central processing unit is busy. As use increases, costs may rise because of higher paging requirements.)
- Reviewing the integrity, confidentiality, and availability of programs and data.
- Verifying that adequate measures have been made to ensure continuity of operations.
- Reviewing the adequacy of the overall security environment.
- Determining the accuracy of program functionality.

AUDIT ALTERNATIVES

It should be noted that resource sharing (i.e., the sharing of common resources with other customers of the vendor) may lead to the vendor's insistence that the audit rights of individual clients be limited. This is reasonable. However, performance review by the internal audit group of the client is only one means of approaching the control requirement. The following alternative measures can be taken to ensure that adequate control can be maintained.

Internal Reviews by the Vendor. In this case, the outsourcing vendor's own internal audit staff would perform the reviews and report their results to the customer base. Auditing costs are included in the price, the auditor is familiar with the operations, and it is less disruptive to the outsourcer's operations. However, auditors are employees of the audited entity; this may limit independence and objectivity, and clients may not be able to dictate audit areas, scope, or timing.

External Auditor or Third-Party Review. These types of audits are normally performed by an independent accounting firm. This firm may or may not be the same firm that performs the annual audit of the vendor's financial statements. In addition, the third-party reviewer may be hired by the client or the vendor. External auditors may be more independent than employees of the vendor. In addition, the client can negotiate for the ability to exercise some control over the selection of the third-party auditors and the audit areas, scope, and timing, and the cost can be shared among participating clients. The scope of external reviews, however, tends to be more general in nature than those performed by internal auditors. In addition, if the auditor is hired by the vendor, the perceived level of independence of the auditor may be impaired. If the auditor is hired by each individual client, the costs may be duplicated by each client and the duplicate effort may disrupt vendor operations.

User-Controlled Audit Authority. The audit authority typically consists of a supervisory board comprising representatives from each participating client company, the vendor, and the vendor's independent accounting firm and a staff comprising some permanent and temporary members who are assigned from each of the participating organizations. The staff then performs audits at the direction of the supervisory board. In addition, a charter, detailing the rights and responsibilities of the user controlled audit authority, should be developed and accepted by the participants before commissioning the first review.

This approach to auditing the outsourcing vendor appears to combine the advantages and minimize the disadvantages previously discussed. In addition, this approach can benefit the vendor by providing a marketing advantage, supporting its internal audit needs, and minimizing operational disruptions.

CONCLUSION

Outsourcing arrangements are as unique as those companies seeking outsourcing services. Although outsourcing implies that some control must be turned over to the vendor, many measures can be taken to maintain an acceptable control environment and adequate review. Some basic rules can be followed to ensure a successful arrangement. These measures include:

- Segmenting the organization's IS activities into potential outsource modules (e.g., by technology, types of processing, or businesses served).
- Using analysis techniques to identify those modules that should be outsourced.

- Controlling technology direction setting.
- Treating outsourcing as a partnership, but remembering that the partner's objective is to maximize its own profits.
- Matching the organization's business needs with the outsource partner's current and prospective capabilities (e.g., long-term viability, corporate culture, management philosophy, business and industry knowledge, flexibility, technology leadership, and global presence).
- Ensuring that all agreements are in writing.
- Providing for continuing review and control.

The guidelines discussed in this chapter should be combined with the client's own objectives to develop individualized and effective control.

VI-4

More Productive User Interfaces

Paul Nisenbaum
Robert E. Umbaugh

THE INTERFACE IS THE USER'S WINDOW to any system. Although standard screen structure, the use of icons and color, and sharper screen quality have all contributed to improvements in the way systems are presented to a wide audience of users, an interface encompasses far more than screen format.

As a mechanism that lets people customize a user interface to their needs, a flexible user interface relates to the total hardware and software system. Most computer hardware and software, either purchased or developed in-house, provides some degree of flexible user interface that allows users to control their computing system. For example, users can adjust the contrast on a display monitor, assign synonyms to commands, create macros, and split the screen for multiple windows.

These and other components of a flexible interface have become expected standard features. Most people probably do not even think about whether the interface is flexible, but if they did not have the ability to customize function keys or set the time to military or AM/PM format, they would no doubt notice its absence.

PC users demand the ability to easily control hardware and software. Even minicomputer and mainframe vendors now offer personal computer-like interfaces. The implementation of client/server systems that let PC users access mainframe information also signifies that people want to control their computers on their own terms. Without a doubt, people want the ability to tailor their computing systems.

Yet, although a system may have a flexible, efficient, and intuitive user interface, the interface may be counterproductive unless it enables users to complete goals, not just subordinate tasks. It is the responsibility of the IS manager and system designer to ensure that gains in efficiency also contribute to the achievement of overall system goals.

NEED FOR GOAL-ALIGNED INTERFACES

The following example illustrates the importance of aligning flexible user interfaces with system goals. An inventory control clerk has the ability to customize the online data entry application and workplace environment in the following ways:

- Designating abbreviations, so that by typing fewer characters and pressing one function key, whole phrases or paragraphs are displayed.
- Changing and intermixing type fonts.
- Changing colors and highlights.
- Rearranging an online form to make entering both data and text easier.
- Adjusting the keyboard and monitor to reduce fatigue and prevent carpal tunnel syndrome.

If the goal is to quickly and accurately itemize the inventory, however, a portable bar code system could be a better solution. Understanding the overall goal, working with the user and others to precisely define what tasks need to be accomplished, and restructuring work steps would have been a better approach to defining the right process and the subsequent system interfaces.

EXAMPLES OF NONPRODUCTIVE FLEXIBILITY

Merely defining a goal first does not guarantee an efficient way of reaching it. Consider the following example of the data center goal of monitoring a workload such as the payroll system.

Workload Monitoring

In this example, monitoring software can display four workload states: not scheduled to run, running on schedule, running behind schedule, and unable to complete on schedule.

To provide maximum user flexibility, hardware and software could be provided that would allow each staff member on each shift to highlight these states with a choice of 1,000 colors. Although this might be beneficial for a single user, each time another person approached the console, the new user would have to determine which color represented a particular state. This user in turn might spend time reconfiguring the colors to his or her liking. The situation may then be repeated when yet another person approaches the console.

Instead of accruing time lost in terms of users figuring out the color scheme and customizing the highlighting, a more efficient solution would be to limit either the color choices or the customization privileges, or to limit a combination of the two with default settings. Although the user interface with 1,000 color choices may be easy to use and flexible, it does not necessarily help meet the overall goal of monitoring the workload.

Navigating from One Panel to Another

Flexible interfaces also let users customize the subordinate task of navigating from one panel to another. Variations of this example can be seen in microcomputer, minicomputer, and mainframe systems. The user can choose from among the following ways to select the name of the destination panel:

- Selecting panel destination choices from a sequence of menus.
- Selecting a labeled push button.
- Typing the destination panel name on the command line.
- Typing the destination panel abbreviation name on the command line.
- Typing a user-created synonym for the destination panel name on the command line.
- Selecting the destination panel from a help panel.
- Selecting an icon (either system provided or user created) representing the destination panel.
- Selecting the destination panel from an index.

Knowing which approach is best in a given situation results from careful analysis and consultation with users. Providing all of these options would probably be overkill.

In both user interface examples described—assigning colors to objects and navigating between panels—an appropriate balance must be reached between too little and too much flexibility. Are the many color selections and the navigation techniques helpful in satisfying the varied desires and needs of users? Do they help users reach stated goals and improve productivity, or do they simply carry a flexible interface to an extreme? If the user has to spend too much time deciding which way to do something or learning a multitude of techniques, productivity may suffer.

CREATING USABLE AND PRODUCTIVE SYSTEMS

Three steps are key to developing systems with beneficial flexible user interfaces:

1. Identifying the users and their characteristics.
2. Identifying user tasks.
3. Involving users continuously.

IDENTIFYING USERS AND THEIR CHARACTERISTICS

The first step in design is to determine who will be using the completed system. Collecting information about these people in an organized fashion helps ensure that fewer items fall through the cracks and that similarities and differences among users become apparent. The system can then be designed to accommodate various user characteristics.

Helpful information includes current and previous job titles or levels, formal and informal education, and experience with various computer systems. Less tangible, but equally important, may be the user's motivation for learning and using new systems.

Collecting User Information

Whether to poll all users or only a representative sample depends on the number of potential users. If 10 people will be using the system, then it probably makes sense to question all of them. For 2,000 users, a sample would be sufficient.

For best results, the sample should include users with various skill levels and job classifications who will actually use the system. It is a mistake to design an interface for the most skilled or most knowledgeable user. A more effective approach is to design a system with the needs of less gifted users in mind and provide enough flexibility for the more advanced users to build their own short-cuts and techniques.

For example, if the payroll manager and IS manager agree on building a new payroll application, they will no doubt have definite ideas about the design. Once the application is up and running, however, they may have only peripheral involvement with it. Within the payroll department, the application administrator and others may be interested in customization and various bells and whistles. But the primary users of the application, the data entry staff, might not have the skill, time, nor desire to explore computer applications. It is therefore essential that the everyday data entry users, as well as the more experienced and the less frequent users, be included in the collection sample.

Exhibit VI—4-1 provides a sample form for collecting information on user characteristics; it can assist the designer and be tailored or used as is. For increased efficiency, it may make sense to combine the step of gathering user characteristics with the next step: identifying current goals and user tasks.

IDENTIFYING USER TASKS

The importance of identifying user tasks should not be underestimated in the development of systems and user interfaces. Special attention needs to be given to collecting this information. User tasks can be as broad as getting the payroll out and as narrow as deleting a looping transaction with one keystroke. It is crucial to identify goals and their subordinate tasks.

Examples of Goals and Subordinate Tasks

The system designer needs to determine the level of detail to be gathered. The first visit to a user may be to gather the higher level tasks, with subsequent visits used to focus on fine details. For example, if the goal is to restore an accidentally

Part 1: User Characteristics	
For project	
Date(s)	
Department	
Observer	
Name	
Job title	
Job level	(entry-level, senior)
Department	
Education	
Degrees	
Additional courses, seminars	
Experience	
Previous job titles	
Current job	(years)
Subject area	(years)
Microcomputer	(years)
Minicomputer	(years)
Mainframe	(years)
Use computer on current job	(daily, quarterly)
Motivation	(reluctant, eager)

Part 2: Current Hardware	
Mainframe or minicomputer terminal type	(monochrome, four color)
Personal computer	(display: monochrome—amber or green, VGA, SVGA; sound; RAM; hard drive; LAN or host emulator cards)
Data entry	(Keyboard, mouse, touch screen, voice)
Printer	(Local, remote, paper type, speed duplex, multiform)
Modem	(Internal, external, baud rate, LAN-connected)
Miscellaneous	(Swivel monitor stand, printer stand, adjustable chair or desk, lighting)

Exhibit VI-4-1. Sample Form for Collecting User Information

Part 3: Current Software Applications		
Application, Brand, and Version	**Frequency Used (Hourly/Quarterly)**	**Expertise (Novice/Expert)**
CAD/CAM		
CASE tools		
Customer support		
Data base		
Desktop publishing		
E-mail		
Financial planning		
Graphic presentation		
Help desk		
Human resources		
Inventory		
Operating system		
Payroll		
Printer management		
Problem management		
Programming (languages)		
Service-level management		
Spreadsheet		
Statistics		
System administrator (name system)		
Tape management		
Utilities Word processing		

Exhibit VI-4-1. *(continued)*

deleted file, the tasks are to phone the data center and ask to have the file restored, locate the tape on which the file resides, and request a tape mount.

If the goal is to print the monthly problem management report, the tasks are: collect the daily and weekly reports from remote machines by uploading to the mainframe, run the job to create the monthly report, set up the printer with preprinted forms, and submit the print job.

Understanding Users' Human Nature

As with other endeavors involving people, things may not always appear as they seem. For reasons that are both conscious and subconscious, people may not be as forthcoming, direct, or honest as might be expected. Personal agendas and corporate politics transcend even the most seemingly cut-and-dried situation.

When approached with questions, some people feel they are being judged and must give a correct answer. Because they do not want to appear ignorant to the observer or their peers or supervisors, their answers may not accurately reflect their job. Other people may not speak candidly for fear of retribution, no matter how much assurance to the contrary they are given. They may have been reprimanded in the past or know of others who have had a similar negative experience. Still other users may emphasize only a small part of their job because that is the area in which they feel most competent. Gaining a more balanced view of some users may necessitate several visits over a few days or weeks.

Many approaches for gathering information about user goals and tasks can be concurrently employed to cover a wide variety of users and their settings. The recommended approaches in the following sections can be adapted or expanded to the particular needs of the designers.

Remaining Open to Surprises. Acute observation of a user without preconceived notions can lead to startling revelations for both the designer and the user. Keeping an open mind early in the design phase avoids costly and time-consuming work later on.

Observing the User or Users on Several Occasions. Personal observation by the designer is the best way to begin to gather information about the user interface. Having personal knowledge of the details of the application is helpful but not absolutely necessary to the experienced observer. Because making a single observation is risky, the designer should try to schedule observations at different times of the day and on different days of the week. In some cases, the system may be suffering from slow response time, the user may be doing nontypical tasks, or personal issues may be affecting the user's performance. When an interface is being designed for a task for the first time, the designer will have to make observations on the manual tasks and use judgment and experience to design a pilot interface for the user to try.

Asking Users Why They Do Tasks in a Certain Way. Designers should listen carefully to what users say about why they perform tasks in a specific way and ask users if they are satisfied with the current way of doing things. If they are not, the designer should ask why or how tasks could be performed differently. Could they be combined or eliminated? And how often are they performed?

To stimulate the user's thinking about options, the designer may opt to describe typical options or demonstrate optional interfaces in a technology demonstration lab. The expense of such labs hinders most companies from using them, but those that do report high levels of success in designing user interfaces that are productive and goal-aligned.

Meeting with a Single User at a Time. Many people are reluctant to speak up when they are part of a group, especially when talking about new technology or new ways of doing things. Meeting with single users often lets the designer probe more deeply into tasks and goals and broadens understanding.

Meeting with a Group of Coworkers Who Share Common Responsibilities. Although there are distinct advantages in meeting with single users, meeting in groups yields valuable information as well. Users stimulate ideas in each other and, perhaps more important, group discussions often indicate areas of disagreement or conflict that the designer should follow up on later.

Meeting with Coworkers of an Employee to Gather Information on the Tasks and Goals of Other Jobs. Although this technique is not always as productive as meeting with the users themselves, it occasionally provides revealing information that can make the difference between success and failure. The designer needs to be a skilled interviewer who can differentiate between comments related to the task and those that reflect personal observations about the person doing the job.

Meeting with Customers—In-House and Out. Meeting with customers is especially critical if the proposed or existing system involves electronic document interchange or is expected to at some point in its life.

Asking if the Task Is Essential to Complete the Specific Goal. In all meetings with users it is important that the designer repeatedly probe whether all the tasks being systematized are necessary to achieving the overall goal. This is especially true when going from manual to computerized tasks for the first time. The older the company the more likely it is that the task is being done because "that's the way we've always done it."

Meeting with Vendors. There are many reasons to meet with vendors, both software and hardware, when considering alternative user interfaces. Vendors are often a good source of information regarding new technology, and they can constructively critique the designers' ideas for proposed interfaces. The IS management team should never allow the vendor to design or dictate the interface, however, unless the vendor is under contract to do so and has special expertise in this area.

Asking the User's Supervisor About the Goals and Tasks of the Job. It is not uncommon for IS staff to design and build a system based on what they think is good input from users only to find that a supervisor or higher management do not like it or, worse yet, were about to restructure jobs and thereby invalidate all or some of the system elements. Checking up the ladder makes good sense.

INVOLVING USERS CONTINUOUSLY

Including users in an iterative design process helps ensure that the final product allows users to meet their overall goals and provides appropriate facilities for users to customize the user interface. The iterative process includes demonstrations of work in progress to users at various developmental stages. Paper and online prototypes can be used initially, with the actual product demonstrated as it evolves. Sessions can be held both for individuals and for groups, with hands-on sessions scheduled when appropriate. The designer needs to constantly monitor user input and adjust the product accordingly.

A usability lab can provide a no-risk opportunity for users to try out a system during the latter stages of development. At the same time, it has the potential to give system builders insight into exactly how users will exercise the system once it goes into production. The usability lab is also often used to stress parts of the system during development in order to identify areas of weakness, missing features, or inappropriate interfaces. The labs can be located in-house or contracted out, and they vary greatly in complexity and cost.

First Iteration

When the designer initially sorts through the collected user characteristics, goals, and tasks, and other information, the overall goals will probably appear fairly obvious. A high-level design based on these goals must be presented to users. Without a confirmation of this system view, the entire project may face an uphill battle. The designer needs to demonstrate that the system is a tool to make the users' jobs more productive.

Continuing the Cycle of User Feedback

As the design evolves, care should be taken to incorporate common areas of concern noted during the examination of the user input. When the users view the next level of design, they should express confidence in their new tool. Designers who listen to and incorporate feedback from users into the next design iteration not only make the users feel part of the process but give them a stake in the eventual adoption and use of the system.

This cyclical process needs to progress quickly so the users do not lose interest. Using rapid prototyping techniques can maximize effort and feedback. Actual scenarios give life to the design, and users can participate by asking specific questions, such as, "What if I selected this field?" or "What happens on the last Friday of the quarter if it is part of a three-day holiday?" Eventually, live data should be used in the user review sessions to help test the system and generate enthusiasm. Once again, feedback about the use of the system needs to be carefully evaluated and, if appropriate, incorporated into the next iteration. Here again, a usability lab may provide an effective and productive tool.

BALANCING FLEXIBILITY AND PRODUCTIVITY

As is the case with almost anything related to systems development, the decision of how much flexibility and how many options to give users involves tradeoffs. Users accustomed to using systems based on dumb terminals—so-called green screens—will find almost any degree of flexibility an improvement. But the step from green screens to intelligent terminals is usually so great that both designers and users can get carried away if some reasonable limitations are not applied.

Improvements that make a user's job go faster and smoother with less strain are easy to justify in most cases because of the inherent power and flexibility already built into most intelligent terminals. It is when further systems development work that results in significant investment of project time and money is needed that some form of cost/benefit analysis is necessary, however informal it might be. Today's emphasis on worker eye strain and injuries to the wrist and arms may make it prudent to provide as much flexibility as the project budget will allow.

The soundest approach therefore may be to allow good judgment and common sense to prevail. The final test is that user management be satisfied that their money is being well spent and that the flexibility provided will pay a return either in direct or indirect benefit to the enterprise.

CONCLUSION

Although the participation of users in the systems development process can clearly lead to better systems, their involvement needs to be carefully managed. It bears repeating that the designer should always remember that people sometimes bring personal agendas into the workplace that may have a negative influence on a seemingly impartial opinion. When involving users in design, designers should work to maintain a positive environment.

One rule to remember is that users may expect their input to be visible in the product. If they do not see their requests in the design, they may reject further participation, become hostile to the project, and discourage others. If they are paying customers, they may even begin buying from other vendors.

Users may become defensive when the tasks they have been doing a certain way turn out to be inefficient or unnecessary. They may contradict or seek to invalidate the statements of others because of office politics. All this is part of the process of designing productive systems, and designers and IS managers must learn how to manage it.

VI-5

Quality Improvement in Systems Development

William J. Egenton
David O'Sullivan

ALTHOUGH QUALITY HAS EMERGED as a formal management function, the discipline is still evolving. Once exclusive to manufacturing departments, quality now embraces functions as diverse as purchasing, engineering, and marketing research and commands the attention of senior management in all industries. Despite the increased interest, quality is a term that remains easily misunderstood. Different companies mean different things when they use the word *quality*, as do different groups within the same firm.

In total quality management (TQM), it is the customer who defines quality. Quality refers to the organization's ability to meet or exceed customer requirements while maintaining a competitive market position. This customer-driven perspective on quality can also be applied to systems development work within the IS department. Even though the development of information systems is significantly different from the manufacture of hardware products, relevant quality assurance (QA) concepts that have their theoretical foundation in systems tried and tested in other industries can be adopted by IS departments. One example is the concept of process improvement as a means of assuring a quality product.

Problems and issues arise as a result of the interaction (or lack of interaction) between IS and its customers in the identification and analysis of product characteristics. Tools and methodologies such as software quality assurance, quality function deployment, structured analysis, and prototyping help to manage some problems, but they are only part of the solution. More important is how these tools can be used in combination to help IS maintain customer interaction throughout the systems development process.

IS departments must produce quality, cost-effective products for increasingly demanding customers. Within an enterprise, all end users requiring information are customers of the IS department. The requirements of the customer are the

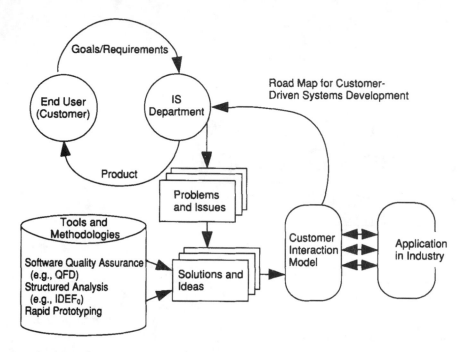

Exhibit VI-5-1. Connection Between QFD and Systems Development Processes

primary goals for the IS department in the development of information systems. IS professionals then manifest these goals in the product.

The Customer Interaction Model is a synthesis of quality function deployment (QFD) and structured analysis. Using this model, a link is developed between QFD and the systems development process (see Exhibit VI-5-1). This approach enables IS management to tailor the customer interaction methodology to its development environment. To test and validate the ideas behind the Customer Interaction Model, this article also describes how the principles are applied in an actual test site.

QUALITY ASSURANCE AND SYSTEMS DEVELOPMENT

Quality assurance in systems development has evolved from inspection methods and procedures into systems of control. QA is characterized by the requirement of conformance to formal procedures. In systems development work, associated problems include:

- The user's difficulty in specifying requirements.
- The intangible nature of information systems.

- The apparent ease of change of software and systems architecture.
- The quantities of information to be handled.
- The difficulty of exhaustively testing information systems.

Accompanying these problems are associated costs. These costs occur right through the systems development life cycle to maintenance. Many of the problems of systems development can be traced to the difficulty of defining user requirements during the analysis phase of the project. Structured techniques and CASE tools attempt to deal with this issue, but further development is still necessary.

A basic information systems framework for requirements definition is the core of this chapter. The quality function deployment methodology is a strategy that recognizes the importance of the customer in the development of new products. This chapter is mainly concerned with the interaction of the QFD process with the systems development process and the early stages of implementation of QFD in an IS department.

QUALITY FUNCTION DEPLOYMENT

QFD is a planning and management methodology used for the design or redesign of products or services. It employs an approach in which customer requirements are translated (deployed) into design and manufacturing requirements in order to deliver products or services that meet or exceed the needs of the customer. It is a way of relying on the voice of the customer to drive the product development process. Japanese industries were the first to formalize and use approaches of quality deployment based on concepts researched by Yoji Ako. When incorporated into systems development work, QFD is a structured and disciplined process that provides a means for the IS staff to identify and carry the customer perspective through each stage of systems development and implementation.

Problem-Solving Tools for Building the Customer View

Planning tools for productivity and quality improvement are an essential feature of QFD. These tools give individuals the ability to contribute to the planning step. Seven tools in particular that have evolved from operations research are now widely employed in strategic planning in organizations. These tools are referred to as the seven new planning tools. The seven new tools include:

- Affinity diagrams.
- Three diagrams.
- Matrix diagrams.
- Matrix data analysis.
- Relational diagrams.
- Program decision process charts.
- Arrow diagrams.

Affinity Diagram. Of the seven planning tools, the one that has the widest application in the QFD methodology is the affinity diagram, which provides a method for building an overall structure of ideas from a set of unstructured ideas. The hierarchy of ideas is built from the bottom up. The affinity diagram organizes the customer's views in a structure that reveals patterns and general areas of focus. The seven new planning tools, including the affinity diagram, have proved to be useful at all levels of management in an organization.

Contextual Inquiry. Vital to the successful design of a new IS product or service is a thorough understanding of the requirements of the customer. Customer requirements must be gathered in a thorough and orderly fashion. A technique called contextual inquiry, developed by Digital Equipment Corporation in the late 1980s, is a means of gathering information from customers about their work practices and experiences. The contextual inquiry approach is based on field research techniques and focuses on interviewing users as they do actual work. Although contextual inquiry is currently used primarily in software development, the technique can be used for the design of other products and services. When the requirements of the customer are gathered, they must be analyzed using the QFD process.

How the QFD Process Works

Quality function deployment covers several aspects of design and development of products, processes, and services—for example, customer requirements, value engineering, reliability, new technology, bottleneck engineering, and other disciplines. These disciplines were integrated using many matrices to analyze relationships between different aspects of product development. Researchers have since developed a more simplified approach to QFD. One approach in particular is dubbed the House of Quality.

The House of Quality approach (see Exhibit VI-5-2) identifies the level of importance of each customer requirement and how well the enterprise is able to meet these requirements. Most important, the House of Quality identifies a number of key areas for breakthrough development of a world-class product, process, or service. For an enterprise implementing QFD as part of product development, this process yields effective results in a relatively short amount of time. The House of Quality analyzes the relationships between the requirements of the customer and the design characteristics of the product being developed. This analysis investigates conflict in product characteristics and allows for a competitive assessment with similar products.

To link the QFD methodology with the systems development process, it is necessary to model this interaction as shown in Exhibit VI-5-3. A modeling technique called $IDEF_0$ is used to illustrate this link. With the $IDEF_0$ methodology, each activity in the system is represented by a box and each activity may be hierarchically deconstructed to greater levels of detail. The $IDEF_0$ methodology also distinguishes between the various types of input to the system. The input is divided into controls, input about customers, and resources. These distinctions enhance the value of the model at greater levels of detail.

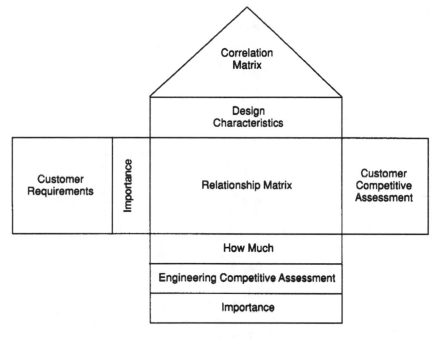

Exhibit VI-5-2. House of Quality

QFD AND SYSTEMS DEVELOPMENT

The systems development life cycle includes systems analysis, design, development, implementation, and maintenance stages. Traditionally, emphasis has been on design and development, with the result that the early stages of customer requirements analysis were neglected. It is the aim of the QFD methodology to correct this weak customer link and carry the requirements of the customer through every stage of product development.

Building the Customer Interaction Model

In the high-level diagram of the Customer Interaction Model shown in Exhibit VI-5-3, the QFD function is identified as a separate function to give it priority as an important component in the development of a quality product that successfully meets customer requirements. QFD is also an ongoing function with many components (including QFD planning and contextual inquiry) that maintains a direct line of communication between the customer and the systems development team throughout the development and maintenance of the product. The controls in the Customer Interaction Model are illustrated as arrows entering the model from the top.

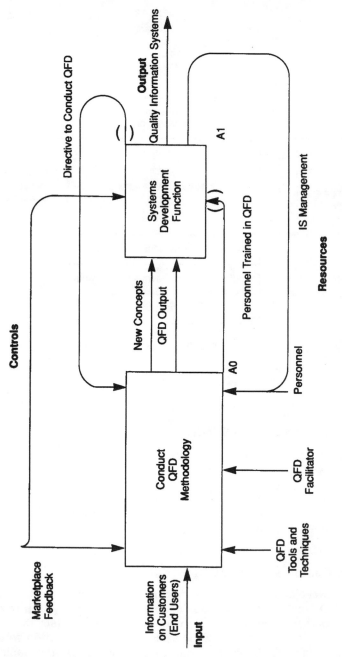

Exhibit VI-5-3. Customer Interaction Model Using the IDEF$_0$ Methodology

Marketplace feedback is the first control and includes customer complaints, field reports, and returns. Information on customers (i.e., end users) is illustrated as a single input to the model. This information includes all customer categories including potential customers.

Among the resources for the QFD function are the QFD tools and technique. The QFD facilitator is a separate resource. The facilitator is a trained person who manages the entire QFD process. The final resource of the QFD function is inherent in any ongoing business and includes personnel, facilities, technology, experience, and equipment. IS management is depicted as an output from the systems development function and feeds into the QFD function as a resource. IS managers and other personnel (including enterprise management) are trained in QFD and subsequently act as a resource to the systems development function. This completes a very important loop in the training of personnel in the principles of the QFD methodology.

There are two other links in the Customer Interaction Model in Exhibit VI-5-3. As a result of the analysis of customer data by the QFD function, ideas and new product concepts may arise. The QFD output is the final link between the QFD and systems development functions of the Customer Interaction Model. The QFD output may be summarized in the form of Pareto diagrams. The output of the model is quality information systems.

The QFD Function Deconstructed

The QFD function may be deconstructed into three activities in the next level of detail (see Exhibit VI-5-4):

- A function to manage customer interaction.
- A facilitator to manage the QFD process.
- The implementation of QFD results in the product development process.

These three activities reflect the three distinct phases that take place in the QFD methodology.

The first function, for managing customer interaction, has two goals:

- Preparation for the QFD process.
- Maintaining a customer link during the product development process.

The QFD facilitator usually conducts an actual QFD planning session for the benefit of those who will be conducting this study. The goals of the QFD planning session are to:

- Decide the focus of the quality function deployment.
- Identify customer categories.
- Identify any competitors.
- Plan and manage customer visits.
- Prepare the logistics of the QFD process.

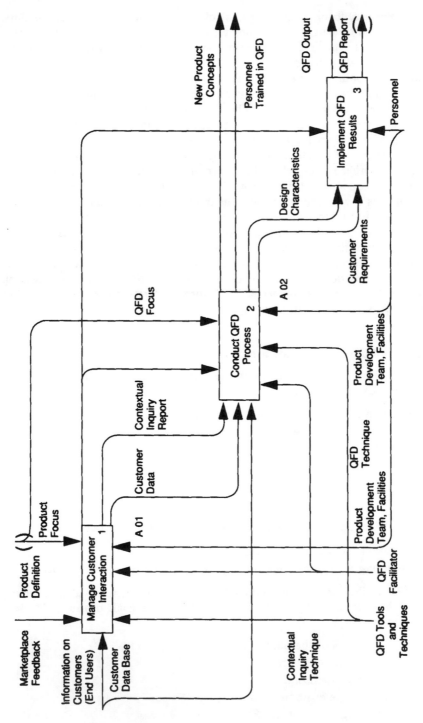

Exhibit VI-5-4. QFD Function Deconstructed

Whereas the QFD planning session may be a one-time occurrence, the customer link must be maintained during the development process to get feedback on progress and other aspects of system design. This element of the customer interaction function results in a closed loop and a continuous communication link between the systems builders and the customers throughout the life cycle.

The second activity of the QFD function is to conduct the actual QFD process and the House of Quality analysis. The final activity is implementing the results of the quality function deployment. The results of QFD are implemented by the systems development team members that participate in the QFD process

Systems Development Function Deconstructed

The systems development function is deconstructed into the four components in Exhibit VI-5-5. The systems development group develops design specifications for the system. It then develops the information system based on design specifications and subsequently tests these products. The systems development group supplies instructions and resources to maintain the products in successful use in the field. These functions are integrated and directed by the managerial function, which directs and administers the total systems development operation.

Input to the systems development function comes mainly from the QFD function of the Customer Interaction Model. The input includes new product concepts and QFD output. The systems development function uses the input to create its output: systems, documentation, support resources, and progress reports to enterprise management. The output of the QFD also influences the product definition at an enterprise management level.

Two resources that are an outcome of the QFD function act as input to the systems development function: the QFD report and the personnel trained in the QFD process. The QFD output is only a summary of the results of the QFD and therefore needs a detailed report to support its content. Reports and new product concepts are essential feedback in order to guide the organization in its development of quality products and services.

QFD: A CORNERSTONE OF TQM

IS Case Study

To test and validate the Customer Interaction Model, it was applied in an industrial test site in the IS department of a US multinational company. A case study is presented that highlights the main points of implementing the QFD methodology.

The responsibility for implementation of the QFD methodology rests with the voice of the customer group, which is a subset of the QA group. The voice of the customer group members consists primarily of IS department members and includes representatives from the entire enterprise.

An examination of the product development process within the IS department was initially done to uncover the weak links between customers and the developers of information systems within the enterprise. Next, a plan was

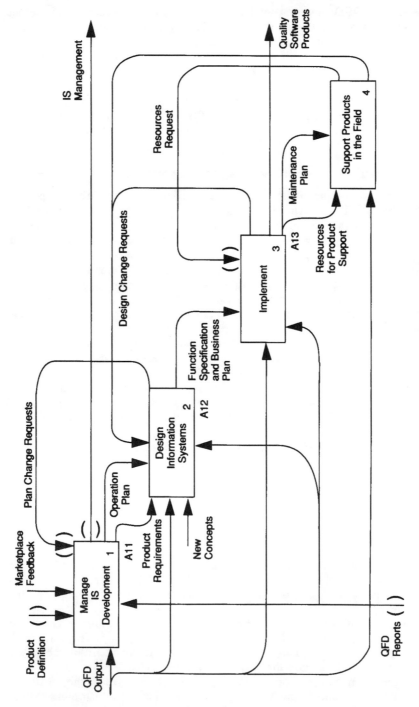

Exhibit VI-5-5. Systems Development Function Deconstructed

prepared for building a customer interaction process in the IS department. Three distinct stages were identified for the adoption of the QFD methodology in the IS department:

- Management breakthrough thinking.
- Awareness and training (phase one implementation).
- Full integration to the systems development process (phase two implementation).

Management Policy. As with the implementation of any process that involves system changes at an enterprise level, corporate management must see and understand the need for such changes. A management policy was published within the enterprise that outlined management's commitment to strengthening the customer integration process within the IS department as a means of improving quality. The first priority of the IS management was to initiate a renewed awareness among IS personnel of the importance of customer interaction in the design process.

Implementing the QFD Methodology. A number of milestones were reached in the implementation of phase one over a 12-month period. An implementation plan was developed by the voice of the customer group, and a study was conducted on the processes and tools necessary for the QFD methodology. The application of each process and tool was identified along with its relevance to the IS department and whether or not training was required. The main points of the implementation plan, developed by the voice of the customer group, are listed in Exhibit VI-5-6.

A document was published within the IS department entitled *Guide to Customer Visits* that outlines important aspects in conducting a contextual inquiry, including recommended forms and questions that may be asked during the interview (see Exhibit VI-5-7). This document helps to achieve a reasonable degree of consistency between different interviewers visiting different sites.

The final part of phase one implementation centered on the QFD process. The most effective approach to training design groups in the QFD process is to use a real-life project. This approach was adopted under the direction of the voice of the customer group. The goal of the pilot QFD was to provide input to an upgrade version of an existing system. Another important aim was to train several in-house QFD facilitators, which enables the IS department to continue with an independent implementation of the QFD methodology. A skilled QFD facilitator from corporate training was assigned to lead the pilot QFD. The voice of the customer group and the QFD facilitator then put a plan together to execute the pilot QFD process.

A planning session for the pilot QFD was scheduled and attended by the facilitator, the development team, and personnel being trained in QFD facilitation. A four-day QFD was then conducted on the system. A system development team emerged, familiar with the QFD process and with tangible input to the upgrade version of the system being examined. Better trained in-house facilitators also emerged. A *QFD Facilitator's Handbook* was published by the IS de-

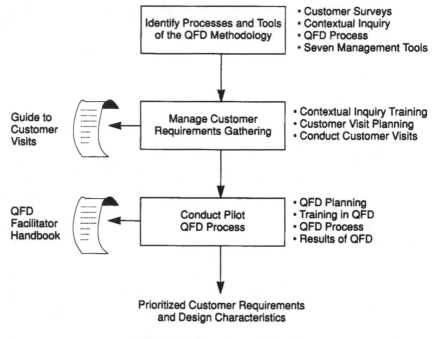

Exhibit VI-5-6. Implementation Plan

partment. The document outlines the QFD process, but more important, the handbook highlights important pointers for resolving conflict within the group and developing interpersonal sensitivity, organization, and motivation skills that facilitate the process. This approach to on-the-job training maintains the momentum of the implementation of phase one within the group.

Phase two implementation is the total integration of the QFD methodology into the systems development process after the IS department is familiar with the QFD process and its impact, based on the pilot project.

CONCLUSION

The Customer Interaction Model described in this chapter was developed in conjunction with the early implementation of the QFD methodology in an IS department. The real-world experience adds value to the model for other IS departments developing a similar customer interaction process in their enterprise.

The Customer Interaction Model is a synthesis of the quality function deployment methodology and the systems development process. The primary goal of the process described is to provide IS managers and their staff with a road map on how best to make the QFD methodology a routine part of the systems development process. The model addresses both social and technical integration issues. The $IDEF_0$ modeling technique was used to create the model because it clearly describes QFD's impact on the systems development process and identi-

Establishing Partnership with the Customer
To establish a partnership, the interviewer should:

- Explain the importance of customer participation in the interview.
- Explain that the customer knows his or her work.
- State that there is no prepared list of questions.
- State that you want the customer to show you what is important to his or her work.
- Ask specific questions about the customer's position in the company, background and education, past experience with computers, experience with a specific product (if appropriate) and what customer liked or disliked about that system.
- Know the customer's full name.

Contextual Inquiry
To uncover design opportunities, the interviewer should:

- Let the customer's work guide the discussion.
- Keep questions open-ended.
- Expand or change focus of necessary.
- Probe for solutions or workarounds.
- Follow-up on comments that contradict your entering assumptions.
- Interpret the work with the customer.

Discuss the Work
When discussing the work, the interviewer should:

- Ask to see an example when customer refers to any kind of work. The interviewer should guide the customer to speak from concrete, ongoing work or recent projects.
- Ask the customer to perform some work using a particular tool that he or she mentions.
- Use questions that keep the conversation focused (e.g., ''What are you doing and why? Is that what you expected? Can you show me?'')
- Highlight important points when taking notes and refer to these points in the wrap-up interview.

Probe Behind Workarounds
Here, the interviewer should:

- Ask about the workarounds that the customer has developed to avoid or solve problems.
- Probe behind the workaround to see how the customer thinks about the structure of his or her work.
- Ask the customer to pay close attention to how and when the tools disrupt work flow.
- Check that his or her assumptions about specific problems are the same as the customer's.

Exhibit VI-5-7. Interview Tips from the Guide to Customer Visits

Design a Solution
To design the best possible solution, the interviewer should:

- Share interpretation and design ideas, stopping to confirm that the interviewer is hearing the customer correctly.
- Ask why the customer requested changes to software.
- Use Post-It notes to represent concepts or design ideas so that the customer can visualize the ideas and validate them.
- When illustrating a prototype, ask the customer to interpret the meaning of an object or display.
- Note what matches and does not match his or her intention; probe behind the mismatches to see how the customer thinks about the work.
- Work with the customer on design alternatives that match the customer's interpretation of work.
- Note how long it takes to explain a concept that the customer does not intuitively understand.
- Note your process for clarifying information.

Wrap Up the Interview
In closing, the interviewer should:

- Summarize the critical issues for the customer.
- Ask if there is anything the customer thinks the interviewer should follow up on or capture.
- Ask about preferred questions or design ideas that did not arise during the interview.
- Give the customer hints or suggestions on system use as appropriate.
- Ask to call or visit again; invite the customer to call with any additional questions or information.
- Thank the customer for participating.

Exhibit VI-5-7. (continued)

---

fies popular tools and techniques, highlighting the stage of the QFD methodology where they are applicable. The structure of the model, in its gradual deconstruction of detail, highlights the need for management commitment for the successful implementation of the QFD methodology.

The QFD methodology is a principal cornerstone in the broad-based pursuit of a TQM culture in an IS department, and when integrated through the use of the Customer Interaction Model, it can help improve operations and effectively orchestrate many of the desired attributes, processes, and tools of TQM. The Customer Interaction Model illustrates that the QFD methodology has great potential in the development of a more holistic approach to information systems design.

VI-6

Improving the Quality of Production Systems

John P. Murray

THE CONSISTENT OPERATIONAL QUALITY of the applications systems within an organization is a matter of great concern to the IS manager. Many managers would like to strengthen the methods they use to ensure the improved quality of those systems. Although considerable progress has been made in the way applications systems are tested and controlled before they are moved to production status, there is room for improvement in many IT installations. Even in well-run installations, flaws can still evade the testing process and become problems in the production environment. Unfortunately, in poorly managed installations, the incidence of such difficulties can seriously affect productivity.

The problems caused by this situation range in difficulty from minor to serious. Minor complications may result in the application in error being corrected and redone. Sometimes that happens quickly, with little disruption. However, if the problem is serious or the application in error cannot be easily corrected or replaced, departments that require the information contained in the failed system to do their work will also experience disruption. Of course, as those departments' inability to complete their work continues, tension levels between them and the IS organization will rise.

A more serious circumstance occurs when the failure of the application affects the customers of the organization. In this case, the entire organization suffers, particularly the IS manager, who must bear responsibility for the failure.

IMPROVING THE DEVELOPMENT CYCLE

A clear trend in the process of developing applications systems is to meet the increasing demand for development in much shorter time frames. At the same time, applications are becoming larger and more complex. As this competitive nature of business increases, the tendency to take shortcuts with the testing process typically increases. In this rush to develop applications as quickly as possible, organizations ignore the well-documented costs of production processing reruns necessitated by—and the unavailability of online system caused by—application errors. The eventual impact these problems and costs can have on the company's outside customers, however, cannot be ignored by the IS manager and are not likely to be overlooked by senior management. The model office is an effective way of discovering these errors before they cause problems. This objective testing procedure ensures the quality of the applications produced by the development group and subsequently the quality of the work the data center provides to its customers.

In addition, more integration of systems, either purchased packages or systems developed in-house, is occurring. Although the concept of systems integration has merit, it also brings a high degree of risk because a problem somewhere along the line carries into other areas of the system. From the perspective of IS management, this phenomenon brings with it new vulnerability. What may have once been considered a small, or even harmless, application error causing a relatively insignificant data center problem now has organizationwide implications.

The model office is essentially a model data center that duplicates the production environment in the actual data center. In this independent facility (i.e., independent from the data center as well as the systems development group), the data center's operations are simulated and proposed applications (e.g., new versions of systems or changes to existing programs and systems) are run in a production-like environment. This provides an unbiased testing phase for these applications. Application errors, which would otherwise result in systems failures that could not only disrupt the entire production environment, but eventually affected the organization's external customers, can thus be identified and corrected. This chapter discusses the potential negative impact of application errors on the data center, examines one company's efforts to avoid the problems associated with these errors by installing a model office, and provides guidelines for establishing such a function.

APPLICATION ERRORS: IMPLICATIONS FOR THE DATA CENTER

The data center operation, like any production line process, operates under tight schedules. The requirement to rerun production work, particularly if the rerun is large, can therefore create problems. Production schedules must be reworked, which incurs additional operations section expenses, and the overtime paid to one, or perhaps several, computer operators and other support personnel adds up.

The additional expense and disruption associated with reworking batch processing cycles is one of the negative issues associated with production processing reruns. Various aspects of the online processing environment must also be considered. In many installations, the existing demand for fast, consistent, and longer online processing sessions is increasing at a rapid pace. The availability of the online system has become an increasingly important business issue—one that users have come to view as a given, much as they expect a dial tone when they pick up the phone.

The expenses associated with the unavailability of the online environment can be high. For example, one medium-sized insurance organization estimates that for each scheduled hour of online processing time the system is unavailable, the organization loses $4,000 in direct salary expense. It is difficult to calculate the actual cost to the organization of more abstract problems associated with system availability, but the cost associated, for example, with the inability to respond to questions from the field about the status of policies submitted for processing or about the status of the processing of claims can add up as well.

Customer service is important for all organizations. Despite the increasingly competitive nature of all areas of business, customer service has been singled out as the critical factor that may differentiate a business from its competitors and ensure its continued existence.

For example, online computer systems are increasingly being linked to automatic voice response telephone systems. In financial industries such as insurance and baking, these systems provide desirable business benefits to the organization's customers. The unavailability of such a system has an immediate and highly visible negative impact on the level of service the organization's customers have come to expect. The IS manager must therefore do everything possible to make certain the quality of the work produced by the data center is as high as possible.

Traditional Techniques for Testing

In most installations, standard applications systems management techniques are already in place. The process in many installations is to attempt to carry out sufficient testing of new or enhanced programs and systems during the applications development stage of the systems life cycle. When the testing is complete, the programs are turned over to the production system. As every IS manager is aware, in that process two items decide the quality of the work turned over to the data center for production processing: the quality of the testing discipline within the development section and the commitment of the people in the applications section to providing high-quality programs and systems. In applications development departments that adhere to the proper disciplines and in which management supports a rigorous testing process, the quality of the work will be high. Unfortunately, for many reasons, including the pressure to reduce IT expenses, installations that have the proper levels of application testing discipline are few.

No organization wants to publicize application systems errors that adversely affect their customers; such potential problems, however, are often found by accident, before the errors move into the organization's product mainstream. Although the preferred method would be to identify errors during the testing stage, it is far more acceptable to accidentally find them than to face the displeasure of customers. Although it is better to catch systems, programming, or operational errors in the user departments than in the customers' offices or homes, it is better yet to catch the errors before they get to the user departments. The best possible circumstance is to catch the errors before they corrupt the IS production processing environment.

No matter how much effort and attention are devoted to preventing applications systems difficulties, some will still occur. However, there are ways to reduce the frequency of the more prevalent causes of systems difficulties. These causes can often be identified and corrected before they result in systems failures. In addition, more often than not, these initial problems require a management solution rather than a technical solution.

INSTALLING A MODEL OFFICE: A CASE STUDY

The model office is a way to provide for much more effective management of application systems. The concept of the model office is straightforward. Data center hardware resources are allocated to the model office project to provide a facility in which new versions of systems, or changes to existing programs and systems, can be processed—essentially tested—in an operating environment that simulates the actual data center's operations. A primary benefit of the model office process is that, properly installed and adequately managed, its use can be taken as an absolute assurance not only that the quality of the work done in the IT department will improve but the improvement will be consistent. Even in very well-managed, tightly controlled installations, slight improvements will occur. In installations that have problems, the improvements produced can be nothing less than dramatic. Although this is a somewhat new concept, sufficient evidence exists to show that it can be cost-effective.

The experience with the model in one organization illustrates the value of the process. This is an organization that experienced considerable growth, almost 200% in three years. Consequently, the organization had to deal with the myriad problems caused by that level of growth.

In three years, the staff in the IT department grew from a total of eight employees to a total of 55. Staffing levels in other areas of the organization grew just as rapidly. Because of the growth of the business, all the existing IT applications systems had to be rewritten to adapt to the growth of the business and the expansion of services to its customers. Because of the processing characteristics of the business, coupled with the organization's explosive growth, the company retained an outside firm to custom design the required systems.

This combination of factors created considerable difficulty for the IT department. That difficulty was compounded because the other departments of the organization also had to deal with their own growth issues. As a result, there was lack of experience in several areas in dealing with the new systems. Problems would arise with newly installed systems components and initially not even be recognized as problems. By the time they were recognized, they were more serious than they would have been in an organization whose staff was more experienced with such a system.

As the process of testing and setting up the new systems moved forward, the strain on both the IT department and the other departments within the organization became intense. The occurrence of problems and subsequent systems rework, once the systems were moved to production, was too high. The progress that was made was not viewed as adequate. Obviously, better methods were necessary to ensure the quality of the applications before they were moved to the production environment.

During this period, it was not unusual (sometimes twice a week) to delay bringing up the online system because the batch updates of the previous night had to be corrected to ensure that the files for the online processing were correct.

Clearly a solution had to be found. The decision, after considerable review of the circumstances and the various alternatives, was to install a model office system as a part of the mainframe processing environment.

The Model Office Process

In this example, a small staff was established to operate the model office function. Staff members came from areas outside the IT department. The model office staff is responsible for operating new or changed programs or systems in the model office environment. The model office can be thought of as the bridge between the development section's test system and the data center's production system.

The purpose of this process is to provide scaled-down online and batch processing environments that mirror the installation's production environment. The model office is not a complete mirror image of the production processing environment. The expense associated with installing a complete mirror system would be prohibitive for most organizations. In addition, because only a rather small part of this organization's production systems would typically require testing in the model office at any given time, a full-scale replication was not required.

In this particular organization, the allocation of the IS resources necessary to accommodate the model office was carefully considered before its installation. Those involved identified the resources that would be required to test the largest system in the installation. On the basis of that determination, a request went to senior management to obtain the required resources and to allocate them to the model office.

The proposal to install the model office was bolstered by an analysis of several processing difficulties. Estimates were made of the costs, both tangible and intangible, associated with these failures. Using these examples, the management of the IT department showed how the problems could have been avoided with the model office. The case was sufficiently strong to convince senior management that the expense associated with the installation of the model office was justified.

When simultaneous requests for model office testing are registered, priorities are determined, and the most urgent system is processed first. Usually, when several systems require testing simultaneously, the model office personnel can make some scheduling adjustments and accommodate each without undue stress.

The process is a front end to the data center production systems. All developments and enhancements of any size are staged and scheduled in the model office exactly as they would be run in the actual production environment. The idea, and the reality, is that the new or changed code gets a final test before becoming live production processing. When the model office staff shows that the programs are operating correctly, those changes required to move the code to the production system are made and the programs or systems are then moved to the production environment.

The movement to the production environment from the model office is a formal process. It does not occur until the manager of the model office has confirmed that the items being tested are ready for movement to production. A benefit of the process is that the decision to move the code from the model office rests with the manager of the model office. Because the decision is removed from the applications development section, the biases of those who developed the system and did the coding are eliminated. The experience of this organization has been that the manager of the model office brings a higher level of objectivity to the process.

The Need for a Model Office

The people in the development section usually have as their mandate the requirement to bring high quality, well-documented applications into production as rapidly as possible. They are also charged with doing that at the price agreed on with the customer departments that have requested the work.

That is the way systems development is supposed to work; unfortunately, of course, things can and do go wrong. When specifications change and users who should help to develop the new or enhanced system become distracted with other work, the project falls behind schedule. As a result, to get the system operational, concessions about minor aspects of the system will be made. Items that were a part of the original specifications become thought of as cosmetic and as such are marked to be attended to at an unspecified later date. Both the operations and user documentation are promised for delivery after the system is installed. This list can be as extensive as the circumstances warrant. The goal evolves into simply getting the system operational as soon as possible.

During applications development, some original goals, such as an emphasis on quality and the delivery of complete, comprehensive documentation, become less important than the delivery of the product. Many data center managers have seen this happen. The problem, from the perspective of the data center manager, is that he or she, along with the departments that requested the project, must live with this less-than-perfect system.

What happens, consequently, is that as projects fall behind and costs rise, those who have requested the project become disenchanted. The members of the application team grow restless and want to move on to the next project. Tension and pressure mount, and as that occurs, some items requiring closer attention, such as testing and documentation, may not receive the complete attention they deserve.

The result can be that programs are turned over to production, often with little notice to the data center, that are less than adequately tested. The new system (or new or changed programs) has been declared operational by the applications section and likely by the customer who has signed off on the project to obtain some relief for his or her problems or to simply get through the process and move on to other pressing issues.

In that circumstance, the data center manager has little recourse but to process the system and hope for the best. The best seldom happens, however, and management must then deal with all the resulting problems. Of course, the IS manager could refuse to run the system; doing that could well be the best solution to the dilemma. However, with both the development section project manager and the user project manager saying they are comfortable with the system as it is, there is little the IS manager can do but comply.

THE MODEL OFFICE MANAGER

The job of the model office manager is not to bring the system into operation or to meet any project deadline. His or her job is to carry out an objective operational analysis and test of the system. This objective perspective results in a more precise assessment of the system and imposes a higher level of reality on the status and the quality of the system than when the development group and a harried customer are the arbitrators of the work.

It becomes a quality issue. The difference, compared with the usual process, is that the manager of the model office is judged on the operational quality of the system. The manager of the model office does not have to worry about such items as implementation dates or the cost of the project. Consequently, he or she can concentrate on one item—quality. The manager of the model office acts as a development watchdog for the user departments as well as IS.

When the system moves to the model office section, that group runs the system exactly as it would in the production environment. All regular systems controls (e.g., balancing, verification of data, error checking, cycle processing, and system restart capabilities) are monitored. Because the system is processed according to the documentation provided by the applications section, the documentation's adequacy is also tested as a part of the process.

When errors occur in the model office environment, members of the model office section work with both the members of the department that requested the system and the members of the development team to identify the problems and make the required corrections. Until the manager of the model office section certifies that the system is correct, it will not be moved to production status in the data center.

The return on investment for the organization used in this example has been considerable. That return is measured in hard dollars, and in more subtle ways as well. The hard dollars are associated with the ability to improve the availability of the online systems as well as a corresponding reduction in the requirement to rerun batch production systems because of design or coding problems. The most important intangible benefit is the improved perception of IS performance throughout the organization. There has also been a noticeable improvement in morale in those departments that use the systems. This is the result of reduced frustration associated with new or changed systems.

JUSTIFYING THE MODEL OFFICE

One justification for the installation of a model office is the potential benefits to IS department customers. The IS function is a customer service function. Three groups of customers should be considered when the organization is thinking about the implications of less-than-high-quality service from IT. The first level is usually referred to as the users. That group consists of employees in other departments of the organization who request the new systems or extensive enhancements to existing systems. The second level consists of the organization's senior management. The third and ultimately the most important group is made up of the customers of the organization.

As the effects of applications systems errors move up the customer level, the degree of pressure experienced by the members of the management team, and particularly the project manager, increases. The absolute worst case occurs when all three groups—department members or users, senior management, and customers—are aware of application weaknesses. It is usually the case that members of the organization's senior management are not very understanding or forgiving of applications systems problems that have found their way to the customer base.

CONCLUSION

IS managers today must do whatever they can to improve the quality of the work they produce. In addition, they must strengthen the perception of the quality of that work by the organization's senior management and the departments that use the data center services. Although there are many ways in which quality improvements can be made, the model office offers considerable benefits.

It is in the IS manager's best interest to do whatever possible to support the installation of the model office within the organization. Working to gain support for the model and approval to implement it and working closely with the manager of the function after it is operational can improve the perception of the value of the work being done by IS. The model office does impose an additional level of processing on the data center and that may require additional work for the staff. The benefits, however, are worth the extra effort it takes to install a model office.

VI-7

Secure Workstations

Leslie S. Chalmers

A MICROCOMPUTER MAY CONTAIN information that should not be available outside the company, outside the department, or to anyone but one or two others. Examples of such information include a negative performance review of an employee or plans for a new marketing campaign. Workstations that contain such information must be protected to ensure the confidentiality of the information.

Some workstations or LANs may run an application that is the system of record for vital company data. (A system of record is a system from which other applications derive their information on particular company activities. It is the one to which everyone refers if there is any inconsistency in the copies and extracts on other systems.) If the system of record were to be modified, it would have to be carefully rebuilt rather than being refreshed from another application or system. Financial systems of record may present opportunities for fraud unless they are properly secured. To do this, it is necessary to ensure that data entries to the application are genuine. Such systems must be protected to ensure the integrity of the information they contain.

Just as a user (whether authorized or not) could modify data on a workstation to contain an incorrect value, that user could also commit fraud by using the workstation to originate a false transaction. A would-be embezzler could also cause a genuine transaction to be modified while in transit, thus producing the same result. To protect the company's assets, the authenticity of both the user and the transactions must be ensured.

Some departments may prepare time-critical documents whose loss would cause a loss of business or violation of government regulations. In some cases, a LAN may be used to process an entire application with the master records residing on the file server. Such systems must be protected to ensure the availability of the information they contain.

Finally, workstations must be protected to ensure the software they run performs only those functions that management intends. Viruses can compromise the utility of a workstation by modifying software to erase files or perform other undesirable functions. Damage may also result if the workstation user is able to modify the functions of the software. For example, in trying to change the appearance of a report, the user might modify a formula in a spreadsheet template, causing the report to contain incorrect information.

DETERMINING WHAT MUST BE PROTECTED

In the best possible world, all workstations would be protected in the same way as a large mainframe. However, because full controls may cost thousands of dollars per workstation, this expense is rarely justifiable.

Not all microcomputers and LANs are alike. Many are used for word processing or simple spreadsheets that carry very little risk. However, others may contain highly sensitive information. It is not advisable to ask the person responsible for a system whether the system should be secured. The answers given may be biased by a need to feel important (i.e., suggesting that a system needs more security than it does) or a wish to avoid the nuisance of security procedures (i.e., suggesting that a system does not need security when it does).

Securing File Servers. LAN file servers should almost always be classified as requiring controls. When a large number of users' data is concentrated on a single system, the system takes on much greater importance than the individual workstations with which it communicates.

Performing Risk Analysis. The only way to ensure that controls are applied where they are needed is to have an objective way of measuring the needs of the various machines. A risk analysis should be performed for every standalone workstation and LAN in the company. Because the analysis is being applied to workstations, it should be done using an automated package that runs on those workstations. It is not necessary to make the analysis detailed or precise; a rating of high, medium, and low risk suffices. Ideally, the process should take no more than an hour or two for the user to complete.

A sophisticated risk analysis program also produces a list of recommended controls based on the responses to the questions. A system with a high need for availability but a low need for confidentiality would be given a different list of required controls than one with the reverse. Users should be required to certify to a central control point that they have performed the analysis using the approved tool and that all recommended controls have been implemented.

Risk analyses should be repeated annually for best results. Over time, users learn more and more about how to use their systems and expand the functions that they perform. Some of these new functions will introduce elements of risk that were not there at the last risk analysis. The only possible exception to this requirement would be a system that had already implemented every control that the analysis might recommend; a fully protected system does not need to be analyzed for protection needs.

RECOMMENDED CONTROLS

Limiting Access to the Workstation

Any system that requires protection must have some method for limiting who may access the system. At the simplest level, this may be done by locating the equipment in a locked room and giving keys only to the authorized users. However, this is often awkward and inefficient.

Other ways to limit access are to use lockable cabinets to contain the equipment, a system password that requires the user to enter a secret code whenever the system is booted, or a lock that prevents the use of the keyboard or the power switch until the lock is removed. These methods are all simple to use and understand, and they can be defeated. They are appropriate for systems with medium or low risks in all protection categories.

Authenticating Users. Any system with a high risk, however, must have a more robust user authentication procedure. Each individual who is authorized to use the computer system at all should be required to provide a personal user identification code and some proof of that identity (usually a password). LANs frequently provide this facility as part of the overall LAN software. Security products for standalone personal computers are available with this feature.

Systems with the highest risk should be protected with sophisticated user authentication. Products have been recently introduced that require microcomputer users to sign on using a token or smart card, small calculator-like devices that compute passwords so that they may be changed with every use.

Ensuring Information Confidentiality

Protecting the confidentiality of information requires ensuring that individuals who are not authorized to see it cannot access it when stored. One approach is to prevent anyone from having access to the system unless cleared to see everything on it. Although this is reasonable for standalone microcomputers, it is probably not practical for a LAN.

LAN Access Control. With later releases of LAN software, vendors have begun to include access control software. This may be sufficient, provided every sensitive file is placed on a protected and controlled file server. In cases in which users may have information stored on their workstation as well as the file server, additional protection at the workstation level should be considered.

Add-on microcomputer security packages can be used to control what each user is allowed to see. A file that is not to be shared by everyone is placed in a protected directory or flagged to show it is protected. Only those users who have been granted read access may then open the file. These packages can be used on both standalone and networked machines.

Data Encryption. The best protection is provided by encryption. This approach means that hardware thieves or those with access to archival media cannot see sensitive files unless they have the secret encryption key. Only an authorized user can cause the information to be decrypted. Such users must, of course, be trained not to do this casually.

Overall Controls. The organization should look for add-on security products that require a user sign-on, limit what the user can see after logging on, and encrypt sensitive data. It does not make sense to consider a product that cannot do all three, given that there are ones on the market that do. The cost of any software is much less than the cost of training staff to install and support it, so the organization should leverage the human investment into providing optimal protection.

Computer equipment is a popular target for thieves. If the hard drive inside the computer contains confidential information, thieves may get a great deal more than they bargained for. Various antitheft devices are available that effectively attach the microcomputer to the desk.

Ensuring Information Authenticity and Integrity

The same controls that are used for maintaining confidentiality may also be used to limit who may create or change the contents of a file. With most add-on security products a user may be granted read access without necessarily being able to create or update the file. In general, the fewer people who can change information, the better.

There are other considerations, however. First, it is vitally important that the software used to process information with high integrity requirements be subject to the same or tougher controls. For example, a user should not be able to easily change the formula in a spreadsheet cell because this could cause the calculation to be incorrect.

Second, the input processing must be designed to prevent false or erroneous information from being accepted into the application files. Data may be edited for such obvious errors as numeric data in an alphabetic field or a field value that is not on a table of acceptable values. It may also be necessary to introduce dual-custody kinds of controls. With such controls, one person performs the initial input and a second person verifies that it is correct. Someone trying to originate a false entry would have to gain the cooperation of the reviewer before being able to complete the transactions.

Ensuring Information Availability

There are many threats to computer hardware today. A user may power down the machine before saving work. A file may be deleted before the user realizes how critical it was. A computer virus may overwrite an entire disk before it is discovered and eliminated. A burglar could steal a computer and all the files stored on it.

Most computer systems are easily replaced with off-the-shelf hardware. What is not easy to replace is the data. The only way to ensure recoverability is to have a reasonably current backup copy of the critical files. With a backup safely stored away from the computer, almost any disaster can be overcome.

Backing Up Data. Users on a LAN should be required to place all critical files on the file server where they can be centrally backed up by the system administrator at the end of the day, but standalone microcomputers must be backed up by their owners. Most users hate making backups. It is time-consuming and unproductive (to the user) to back up a 40M-byte hard drive every day. As drive capacities grow, the time to back up a fairly full disk can become intolerable to even the most dedicated.

Several products can ease this burden. First, a backup software product can speed the process while reducing the number of diskettes needed. More sophisticated ones provide incremental or differential backups, which means backing up only those files that have changed since the last backup. This can make it possible for the user to spend only a few minutes a day at this activity with longer times reserved for doing a full system backup.

Faster, removable media may be used for the actual backup copy. Tape streaming devices and removable hard disks are faster than diskettes, so the time to write the backup copy is decreased. With larger capacities than diskettes, they also reduce the time required to change media when the backup is too big to fit on one volume.

Delete Protection. It is also recommended that utility software be installed on every system. A file is not completely deleted at first and may be easily recovered with the right tools.

Disaster Recovery. Company disaster recovery plans should include provisions for the recovery of critical distributed systems of all sizes. Sources need to be identified who can supply and install replacement equipment. A recovery location must be established before rather than after a disaster. Users should be trained and practice how to recover their critical business functions on a brand new machine. Many people who routinely create backups have never tried to use those copies for a complete system recovery.

Ensuring Systems Utility

When appropriate, there should be software life cycle controls very similar to those for mainframe programs. Software should be clearly specified, carefully developed, thoroughly tested, and well documented before it is placed in use. This concept is difficult to sell to those who consider the microcomputer a way to free themselves from those bureaucrats in the IS department. Still, it is the correct level of control for very high risk systems.

At the very least, a security package should be used to prevent the modification of software. Normal system settings should grant individuals execute-only authority to stored software. Ordinary users should not be able to change software that has been identified as being the production copy. They should also be provided with a mechanism that will ensure they do not accidentally execute other versions. For example, on a DOS computer, there could be a write-protected BAT file in the first directory on the PATH command that changes the system to the appropriate, protected directory before issuing the command to execute the program.

Computer Viruses

Computer viruses are a problem that are, so far, unique to microcomputers. Security products that limit users to execute-only authority may prevent virus infections, too. However, some viruses, such as boot-sector viruses, may elude this kind of protection, which was not designed with viruses in mind. Specialized antivirus software should prevent most virus attacks against the system.

Because new viruses are being created that even the most effective antivirus software cannot stop, users need to be taught how to avoid bringing infections into the office in the first place. In general, software should be purchased from reputable sources in shrink-wrapped packages. All other software should be banned. Even though bulletin boards contain much useful software that has been screened to make sure it is free of viruses, users may introduce the viruses themselves in their subsequent handling of that software.

Media should never be carried home and placed on a home system if other members of the household have access to the system. Regardless of how conscientious the employee may be, others may have brought home an infection that is carried back to the office by the employee.

Copyright Infringement

Software is protected by federal copyright laws. Because illegal copying is so common and easy, users may not know this simple fact. A company that does nothing to discourage copying may find itself answering charges brought by software publishers. To counter this threat, a company policy prohibiting copying must be developed and users educated about it. In addition, systems should be periodically audited to identify what software is installed.

The Software Publishers Association (SPA) provides a product, SPAUDIT, for free that will scan a system looking for telltale software product footprints, just as a virus scanner looks for viruses. SPAUDIT is designed to identify the software products of the companies that belong to SPA. This audit is not complete, but because most popular products come from SPA members, SPAUDIT provides the first step in conducting a complete audit. A user who has pirated products identified by SPAUDIT has probably pirated other programs as well.

END-USER SECURITY AWARENESS

No controls will work unless the end user is committed to protecting the system. The fanciest backup equipment and software in the world cannot install and run itself. Users who take home diskettes compromise the protection of encryption when they decrypt the information for copying. Poorly implemented add-on security software is as bad as no security.

Users must be taught about the risks they face. They may not know about all the ways they can get into trouble. Then they must be given instruction in the risk analysis process so they can determine for themselves how great a risk they face. Once they have been motivated, they will be much more likely to comply with security procedures.

Importance of Backups. Users must be taught to perform file backups and given detailed instructions on how to run the hardware and software they use for making copies. Rules for how often to make backups and where to store them must be explained. There have been instances in which users who were instructed to make backup copies of their disks were found to have made photocopies of them. Therefore nothing can be assumed about their understanding of this procedure. A check should be made after a month or two to see if they are following the procedure as specified.

Education About Viruses. Users also need an explanation of how viruses are spread and how to protect against them. If antivirus software is available, users must be shown how to install and use it. They should be told how they may notice a virus even if there is no antivirus software on their systems, and finally who they are to call or what they are to do if they detect a virus on their systems.

Education About Copyright Laws. Users must be made to understand how to copying of software puts the company at risk for copyright infraction lawsuits. If they are unsure whether it is permissible to copy a software product, they should not. Users must also be informed of the disciplinary action that will result if they do not comply with company policy.

Handling Diskettes. Handling diskettes is less of an issue today with the smaller, plastic-encased diskettes. Anyone with an older machine must be taught these lessons, however. Although diskettes are not likely to be damaged by bending or touching, the threats of heat and demagnetization remain. Diskettes must not be stored near any magnet or potential magnetic field. Older telephones with ringers can destroy the contents of a diskette when a call comes in; the ringer is controlled by a powerful magnetic force that can erase the entire diskette. Magnets used for holding display items should be stored well away from all diskettes. Overheating will eventually result in warping and other damage.

CONCLUSION

Implementing effective workstation controls involves three phases. First, an analysis is needed to determine which systems are at risk and for what. Next, the appropriate technical security tools must be selected and implemented on the systems that need them. Finally, users must be educated about these tools and the additional steps they must take to ensure the availability of the equipment and the control of the data it contains.

VI-8

Software Quality and Auditing

David C. Chou
Amy Y. Chou

SOFTWARE QUALITY is one of the important ingredients of competitive advantage in the software industry. Software users ask for the commitment of software development companies to providing quality software. Because only zero-defect products can survive in today's competitive software market, software industries manage their software quality process to meet users' needs.

Delivering high-quality software is also paramount in internal software shops that design, generate, implement, and maintain their own software products. Here customer satisfaction relies on the following software attributes: quality (i.e., accommodating requirements specified by the customer), support (i.e., providing software services and assistance whenever needed), reliability (i.e., delivery of defect-free software), timeliness (i.e., delivery of software on schedule and on time), and cost (i.e., providing an affordable price to the customer).

The software quality assurance (SQA) process guides IS shops in delivering the high-quality software customers demand. The SQA process is a technique of achieving software quality through the software engineering process. SQA activities include adopting analysis, design, coding, and testing methods and tools; applying formal technical reviews during each software engineering step; using a multitiered testing strategy; controlling software documentation and the changes made to it; implementing a procedure to ensure compliance with software development standards; and utilizing measurement and reporting mechanisms.

Two software process control methods are widely used in IS shops. They are total quality management (TQM) and the Capability Maturity Model (CMM). The International Standards Organization (ISO) has initiated another important SQA process for the European community—ISO 9000. The quality standards of ISO 9000 and its certification process ensure the quality conformance of an IS shop's product. An auditing process is needed for all three SQA methods to determine whether the elements in the quality system conform with stated quality objectives.

The ISO 9000 series is universally applicable to any total quality system, and the TQM or CMM program can be integrated into the ISO 9000 certification process. This chapter discusses how the software quality process, methodologies, and auditing process ensure software quality in the IS shop and reviews a framework and tool for implementing the ISO 9000 auditing process.

SQA PROCESSES AND METHODS

Software quality assurance (SQA) is based on three important premises:

1. Software requirements are the foundation from which quality is measured.
2. Specified standards define a set of development criteria that guide the manner in which software is engineered.
3. There is a set of implicit requirements that often goes unmentioned.

The main goal of producing a quality product is to ensure that customers are satisfied about the product delivered. Although building a software product to meet customer requirements is not a difficult job, accommodating changes in the requirements and replying to problems in a timely manner are enigmatic tasks. Customers might want to add functionality or make other changes to reflect a changing business environment. These explicit and implicit requirements create uncertainty for the SQA process.

SQA Activities

SQA activities include the following seven tasks:

- Application of technical methods.
- Conducting of formal technical reviews.
- Software testing.
- Enforcement of standards.
- Control of change.
- Measurement.
- Record keeping and recording.

SQA begins with a set of technical methods and tools that help the system analyst and designer build high-quality software. The quality assessment process is implemented through a formal technical review procedure. The formal technical review is a meeting conducted by technical staff to uncover quality problems. Software testing includes a series of test case design methods that help ensure effective error detection. However, testing is not as effective as it should be for all classes of errors.

If formal standards do exist, an SQA activity must be conducted to ensure that they are being followed. Standards may be dictated by customers or mandated by regulations, although sometimes they are self-imposed by IS shops.

The change control process, a part of software configuration management, contributes directly to software quality by formalizing requests for change, evaluating the nature of change, and controlling the impact of change.

Measurement is another integral part of SQA. Various software metrics on technical and managerial measures must be collected for analytical purposes. Record keeping and reporting provide the vehicle for the collection and dissemination of software quality information.

Identifying Defects

The best way to identify software defects is through the use of walkthroughs, reviews, and inspections. The most important quantifiable measure of goodness is the defect-discovery rate. Most SQA activities, including technical review, testing, standards enforcement, change control, measurement, and record keeping and reporting, are fulfilled through the three methods of walkthroughs, reviews, and inspections.

Walkthroughs. Walkthroughs allow the developers, or authors, of a specific system to present their design in front of the design team and other reviewers. This process helps the other members of the team know more about the project. The reviewers' job is to spot problems or errors in the product under review.

Reviews. Two kinds of reviews are conducted: management reviews and technical reviews. Management reviews provide a means for managers to track the development progress, identify problems, recommend solutions, and ensure proper allocation of resources. These reviews establish a forum between the project team and management. Technical reviews are used to examine the project's development process and planning documentation. This includes the development plan, SQA plan, verification and test plans, and configuration management plans.

Inspections. Software inspection is a method of static testing to verify that software meets its requirements, including external product requirements and internal development requirements, and their standards. The benefits of software inspection include defect reduction, defect prevention, and cost improvement. In contrast to inspections, walkthroughs do not usually follow a process that is repeatable or collect data, and hence the process cannot reasonably be studied and improved. The defect-detection efficiencies of walkthroughs are much lower than those of inspections.

The inspection process comprises the following six operational stages:

1. *Planning.* The planning stage ensures that materials to be inspected meet inspection entry criteria and the availability of the right participants and suitable meeting places and times.
2. *Overview.* Group education is held during this stage, and assignments are made to participants involved in the inspection process.

3. *Preparation.* During the preparation stage, participants learn the material and prepare to fulfill their assigned roles.

4. *Inspection.* The inspection stage comprises the defects-finding process.

5. *Rework.* During the rework stage, the author reworks all defects.

6. *Follow-up.* In this stage, the inspection moderator or the entire inspection team verifies all fixes made by the author. The ideal result is that no secondary defects are introduced.

The inspection process aims to identify defects in the software development process. Inspections are an element of a formal process where an intermediate product—such as a requirements document, design document, or code element—is examined for defects by the producer's peers. The inspections should ensure conformance with applicable specifications and standards, identify logical defects, and identify problems with the internal and external interfaces.

TQM AND CMM

Two software process control methodologies are widely used in IS shops: total quality management and the Capability Maturity Model. The SEI's Software Maturity Framework (SMF) produced the Capability Maturity Model in 1991, which is used to guide the project management control and software quality control processes. Neither methodology provides specific implementation tools.

The functions, strengths, and weaknesses of TQM and CMM are discussed in the following sections.

THE TQM MODEL

TQM has three main components: quality planning, problem-solving, and process management. The TQM process includes the following phases:

1. *Quality planning.* During this phase, IS personnel incorporate the voice of the customer and the voice of the business in establishing IS quality objectives. These objectives drive an implementation plan for quality improvement activities.

2. *Problem solving.* Based on the objectives, staff are trained in process management and the problem-solving process, initiating quality teams. If the process is measured and defined, a quality improvement team is initiated to improve the process. If the process is poorly defined or understood, process management is used to define and measure the process. Here benchmarking helps establish proven processes for software. As the team moves through the problem-solving process, the quality council reviews its progress and assists it in remaining on track.

3. *Process management.* Once the team has a proven solution to a quality problem, team members need to develop a process management system to standardize it. Then the quality council and IS management act to multiply

the benefits by replicating the improved process throughout the IS shop. All three activities fit together in a system of continuous improvement.

SEI'S CAPABILITY MATURITY MODEL

The five levels in SEI's CMM for managing the software process are similar to the process maturity levels of SMF. The five levels are:

1. *Initial.* In the initial level, until the process is under statistical control, orderly progress in process improvement is not possible. Although there are many degrees of statistical control, the first step is to roughly predict schedules and costs.
2. *Repeatable.* In the repeatable level, the organization has achieved a stable process with a repeatable level of statistical control by initiating rigorous project management commitments, costs, schedules, and changes.
3. *Defined.* Here, the organization has defined the process as a basis for consistent implementation and better understanding. At this point, advanced technology can be introduced.
4. *Managed.* Organizations at the managed level have initiated comprehensive process measurements and analysis. This is when the most significant quality improvements begin.
5. *Optimizing.* At the optimizing level, the organization has a foundation for continuing improvement and process optimization.

The five levels were selected for the following reasons:

- They represent historical phases of evolutionary improvement.
- They provide achievable improvement steps in a reasonable sequence.
- They suggest interim improvement goals and progress measures.
- They provide immediate improvement priorities once an organization's status in this framework is known.

Key Process Areas of CMM Levels

Each CMM level except for the first includes key process areas (KPAs) on which organizations must concentrate in order to raise their software processes to that level. KPAs also serve as the threshold for achieving certain maturity levels.

The required KPAs for levels two through five are as follows:

- Level 2:
  - Software configuration management.
  - Software quality assurance.
  - Software subcontract management.
  - Software project tracking and oversight.
  - Software project planning.
  - Requirements management.

- Level 3:
  - Peer reviews.
  - Intergroup coordination.
  - Software project engineering.
  - Integrated-software management.
  - Training program.
  - Organization process definition.
  - Organization process focus.
- Level 4:
  - Software quality management.
  - Quantitative process management.
- Level 5:
  - Process-change management.
  - Technology-change management.
  - Defect prevention.

Each KPA is subdivided into goals, commitment to perform, ability to perform, activities performed, measurement and analysis, and verification of implementation. Each area except goals is further defined by specific statements applicable to the area that are used to judge whether the specific software project meets the expressed criteria. For statements to be considered met, hard evidence demonstrating achievement of statement intent must be provided. An evaluation team requires hard evidence when judging the CMM level of a company.

The following sections discuss the two reviews conducted in the CMM process: the software process assessment and the software capability evaluation.

The Software Process Assessment

A software process assessment is launched by an IS shop to help improve its software development practices. This assessment is generally conducted by six to eight of the organization's senior software development professionals and by one or two coaches from the SEI or from an SEI-licensed assessment vendor. The assessment is conducted in six phases:

1. *Selection phase.* During the selection phase, the organization is identified as an assessment candidate, and the qualified assessing organization conducts an executive-level briefing.

2. *Commitment phase.* This phase encompasses the organizations' commitment to the full assessment process as evidenced by the signing of an assessment agreement by a senior executive.

3. *Preparation phase.* In the preparation phase, the organization's assessment team receives training, and the on-site assessment process is fully planned. All assessment participants are identified and briefed, and the maturity questionnaire is completed.

4. *Assessment phase.* The on-site assessment is conducted in about one week. Then the assessment team meets to formulate preliminary recommendations.

5. *Report phase.* In the report phase, the entire assessment team helps prepare the final report and presents it to assessment participants and senior management. The report includes team findings and recommendations for actions.

6. *Assessment follow-up phase.* During this phase, the assessed organization's team, with guidance from the assessment organization, formulates an action plan. After approximately 18 months, the organization should conduct a reassessment to review progress and sustain the software process improvement cycle.

The Software Capability Evaluation

After the software process assessment, an organization establishes the CMM level it has reached. The organization then pursues a software capability evaluation (SCE), which is typically conducted by an outside organization such as the government or a software contractor.

Organizations that are candidates for an SCE first complete a maturity questionnaire. An evaluation team visits the organization and uses the maturity questionnaire to help select representative practices for a detailed examination.

The examination consists of interviewing the organization's personnel and reviewing the organization's software development-related documentation. It focuses on three important premises:

1. The proposed processes will meet the acquisition needs.
2. The organization will actually install the proposed processes.
3. The organization will effectively implement the proposed processes.

The SCE is a judgmental process, and it is mandatory that all organizations for a single project be evaluated consistently. The SEI believes that the SCE method provides the necessary consistent criteria and method.

COMPARISON OF TQM AND CMM

TQM and CMM are the two important software quality methodologies used in IS shops both in the private and public sectors. However, CMM has been adopted by the US government, especially in the Department of Defense (DOD), to evaluate contractors' capability maturity and as a qualification indicator for receiving and continuing DOD contracts.

Similarities

The process life cycles of TQM and CMM include all aspects of project planning, analysis, design, and development. For example, level 2 of CMM (i.e., repeatable) allows software projects to be repeatedly delivered on schedule with reasonable quality. At level 3 (i.e., defined), the software process is defined, trained, and followed by software engineers.

TQM's process management mechanism can be used at the repeatable level to define software processes. Once these processes are defined, IS shops can move to the next level (i.e., managed), where they begin to measure and analyze the process and its products. In addition, TQM's process management can help identify the measurements necessary to determine the stability and capability of the software process.

At the highest level, the optimizing level, all software engineers apply project management tools and principles of TQM to improve the software process. TQM's problem-solving mechanism can be used at all levels of CMM to identify and reduce the causes of customer dissatisfaction. Also, TQM's quality-planning mechanism links these project management tools and processes together to achieve breakthrough improvements in productivity and quality.

Dissimilarities

TQM's principles may affect all of an organization's projects, because TQM is a comprehensive mechanism for an organization's overall quality management. CMM, however, affects only software development projects. For that reason, CMM ignores the cultural dimension of organizational processes.

Weaknesses

Both the TQM and CMM approaches to software quality control are subject to failure from several inherent weaknesses:

- Lack of expertise in a particular application domain.
- Poorly implemented statistical process control.
- Confusion of a process's infrastructure and activity dimensions.
- Lack of recommended tools, methods, or software technologies.
- Lack of an automated implementation tool, particularly one supported by artificial intelligence (AI).

To remedy the flaws inherent in both methodologies and improve the effectiveness and efficiency of the quality control process, a competent, AI-enhanced CASE (computer-aided software engineering) auditing tool should be developed.

THE ISO 9000 SERIES AND ITS CERTIFICATION

The ISO 9000 series is a set of quality assurance standards. The first standard in the series, ISO 9000, is a guide to the other four standards. ISO 9001 is a standard for quality assurance in the design/development, production, installation, and servicing areas. It applies to companies that design as well as manufacture products. The ISO 9002 standard applies to quality assurance in the prevention, detection, and correction of problems during production and installation. ISO 9003 is a standard for quality assurance in final inspection and testing. It applies mainly to manufactured products. ISO 9004 is technically not a standard because it contains guidelines on quality management and quality system elements.

ISO 9001, 9002, and 9003 must be certified through an external audit. ISO certification means only that a company has a system in place that enables it to meet the quality standards it established for itself. Because ISO 9000 does not specify quality criteria, each certified company sets its own quality standards. As long as companies conform to their own quality criteria, they retain ISO registration. ISO certification is not given to products or companies; it applies only to individual plant sites. It therefore does not guarantee that a company produces quality products. Rather, it certifies that a system of policies and procedures is in place to make the manufacture of quality products possible.

The ISO 9000 series is a uniform, consistent set of procedures, elements, and requirements for quality assurance that are universally applicable to any total quality system. A TQM or CMM program can be integrated into the ISO 9000 certification process; successful programs can guarantee an eminent ISO 9000 certification.

US companies obtain ISO 9000 certification for four reasons:

- The European community legally requires suppliers of certain regulated products to have ISO 9000 certification. This requirement has a significant impact on American companies doing business overseas.
- ISO 9000 certification enables companies to compete for business from individual customers (both foreign and domestic) that contractually require their suppliers to be certified. The North Atlantic Treaty Organization and the US Department of Defense also require suppliers to obtain ISO 9000 certification.
- Certification enables companies to differentiate themselves from noncertified competitors.
- Most important, many managers insist that the process of creating, documenting, and establishing the controls for a quality system is a catalyst for improving quality.

THE AUDITING PROCESS

One way of assessing the quality control process is through auditing. Audits should be carried out to determine that the various elements within a quality system meet stated quality objectives.

A quality audit is defined by ISO as a "systematic and independent examination to determine whether quality activities and related results comply with planned arrangements and whether these arrangements are implemented effectively and are suitable to achieve objectives."

Audits are generally initiated for one or more of the following reasons:

- To initially evaluate a potential supplier.
- To verify that an organization's own quality system continues to meet specified requirements and is being implemented.

- To verify that the quality system of a supplier in a contractual relationship continues to meet specified requirements and is being implemented.
- To evaluate an organization's own quality system against a quality system standard.

An auditing process is thus initiated for internal or external purposes. It contrasts with other SQA methods, such as walkthroughs, reviews, and inspections, which are initiated purely for internal quality management purposes.

Auditors are responsible for many duties, including:

- Complying with the applicable audit requirements.
- Communicating and clarifying audit requirements.
- Planning and carrying out assigned responsibilities effectively and efficiently.
- Documenting the observations.
- Reporting the audit results.
- Verifying the effectiveness of corrective actions taken as a result of the audit.
- Retaining and safeguarding documents pertaining to the audit: submitting such documents as required, ensuring documents remain confidential, and treating privileged information with discretion.
- Cooperating with and supporting the lead auditor.

The audit requirements and standards are set by internal and external sources. Most internal requirements and standards focus on maintaining product quality and meeting customer needs. The external sources of quality standards are usually identified by quality agencies, such as the International Standards Organization (ISO), American Society of Quality Control (ASQC), and Institute of Electric and Electronic Engineering (IEEE), and by such government agencies as the Food and Drug Administration (FDA), Federal Aviation Administration (FAA), Department of Energy (DOE), Department of Defense (DOD), and Environmental Protection Agency (EPA).

FRAMEWORK FOR SOFTWARE QUALITY AUDITING

Preparation

A software company begins its certification process by selecting the most appropriate ISO 9000 standard (i.e., ISO 9000, ISO 9001, ISO 9002, ISO 9003, and ISO 9004). After determining which ISO standard applies to its operations, the company seeks an ISO guidebook, such as the *ISO 9000 Handbook* or *ISO 9000: Preparing for Registration*, for preparing the process. These books provide guidance on documenting the work performed at every function affecting quality. They also identify the installing mechanisms to ensure that employees follow through on the documented procedures. These books are available from the American National Standards Institute (ANSI).

A company seeking ISO 9000 certification should form an internal team to identify its own software quality assurance criteria, standards, and procedures to implement its quality system, such as TQM or CMM. The entire company should follow the SQA methodology to analyze, design, and develop software. All software production should be based on the quality system and all related activities should be documented.

A qualified external audit registrar is invited to inspect the company's quality system (for information on registrars accredited in the US, contact the Registrar Accreditation Board in Milwaukee WI). This auditor conducts the auditing process and checks the actual practices and records to ensure that they are in compliance with the ISO quality system. If the results are appropriate, a compliance certificate is awarded. If any nonconformances are found during an audit, they must be corrected before certification can proceed. The certificate can be renewed every three to four years.

Stages of the Process

Auditing is divided into four stages: preparation, performance, reporting, and closure. The preparation stage starts from the decision to conduct an audit and includes the activities of team selection and on-site information gathering. The performance stage begins with the on-site opening meeting. It includes information gathering and information analysis through interviews and examination of items and records. The reporting stage encompasses translation of the audit team's conclusions into a tangible product. It includes the exit meeting with managers and publication of the formal audit report. The closure stage deals with the actions resulting from the report and the recording of the entire effort. For audits resulting in the confirmation of some weakness, the closure stage includes tracking and evaluating the follow-up actions taken to fix the problem and prevent its recurrence.

Corporate executives should review the effectiveness of the certification process. The results of certification guide them in setting up strategies for improving software quality and productivity.

A TOOL FOR THE AUDITING PROCESS

A CASE tool is a software package used to automate the systems analysis and design process. An integrated-CASE (I-CASE) tool integrates all aspects of software development activity, including project management activities. It helps automate systems analysis, design, and development. Its code generator generates usable programs for systems implementation. Because the software quality assurance process requires computation of the metrics for software validation and verification, an AI-based CASE tool is well suited to making the software quality and auditing processes more efficient.

A CASE tool for the software quality control process is expanded by adding in the auditing capability for meeting the following objectives:

- Monitoring SQA standards and procedures.
- Housing TQM or CMM operational standards in its repository for the auditing process.
- Generating assessment process harmony.

- Substituting paper documents with electronic data.
- Automating the generation audit reports.
- Tracking weaknesses found during auditing.
- Automating the auditing process.

To include the capability of implementing the CMM methodology, the knowledge base of an AI-enhanced CASE tool must contain key process areas (KPAs) separated by maturity levels. If the company chooses the TQM method, the various elements for the Plan-Do-Check-Act (P-D-C-A) cycle must be included in the knowledge base.

The measurements and metrics collected from software process assessments and software capability evaluations are also stored in the knowledge base for future usage. Any weakness found during the auditing process is recorded for correction purposes.

SUMMATION

Software quality assurance is a technique of achieving software quality through the software engineering process. The SQA process guides the IS shop in delivering the high-quality software demanded by customers. It is one of the important ingredients for gaining competitive advantage in the software industry. Total quality management and the capability maturity model are the two software process control methodologies used in the software industry. Neither provides specific implementation tools.

Software quality auditing is the process of determining whether the various elements within a quality system meet stated quality objectives. It is one way of assessing the quality control process in the IS shop. The ISO 9000 series is a uniform, consistent set of procedures, elements, and requirements for quality assurance that can be applied universally to any total quality system. Successful integration of the TQM or CMM methodology into the ISO 9000 process guarantees an eminent ISO 9000 certification that enhances the competitive advantage of US companies doing business in the international software market.

An AI-enhanced CASE tool automates software analysis, design, and coding and the performance of the SQA and auditing processes. It stores various data in its repository, such as business goals, business strategies, operational models, programming structures and codes, design specifications and requirements for the system, SQA criteria and methodologies (such as TQM and CMM), and auditing criteria and procedures for ISO 9000 certification. As a comprehensive software quality and auditing tool, it further enhances the competitive capability of the IS shop.

VI-9

Benchmarking the IT Function

Howard A. Rubin

EFFORTS TO IMPROVE QUALITY and increase efficiency are often based on benchmarking, a technique that companies use to compare themselves against other companies and identify best practices. In IT organizations, the use of benchmarking as a proactive tool for performance assessment and goal setting is rare but on the rise.

There are two fundamental reasons why this technique has not been widely applied in IT. First, benchmarking requires measurement, and few IT organizations have mature measurement programs. Second, there is a limited understanding of the key aspects of IT performance to be benchmarked.

One source of the problem is that measures in the world of IT are thought to be overhead and are not evaluated in terms of the value they add. In addition, those who have attempted to measure usually get caught up in trying to find the best measure instead of a suitable one. Finally, IT is under tremendous pressure to reduce expenses while producing more work. Most IT executives believe that to establish a baseline they have to impede progress before moving on again. If they slow down, they are unemployed. In short, baselining and benchmarking in IT are generally resisted for all of the wrong reasons.

USING BENCHMARKING TO SET PERFORMANCE TARGETS

There are some documented cases of IT organizations that have invested the time and energy to focus their benchmark efforts by identifying key areas of performance, measuring them, and then comparing themselves against others using the same rigorous measurement criteria. Here are two examples that illustrate the results of properly applying benchmarking to the IT organization.

Case 1. A bank's IT organization seeking to lower costs while strengthening the business alignment of IT. Bank X focused its internal measurement efforts on characterizing the yield of the support costs of its work prioritization process, the support process cycle time, the application quality, organizational structure, and new development productivity. By applying the same measures to a broad-based group of organizations with exemplary IT workflow management and support, the following findings became evident:

- Approximately 60% of the application enhancement requests being approved in the process had no business value.
- It was taking longer to approve work to be done than to actually do it.
- The organization had a hierarchical structure with seven levels; peer companies only had three layers.
- Support costs were 40% higher than would be expected and applications were almost twice as defect-prone compared to equivalents at other companies.
- Almost three times more staff was allocated to support the applications portfolio than was the average in the benchmark group.

Using the insights and observations obtained from the benchmarking project as a basis for setting IT performance targets, Bank X redeployed approximately 55% of its IT staff from support to new development while decreasing support cycle time by approximately 40%. The time frame for this radical change: 10 months.

Case 2. A multinational chemical company wanting to become more responsive to environmental regulations while improving the quality of its applications portfolio. Company Y focused on its applications maintenance process complexity, productivity, and cycle time. Simultaneously, it assessed its entire portfolio from the vantage points of technical quality and functional quality. It also determined its current return-on-investment for its extensive suite of support and quality assurance tools. By doing a benchmark analysis with peer companies and vendors of packages in its industry, Company Y discovered that:

- Its maintenance processes were more serial in nature and had more bottlenecks than those in the comparison group.
- Technical quality of its applications was about 22% better than others, but functional quality was 35% worse. Customers were not benefiting from these technically high-quality systems.
- Tool penetration was low. The high technical quality of the applications could be attained with roughly 67% of the current level of effort if the toolsets were used as intended.

Within three months, Company Y had a program in place that reduced the cycle time of enacting a regulatory change from an average of four months to three weeks. In addition, specific functional quality targets were being set for each application. Skills upgrading and retraining of the staff was under way, with a target of increasing support productivity by 25%.

These two examples illustrate how benchmarking can be effective when the results are used for target setting and action. Benchmarking is not a passive activity, but one in which quantifiable findings must be turned into measurable gains. Benchmarking is also a continuous process.

BASELINING: THE LINK TO BUSINESS VALUE

Perhaps the biggest challenge facing IT organizations is to establish the link between their work efforts and business value. The starting point for making this happen is obtaining a total picture of the IT organization and relating it to the production of value for the business.

The key element in making the connection is the creation of an IT baseline—a quantitative view of where the IT organization is today—focusing on six basic factors that are the drivers of IT business value:

- The applications and project inventory.
- The management practices portfolio.
- The delivery process methodology.
- The technology infrastructure.
- The business-IT interface.

Once complete, the baseline provides a framework for comparing an organization against competitive benchmarks and a context for improvement and innovation.

Among the criticisms of baselining is that it is costly, time-consuming, and involves too much introspection. Resistance to measurement is quite widespread. Less than 10% of software-producing organizations worldwide have any kind of ongoing measurements program. Only one out of six organizations that start measurement programs is successful.

To be a success, a measurement program must be perceived by all to add value to the organization's processes and products, must supply information used for decision making and organizational learning (e.g., continuous process improvement), and must serve as a common basis for communication both within the IT organization and throughout the business itself. This definition of success fits with analyses undertaken by the Software Engineering Institute (SEI) of Carnegie Mellon University in Pittsburgh.

The SEI methodology classifies an organization's software process maturity into one of 5 levels. Level 1 is the Initial level, where there is no formalization of the software process; Level 5, by contrast, is the Optimizing level, where methods, procedures, and metrics are in place with a focus on continuous improvement. According to SEI studies, 80% to 86% of organizations in the US are at Level 1.

GETTING READY TO BENCHMARK

The ability and capability to measure is a prerequisite for benchmarking. The first step is to carefully assess the IT organization's "measurement readiness."

Answers to the following questions can be a basis for setting the direction for the IT organization's measurement strategy; each answer is scaled from 0 to 5:

1. How intense is the organization's desire to improve its performance? 0—No desire; 5—Intense desire.

2. Is the enterprise willing to invest time and money to improve systems performance with measurement? 0—No; 5—The organization already allocates funds and people.

3. What is the level of the systems skill inventory for using metrics? 0—None; 5—Already in widespread use.

4. To what extent are measurement concepts known and understood by the systems staff? 0—No staff members have this knowledge; 5—Staff is 100% trained.

5. Is the systems culture opposed to using measurements at the organizational and individual level? 0—100% opposed; 5—Eager to implement.

6. To what extent is a support structure in place to foster measurement practices and perform metrics technology transfer? 0—No infrastructure; 5—An in-place team exists.

7. Are tools and repositories for acquiring and analyzing metrics data in place? 0—No; 5—A full suite and warehouse are available.

8. Does the systems organization understand its role in business processes? 0—No; 5—Yes, and it is documented and tracked through metrics.

Readers can chart their answers to these questions on Exhibit VI-9-1.

Actions for Beginners

If the answer to these questions is at the low end of the scale, the IT organization's measurement readiness is quite low. A good starting point is to contact professional societies so that experiences in measurement can be shared and exchanged. Readers can make contacts through the International Function Point User Group, the IEEE Computer Society, the Quality Assurance Institute, the Gartner Group, and at seminars offered by Digital Consulting and Software Quality Engineering.

Actions for Improving Readiness

If the IT organization's readiness is somewhat higher according to the readiness profile in Exhibit VI-9-1, several actions should be taken in parallel.

First, IS management should reinforce the measurement infrastructure by acquiring automated tools to help with collection and analysis, then build a rapid baseline using the structure outlined in this chapter and produce a measurement program design. Next, it should embark on a 30-day mission to collect and analyze information using the 80/20 rule as a guide. The metrics findings for the baseline should be consolidated into a single page and should paint a picture of the six driver areas.

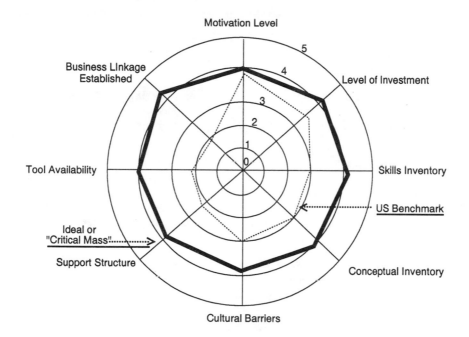

Exhibit VI-9-1. Measurement Readiness Profile

The baseline then has to be related to published benchmarks, translating the opportunities uncovered into business terms that will provide value from the efforts. This step brings focus to all the work and provides a platform for transformation and the ongoing use of measurement.

BUILDING A COMPLETE INFORMATION TECHNOLOGY BASELINE

IT productivity and quality measures do not in themselves provide a complete picture of IT performance. Establishing a baseline as a prelude to benchmarking goes beyond just productivity and quality. A complete baseline involves assessing an IT organization's current portfolio of projects and applications, its human resources and organizational structure, its management practices and processes, the technology infrastructure, and most important, business factors that drive the computing function.

The typical time it should take to construct such a baseline is approximately 30 to 60 days with a dedicated team of no more than three individuals (including consultants). The goal is to create a workable and useful organization profile rather than accomplish 100% complete data acquisition—the 80/20 rule applies here.

Six Baseline Components

A divide-and-conquer approach is needed for assessing the baseline performance of an organization; this means viewing the baseline as containing six key segments that can be combined into a single comprehensive picture.

Applications and Project Portfolio Baseline. The work performed by the software side of a typical IT organization takes the form of creating new software applications or modifications to existing applications. This baseline segment creates an inventory of the applications and current projects as they exist today "as is." Key descriptive information and metrics for each existing system and project underway includes:

- Demographics (e.g., age, language, implementation date, technology platform, and tools and techniques used).
- Financial history (e.g., cost to build, cost to maintain, cost to use, and cost to operate).
- Size (e.g., lines of code or function points counts).
- Support information (e.g., number of people on staff, number of requests, and average request size).
- Quality attributes, such as:
  - Rating of functional quality by the user (i.e., the ability to support user requirements in terms of functionality, accuracy, reliability, and data quality).
  - Rating of technical quality by systems staff (i.e., design strength, complexity, architecture, maintainability, portability, and interoperability).
  - Problem history.
  - Defects found per line of code or function point.
- Productivity attributes, such as:
  - Support ratios (e.g., lines of code or function points per support staff member).
  - Original delivery rate (e.g., lines of code or functions points per team member per month).

Systems Organization and Human Resources Baseline. This baseline segment provides a profile of the people side of the IT equation and the current organizational structure, including:

- Organizational chart (functional).
- Average managerial span of control.
- Human resources profile (e.g., skills inventory, educational inventory, training history, team and individual profiles such as Meyers-Briggs).

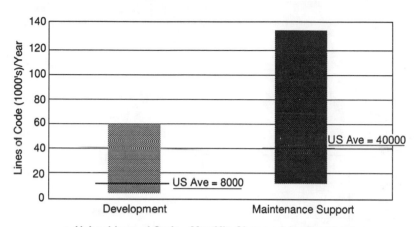

a. Using Lines of Code—Max/Min Observed Against the Average

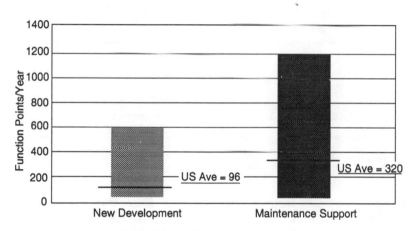

b. Using Function Points—Max/Min Observed Against the Average

Exhibit VI-9-2. Rating Development and Support Productivity

- Work distribution (i.e., percentage of people and dollars expended on development versus support).

Ultimately, this effort attempts to answer one question: Does the IT organization have the right resources to support the business today and into the future?

The production of an "organizational readiness" footprint, to determine the ability to assimilate new technology, is a major baseline output. This parallels the

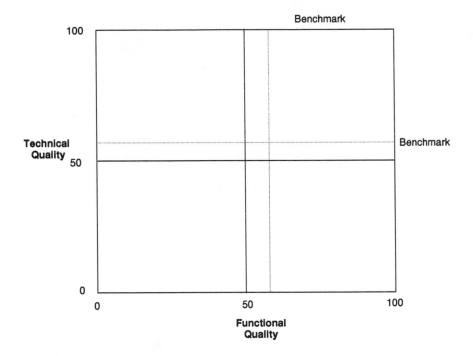

Exhibit VI-9-3. Application Portfolio Characteristics

measurement readiness footprint used earlier (and shown in Exhibit VI-9-1) but concentrates on software technology instead of measurement. Assessment questions include:

1. How intense is the organization's desire to improve its performance? 0—No desire; 5—Intense desire.
2. How much is the organization willing to invest to improve its performance? 0—No investment; 5—Up to $100,000/professional.
3. What is the current level of the systems skills inventory in software engineering? 0—Abstractions and models not used at all; 5—Formalization and models used by all.
4. To what extent are basic software engineering concepts known and understood by the systems staff? 0—No staff members have been exposed to software engineering principles; 5—Staff is 100% trained.
5. Is the systems culture averse to using new tools, techniques, or innovations? 0—100% opposed; 5—Eager to implement.
6. To what extent is a support structure in place to foster measurement soft-

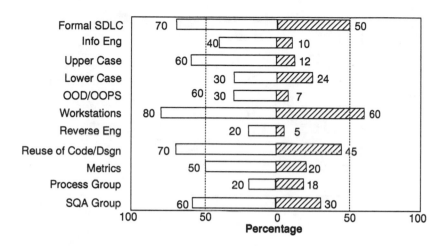

KEY:
▢ % Using
▨ % Penetration

Exhibit VI-9-4. Tool/Technique Inventory: US Data

ware engineering technology transfer? 0—Not in place; 5—An in-place team of critical mass exists.

7. What is the current software engineering platform? 0—Dumb terminals; 5—Client/server workstations.

8. What is the development and support split? 0—0% versus 100%; 5—100% versus 0%.

The results should be plotted on a circular scale similar to the measurement assessment. This time, however, it is necessary to plot where the IT organization should be in regard to either a particular technology (e.g., client/server) or the overall software process (e.g., SEI Level 1 through 5). The gaps that become apparent are those that need to be filled to transform an organization to where it should be.

Management Practices Baseline. The focus of this baseline segment is on how the existing resources perform work. It means gathering and summarizing answers to basic questions about management practices:

- How are planning and prioritizing done?
- How does the organization translate requests into systems?
- How well defined is the systems development life cycle?

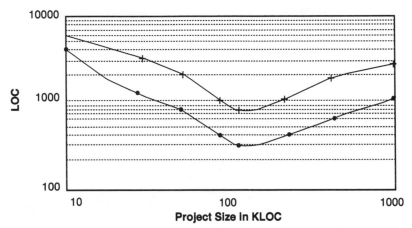

KEY:
+ Efficient
● Average

**a. Productivity in Thousands Lines of Code
(KLOC) per Staff Month**

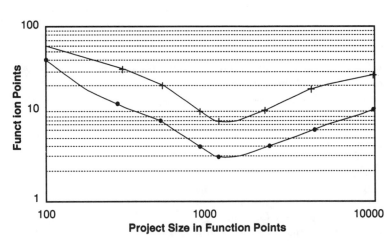

KEY:
+ Efficient
● Average

b. Productivity in Function Points per Staff Month

Exhibit VI-9-5. Rating Effort Productivity

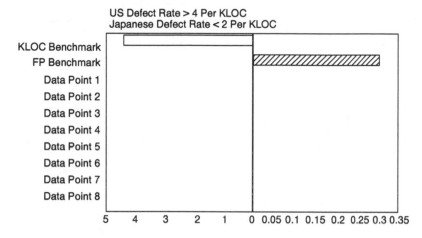

US Defect Rate > 4 Per KLOC
Japanese Defect Rate < 2 Per KLOC

Per KLOC Benchmark/Per Function Point

KEY:
☐ Per KLOC
▨ Per Function Point

Exhibit VI-9-6. Defect Density

- What is in the organization's tool inventory and what is actually used (i.e., tool penetration)?
- What is the organization's current software process maturity level? (This entails performing a formal or informal SEI assessment.)

This information is typically gathered through interviews and workshops conducted to assess the rigor, actual end use, and effectiveness of these practices themselves.

Delivery Process Baseline. This is the baseline segment in which specific representative development and maintenance projects are examined in detail to assess schedule and effort productivity and quality. However, it is often necessary to go beyond these issues and assess other factors likely to affect productivity and quality.

By determining delivery and support rates at the project level and comparing them to external benchmarks, an organization can create a framework to quantitatively assess the impact of potential changes and identify opportunities. Furthermore, the framework provides a clear basis for understanding the impact that tools potentially have on the overall delivery rate and product/process quality.

Typical metrics collected at the project level include:

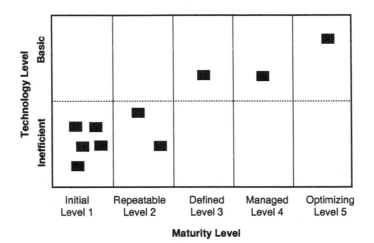

Note: *With the exception of the Level 5 point, each square represents 10 companies*

Exhibit VI-9-7. Software Process Maturity (US)

- Lines of code or function points per professional by project size.
- Lines of code or function points per work month by project size.
- Defects per line of code or function discovered before and after implementation.
- Percentage of defects originating from each life cycle phase.
- Point in life cycle where defects were found.
- Percentage of defects removed by phase.

Technology Infrastructure Baseline. This segment of the baseline identifies current and proposed delivery and production environments. If a company is, for example, wrestling with the possibility of shifting from mainframe to workstation-based development to reduce cost while increasing productivity, the practical steps that must be taken to accomplish this transition are evaluated.

Business Factors Baseline. Perhaps the most important component of any baseline is mapping the link between the software engineering function and the business's performance itself. Executive interviews with major systems customers have to be undertaken to develop an understanding of the mix of internal and external factors that may be changing the business. Understanding the volatility of the customers' environment sets the stage for examining how the systems organization is aligned to support the business. In addition, interviews with key

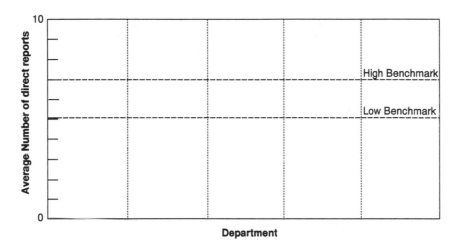

Exhibit VI-9-8. Span of Control

end users should be used to identify projects not currently being worked on that could provide a measurable difference in the way the business is run.

The essence of this baseline segment is to discover how business value is created by the systems area from the vantage point of the business. This step also sets the stage for defining business-value-level metrics.

The results of this baseline segment should be used to create a table that separates all the people interviewed into peer group audiences. For each audience the cross-reference table shows which performance assessment areas are essential. Another table can then be constructed to link each performance area to the measurements that support it. The resulting set of tables is essentially the IT organization's measurement program design document.

Making the Connection to Business Value

The baseline process illustrates a metrics dashboard. With proper instrumentation, it can be used to monitor and manage organizational performance as well as to clearly identify the system's contribution to business value.

The exact business measures can be derived by extending the business factors baseline segment. If external business customers and internal business customers are included in addition to the IT audiences, a complete dashboard framework will be the result, showing what performance improvement looks like to each constituency and what the suitable indicators (metrics) are. Using this as a framework, two types of dashboards should be constructed—one containing the navigation gauges, the other containing the destination gauges used to declare success.

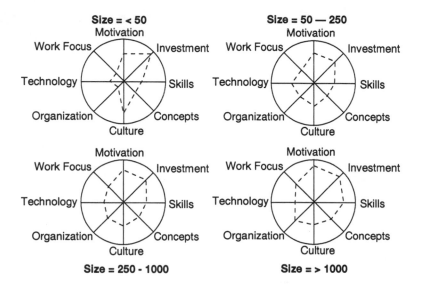

Exhibit VI-9-9. Organizational Readiness Profits

USING EXTERNAL BENCHMARKS: A STEP-BY-STEP APPROACH

If the IT baseline is constructed following the guidelines given, it then becomes possible to assess IT performance against best-in-class benchmarks. Exhibits VI-9-2 to VI-9-11 are tools to help readers create a baseline/benchmark snapshot. The steps to construct a snapshot of an IT organization are detailed next.

Applications Portfolio. Using Exhibit VI-9-2a (lines of code) or VI-9-2b (function points), mark a point for each application for which a support ratio is computed per support professional. Using Exhibit VI-9-3, fill in the percentage of the total portfolio that is in each of the following categories:

- Low functional quality (FQ) and low technical quality (TQ).
- High FQ and low TQ.
- Low FQ and high TQ.
- High FQ and high TQ.

Technology Infrastructure. Using the inventory of tools and techniques in Exhibit VI-9-4, indicate what percentage of the target audience is properly employing each tool and technique in the intended manner at least 80% of the time.

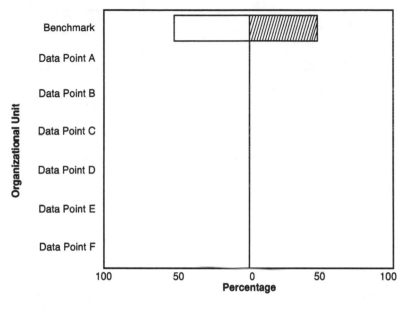

KEY:

☐	Development % (US 52%)
▨	Maintenance % (US 52%)

Exhibit VI-9-10. Development Versus Maintenance

Delivery Process. Mark a point on Exhibit VI-9-5a (lines of code) or VI-9-5b (function points) for each project that has been assessed in terms of delivered lines of code or function points per person month. Then, mark a point on Exhibit VI-9-6 that represents either the average number of postimplementation defects detected per function points or lines of code.

Management Practices. Using Exhibit VI-9-7, place an X in the segment that most clearly relates the IT organization's SEI process maturity rating.

Organization and Human Resources. On Exhibit VI-9-8, mark the point that shows the average span of control, then answer the organizational readiness assessment questions and plot the results on Exhibit VI-9-9. Using Exhibit VI-9-10, indicate the percentage of resources allocated to maintenance and development. At a more detailed level use Exhibit VI-9-11 to categorize the work as corrective, adaptive, and perfective maintenance.

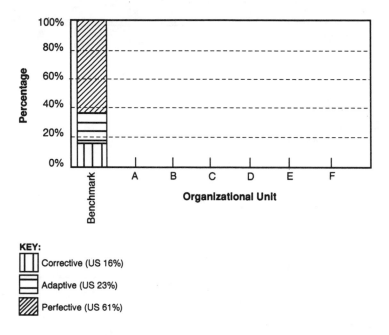

Exhibit VI-9-11. Maintenance Distribution

When used together, Exhibits VI-9-2 through VI-9-11 provide a snapshot of conditions in the IT department. Though this picture does not take into account industry specifics, IT managers can still use the worksheets to get value from their baseline through comparisons with the benchmarks provided.

CONCLUSION

Benchmarking clearly provides a context for assessing IT performance, business contribution, and competitive positioning. The real issue for IT organizations is how to transform performance and produce business value through computing:

In the context of benchmarking, metrics are a core competency that an organization must develop to promote learning and continuous improvement. The road to a learning organization is a difficult and complex one. It starts with the basics—knowing where the business is, where it is going, and how it is going to get there. These issues are the essence of the benchmarking process.

IT organizations must transform their performance as business transforms. The benchmarking steps and worksheets in this chapter can be a tool and catalyst for making this change happen in the IT department.

VI-10

Improving Quality with Software Metrics

Ian S. Hayes

TO SUCCESSFULLY MANAGE INFORMATION SYSTEMS in the cost-conscious 1990s and beyond, IS managers must become more effective in delivering their services. They should be able to identify and replicate practices that result in higher productivity, lower costs, and higher quality and correct or eliminate ineffective practices. This process requires a continuous flow of information to determine the quality of the IS activities being managed. Subjective guesses are not accurate enough and are open to dispute. Solid, objective measurements are required.

This chapter discusses how to use software metrics as a basis for making management decisions. It describes the sources and methods of collection for the data presented and demonstrates how cost and frequency-of-change information can be used with technical metrics to provide comparisons of cost for quality and productivity. The chapter also uses historical data collected from over 12,000 programs in over 65 companies to illustrate the current state of IS software and conversely to identify the desirable characteristics in a program or system. Programming standards are recommended, based on the metrics described, and steps for implementing a software measurement program are outlined.

WHY MEASURE?

Most IS departments do not have formal methods for measuring their programming and maintenance activities. As a result, the management of these departments would be hard pressed to answer questions such as:

- Has quality or productivity improved or declined in the last five years? By how much?
- Which programs or systems require a disproportionate amount of programming resources to maintain?
- What are the desirable characteristics in a program or system?

- Does a new software tool or development technique actually improve quality or productivity?
- Does increased quality actually lead to decreased cost?

All these questions, and more, could be answered through the implementation of a formal measurement program.

A comprehensive measurement program would combine software quality metrics, frequency, and quantity of change information with cost or effort information to provide an accurate picture of current programming practices. This information would be used to:

- *Define Quality.* Which practices are effective? What are their characteristics? The answers to these questions provide a real-world definition of quality. For example, small programs cost far less to maintain than large programs. Therefore, small is beneficial in programming.

- *Quantify Quality.* Once a definition of quality has been identified, the IS environment can be measured to determine the volume of characteristics. For instance, how many of the programs in a general ledger application are too large to maintain effectively?

- *Set a Baseline.* Each compilation of measurement information provides a baseline of comparison for future runs. This baseline can be used for identifying trends, setting improvement objectives, and for measuring progress against those objectives. Continuing the example, after one year of improvement activities, what percentage of the general ledger programs are still too large?

Combining cost and effort data with software quality metrics is particularly valuable for management decision making. It enables IS managers to determine relative productivity rates and to make cost-quality comparisons. Productivity rate information helps identify productivity bottlenecks and provides a foundation for accurate task estimation. Cost and quality information provides a basis for cost justifying quality improvement activities.

Maintaining a historical data base of key measurement data enables long-term validation of the decisions made from the measurement data. Quality and productivity trends can be determined, and the use and collection of the metrics information can be fine tuned if necessary. The benefits can be periodically reanalyzed to ensure that they are matching their cost-justification projections. Finally, as the base of historic data grows, it becomes more valuable for verifying task estimates.

METHODS USED IN THIS STUDY

There are three major categories of data used in this chapter: technical quality data, cost and effort data, and frequency-of-change data. This data is gathered through a variety of automated and manual methods. Much of the data presented here was gathered from Keane Inc.'s reengineering consulting practice. Keane's analysis combines its ADW/INSPECTOR metrics with other statistics, such as programmer cost, maintenance effort by task, frequency of change, and failure rates, to assist IS management in making strategic decisions about its application portfolio.

Technical Quality

The software metrics data presented in this chapter is a combination of statistics that have been gathered over five years by both Language Technology, Inc. (now a part of KnowledgeWare) and Keane, Inc. The software quality statistics are derived primarily from a data base of more than 150 software metrics collected using an automated COBOL analysis tool. This data base was originally developed by Language Technology from data contributed by more than 65 customers. The data base encompasses well beyond 12,000 COBOL programs consisting of 23 million lines of code.

Cost and Effort

Cost and effort data are essentially equivalent. For example, effort can be converted to cost by multiplying the number of hours expended by the average cost rate for the programmers. Gathering this data varies widely from company to company. Common sources are time-accounting systems, budgets, or task-tracking systems. Ideally, programmer time is charged to specific projects and is broken down by type of task. This is rarely available, so in most cases, specific cost allocations have to be extrapolated from available information.

Frequency of Change

Determining productivity rates and isolating the most volatile, and hence high-payback, areas of an application system require the availability of frequency- and volume-of-change information. This type of information can take many forms, depending on the company's practices. For example, if the company measures function points, volume of change is the number of function points added, deleted, or modified over a period of time. Another method is to track user change requests. These methods are less accurate from a programmer's point of view, but they have the advantage of having more meaning to the business areas. More commonly, however, this data is derived from the company's library management system. If set up properly, the library management system can provide reports on the number of changes made for each module contained within it, and often it can state the size, in lines of code, for each of those changes.

METRICS AND COST

Many companies are content with using only technical metrics to evaluate their programming and maintenance practices. Although this gives them the ability to evaluate the quality of their practices, it does not allow the correlation of quality to cost. This correlation provides a wide range of information that is invaluable to IS management. Some key examples of analyses that are accomplished in conjunction with cost are:

- Productivity measurement.
- Comparison of applications by effectiveness.
- Establishing the value of quality.
- Isolation of areas of high cost and high payback.
- Task estimate validation.
- Cost justification.

Examples of these analyses are illustrated in the following sections.

Productivity Measurement

Productivity measurement is a sensitive topic among those who are to be measured. Given the level of accuracy of the numbers available in most companies, productivity comparisons are best done at the application level rather than at the individual level. The basis for measuring productivity is effort per unit of work. Using cost in place of effort is equally valid. Potential units of measure include function points, statements, lines of code, change requests, or any other unit of work routinely measured by the company. Each has its own strengths and weaknesses, but any unit works for gross measures. The best unit that is easily available should be used because 80% of the value can be derived with the first 20% of the effort. A typical unit used by Keane is the number of lines of code changed per year. *Changed,* in this case, is defined as added, modified, or deleted. This number has the advantage of being relatively easy to obtain through ADW/INSPECTOR and a library management tool.

Taking the overall cost or effort expended on maintenance for a given application and dividing it by the volume of changes leads to a productivity factor. For example, application A may have a productivity factor of $10 per line of code changed. Improvements in productivity are reflected as a corresponding drop in the per-unit cost.

Comparison of Applications by Effectiveness

As Exhibit VI-10-1 illustrates, the productivity factor may be used to compare applications against each other. Five applications that were measured during a portfolio analysis are ranked against each other by their relative productivity factors. This relative ranking provides a challenge to management to determine why productivity is better for one application than another. Is it due to technical quality, personnel, or maintenance records? Answering these questions has inestimable value. If the factors causing low costs on ACP could be duplicated on CSA, there would be an almost four-fold increase in productivity.

Establishing the Value of Quality

Exhibit VI-10-2 is a combination of the productivity statistics in Exhibit VI-10-1 and average application quality statistics. The quality is measured by

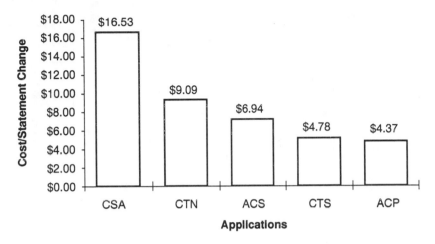

Exhibit VI-10-1. Comparison of Applications by Effectiveness

the average ADW/INSPECTOR composite score for all the programs within the application. This metric will be explained in more detail later in this chapter; in short, the score ranges from 0 to 100, with 100 being ideal.

The combination of cost and quality data as illustrated in Exhibit VI-10-2 demonstrates the strong correlation between application quality and the cost of maintenance. Even small increases in average quality generally result in measurable reductions in the cost of maintenance. This correlation between application quality and the cost of maintenance is typical when multiple applications within a company are measured. Performing this type of correlation is valuable for developing actual dollar benefit estimates for quality improvement efforts.

Isolation of Areas of High Cost and High Payback

Some programs within an application consume a much higher proportion of maintenance resources than other programs. These programs may be isolated by distributing the cost and change data down to the individual program level. Whereas programmers can generally identify the top one or two programs requiring the most maintenance effort, the overall distribution of maintenance cost and effort is often surprising. Interestingly, between 3% and 5% of the programs in a given application account for 50% to 80% of the overall maintenance effort. Those numbers are remarkably consistent across many companies. Exhibit VI-10-3 illustrates this cost distribution for the five applications used in the previous examples. Twenty programs, or approximately 3% of the total number of programs, account for 50% of the overall cost of maintenance.

To develop this graph, yearly maintenance costs were allocated to each program within the applications. The programs were sorted by cost, and the graph

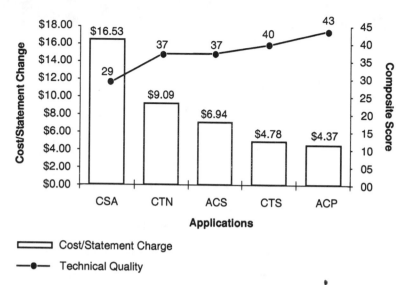

Exhibit VI-10-2. The Impact of Quality on Maintenance Costs

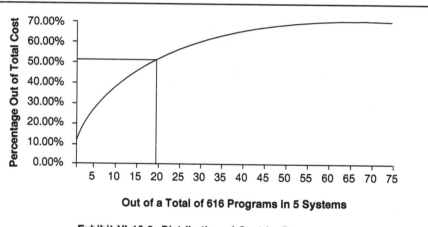

Out of a Total of 616 Programs in 5 Systems

Exhibit VI-10-3. Distribution of Cost by Program

line charts the accumulation of these costs (i.e., from the left, point 1 is the percentage cost of the most expensive program; point 2 is the sum of the two most expensive programs). Knowing where effort is actually expended allows improvement activities to be targeted to those programs in which the value would be the highest. Of the programs in this example, 28% had no maintenance effort expended on them at all. Consequently, they would have no payback if improved.

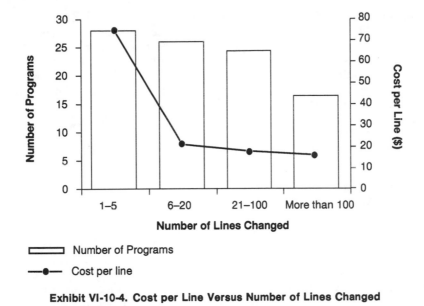

Number of Programs

Cost per line

Exhibit VI-10-4. Cost per Line Versus Number of Lines Changed

Task Estimate Validation

Keeping cost statistics and in particular developing a productivity factor is valuable for use in estimating the effort and cost of maintenance changes. Once the approximate size of a maintenance change is known, it can be multiplied against the productivity factor to provide an approximate scope of effort. This may be used directly or as a validation for other estimation methodologies. If cost or effort data is correlated against other factors, such as size or technical quality, these factors can be used as modifiers when estimating. For example, Exhibit VI-10-4 illustrates the cost per statement changed based on the number of statements changed in a particular modification for a client.

As would be expected, maintenance modifications that encompass only a few lines of code are more expensive per statement changed than larger modifications. This reflects the overhead for finding, inserting, testing, and placing the modification into production. The overhead factors are generally uniform relative to the size of the modification. Knowing this curve allows the calculation of an adjustment factor that can be applied to estimates for small changes.

Cost Justification

Using methods similar to the estimation technique described above, cost and productivity factors can be used for developing cost/benefit analyses for justifying new productivity tools or methodologies. For instance, the comparison of the five applications in Exhibit VI-10-2 can easily be used to justify quality improvement efforts. The only additional piece of necessary information is the total volume of changes. In the case of application CTN, this volume is about 114,000 statements per year. This application costs $9.09 per statement maintained, whereas application ACP costs $4.37. If the quality of CTN could be improved to meet or exceed that of ACP, one would expect that its cost of maintenance would be similar to ACP. Thus, if the improvement activities resulted in a per-statement drop in cost of $4.72 ($9.09 - $4.37), the anticipated benefit would be about $538,000 per year (114,000 × $4.72). Comparing this benefit to the cost of making the necessary improvements would provide the cost justification.

METRICS AND QUALITY

Cost metrics describe the value of a particular activity, whereas technical metrics evaluate the quality of its implementation. This industry has over 30 years of experience in the effectiveness of programming practices, yet practitioners still argue about the merits of such advances as GO TO-less programming. The metrics in the Language Technology data base confirm most theories on effective programming. This data should be used to define the standards by which programming is performed. Even though the data is based on COBOL programs and programmers, the insights are not limited to that environment. Human abilities for handling complexity, for example, are not language, or even programming, dependent.

Background

Some important definitions are necessary before embarking on a discussion of the metrics and their meanings. Perhaps the most important definition is that of the ADW/INSPECTOR composite score. As its name implies, this metric is a composite of a number of other measures. A 0-to-100 scale is used to provide a quick ranking of programs by quality. This metric has a consistently high correlation with other measures of quality, and it is an accurate predictor of relative maintenance effort, and therefore cost, between programs. The composite score is calculated as follows:

- Degree of structure (based on McCabe's essential complexity): 25%.
- Degree of complexity (based on McCabe's cyclomatic complexity): 25%.
- Number of ALTER verbs: 10%.
- Number of GO TO (i.e., non-exit) verbs: 10%.
- Number of fall throughs (i.e., to non-exit statements): 10%.
- Number of active procedure exits: 10%.
- Number of recursive procedures: 10%.

A perfect program receives 100 points (totally structured, not complex, and no ALTERs or GO TOs).

Other important definitions are:

Metric	Industry Average
Composite Score	60
Unstructured	43%
Complex	47%
Lines of Code	1,902
Statements	666
GO TOs	52
Fall Throughs	23
Recursion	3
Active Exits	12

Exhibit VI-10-5. Characteristics for the Average COBOL Program

- *Active Procedure Exits.* These result when a PERFORMed procedure within a COBOL program is exited by a GO TO while still active from a PER-FORM. Unpredictable program behavior can result when these procedures are reached via another control path in the program.
- *Recursive Procedures.* These are the result of a PERFORMed procedure being re-PERFORMed before completing a previous PERFORM. It too can cause unpredictable behavior.
- *McCabe Metrics.* These will be discussed in the following sections.
- *Lines of Code.* These are the specific number of lines of 80 column COBOL text in a program. Statements are the count of COBOL procedural verbs, such as MOVEs or ADDs.

Industry Averages

A major advantage of having a large multicompany data base of metrics is the ability to compare a specific company's application software with the rest of the industry. This provides a relative measure for that software against its "average" peers.

Exhibit VI-10-5 contains the characteristics for the average COBOL program. The characteristics of the average program may surprise many programmers: It is small in size and its quality is high. These effects are the result of mixing older COBOL programs with more newly developed code. As with averages in general, this obscures the very different characteristics of these two distinct categories.

This distinction between new and old code is illuminated when the programs within the data base are distributed by composite score. This distribution is shown in Exhibit VI-10-6.

The score distribution in Exhibit VI-10-6 is bimodal (i.e., there are two separate peaks). The peak at the far left is that of the average old application; its

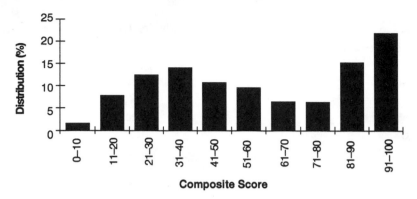

Exhibit VI-10-6. Characteristics for the Average COBOL Program

average program is in the 35-point range for the composite score. The peak at the far right mostly represents newer programs. Many new programs are created in conformance with rigorous structured programming standards and tend to be high in quality.

Over time, however, this quality tends to degrade through the effects of multiple coding changes unless specific quality efforts are in place. This decline is made apparent by the number of programs in the 81-to-90-point range.

The differentiation between the newer code and the older code becomes even more clear when the data is categorized by structure and complexity.

Exhibit VI-10-7 divides programs into four quadrants using unstructured percentages and complex percentages. The newer programs tend to be concentrated in the lower left quadrant. They are structured (i.e., under 25% unstructured), and not complex (under 25% complex). Whereas 30% of the programs fall into this desirable category, they make up only 11% of the total number of COBOL statements, the measure of actual code content.

Conversely, the older programs tend to be concentrated in the upper right quadrant and are both unstructured and complex. The upper right quadrant contains the worst 50% of the programs. Using statements as the measure, those programs make up 74% of the physical volume of COBOL code.

Even more instructive is selecting the average program from each quadrant. These averages are illustrated in Exhibit VI-10-8. In this exhibit, overlaps are the same as active procedure exits, and procedure size is the number of lines of code in the COBOL program's procedure division.

Structured/Noncomplex. As mentioned earlier, these are the desirable programs. They have a very high average composite score of 93 points and a very low unstructured percentage and complex percentage. This is particularly interesting given the 25% cutoff for unstructured percentage and complex percentage; the low numbers are well below the expected average of about 12.5% for an even

	Noncomplex	Complex
Unstructured	678 Programs 5% of Total 347,083 Lines 1% of Total 89,864 Statements 1% Total	6,285 Programs 50% of Total 16,867,023 Lines 71% of Total 6,131,621 Statements 74% Total
Structured	3,769 Programs 30% of Total 3,070,197 Lines 13% of Total 906,110 Statements 11% Total	1,772 Programs 14% of Total 3,501,542 Lines 15% of Total 1,200,590 Statements 14% Total

25% (Unstructured/Structured axis)

Noncomplex 25% Complex

Exhibit VI-10-7. Categorizing Code by Structure and Complexity

Metric	Structured Noncomplex	Unstructured Noncomplex	Structured Complex	Unstructured Complex
Composite Score:	93	63	78	34
Unstructured %:	1	66	4	77
Complex %:	3	1	50	78
Size in Location:	815	512	1976	2684
Statements:	240	133	678	976
Procedure Size:	451	236	1227	1565
Go Tos:	1	9	5	100
Fall Throughs:	1	5	3	43
Recursion:	0	0	0	5
OverLaps:	0	1	0	14
Deadcode:	6	10	9	36

Exhibit VI-10-8. Average Programs per Structure and Complexity

distribution. Size is an important factor in the quality of these programs. Desirability in programs is achieved through the combination of small size, high structure, and no GO TOs. Despite the single GO TO in the average, the majority of the programs in this category have no GO TOs at all.

Unstructured/Noncomplex. These are the rarest programs in the data base sample due to the difficulty of creating an unstructured program without measurably increasing its complexity. This is generally possible in only very small programs, which is demonstrated by the fact that these programs make up 5% of the total in number but only 1% of the physical volume in statements.

These programs are surprisingly unstructured given their very small size. This lack of structure is caused by the increase in the average number of GO TOs and fall throughs. The drop in quality reflected in this quadrant results in an increase in the number of lines of dead code. Dead or inexecutable code is generally introduced accidentally when a poorly understood program is modified. These programs could easily be improved by simply running them through a structuring tool.

Structured/Complex. This quadrant contains programs that are highly complex despite being structured. Some of these programs are necessarily complex due to the nature of the tasks they are performing; others are needlessly complex because of poor design. In either case, studies have shown that both error rates and testing effort increase as complexity increases. Complexity is highly correlated to program size. This is demonstrated by the significantly larger size of these programs as compared to the programs in the noncomplex categories. The presence of additional numbers of GO TOs and fall throughs also results in an increased complexity. The best method for reducing complexity in a structured program is to subdivide complex portions of that program into smaller noncomplex portions whenever possible. Breaking larger programs into multiple smaller modules also reduces complexity.

Unstructured/Complex. The programs in this quadrant are the classic old COBOL programs. The composite score drops dramatically as the result of increased complexity and decreased structure. Size has increased significantly, making these programs even more difficult to understand. Particularly alarming is the massive increase in poor coding practices, such as GO TOs, fall throughs, and recursion. These are averages, however; there are programs that are considerably worse in this category. This major drop in quality results in programs that are very hard to maintain and have significantly higher failure rates. This is reflected in the increased number of lines of dead code. Programs in this category are candidates for reengineering improvement activities to lower the cost and effort of maintenance.

Program Size Factors

Perhaps the most critical determinate of a program's quality and maintainability is its size. It is theoretically possible to produce a large, well-structured, noncomplex program; however, analysis of the statistics in the metrics data base show that this is an extremely rare occurrence. In fact, size in statements has a strong correlation with every negative programming practice. Size is not a component of the composite score, but as Exhibit VI-10-9 illustrates, the composite score increases as the size decreases.

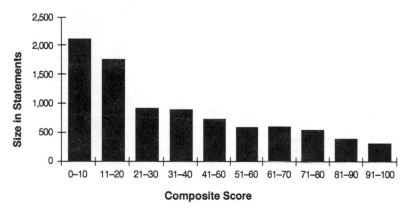

Exhibit VI-10-9. Distribution by Size

In fact, programs in the ideal composite score range of 93 to 100 points average only 240 statements in size. This size appears to be a boundary in program understanding. It appears consistently in the analysis results from company to company. In one client engagement, multiple years of ADW/INSPECTOR metrics were available, allowing the comparison of the effects of maintenance changes over time. When compared by composite score, programs under 240 statements in size either retained their score through multiple maintenance changes or actually improved in score. Conversely, programs over 240 statements generally declined in composite score. This effect appears to be caused by the relationship of complexity and size. Exhibit VI-10-10 is a graph of size, structure, and complexity correlation.

Again, although structure and complexity should be independent of size, they are highly related in practice. The sheer size of large programs makes them difficult to understand fully and often forces programmers to make localized program modifications that actually increase complexity and degrade structure. This tendency accelerates over time until the large program becomes unmaintainable.

As would be expected, size is related to the cost of maintenance. The graph in Exhibit VI-10-11 is from another Keane portfolio analysis engagement. The bars represent the number of programs in each size category. The graph line shows the total maintenance expenditure in dollars for all the programs in each respective column.

The highest expenditure of dollars is on the smallest number of programs, those with over 1,000 statements in size. The second-highest total expenditure is for the programs in the under-250-statement category; however, as the bar indicates, they make up a disproportionate number of the programs. This data is more dramatic when the average maintenance cost per program is graphed in Exhibit VI-10-12.

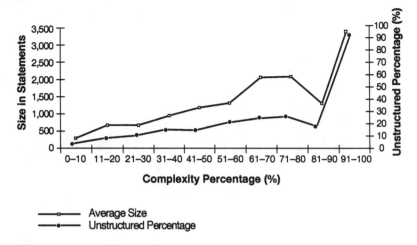

Average Size
Unstructured Percentage

Exhibit VI-10-10. Size and Structure Compared to Complexity Percentage

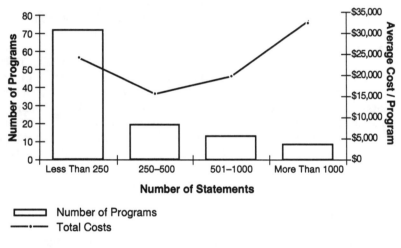

Number of Programs
Total Costs

Exhibit VI-10-11. Distribution of Expenditures by Program Size

McCabe Metrics

The metrics of T.J. McCabe are widely used to measure the complexity and structure of programs. Numerous studies have shown that these metrics are accurate predictors of program defect rates and program understandability. There are two separate metrics: cyclomatic complexity and essential complexity. Both metrics are based on measuring single-entry, single-exit blocks of code. In a structured COBOL program, these blocks of code are individual paragraphs. In

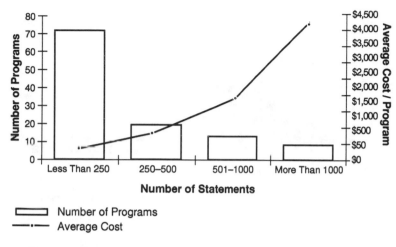

Number of Programs

Average Cost

Exhibit VI-10-12. Distribution of Programs and Costs by Program Size

a convoluted, unstructured program, the entire program may be one single-entry, single-exit block.

Cyclomatic Complexity. This is the measure of the number of test paths within a given single-entry, single-exit block of code. The number of defects in a block of code greatly increases when the number of test paths in that block exceeds 9. This is based on the theory that the average person can assimilate 7 plus or minus 2 (i.e., $7+/-2$) pieces of detail at a time in short-term memory. In COBOL programs, the number of test paths can be estimated by counting the number of IF statements in the block of code being measured. Adding 1 to this number gives the total number of unique test paths in that block of code. Therefore, as long as there are no more than 9 IF statements in each single-entry, single-exit block of code, it will meet the McCabe standards.

The $7+/-2$ levels of detail principle represented by cyclomatic complexity comes through in Exhibit VI-10-13. This graph from the metrics data base shows the distribution of the number of statements per paragraph across the entire data base.

The standard deviation in this graph is $8+/-4$, showing that programmers naturally tend to limit themselves to the McCabe cyclomatic constraints.

Essential Complexity

Essential complexity measures the degree of structure in a block of code. Essential complexity is measured by reducing all structured constructs (i.e., conditions, sequence, and loops) out of a given block of code, then measuring the remaining complexity. If no complexity remains, the piece of code is

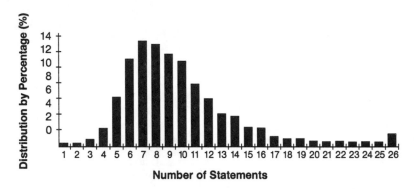

Exhibit VI-10-13. Distribution by Number of Statements per Paragraph

structured. Any remaining complexity is unessential (i.e., it could be removed by structuring that piece of code). Unstructured percentage is the sum of the number of unstructured paths divided by the total number of paths in the program. Essential complexity is used to measure if a particular program is in compliance with structured standards. It also predicts the ease with which a program can be modularized. Exhibit VI-10-14 illustrates this principle in an unstructured program.

The program in the real life portion of the illustration is primarily unstructured. The first two and last two paragraphs are single-entry, single-exit, but the entire middle section of the program is so interwoven with control flow that it is effectively one large block. The structure chart represents a view of the program if it were imported into a CASE tool. The middle block in the structure chart, shown as paragraph Initial-Read, actually contains all the code from the convoluted middle section, as shown in Behind the Scenes. If this program were to be modularized for reusability, this large block of code could not be easily subdivided unless it was structured first. Ideal programs should be totally structured as measured by McCabe's essential complexity.

GO TOs and GO TO EXITs

Despite the advent and supposed acceptance of structured programming, GO TOs are still used. ADW/INSPECTOR can differentiate between GO TOs and GO TO EXITs, which allows them to be examined separately. GO TO EXITs can theoretically be structured, so they are not penalized in the composite score. When GO TO non-exits are used within a program, they greatly increase its complexity. This is shown in Exhibit VI-10-15.

This complexity increases disproportionately as the number of GO TOs increases. Exhibit VI-10-16 shows the effect of GO TOs on the Composite Score.

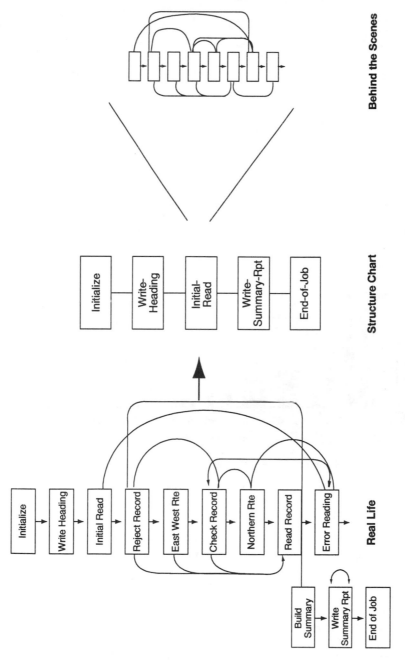

Behind the Scenes

Structure Chart

Real Life

Exhibit VI-10-14. The View Within the CASE Tool

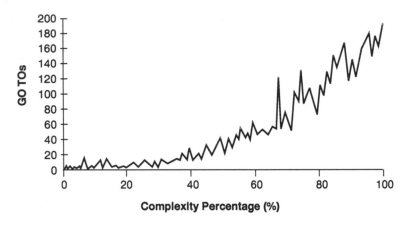

Exhibit VI-10-15. The Effect of GO TOs on Complexity

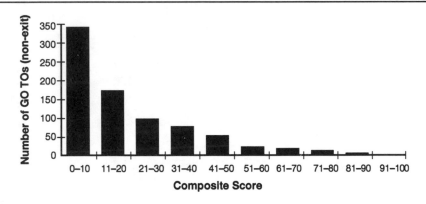

Exhibit VI-10-16. GO TOs by Composite Score

The number of GO TO non-exits accounts for 10 points of the composite score. As can be seen from this graph, their effect on program quality is far beyond those 10 points.

Many programmers argue in favor of the use of GO TO EXITs. This construct was originally used to reduce nesting of IF statements within paragraphs. This is no longer necessary in structured programming. Structuring tools can be used to automatically control nesting levels, and the use of structured CASE constructs eliminates much of the complexity involved in the nesting of related conditions. If implemented correctly, GO TO EXITs should have no effect on structure and complexity. In practice, this is not true. Exhibit VI-10-17 shows the number of GO TO EXITs graphed against complexity.

Although the correlation is not as strong with GO TO non-exits, GO TO EXITs increase program complexity. This appears to be the result of two factors. First,

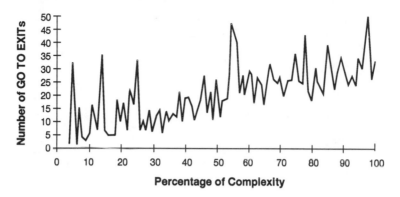

Exhibit VI-10-17. The Effect of GO TO EXITs on Complexity

GO TO EXITs are legitimate as long as they go to the correct exit. Unfortunately, their very existence invites accidental misuse. Second, the presence of any GO TOs tends to beget other GO TOs. For these reasons, it makes the most sense to avoid the use of any GO TOs when programming.

ALTERs

The ALTER verb is rarely seen in COBOL nowadays. A vestige from the assembly programming days, it appears only in some of the oldest programs. When ALTERs are present, their negative effect on quality amply justifies their removal. This may be done manually or automatically with a structuring tool. As with GO TOs, ALTERs comprise 10 points of the composite score. Again, however, their effect extends far beyond those 10 points. This is shown in Exhibit VI-10-18.

The presence of any ALTERs within a program tends to push it below 30 points in the composite score. ALTERs also greatly increase the number of test paths within the program. The worst example in the data base was a 1,500-line program that contained 508 ALTER statements. It had 4,718 separate test paths as measured by McCabe's cyclomatic complexity.

METRICS AND STANDARDS

With over 30 years of industry experience in COBOL programming, one would expect that the characteristics that make one program easier to maintain than another would be well known and incorporated into IS practices. Unfortunately, that has not been the case, in part because of the lack of agreement about what makes a desirable program. The data presented in this chapter should shed some light on some of the most important characteristics.

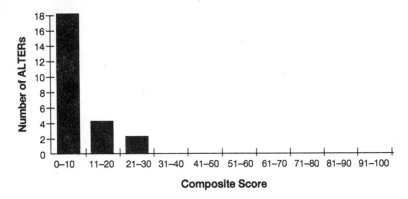

Exhibit VI-10-18. ALTERs by Composite Score

The maintainability of COBOL programs can be greatly enhanced, thereby lowering the cost and effort of maintenance, by following a few simple standards. The key to excellence in COBOL programs is understandability. If a program can be easily understood, it can be modified quickly, can be tested more thoroughly, and will be less likely to contain defects.

When all the data is analyzed, the three crucial characteristics that lead to understandability are size, structure, and modularity.

- *Size.* Once a program exceeds a certain size, it becomes difficult to understand and maintain just due to sheer mass. Changes are made to large programs without understanding the context of the whole program. This introduces poor coding practices and increases complexity. Large programs tend to contain many functions. Modifications to any of these functions require analysis and testing of the entire program, further increasing maintenance effort. Thus, small is better in programming. As the data in this chapter indicates, the ideal size limit appears to be 240 or fewer COBOL statements.

- *Structure.* Programs consisting of well-organized, single-function paragraphs are the easiest to understand and maintain. Strictly following the dictates of structured programming helps ensure that programs meet these standards. Further, if a program is structured, by definition it does not contain any poor coding constructs, such as GO TOs, fall throughs, or recursion.

- *Modularity.* Each paragraph in the program should consist of a single business function, and it should have only one entry and one exit. Further, the business functions must be modularized such that each paragraph consists of 8± 4 COBOL statements. If the program is structured, this rule ensures that the program is not complex by the McCabe standard. This simplifies testing, enables the location and reuse of business functions, and enhances the value of importing the program into a CASE environment.

Metric	Recommended Standard
Composite Score	93–100
Unstructured	0%
Complex	Less than 25%
Lines of Code	750 or Less
Statements	240 or Less
Statements per Paragraph*	8–12
GO TOs	0
Fall Throughs	0
Recursion	0
Active Exits	0

*Average Standards per Paragraph is calculated by dividing the total number of procedure statements in the program by the total number of paragraphs.

Exhibit VI-10-19. Optimum Metric Values for Measuring COBOL Compliance

Each of the three characteristics described above should be incorporated into IS coding standards for both new development and maintenance. Adherence to these standards can be easily measured with an analysis tool like ADW/INSPECTOR using the metrics discussed in this chapter. Exhibit VI-10-19 contains the optimum metric values for measuring compliance with these recommended COBOL programming standards.

For new development, these standards can be directly applied as acceptance criteria for newly written programs. Following these standards ensures that these new programs are easy to modify in the future.

For maintenance and enhancement projects, these standards become targets. At the very minimum, programs should be measured after each modification to ensure that quality is maintained. Ideally, programmers can attempt to slowly improve programs to get closer to these standards each time they make a maintenance change.

Finally, these standards are a goal for reengineering or other improvement activities to existing code. These efforts should be targeted at only the most highly maintained 3% to 5% of the programs to ensure payback. Attempting to reengineer all code to this standard would be cost prohibitive, and it would not provide any benefits on rarely maintained programs.

CONCLUSION

A software measurement program can provide IS managers with valuable insights on how to best manage their scarce resources. As the examples in this chapter demonstrate, metrics can be used to identify specific standards and methods that save money and resources. They can pinpoint when to apply most effectively that knowledge. They can be used to estimate the effort for programming tasks and to quantify the benefits of improvement tasks.

Gaining these benefits requires implementing a software measurement program. Some basic steps are:

Defining Objectives. Define why a measurement program is needed and what the specific types of questions to be answered are. This identifies what type of data is needed and the scope of the effort.

Identifying Data Sources. A check should be made of the existing data. Is the data complete as required? What data can be collected automatically? What data must be collected manually? Some examples of each type of data should be collected to check its ease of collection, accuracy, and completeness.

Obtaining Tools. If the company does not already have an analysis tool, it is time to get one. Some tools, such as project tracking and library management software, may be in house but may require specific option settings to collect the necessary data.

Defining Reports. Copies of the reports and graphs that will be used as output of the measurement program should be mocked up. Distribution frequency should be decided on. Displaying these examples is a key to getting buy-in on the project.

Pilot Testing. Results from the preceding steps should be tested on a pilot set of applications. It is best to use two or three diverse types of applications to ensure that any potential problems are caught. Training requirements should be identified for those involved in the project.

Tuning the Results. Any of the collection methods, metrics, and reports should be fine tuned, using the results of the pilot test. This information should be used to estimate effort for the final roll out.

Developing a Roll-Out Plan. A plan should be developed to roll out the measurement program across the organization, making sure to include sufficient training.

Implement. The process is put in place. Data should be reexamined periodically and tuned accordingly as new information is received. Results should be saved in a data base to allow for comparisons over time.

About the Editor

ROBERT E. UMBAUGH is principal consultant and head of Carlisle Consulting Group, an affiliated consulting firm specializing in productivity improvement in IS and the strategic application of technology. He is a consulting editor for Auerbach Publications and served for many years as chief information officer for Southern California Edison. As an adjunct professor of information systems at Claremont Graduate School (Claremont CA) and visiting lecturer at other schools, Umbaugh has helped educate many of today's IS managers. He can be reached at (717) 245-0825 or by mail at 700 West Old York Rd., Carlisle PA 17013.

Index